Research Guide to the Arid Lands of the World

by Stephen T. Hopkins and Douglas E. Jones

with the technical assistance of
John A. Rogers

ORYX PRESS
1983

The rare Arabian Oryx is believed to have inspired the myth of the unicorn. This desert antelope became virtually extinct in the early 1960s. At that time several groups of international conservationists arranged to have 9 animals sent to the Phoenix Zoo to be the nucleus of a captive breeding herd. Today the Oryx population is over 400 and herds have been returned to reserves in Israel, Jordan, and Oman.

Copyright © 1983 by Stephen T. Hopkins and Douglas E. Jones

Published by The Oryx Press
2214 North Central at Encanto
Phoenix, Arizona 85004

Published simultaneously in Canada

All rights reserved
No part of this publication may be reproduced or transmitted in any form or by any means, electronic or mechanical, including photocopying, recording, or by any information storage and retrieval system, without permission in writing from The Oryx Press

Printed and Bound in the United States of America

Library of Congress Cataloging in Publication Data

Hopkins, Stephen T.
   Research guide to the arid lands of the world.

   Includes indexes.
   1. Arid regions—Bibliography. 2. Arid regions—
Research—Bibliography. I. Jones, Douglas E., 1948–
II. Title.
Z6004.A7H66   1983   [GB611]   016.33373   83-42500
ISBN 0-89774-066-1

TABLE OF CONTENTS

Introduction — v
Acknowledgments — vi
How to use this Sourcebook — vi

Part One: Geographic Regions

Drylands of The World	1
Oceans and Coasts	6
The Third World	7
Europe	11
Mediterranean Europe	12
Cyprus	13
France	15
Gibralter	17
Greece	18
Italy	21
Malta	23
Portugal	24
Spain	26
Eastern Europe	28
Bulgaria	29
Hungary	31
Romania	33
Yugoslavia	35
The Atlantic Ocean and Islands	37
Azores Islands	38
Canary Islands	39
Cape Verde Islands	40
Madeira Islands	42
St. Helena and Ascension Islands	43
Africa	44
Northern and Western Africa	53
Algeria	57
Benin	59
Cameroon	60
Central African Republic	62
Chad	63
Egypt	65
The Gambia	68
Ghana	70
Guinea	72
Guinea-Bissau	73
Ivory Coast	74
Libya	75
Mali	78
Mauritania	80
Morocco	82
Niger	85
Nigeria	87
Senegal	90
Sudan	92
Togo	95
Tunisia	96
Upper Volta	99
Western Sahara	101
Eastern and Southern Africa	103
Angola	105
Botswana	107
Ethiopia	109
Jibuti	111
Kenya	113
Lesotho	116
Malawi	118
Mozambique	120
Namibia	122
Rwanda	124
Somalia	125
South Africa	127
Swaziland	132
Tanzania	134
Uganda	137
Zaire	139
Zambia	140
Zimbabwe	142
The Indian Ocean and Islands	145
Madagascar	146
Mauritius	148
Reunion	149
Seychelles	150
Socotra	151
Asia	152
The Middle Est	155
Afghanistan	159
Bahrain	162
Iran	163
Iraq	166
Israel	168
Jordan	171
Kuwait	173
Lebanon	174
Oman	176
Qatar	178
Saudi Arabia	179
Syria	181
Turkey	183
United Arab Emirates	185
Yemen (North)	186
Yemen (South)	188
South and Southeast Asia	191
Burma	192
India	194
Indonesia	199
Pakistan	201
Sri Lanka	204

Table of Contents

Thailand	206	New Mexico	282
Vietnam	207	North Dakota	283
Northern and Eastern Asia	209	Oklahoma	284
China	210	Oregon	284
Mongolia	214	South Dakota	285
Union of Soviet Socialist Republics	215	Texas	285
Australia and the Pacific	223	Utah	286
Australia	225	Washington	287
Galapagos Islands	233	Wyoming	287
Hawaiian Islands	234	The Caribbean Islands	291
Kiribati	236	Bahama Islands	293
Marquesas Islands	237	Cuba	294
New Caledonia	238	Dominican Republic	296
New Zealand	239	Haiti	297
The Americas	240	Jamaica	299
North and Central America	245	Netherlands Antilles	300
Canada	248	Puerto Rico	302
El Salvador	253	Trinidad	304
Guatemala	254	Turks and Caicos Islands	305
Mexico	256	Virgin Islands	306
The United States	261	South America	309
Arizona	273	Argentina	311
California	276	Bolivia	314
Colorado	278	Brazil	316
Idaho	279	Chile	319
Kansas	279	Colombia	321
Minnesota	280	Ecuador	323
Montana	280	Paraguay	325
Nebraska	280	Peru	327
Nevada	281	Venezuela	330

Part Two: Subjects of Drylands Research

Bibliography and Information Science	334	Agricultural Economics	353
Geography	335	Energy	354
Science and Technology	337	Solar Energy	355
Earth Sciences	338	Geothermal Energy	355
Geology	339	Wind Energy	355
Climatology	340	Biomass Energy	356
Hydrology	342	Petroleum	356
Soils	345	Human Geography	356
Biology	346	Demography	357
Botany	347	Anthropology	358
Zoology	347	Medicine	359
Agriculture	348	Urban Geography	360
Irrigation	350	Architecture	360
Plant Culture	350	Economic Development	361
Animal Culture	352	Historical Geography	363

Appendix A: The Areal Extent of the Dryland of the World ... 365
Appendix B: Sources used in the Preparation of this Guide ... 367

Author Index 369
Subject Index 379

INTRODUCTION

The arid lands of the world fall within the territories of over 110 national governments and are the homes of roughly 700 million people. For many of these people, the dryland environments offer an extremely limited resource base, made worse as expanding populations place greater and greater pressure on the land. The seasonally-dry forest, the grassland steppe, and the true desert are all scarce in water, soils, plant matter, and energy, yet man has repeatedly brought these areas to bloom and, in the process, has created some of his earliest and most enduring civilizations. The challenge for us, today, is just as significant; the drylands of the world are growing and their resources—however limited—are becoming more and more valuable. A new and deeper understanding of these environments and their potential is now possible, indeed, essential to us all.

The present Sourcebook is a first attempt at gathering and evaluating bibliographies, directories, abstracting journals, statistical sources, online databases, atlases, and gazetteers deemed most useful to researchers in the physical and human geography of the world's drylands. We have chosen to emphasize sources of current information, though major retrospective bibliographies are also included. It should be mentioned that the diversity of places and subjects relevant to arid lands research makes it impossible for a volume even of this size to be comprehensive in all areas; rather, we have designed this Sourcebook as an *interdisciplinary* tool of greatest value to researchers *outside* their professional specialities.

A word or two should be said about the history and creation of this volume. In 1980 the University of Arizona Library received a grant from the U.S. Department of Education to strengthen its already considerable holdings of arid lands materials and to make the bibliographic records of its acquisitions available to other libraries through the OCLC cataloging network. A further provision of the grant called for the creation of a research guide; this volume is the result. The authors gratefully acknowledge the assistance of the Library Education, Research and Resources Branch, U.S. Department of Education, and its Chief, Mr. Frank A. Stevens, for making the Title II-C grant available.

Users of the Sourcebook are encouraged to comment on errors, omissions, and the organization of this guide to the following address:

Douglas Jones
Science-Engineering Library
University of Arizona
Tucson, AZ 85721

ACKNOWLEDGMENTS

Many hands and minds at the University of Arizona contributed to the creation of this work. *Stephen Hopkins* (acquisitions librarian) served as general editor, annotator of citations, author of the geographical summaries, and indexer. *Douglas Jones* (science librarian) conceived the original idea for the Sourcebook, contributed annotations, conducted online literature searches, and supervised the final preparation of the manuscript. *John A. Rogers* (doctoral candidate, School of Renewable Natural Resources) developed the data on the extent of the world's drylands. *Linda Weingarten* and *Erika Kreider* assisted the general editor in the tedious tasks of verifying citations and editing the text. *Michael McAnnis* helped verify citations and scan maps for the areal data. *Lorene Moore* also helped verify citations.

Additional assistance came from the staff of the University of Arizona Office of Arid Lands Studies, especially *Justin Wilkinson*, *Barbara Hutchinson*, and *Elaine Cook*. The maps were prepared by *Paul Mirocha* of the Office of Arid Lands Studies. *Janet and Bill Crider* of Wordsmith Graphic Communications, Tucson, Arizona, computerized the database, and formatted and typeset the manuscript.

Special thanks are given to the arid lands researchers of the world—some anonymous, some renowned—who responded to our call for documents, reports, and publication lists and in so doing contributed to the success of the Arid Lands Project.

HOW TO USE THIS GUIDE

The reference sources listed in this Sourcebook can be located in several ways.

Access by Place. Since most reference sources cover specific geographical areas of the earth's surface, the largest share of this Sourcebook is a geographical arrangement of entries comprising Part One. The Table of Contents on pages iii–iv presents a synopsis of the arrangement and an index map on page viii provides a visual aid.

Access by Subject. Reference sources covering subjects with no particular geographical emphasis are found in Part Two. Subjects with some sort of geographical limitation are found under the appropriate place name in Part One. The Subject Index on pages 379-391 provides an alphabetical listing of subjects to all entries in both Parts One and Two.

Access by Author. The Author Index lists authors alphabetically on pages 369-377.

Online Databases. In addition to printed indexes, abstracting journals, and statistical sources, over 55 major computerized databases are included in this Sourcebook. Since most are worldwide in coverage they are usually cited under the appropriate subject headings in Part Two. Although traditionally the online databases have been simply computerized alternatives to standard print sources (thus duplicating their information) recently the online versions have begun carrying data not found in print; in addition, a number of sources are now available only online. Information on the vendors of each database is provided; consult the *Directory of Online Databases* (entry 2713) for more complete information.

PART ONE:

GEOGRAPHIC REGIONS

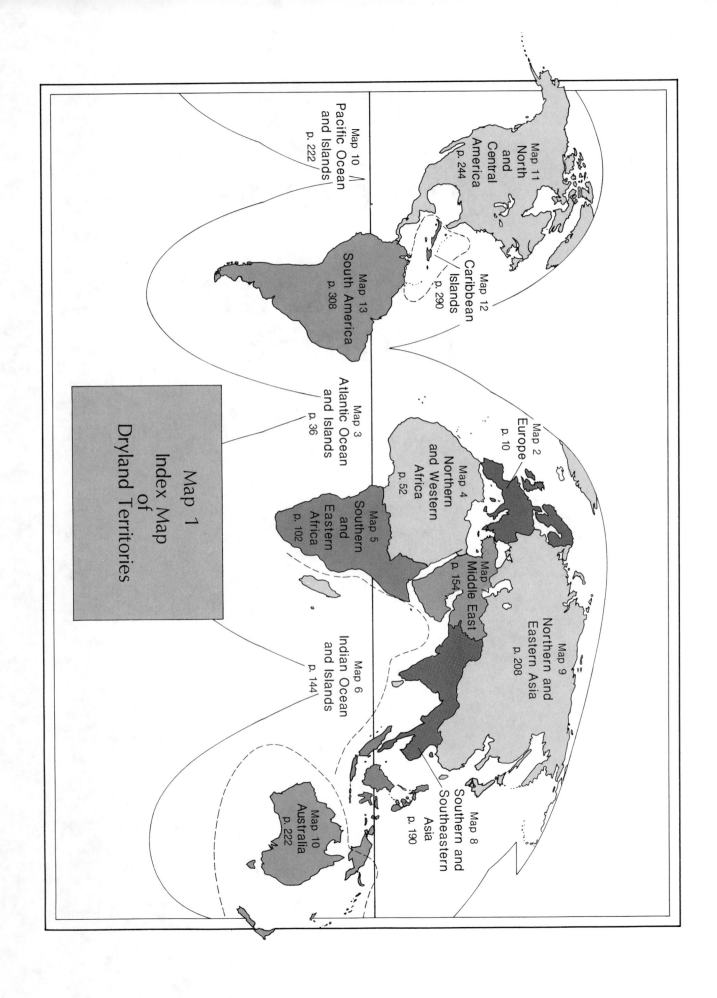

DRYLANDS OF THE WORLD

As shown on the chart below, the drylands of the world comprise some 40 percent of the earth's land surface. Such a large area, of necessity, includes a variety of environments, from grassland steppe to scrub forest to barren rock and sand. Whatever the composition of the plant cover or the range of temperatures or the variations in topography, the overriding factor is *aridity*, defined most simply as the existence of a moisture deficit. This means that the drylands of the world experience during all or part of the year a period when evaporation exceeds precipitation, a period when all life in such lands must adapt in some way to reduced supplies of water or face death from dehydration. The varying degrees of aridity and the extent to which land is susceptible to the erosive processes of desertification are explained in Appendix A on page 365.

The entries below provide reference sources on the world's drylands in general.

Dryland areas of THE WORLD (INCLUDES ANTARCTICA)

DESERTIFICATION RISK	ARIDITY									
	Hyperarid		Arid		Semiarid		Subhumid		Aridity Totals	
	km²	%	km²	%	km²	%	km²	%	km²	%
Very High	By definition, desertification does not exist in hyperarid regions.		1,173,497	0.8	2,017,892	1.3	165,252	0.1	3,356,641	2.2
High			14,856,563	9.9	2,798,457	1.9	615,282	0.4	18,270,302	12.2
Moderate			2,370,746	1.6	11,803,973	7.9	3,578,544	2.4	17,753,263	11.9
Desertification Totals			18,400,806	12.3	16,620,322	11.1	4,359,078	2.9	39,380,206	26.3
No Desertification	9,181,052	6.1	859,955	0.6	1,085,988	0.7	10,514,060	7.0	21,641,055	14.4
Total Drylands	9,181,052	6.1	19,260,761	12.9	17,706,310	11.8	14,873,138	9.9	61,021,261	40.7
Non-dryland									88,699,184	59.3
Total Area of the Territory									149,720,445	100.0

RESEARCH GUIDES

1   Dickson, Bertram Thomas, ed. *Guide book to research data on arid zone development.* Paris: Unesco, 1957. (Arid zone research 9) 191 pp.

Covers physical and human geography with chapter bibliographies.

2   McGinnies, William G.; Goldman, Bram J.; and Paylore, Patricia, eds. *Deserts of the world: an appraisal of research into their physical and biological environments.* Tucson AZ: University of Arizona Press, 1968. 788 pp.

A major research guide covering the physical geography of hyperarid, arid, and semiarid lands of the world. Includes extensive annotated bibliographies, which are listed separately in this Guide.

3   Vogel, Harvey. *An inventory of geographic research of the humid tropic environment.* Natick MA: U.S. Army Natick Laboratories, 1966. 2 vols.

Although from the title one might conclude that this bibliography has something to say about anything *except* arid lands, in fact, for some semiarid and subhumid regions, the boundaries overlap. Within these limits this set is an excellent research summary. Prepared by Texas Instruments, Inc., under contract.

GENERAL BIBLIOGRAPHIES

4   "Bibliographies." *Arid lands abstracts* 4 (1973): items 184-271.

5   Paylore, Patricia. "Arid lands information in United States government-sponsored indexing tools: What? Where? When? How?" In *Food, fiber and the arid lands*, pp. 407-420. Edited by William G. McGinnies, Bram J. Goldman, and Patricia Paylore. Tucson AZ: University of Arizona Press, 1971.

Drylands of the World/6-21

6 Paylore, Patricia. "Bibliographical sources for arid lands research." In *Arid lands in perspective*, pp. 247-287. Edited by William G. McGinnies and Bram J. Goldman. Tucson AZ: University of Arizona Press, 1969.

Annotated.

7 Paylore, Patricia. *A bibliography of arid lands bibliographies*. Natick MA: U.S. Army Natick Laboratories, 1967. (NTIS: AD 663-843) 80 pp.

Contains 362 annotated citations.

8 Paylore, Patricia, comp. *Desert research: selected references 1965-1970*. Natick MA: U.S. Army Natick Laboratories. Earth Sciences Laboratory, 1969-1970. 2 vols. (Technical Reports 70-24-ES and 71-20-ES; NTIS: AD 703-884 and AD 723-062).

Volume I covers 1965-1968, 410 pages; Volume II covers 1966-1970, 169 pages. Contains a total of 2,073 citations.

9 Sherbrooke, Wade C., and Paylore, Patricia. "The desert—for children." *Arid lands newsletter* 14 (1981):15-22.

Annotated listing covers juvenile works in several languages.

10 Templer, Otis Worth, comp. *The geography of arid lands: a basic bibliography*. Lubbock TX: Texas Tech University. International Center for Arid and Semi-Arid Land Studies, 1978. (ICASALS publication 78-1) 101 pp.

Unannotated.

INDEXES AND ABSTRACTS

11 *Arid lands abstracts*. Tucson AZ: University of Arizona. Office of Arid Lands Studies, 1972-76.

A first attempt at establishing an ongoing abstracting journal for arid lands research. Succeeded by *Arid Lands Abstracts* (1980-) which became *Arid Lands Development Abstracts* in 1982 (entry 12).

12 *Arid lands development abstracts*. Farnham Royal, U.K.: Commonwealth Agricultural Bureaux, 1980-1982.

Formerly: *Arid Lands Abstracts* (1980-82). Prepared by the staff of the Arid Lands Information Center, Office of Arid Lands Studies, University of Arizona. Available online.

DIRECTORIES

13 Goudie, Andrew S. *Directory of arid zone research in the United Kingdom*. Oxford, U.K.: Oxford University. School of Geography, 1971. 18 pp.

Lists names, addresses, and specialties of some 85 researchers.

14 Paylore, Patricia. *Arid lands research institutions: a world directory*. 2nd ed. Tucson AZ: University of Arizona Press, 1977. 317 pp.

Each entry includes a variety of information including address, scope of interest, staff, facilities, and publications. Partially updated by *Inventory of arid lands research institutions and their sources of financial support* (1982) (entry 15).

15 Taylor, Suzanne N.; Mortensen, B. Kim; and Dunford, Christopher. *Inventory of arid lands research institutions and their sources of financial support*. Tucson AZ: University of Arizona. Office of Arid Lands Studies, 1982. 133 pp.

Prepared for United Nations Environment Programme (Nairobi). Serves as a partial updating of Paylore, *Arid Lands Research Institutions* (entry 14). Coverage is worldwide.

DESERTIFICATION

16 *Desertification: its causes and consequences*. Oxford, U.K.: Pergamon, 1977. 448 pp.

Prepared by the staff of the United Nations Secretariat Conference on Desertification held in Nairobi, Kenya. Contains highlights of the Conference.

17 Paylore, Patricia, ed. *Desertification: a world bibliography*. Tucson AZ: University of Arizona. Office of Arid Lands Studies, 1976. 644 pp.

Prepared for 23rd International Geographical Congress, Moscow, 1976. Contains 1,645 annotated citations.

18 Paylore, Patricia, and Mabbutt, J. A., eds. *Desertification: world bibliography update 1976-1980*. Tucson AZ: University of Arizona. Office of Arid Lands Studies, 1980. 196 pp.

Updates entry 17. Contains some 500 annotated citations.

19 Reining, Priscilla. *Handbook on desertification indicators based on The Science Associations' Nairobi Seminar on Desertification*. Washington DC: American Association for the Advancement of Science, 1978. (AAAS publication 78-7) 141 pp.

20 Sherbrooke, W. C., and Paylore, Patricia. *World desertification: cause and effect; a literature review and annotated bibliography*. Tucson AZ: University of Arizona. Office of Arid Lands Studies, 1973. (Resource information paper 3; NTIS: PB 228-100) 168 pp.

Contains 252 citations.

21 *United States sources of information in the area of desertification*. Washington DC: U.S. Environmental Protection Agency, 1977. (EPA 840-77-008) 23 pp.

Prepared by the staff of the U.S. International Environmental Referral Center (USIERC).

22 **Walker, Alta Sharon, and Robinove, Charles Joseph.** *Annotated bibliography of remote sensing methods for monitoring desertification.* (Reston VA?): U.S. Geological Survey, 1981. (Circular 851) 25 pp.

Covers changes in arid and semiarid lands and their monitoring by remote sensing; ignores use of remote sensing for inventory of arid lands.

REMOTE SENSING

23 *Conference on Remote Sensing in Arid Lands. Proceedings.* Tucson AZ: University of Arizona. Office of Arid Lands Studies, 1971-. Annual.

The first conference in 1970 was a conceptual meeting to establish the Arizona Ecological Test Site (ARETS), and is the only Conference for which no *Proceedings* were produced.

24 **McGinnies, William G.** *An annotated bibliography and evaluation of remote sensing publications relating to military geography of arid lands.* Natick MA: U.S. Army Natick Laboratories. Earth Sciences Laboratory, 1970. (Series ES-61. Technical report 71-27-ES; NTIS: AD 723-061) 103 pp.

Contains 355 references.

CLIMATOLOGY

25 *Climatology: reviews of research.* Paris: Unesco, 1958. (Arid zone research 10) 190 pp.

26 **Reitan, Clayton H., and Green, Christine R.** "Appraisal of research on weather and climate of desert environments." In *Deserts of the world.* pp. 21-92. Edited by William G. McGinnies. Tucson AZ: University of Arizona Press, 1968.

HYDROLOGY

27 **Bowden, Charles.** *The impact of energy development on water resources in arid lands: a literature review and annotated bibliography.*

See entry 56.

28 **Keith, Susan Jo.** *Impact of groundwater development in arid lands: a literature review and annotated bibliography.* Tucson AZ: University of Arizona. Office of Arid Lands Studies, 1977. (Resource information paper 10; NTIS: PB 276-908) 139 pp.

Contains 188 references.

29 **Lustig, Lawrence K.** "Appraisal of research on geomorphology and surface hydrology of desert environments." In *Deserts of the world*, pp. 95-283. Edited by William G. McGinnies. Tucson AZ: University of Arizona Press, 1968.

30 *Moisture utilization in semi-arid tropics: summer rainfall agriculture project.* Boston MA: G. K. Hall, 1977. (Bibliographies and guides in African studies) 401 pp.

31 *Reviews of research on arid zone hydrology.* Paris: Unesco, 1953. (Arid zone programme 1) 212 pp.

32 **Rhoades, Marjorie.** *Water and soil in arid regions (WASAR): an index to selected materials in Colorado State University Libraries.* Ft. Collins CO: Colorado State University. University Libraries, 1977. 3 vols.

Contains some 3,800 unannotated citations covering soils, water, crops, food, and nutrition, emphasizing the developing nations. The WASAR database is computer-searchable: contact the publisher for details.

33 **Simpson, Eugene S.** "Appendix: general summary of the state of research on ground-water hydrology in desert environments." In *Deserts of the world*, pp. 727-744. Edited by William G. McGinnies. Tucson AZ: University of Arizona Press, 1968.

34 *Utilization of saline water, reviews of research.* 2nd ed. Paris: Unesco, 1956. (Arid zone research 4) 102 pp.

35 *Water resources in arid and semiarid regions, 1964-Dec. 1981 (citations from the NTIS database).* Springfield VA: U.S. National Technical Information Service, 1981. (NTIS: PB 81-805-632) 274 pp.

SOILS

36 *Bibliography on desert pavement (1965-1936).* Harpenden, U.K.: Commonwealth Bureau of Soils, 196-. 7 pp.

Contains 28 citations.

37 **Dregne, Harold E.** "Appraisal of research on surface materials of desert environments." In *Deserts of the world*, pp. 287-377. Edited by William G. McGinnies. Tucson AZ: University of Arizona Press, 1968.

38 **Rhoades, Marjorie.** *Water and soil in arid regions (WASAR): an index to selected materials in Colorado State University Libraries.*

See entry 32.

BIOLOGY

39 **Hawkes, Clifford L.** *Aquatic habitat of wetlands, ponds and lakes of semiarid regions: an annotated bibliography of selected literature.* Ft. Collins CO: U.S. Forest Service. Rocky Mountain Forest and Range Experiment Station, 1980. (NTIS: PB 81-131-682) 150 pp.

Covers period 1905-1978.

40 *Indice de proyectos en desarrollo en ecologia de zonas aridas/Index of current research in arid zones ecology/Index des projects [sic] en developpement sur l'ecologie des zones arides.* Xalapa, Mexico: Instituto de Investigaciones sobre Recursos Bioticos A.C., 1978-. Irreg.

In English, French, and Spanish.

BOTANY

41 **McGinnies, William G.** "Appraisal of research on vegetation of desert environments." In *Deserts of the world*, pp. 381-566. Edited by William G. McGinnies. Tucson AZ: University of Arizona Press, 1968.

42 **Pederson, B. O., and Grainger, A.** "Bibliography of Prosopis." *International tree crops journal* 1:4 (1981):273-286.

Covers period 1904-1980. *Prosopis* (the mesquite) is distributed in drylands worldwide.

43 *Plant ecology: reviews of research.* Paris: Unesco, 1955. (Arid zone research 6) 377 pp.

In English with French summaries.

44 *Plant-water relationships in arid and semi-arid conditions, reviews of research.* Paris: Unesco, 1960. (Arid zone research 15) 225 pp.

Contains 297 citations.

ZOOLOGY

45 "Desert animals." *Arid lands abstracts* 5 (1974).

Contains 212 annotated references covering wildlife in desert biomes. Livestock is not considered.

46 *Human and animal ecology; reviews of research.* Paris: Unesco, 1957. (Arid zone research 8) 244 pp.

47 **Lowe, Charles H.** "Appraisal of research on fauna of desert environments." In *Deserts of the world*, pp. 569-645. Edited by William G. McGinnies. Tucson AZ: University of Arizona Press, 1968.

Discusses taxonomy, distribution, and general ecology of native fauna in 13 major deserts of the world. Does not examine information directly on the Takla-Makan desert of China. Extensive partially annotated bibliography.

AGRICULTURE

48 **Illes, Doris.** *Resource directory for agricultural research in semi-arid regions.* Boston MA: G. K. Hall, 1979. (Bibliographies and guides in African studies) 757 pp.

Contains some 1,500 unannotated citations indexed by subject, geographical area, personal name and/or organization.

49 *Moisture utilization in semi-arid tropics: summer rainfall agriculture project.*

See entry 30.

50 **Rhoades, Marjorie.** *Water and soil in arid regions (WASAR): an index to selected materials in Colorado State University Libraries.*

See entry 32.

51 **Riehl, S. K.,; Kinch, M.,; and Baker, R.** *Dryland agriculture bibliography: a list of materials on agriculture in semiarid temperate regions.* Washington DC: U.S. Agency for International Development, 1976. (PN-AAC-775) 169 pp.

52 **Riehl, S. K. et al.** *Bibliography of dryland agriculture.* 3rd ed. Corvallis OR: Oregon State University, 1980. 2 vols.

IRRIGATION

53 **Casey, Hugh E.** *Salinity problems in arid lands irrigation: a literature review and selected bibliography.* Tucson AZ: University of Arizona. Office of Arid Lands Studies, 1972. (Resource information paper 1; NTIS: PB 214-172) 300 pp.

Contains 986 citations.

54 **Porto, Everaldo Rocha; Silva, Aderaldo de Souza; and Luz, Maria Cira Padilha da.** *Bibliografia sinaletica sobre a pequena irrigacao "nao convencional" no tropico semi-arido [synalectic bibliography on non-conventional supplemental irrigation in the semi-arid tropics].* Brasilia: EMBRAPA. Dept. de Informacao e Documentacao, 1980. 122 pp.

RANGE MANAGEMENT

55 "Burning as a tool for arid range management." *Arid lands abstracts* 2 (1972):134-221.

ENERGY

56 **Bowden, Charles.** *The impact of energy development on water resources in arid lands: a literature review and annotated bibliography.* Tucson AZ: University of Arizona. Office of Arid Lands Studies, 1975. (Resource information paper 6; NTIS: PB 240-008) 277 pp.

Contains about 300 references and supplemental references.

57 **Duffield, Christopher.** *Solar energy, water, and industrial systems in arid lands: technological overview and annotated bibliography.* Tucson AZ: University of Arizona. Office of Arid Lands Studies, 1978. (Resource information paper 12; NTIS: PB 285-129) 151 pp.

Covers the interrelationships between solar energy, power generation, and water supplies in the drylands.

GEOTHERMAL ENERGY

58 **Duffield, Christopher.** *Geothermal technoecosystems and water cycles in arid lands.* Tucson AZ: University of Arizona. Office of Arid Lands Studies, 1976. (Resource information paper 8; NTIS: PB 263-091) 202 pp.

Includes annotated bibliography.

59 *Exploration and exploitation of geothermal resources in arid and semiarid lands: a literature review and selected bibliography.* Tucson AZ: University of Arizona. Office of Arid Lands Studies, 1973. (Resource information paper 2; NTIS: PB 218-830) 119 pp.

Contains 102 annotated references.

HYDROPOWER

60 **Cathcart, Richard Brook.** *Evaporative sea basin power-drop sites.* Monticello IL: Vance Bibliographies, 1979. (Public administration series bibliography P-328) 12 pp.

A narrative summary with unannotated references covering potential sites in the Persian Gulf, the Gulf of California, the Dead Sea, and the Qattara Depression in Egypt.

ECONOMIC DEVELOPMENT

61 *Alternative strategies for desert development: proceedings of an international conference held in Sacramento CA, May 31-June 10, 1977.* New York: Pergamon, 1982. 4 vols.

MILITARY AFFAIRS

62 **Brooks, Walter R.** *Desert testing environmental bibliography.* Yuma AZ: U.S. Army. Yuma Proving Ground, 1980. (NTIS: AD A082-347/0) 470 pp.

Includes some 900 citations on environmental factors that influence the design and testing of military equipment in arid lands. Emphasis on the Sonoran Desert near Yuma, Arizona, but coverage is worldwide.

63 **Greyeris, Harry A.** *Desert environmental handbook: first edition.* Springfield VA: U.S. National Technical Information Service. Yuma Proving Ground, 1977. (NTIS: AD A-048-608/4ST) 141 pp.

A compendium of information on testing and maintenance of equipment in arid lands.

64 **Miller, Lester L., Jr.** *Desert operations: a bibliography.* Ft. Sill OK: U.S. Army. Field Artillery School, 1979. (NTIS: AD A066-542/2ST) 11 pp.

Covers citations on desert operations as applied to warfare.

OCEANS AND COASTS

The relationships between dryland and ocean are surprisingly close and complex. The oceans drive the weather engine that creates the world's climate belts; the driest lands on earth are, in fact, stretches of narrow coastline washed by cold currents of the eastern margins of several oceans. A great geographical irony places much of the world's drylands within easy reach of the sea, making the efficient extraction of freshwater one of the primary goals of dryland technology. Because cold currents often bring great quantities of plankton some of the world's richest fishing grounds are found along dryland coasts. For dryland nations fortunate enough to possess a seaboard the resource base is vastly expanded.

The entries below and under later sections on the Atlantic, Indian, and Pacific Oceans provide a number of reference sources on oceanography and coastal deserts generally.

ONLINE DATABASES

65 *ASFA (Aquatic sciences and fisheries abstracts)*. Rome: U.N. FAO, 1978-. Updated monthly.

Corresponds to printed ASFA source. Covers oceanography, marine biology, living and non-living resources of the sea. Vendors: DIALOG, DIMDI, QL Systems.

INDEXES AND ABSTRACTS

66 *Oceanic abstracts with indexes*. La Jolla CA: Pollution Abstracts, 1972-. 6 per year.

ENCYCLOPEDIAS

67 Parker, Sybil P., ed. *McGraw-Hill encyclopedia of ocean and atmospheric sciences*. New York: McGraw Hill, 1980. 580 pp.

COASTAL DESERTS

68 Amiran, David H. K., and Wilson, Andrew H., eds. *Coastal deserts: their natural and human environments*. Tucson AZ: University of Arizona Press, 1973. 207 pp.

Based on the papers from a symposium held in Peru (1967), sponsored by the International Geographical Union, Unesco, and the Government of Peru.

69 Meigs, Peveril. *Geography of coastal deserts*. Paris: Unesco, 1966. (Arid zone research 28) 140 pp.

70 Schreiber, Joseph F., Jr. "Appraisal of research on desert coastal zones." In *Deserts of the world*, pp. 649-724. Edited by William G. McGinnies. Tucson AZ: University of Arizona Press, 1968.

GEOLOGY

71 McGill, John T. *Selected bibliography of coastal geomorphology of the world*. Los Angeles CA: University of California and the Office of Naval Research. Geography Branch, 1960. 50 pp.

Contains 933 citations.

72 Richards, Horace G., and Fairbridge, R. W. *Annotated bibliography of quaternary shorelines, 1945-1964*. Philadelphia PA: Academy of Natural Sciences, 1965. (Special publication 6) 280 pp.

Supplement 1 (1970), covers period 1965-1969 (Special publication 10), 240 pp.; *Supplement* 2 (1974) covers period 1970-1973 (Special publication 11), 214 pp.

CLIMATOLOGY

73 *Guide to NOAA's computerized information retrieval services*.

See entry 2820.

BIOLOGY

74 *Atlas of the living resources of the sea*. 3rd ed. Rome: FAO, 1972. Various pagings.

In English, French, and Spanish. Presents three series of maps: maps of geographical distribution and present state of ocean- living resources; characteristic examples of fish migration; and regional maps showing distribution and abundance of main stocks.

THE THIRD WORLD

The Third World is a convenient designation for the nations whose economies have not yet developed strong manufacturing industries and whose labor supply is largely agrarian. These "underdeveloped" or "developing" nations often face staggering economic difficulties and all the social and political disruption that accompanies poverty, famine, high birth rates, illiteracy, lack of resources, and crushing foreign debt. The dryland nations are almost always members of the Third World, even though aridity alone does not necessarily weaken a national economy. Aridity does, however, limit subsistence agriculture, population density, and economic growth unless tremendous infusions of capital and technological expertise are supplied. The great challenge to the comparatively wealthy nations of the First and Second Worlds will be their ability and willingness to share their wealth and knowledge and accept the peoples of the Third World as more than suppliers of raw materials, customers for consumer goods, or occupiers of strategic military positions.

The entries below provide reference sources on the Third World generally, covering both dryland and non-dryland nations.

RESEARCH GUIDES

75 Slamecka, Vladimir, and McCarn, Davis B. *The information resources and services of the United States: an introduction for developing countries.* Washington DC: U.S. Dept. of State, 1979. 50 pp.

Designed to assist persons in developing countries in accessing, interpreting, and using American information sources. Introduction to United States resources, and tables listing major computer-based services in the United States, are listed alphabetically and by subject. In English, with multilingual introduction.

THESES AND DISSERTATIONS

76 Sims, Michael. *United States doctoral dissertations in Third World studies, 1869-1978.* Waltham MA: Crossroads, 1981. 450 pp.

ENCYCLOPEDIAS

77 Kurian, George Thomas. *Encyclopedia of the Third World.* Rev. ed. New York: Facts on File, 1982. 3 vols.

Good nation-by-nation summaries of basic geographic, economic, and political information.

DIRECTORIES

78 Duffy, James; Hevelin, John; and Osterreicher, Suzanne. *International directory of scholars and specialists in Third World studies.* Waltham MA: Crossroads, 1981. 563 pp.

79 Duffy, David, and Jacobs, Barbara. *Directory of Third World studies in the United States.* Waltham MA: Crossroads, 1981. 463 pp.

Contains formal programs in Third World Studies as well as colleges and universities which offer at least a few courses on the Third World.

INFORMATION SCIENCE

80 Huq. A. M. Abdul. *Librarianship and the Third World: an annotated bibliography of selected literature on developing nations, 1960-1975.* New York: Garland, 1977. (Garland reference library of social science 40) 372 pp.

HYDROLOGY

81 Stow, D. A. V. *Preliminary bibliography on groundwater in developing countries, 1970 to 1976.* St. John's, Canada: Association of Geoscientists for International Development, 1976. (Geosciences in international development report 4) 305 pp.

AGRICULTURE

82 *Agricultural research in developing countries.* 1978 ed. Rome: FAO, 1978. 3 vols.

Gives addresses and basic information on various governmental agencies, universities, and institutions.

AGRICULTURAL ECONOMICS

83 Siddiqi, Akhtar Husain. *Planning policies, strategies, and performance in the agricultural sector of developing countries: a selected annotated bibliography.* Monticello IL: Vance Bibliographies, 1982. (Public administration series P-982) 42 pp.

AGRICULTURAL EXTENSION

84 de Vries, C. A. *Agricultural extension in the developing countries: a bibliography.* Wageningen, The Netherlands: International Institute for Land Reclamation and Improvement, 1968. (Bibliography 7) 125 pp.

Annotated.

Third World/85-99

AGRICULTURAL MARKETS AND COOPERATIVES

85 *Research register of studies on cooperatives in developing countries and selected bibliography.* Budapest: Co-operative Research Institute, [n.d.] Annual.

Annotated.

86 Riley, Peter, and Weber, Michael T. *Food and agricultural marketing in developing countries: an annotated bibliography of doctoral research in the social sciences, 1969-79.* East Lansing MI: Michigan State University. Dept. of Agricultural Economics, 1979. (MSU rural development series, working paper 5) 49 pp.

SOLAR ENERGY

87 Eggers-Lura, A. *Solar energy in developing countries: an overview and buyers' guide for solar scientists and engineers.* Oxford, U.K.: Pergamon, 1978. 206 pp.

MEDICINE

88 Akhtar, S. et al. *Low-cost rural health care and health manpower training: an annotated bibliography with special emphasis on developing countries.* Ottawa: International Development Research Centre, 1975-1981. 7 vols.

89 Singer, Philip, and Titus, Elizabeth A. *Resources for Third World health planners: a selected subject bibliography.* Owerri, Nigeria; Buffalo NY: Trado-Medic Books, 1980. 155 pp.

URBAN GEOGRAPHY

90 Bovy, Philippe H., and Gakenheimer, Ralph A. *Urbanization and urban transport planning in developing countries: a selected bibliography.* Monticello IL: Council of Planning Librarians, 1975. (CPL exchange bibliography 895) 26 pp.

Unannotated.

91 Brunn, Stanley D. *Urbanization in developing countries: an international bibliography.* East Lansing MI: Michigan State University. Latin American Studies Center, 1971. (Research report 8) 693 pp.

Over 7,000 entries dealing with all facets of urbanization in Latin America, Africa, and Asia. Multilingual.

92 Buick, Barbara. *Squatter settlements in developing countries: a bibliography.* Canberra, Australia: Australian National University. Research School of Pacific Studies, 1975. (Aids to research series A/3) 158 pp.

Covers publication from 1950 through 1973. Multilingual references.

93 Hallaron, Shirley Anderson. *Urbanization in the developing nations: a bibliography compiled for the 1960's and 1970's.* Monticello IL: Council of Planning Librarians, 1976. (CPL exchange bibliography 1181) 46 pp.

Unannotated.

ARCHITECTURE

94 *Architecture in developing countries: a bibliographic guide to local self-assistance programs.* Monticello IL: Vance Bibliographies, 1981. (Architecture series bibliography A505) 10 pp.

Unannotated. Prepared by Coppa and Avery Consultants.

HOUSING

95 *L'habitat defectueux dans les pays en voie de developpement [substandard housing in developing countries].* Paris: France. Secretariat des missions d'urbanisme et d'habitat. Service documentation-information, 1974. (Bibliographie 10).

Annotated.

96 *L'habitat du grand nombre dans les pays en developpement [social housing in developing countries].* Paris: France. Secretariat des missions d'urbanisme et d'habitat. Service documentation-information, 1973-1974. (Bibliographie 9).

97 Mackenzie, Donald R., and Kerst, Erna W. *A bibliographic overview of housing in developing countries: annotated.* Monticello IL: Council of Planning Librarians, 1977. (CPL exchange bibliography 1225-1226-1227) 281 pp.

ECONOMIC DEVELOPMENT

98 Auerbach, Devoira et al. *Regional plans of developing countries: an annotated bibliography.* Giessen, Federal Republic of Germany: Justus-Liebig-University Giessen. Zentrum fur Regionale Entwicklungsforschung, 1980. 414 pp.

99 Bauer, A. J. H. *Rural development in the Third World, 1970-1977.* Wageningen, The Netherlands: International Institute for Land Reclamation and Improvement, 1980. (Bibliography 17) 191 pp.

Briefly annotated. Covers the modernization of the rural environment, focusing on socio-economic patterns. Rural development is characterized as that based on deliberate planning, with the use of modern technology, and serving to benefit the rural population as a whole. Special attention to small peasants, rural women, project organization. Includes Japan and Israel. Omits socialist countries.

100 *The developing areas; a classed bibliography of the Joint Bank-Fund Library, Washington, D.C..* Boston: G. K. Hall, 1976. 3 vols.

Produced by the Joint Library of the International Monetary Fund and the International Bank for Reconstruction and Development.

101 *Development guide: a directory of non-governmental organisations in Britain actively concerned in overseas development and training.* 3rd ed. London: Allen & Unwin, 1978. 216 pp.

Contains entries for 198 organizations with detailed accounts of their facilities. Educational establishments, commercial organizations, foundations, and social-cultural organizations excluded. Produced by Overseas Development Institute, London.

102 *Devindex.* Ottawa: International Development Research Centre, 1977-. Annual.

Annotated index to literature on economic and social development. Includes English and French publications.

103 **Geiger, H. Kent.** *National development 1776-1966: a selective and annotated guide to the most important articles in English.* Metuchen NJ: Scarecrow, 1969. 247 pp.

Extensively annotated with rating system. Material published through 1966 on cultural, economic, political and social aspects of development.

104 *International development and the human environment: an annotated bibliography.* New York: Macmillan Information, 1974. 334 pp.

Extensive annotations. Most material published between 1968 and 1972. Considers environmental deterioration of the developed world and possibilities open to Third World development.

105 **Joseph, Richard M., Jr.** *Budgeting in Third World countries: an annotated bibliography.* Monticello IL: Vance Bibliographies, 1982. (Public administration series bibliography P-893) 43 pp.

Extensive annotations.

106 **Powelson, John P.** *A select bibliography on economic development, with annotations.* Boulder CO: Westview, 1979. (Westview special studies in social, political, and economic development) 450 pp.

Contains books and journal articles published since 1970.

107 **Rondinelli, Dennis A., and Palia, Aspy P.,** eds. *Project planning and implementation in developing countries: a bibliography on development project management.* Honolulu HI: East-West Center. Technology and Development Institute, 1976. 174 pp.

108 **Schumacher, August.** *Development plans and planning: bibliographic and computer aids to research.* London: Seminar Press, 1973. (International bibliographical and library series 3) 195 pp.

Contains bibliographies relevant to development plans, documentation centers, and technology.

TRANSPORTATION

109 **Mahayni, Riad C.** *Transportation planning in Third World countries: an annotated bibliography.* Monticello IL: Council of Planning Librarians, 1976. (CPL exchange bibliography 1108) 26 pp.

Extensive annotations.

TOURISM AND RECREATION

110 **de Burlo, Charles.** *The geography of tourism in developing countries: an annotated bibliography, with special reference to sociocultural impacts.* Monticello IL: Vance Bibliographies, 1980. (Public administration series bibliography P-546) 28 pp.

Extensive annotations.

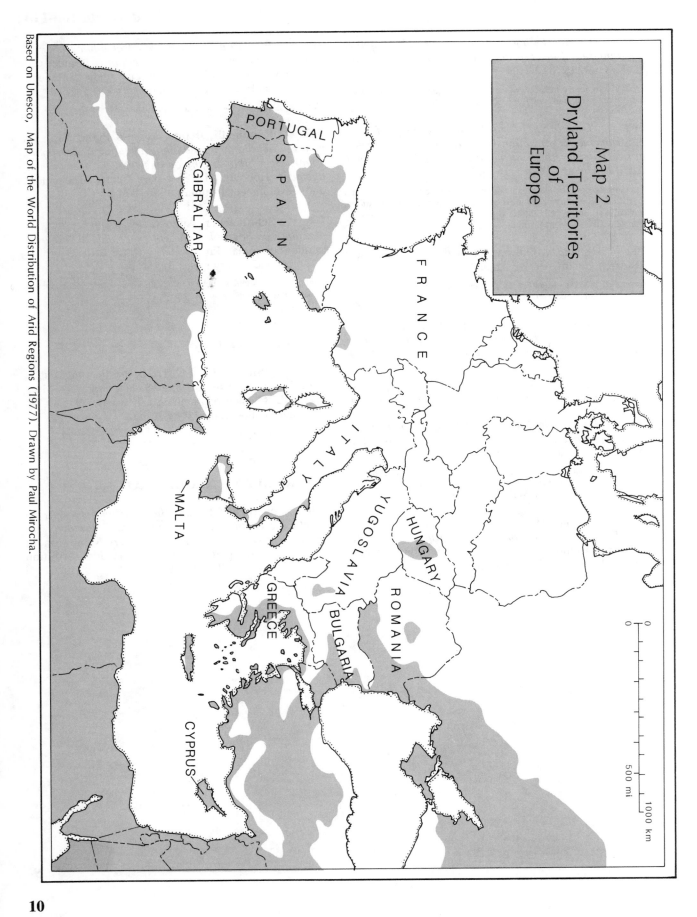

EUROPE

Most of Europe lies in the path of cool, wet maritime air masses that sweep inland from the North Atlantic; only along the Mediterranean, where dry tropical maritime air masses invade during the summer, do extensive dryland areas occur. Isolated drylands also exist in Eastern Europe due to inland rainshadow effects or Mediterranean-type climates extending along the Black Sea coast.

For organizational purposes, the Soviet Union is considered later under Northern and Eastern Asia.

Dryland areas of EUROPE (EXCLUDES U.S.S.R.)

DESERTIFICATION RISK	ARIDITY									
	Hyperarid		Arid		Semiarid		Subhumid		Aridity Totals	
	km²	%	km²	%	km²	%	km²	%	km²	%
Very High	By definition, desertification does not exist in hyperarid regions.		2,488	0.1	63,196	1.3			65,684	1.4
High					70,660	1.5			70,660	1.5
Moderate					4,956	0.1	10,812	0.2	15,768	0.3
Desertification Totals			2,488	0.1	138,812	2.9	10,812	0.2	152,112	3.2
No Desertification					36,407	0.7	486,023	10.1	522,430	10.8
Total Drylands			2,488	0.1	175,219	3.6	496,835	10.3	674,542	14.0
Non-dryland									4,146,817	86.0
Total Area of the Territory									4,821,359	100.0

DIRECTORIES

111 Anderson, Ian Gibson, ed. *Directory of European Associations/Repertoire des associations europeenes/Handbuch der europaeischen Verbaude.* 3rd ed. Detroit: Gale Research; Beckenham, U.K.: CBD Research, 1981. 2 vols.

Part 1 covers national industrial, trade and professional associations. Part 2 covers national learned, scientific and technical societies. In English, French, and German; essential instructions also in Dutch, Italian, and Spanish. Includes continental Europe, Iceland, Malta, and Cyprus; excludes the United Kingdom and the Republic of Ireland. Items are arranged by subject. Information for each of 5,000 organizations includes name(s), address, membership, stated purpose, and publications. Includes tri-lingual index of subjects plus index of organizations.

GAZETEERS

112 *Europe and U.S.S.R.: official standard names approved by the United States Board on Geographic Names.* Washington DC: U.S. Board on Geographic Names, 1971. (Gazeteer supplement 118) 151 pp.

List of 1,770 place-name corrections with latitude/longitude coordinates. Includes European and Asiatic U.S.S.R. Does not include Malta.

CLIMATOLOGY

113 *Climatic atlas of Europe.* Geneva: World Meteorological Organization/Unesco/Cartografia, 1970-. 28 maps.

In English, French, Russian, and Spanish. Covers data from 1931-1960, with base maps at 1:10,000,000. Covers Europe, the Soviet Union to the Urals, and the Middle East of Turkey, Syria, Lebanon, Israel, and Jordan.

114 Thran, P., and Broekhuizen, Simon. *Agroclimatic atlas of Europe.* Wageningen, The Netherlands; Amsterdam: PUDOC; Elsevier, 1965. Looseleaf.

In English, German, French. Base map at scale 1:10,000,000. Covers Europe and European Soviet Union, with sporadic coverage of northern Morocco, Algeria, Tunisia. Forms volume 1 of *Agro-ecological atlas of cereal growing in Europe.*

BOTANY

115 Kuechler, August Wilhelm. "Europe." In *International bibliography of vegetation maps*, vol. 2. Lawrence KS: University of Kansas Libraries, 1966. (Library series 26) 584 pp.

Part II of a four volume set. (The set covers North America; Europe; U.S.S.R., Asia, and Australia; and Africa, South America and world maps.)

ZOOLOGY

116 Harrison, Colin. *An atlas of the birds of the Western Palaearctic.* Princeton NJ: Princeton University Press, 1982. 322 pp.

Covers Europe, Northern Africa, and the Middle East. Distribution maps are small-scale but detailed. Includes extensive text and short bibliography.

AGRICULTURE

117 *Indices of agricultural and food production for Europe and the U.S.S.R..* Washington DC: U.S. Dept. of Agriculture, 1979. (Statistical bulletin 620) 42 pp.

Covers average production from 1961-1965; annual production from 1969-1978.

118 Thran, P., and Broekhuizen, Simon. *Atlas of the cereal-growing areas in Europe.* Wageningen, The Netherlands; Amsterdam: PUDOC; Elsevier, 1969. 157 pp.

In English, German, French. Covers Europe and European Soviet Union. Forms volume 2 of *Agro-ecological atlas of cereal growing in Europe.*

HUMAN GEOGRAPHY

119 Gabrovska, Svobodozarya. *European guide to social science information and documentation services.* Oxford; New York: Pergamon Press, 1982. 234 pp.

A directory of social science institutions—primarily libraries and other documentation centers—in Europe and the U.S.S.R.

ECONOMIC DEVELOPMENT

120 *Sources of European economic information.* Epping, U.K.: Gower Press, 1974. 343 pp.

In English, French, and German. Covers Western Europe.

121 Whiteside, R. M., ed. *Major companies of Europe - 1982.* London: Graham and Trotman Ltd., 1982. 2 vols.

Volume one deals with companies in the European Economic Community; volume two with companies in Western Europe outside the EEC.

Mediterranean Europe

The lands surrounding and emerging from the Mediterranean Sea display in classic form the climate-type known on all continents as "Mediterranean"; this type is defined by coastal areas exposed to westerly winds that bring wet polar air masses in the winter and dry hot air masses in the summer. Although summer drought is a requirement of the type, sufficient precipitation can exist to place some areas outside the dry zone; otherwise, the aridity range is usually subhumid to semiarid since the dry period falls when evapotranspiration is highest.

CLIMATOLOGY

122 Gleeson, Thomas Alexander. *Bibliography of the meteorology of the Mediterranean, Middle East and South Asian areas.* Gainesville FL: Florida State University. Dept. of Meteorology, 1952 (Scientific report 1 appendix) 37 pages.

Annotated. Covers the Mediterranean Sea, southern Europe, west, north, east Africa, southwest Asia, India, the Indian Ocean, Burma, and Thailand.

HYDROLOGY

123 Beavington, C. F., and Williams, J. B. *Bibliography of water resources of Commonwealth countries in the Caribbean and Mediterranean.* London: U.K. Commonwealth Geological Liaison Office, 1980. (Special liaison report) 59 pp.

ANTHROPOLOGY

124 Sweet, Louise Elizabeth, and O'Leary, Timothy J. *Circum-Mediterranean peasantry: introductory bibliographies.* New Haven CT: Human Relations Area Files Press, 1969. 106 pp.

HISTORICAL GEOGRAPHY

125 Hammond, Nicholas G. L., ed. *Atlas of the Greek and Roman world in antiquity.* Park Ridge NJ: Noyes Press, 1981. 56 pp.

A good recent historical atlas of the ancient Mediterranean, covering the period from the Neolithic to the sixth century A.D. Includes gazetteer.

CYPRUS

(formerly a colony of the United Kingdom)
Kypriaki Dimokratia (Greek portion)
Kibris Cumhuriyeti (Turkish portion)

Cyprus lies equidistant from the coasts of Turkey and Syria and is often included in the Middle East rather than Europe. Ethnically it is also split: Turkish Cypriots (backed by a Turkish occupying army since 1974) control the northern 40 percent, while Greek Cypriots live in the southern remainder. Separate governmental administrations exist for each.

The island is subhumid and drier overall than any other large island in the Mediterranean. Irrigated agriculture is common on the Mesaoria Plain that forms the center of the island; grapes, citrus, and potatoes are the principal export crops. From ancient times mining the copper and iron of the Troodos Massif on the southern half of the island has been a major industry; together with asbestos and chromite these are important exports. Despite great potential, tourism has developed only slowly, in part because of the many years of political violence.

Dryland areas of CYPRUS

DESERTIFICATION RISK	ARIDITY									
	Hyperarid		Arid		Semiarid		Subhumid		Aridity Totals	
	km²	%	km²	%	km²	%	km²	%	km²	%
Very High	By definition, desertification does not exist in hyperarid regions.									
High										
Moderate										
Desertification Totals										
No Desertification								9,251	100.0	9,251	100.0
Total Drylands								9,251	100.0	9,251	100.0
Non-dryland										
Total Area of the Territory									9,251	100.0

HANDBOOKS

126 Burge, Frederica M., ed. *Cyprus: a country study*. 3rd ed. Washington DC: U.S. Dept. of Defense, 1980. (American University foreign area studies) 306 pp.

Narrative summary of the nation's history, politics, geography, economy, social life; includes unannotated bibliography.

GEOLOGY

127 Avnimelech, Moshe A. *Bibliography of Levant geology, including Cyprus, Hatay, Israel, Jordania, Lebanon, Sinai, and Syria*.

See entry 1157.

128 *Current bibliography of Middle East geology*.

See entry 1158.

129 Haralambous, Diomedes. *Geologike kai physikogeographike bibliographia tes Hellados/ Geological and physicogeographical bibliography of Greece*.

See entry 157.

130 Haralambous, Diomedes. *Geoscience and natural science: bibliography of Greece*.

See entry 158.

131 Pantazis, Th. M. "A revised bibliography of Cyprus geology." Cyprus Geological Survey. *Bulletin* 2 (1969):57-81.

132 **Ridge, John Drew.** "Cyprus." In *Annotated bibliographies of mineral deposits in Africa, Asia (exclusive of the USSR) and Australasia*, pp. 223-229. Oxford U.K.: Pergamon, 1976.

CLIMATOLOGY

133 **Taha, M. F. et al.** "The Climate of the Near East."

See entry 1165.

ANTHROPOLOGY

134 **Allen, Peter Sutton, and Bialor, Perry A.** "Bibliography of anthropological sources on modern Greece and Cyprus."

See entry 166.

ECONOMIC DEVELOPMENT

135 *Quarterly economic review of Lebanon, Cyprus.*

See entry 1324.

FRANCE
Republique Francaise

France's drylands are confined to two small subhumid regions, one at the head of the Gulf of Lyon and one along the low east coast of Corsica.

The Midi. France's Mediterranean coast displays some features typical of the Basin while others are more unusual. The drylands themselves are confined to the area at and around the delta of the Rhone River; this stream is one of few in the Basin which is strong enough to create a sizeable alluvial plain. Here can be found irrigated fodder crops, livestock, and extensive plantings of rice. Elsewhere, on the slopes, olives and vines are common. The whole Midi is a very desirable tourist destination, though the famous Riviera is to be found in the wetter area to the east.

Corsica. This rocky island possesses a subhumid rainshadow dryland on its only large lowland, the Plaine d'Aleria along the east coast. Though once plagued by malaria, after the Second World War the region was cleared of mosquitoes and now contains considerable tracts of irrigated land that produce a variety of typical Mediterranean crops.

Due to the relatively small importance of these drylands to the nation, only a few selected references are provided.

Dryland areas of FRANCE

DESERTIFICATION RISK	ARIDITY									
	Hyperarid		Arid		Semiarid		Subhumid		Aridity Totals	
	km²	%	km²	%	km²	%	km²	%	km²	%
Very High	By definition, desertification does not exist in hyperarid regions.									
High										
Moderate										
Desertification Totals										
No Desertification							9,247	1.7	9,247	1.7
Total Drylands							9,247	1.7	9,247	1.7
Non-dryland									534,718	98.3
Total Area of the Territory									543,965	100.0

GENERAL BIBLIOGRAPHIES

136 **Chambers, Frances.** *France*. Oxford U.K.; Santa Barbara CA: Clio Press, 1980. (World Bibliographical series 13) 177 pp.

A relatively brief bibliography of 542 annotated citations limited to the English language. Most major subjects are touched.

GOVERNMENT DOCUMENTS

137 **Westfall, Gloria.** *French official publications*. New York: Pergamon Press, 1980. (Guides to official publications 6) 209 pp.

Narrative discussion of French government documents, with bibliographies.

ATLASES

138 *Grand atlas de la France [grand atlas of France]*. Paris: Reader's Digest S.A.R.L., 1969. 224 pp.

In French. Covers continental France and Corsica only.

GAZETTEERS

139 *France: official standard names approved by the United States Board on Geographic Names*. Washington DC: U.S. Board on Geographic Names, 1964. (Gazetteer 83) 2 vols. 1381 pp.

List of 45,000 place-names with latitude/longitude coordinates. Covers France, Corsica, and Monaco.

GEOLOGY

140 Gouvernet, C.; Guiell, G.; and Rousset, C. *Provence*. 2nd ed. Paris; New York: Masson, 1979. (Guides geologiques regionaux) 240 pp.

In French.

HYDROLOGY

141 *Bibliographie hydrogeologique de la France, 1968-1977 [hydrogeologic bibliography of France]*. Orleans: France. Bureau de Recherches Geologiques et Minieres, 1977. (Memoires 93) 97 pp.

In French. Annotated.

CLIMATOLOGY

142 *Carte climatique detaillee de la France [detailed climatic map of France]*. Paris: France. Centre National de la recherche scientifique. Section de geographie, 1970-.

In French. Map series. The Marseille and Corsica sheets depict France's subhumid drylands. Scale 1:250,000.

AGRICULTURE

143 *AGREP: permanent inventory of agricultural research projects in the European Communities.*

See entry 177.

ECONOMIC DEVELOPMENT

144 *KOMPASS: repertoire general de la production francaise [general register of French production]*. Paris: Societe Nouvelle d'Editions pour l'Industrie, 1923-. Annual.

In French with summaries in French, English, German, and Spanish. Contains general economic and financial information on 55,000 French and 19,000 non-French firms.

145 *Quarterly economic review of France*. London: Economist Intelligence Unit, 1976-. Quarterly.

Includes summary of political and economic news as well as charts and statistics of selected economic indicators. Annual supplement.

GIBRALTER
The City of Gibralter
(presently a territory of the United Kingdom)

The narrow peninsula of Gibralter projects southward into the Mediterranean just inside the Strait of Gibralter. Though attached to Spain by a flat isthmus, "The Rock" has been a British naval base since its capture in 1704 and its inhabitants are overwhelmingly pro-British. The Rock itself is a limestone mountain rising to some 470 meters (1,400 feet) above the sea, riddled with caves, with a lower tableland skirting the base of the peak on two sides. The settlement of Gibralter lies below the west face alongside an open harbor protected by artificial moles. With a typically Mediterranean subhumid climate, The Rock has a sizeable summer water deficit that must be met with a rainwater catchment system on the eastern slopes leading to deep cisterns within The Rock itself. Save for small home gardens there is no agriculture; the economy rests solely on its functions as a tourist stop, a transshipment center for goods, a provisioning port for visiting ships, a site for commercial ship-repair, and a large military presence.

Dryland areas of GIBRALTER

DESERTIFICATION RISK	ARIDITY									
	Hyperarid		Arid		Semiarid		Subhumid		Aridity Totals	
	km²	%	km²	%	km²	%	km²	%	km²	%
Very High	By definition, desertification does not exist in hyperarid regions.									
High										
Moderate										
Desertification Totals										
No Desertification								6.5	100.0	6.5	100.0
Total Drylands								6.5	100.0	6.5	100.0
Non-dryland										0.0	0.0
Total Area of the Territory										6.5	100.0

GENERAL BIBLIOGRAPHIES

146 Green, Muriel M. *A Gibralter bibliography.* London: University of London. Institute of Commonwealth Studies, 1980. 108 pp.

CLIMATOLOGY

147 Wallace, J. Allen, Jr. *Bibliography on climate of Malta and Gibralter.* Washington DC: U.S. Weather Bureau, 1958. (NTIS: AD 664-695) 55 pp.

Annotated bibliography of 120 items.

GREECE
Elliniki Dimokratia
[The Hellenic Republic]

Greece consists of a series of rocky peninsulas deeply scored by bays of the sea. To the east and south of this mainland a large number of equally rugged islands stretches across the Aegean Sea to the Turkish coast. The southern portion of the Aegean and its islands and nearby mainland form the semiarid dryland center with pockets of subhumid territory appearing around the perimeter of the basin and inland. Only in the northwest—ancient Macedonia—is the land well-watered.

The resource base is poor. The natural sclerophyllous scrub vegetation is over-grazed and of little value. Forests have disappeared from the upper slopes. Patches of deeper soils along the narrow, disconnected coastal plains are intensely tilled for cereals, cotton, sugar beets, and irrigated vegetables; tree crops of olives, fruits, and vines flourish on the slopes. Despite generally high yields, so little land can be properly cultivated that grazing of goats and sheep remains a mainstay of the local economies. Fishing is an expanding industry and mineral resources have long contributed to the nation's limited exports. Although food and fuel must both be imported, Greece survives on a strong tourist trade, a large shipping fleet, a variety of manufactured goods, and on remittances sent home by migrants abroad.

Dryland areas of GREECE

DESERTIFICATION RISK	ARIDITY									
	Hyperarid		Arid		Semiarid		Subhumid		Aridity Totals	
	km²	%	km²	%	km²	%	km²	%	km²	%
Very High	By definition, desertification does not exist in hyperarid regions.									
High										
Moderate										
Desertification Totals										
No Desertification					19,534	14.8	50,551	38.3	70,085	53.1
Total Drylands					19,534	14.8	50,551	38.3	70,085	53.1
Non-dryland									61,901	46.9
Total Area of the Territory									131,986	100.0

GENERAL BIBLIOGRAPHIES

148 Horecky, Paul Louis. *Southeastern Europe: a guide to basic publications.*

See entry 208.

ENCYCLOPEDIAS

149 *Megali elliniki enkyklopaideia [great Greek encyclopedia].* 2nd ed. Athens: O Phoinix, 1956-1965. 28 vols.

In Greek. A general reference national encyclopedia.

HANDBOOKS

150 Keefe, Eugene K. *Area handbook for Greece.* 2nd ed. Washington DC: U.S. Dept. of Defense, 1977. (American University foreign area studies) 284 pp.

Narrative summary of the nation's history, politics, geography, economy, social life; includes unannotated bibliography.

DIRECTORIES

151 Wheeler, Stella E. L., and Shah, Khawaja T., eds. *Greece, Turkey, and the Arab states.* 2nd ed. Guernsey, U.K.: Francis Hodgson, 1976. (Guide to world science 12) 236 pp.

Provides descriptions of scientific research activities and organizations. Covers Greece, North Africa (Morocco, Algeria, Tunisia, Libya, Sudan, Egypt), and the Middle East (Turkey, Syria, Jordan, Lebanon, Saudi Arabia, the Gulf States, Kuwait, the Yemens).

STATISTICS

152 McCarthy, Justin, Jr. *The Arab world, Turkey, and the Balkans (1878-1914): a handbook of historical statistics.*

See entry 1154.

153 *Statistiki epiteris tis Ellados [statistical yearbook of Greece].* Athens: Greece. National Statistical Service, 1930-. Annual.

In Greek.

GEOGRAPHY

154 Gerakis, Pantazis A. "Greece." In *Handbook of contemporary developments in world ecology*, pp. 157-183. Edited by Edward John Kormondy and J. Frank McCormick. Westport CT: Greenwood, 1981.

A good narrative summary of ecological research in Greece. Includes short annotated bibliography.

ATLASES

155 Kayser, Bernard et al. *Oikonomikos kai koinonikos atlas tes Hellados/Economic and social atlas of Greece.* Athens: National Statistical Service of Greece, 1964. 508 maps. (Looseleaf).

In Greek, English, and French. Most data refers to 1961.

GAZETTEERS

156 *Greece: official standard names approved by the United States Board on Geographic Names.* Washington DC: U.S. Board on Geographic Names, 1955. (Gazetteer 11) 404 pp.

List of 15,600 place-names with latitude/longitude coordinates. Covers Greece, Crete, and the Aegean Islands.

GEOLOGY

157 Haralambous, Diomedes. *Geologike kai physikogeographike bibliographia tes Hellados/ Geological and physicogeographical bibliography of Greece.* Athens: Greece. National Institute of Geological and Mining Research, 1961. 236 pp.

In English and Greek. Continued by Haralambous (1975), entry 158. Covers Greece and Cyprus for the years 1500-1959.

158 Haralambous, Diomedes. **Geoscience and natural science: bibliography of Greece.** Athens: Greece. National Institute of Geological and Mining Research, 1975. 192 pp.

In English and Greek. Covers Greece and Cyprus for the years 1960 to 1973. Continues Haralambous (1961), entry 157. Volume two of *Geologike kai physikogeographike bibliographia tes Hellados/Geological and physicogeographical bibliography of Greece.*

CLIMATOLOGY

159 Furlan, D. "The climates of southeast Europe." In *Climates of central and southern Europe*, pp. 185-235. Edited by C. C. Wallen. Amsterdam, The Netherlands: Elsevier Scientific Publishing Co., 1977. (World survey of climatology 6) 248 pp.

Covers Greece, Romania, Bulgaria, Yugoslavia. A narrative summary of these nations' climates, with maps, charts, and an unannotated bibliography.

BOTANY

160 Davis, Peter Hadland, and Edmondson, J. R. "Flora of Turkey: a floristic bibliography."

See entry 1372.

161 Lavrentiades, G. J. "Bibliographia phytosociologica: Graecia" [bibliography of plant ecology: Greece]. *Excerpta botanica: sectio B: sociologica* 7:2 (1966): 105-108.

Unannotated. Contains 39 items.

AGRICULTURE

162 *AGREP: permanent inventory of agricultural research projects in the European Communities.*

See entry 177.

HUMAN GEOGRAPHY

163 Clogg, Mary Jo, and Clogg, Richard, comp. *Greece.* Oxford, U.K.; Santa Barbara CA: Clio Press, 1980. (World bibliographical series 17) 224 pp.

A good recent annotated bibliography covering the social sciences; 830 citations. Limited to English language titles with few exceptions. Ignores physical geography.

164 Dimaras, C. Th.; Koumarianou, D.; and Droulia, L. *Modern Greek culture: a selected bibliography (in English, French, German, Italian).* 4th ed. Athens: National Hellenic Committee of the International Association for South Eastern European Studies, 1974. 119 pp.

Covers geography, history, language, literature.

MIGRATION

165 Vlachos, Evan. *An annotated bibliography on Greek migration.* Athens: Social Sciences Centre, 1966. (Research monographs on migration 1) 127 pp.

ANTHROPOLOGY

166 **Allen, Peter Sutton, and Bialor, Perry A.** "Bibliography of anthropological sources on modern Greece and Cyprus." *Modern greek society: a newsletter* 4:1 (1976): 6-60.

ECONOMIC DEVELOPMENT

167 *Quarterly economic review of Greece*. London: Economist Intelligence Unit, 1976-. Quarterly.

Includes summary of political and economic news as well as charts and statistics of selected economic indicators. Annual supplement.

ITALY
Repubblica Italiana

Italy's long peninsula, large islands, and central position make it the most Mediterranean of nations. As the homeland of a magnificent ancient empire, it has exercised its influence throughout the Basin and surrounding lands; from Spain, across North Africa, and deep into the Middle East the remains of Roman agriculture and engineering testify to the industry and skill these people brought to the problems of making the drylands bloom. Italy itself is generally wet save for three regions.

Southern Peninsula. Where the heel, instep, and toe of the Italian boot face eastward can be found the peninsula's subhumid dryland. The area corresponds to the *regioni* of Apulia, Basilicata, and Calabria, and consists of a lowland plain in the first region and much more mountainous country in the second and third. Vineyards are important across the lowland while small coastal pockets of irrigated soil support fruit and vegetable crops; upland areas are given over to sheep and goats. Hydropower sites in the mountains of Calabria have been developed while shipbuilding and other industries are found in the coastal towns.

Sicily. The Mediterranean's largest island is a triangular bridge between Italy and North Africa, of importance since ancient times as a source of grain, minerals, and a site for strategic military bases. Today it is relatively poor agriculturally but does possess hydropower sites, mineral wealth (sulfur and oil), and petroleum refineries. Agriculture is concentrated in coastal lowlands where citrus, grapes, and irrigated fruits and vegetables are produced.

Sardinia. This island is Corsica's larger southerly companion and, like Corsica, is relatively underdeveloped compared to similar mainland areas. The typically Mediterranean agricultural pattern prevails; mineral wealth (lead, zinc, coal) is also exploited.

Dryland areas of ITALY

DESERTIFICATION RISK	ARIDITY									
	Hyperarid		Arid		Semiarid		Subhumid		Aridity Totals	
	km²	%	km²	%	km²	%	km²	%	km²	%
Very High	By definition, desertification does not exist in hyperarid regions.									
High										
Moderate										
Desertification Totals										
No Desertification								54,525	18.1	54,525	18.1
Total Drylands								54,525	18.1	54,525	18.1
Non-dryland										246,720	81.9
Total Area of the Territory										301,245	100.0

HANDBOOKS

168 **Keefe, Eugene K.** *Area handbook for Italy.* Washington DC: U.S. Dept. of Defense, 1977. (American University foreign area studies) 296 pp.

Narrative summary of the nation's history, politics, geography, economy, social life; includes unannotated bibliography.

DIRECTORIES

169 **Moore, Howard, ed.** *Italy.* 2nd ed. Guernsey, U.K.: Francis Hodgson, 1975. (Guide to world science 4) 187 pp.

Provides descriptions of scientific research activities and organizations.

GEOGRAPHY

170 *Collana di bibliografie geografiche delle regioni Italiane [bibliographies of the geography of Italy]*. Napoli, Italy: R. Pironti, 1959-.

In Italian. Classified annotated bibliography with sections on geology, geomorphology, vulcanism and earthquakes, hydrology. Volumes 7- issued by the Consiglio nazionale delle ricerche, Comitato per le scienze storiche, filologiche e filosofiche.

171 **Ravera, Oscar.** "Italy." In *Handbook of contemporary developments in world ecology*, pp. 223-228. Edited by Edward John Kormondy and J. Frank McCormick. Westport CT: Greenwood, 1981.

A very short narrative summary of ecological research in Italy. Lacks bibliography.

GAZETTEERS

172 *Italy and associated areas: official standard names approved by the United States Board on Geographic Names*. Washington DC: U.S. Board on Geographic Names, 1956. (Gazetteer 23) 369 pp.

List of 28,900 place-names with latitude/longitude coordinates. Covers Italy, Vatican City, San Marino, and Sardinia.

GEOLOGY

173 *Bibliografia geologica d'Italia [geological bibliography of Italy]*. Napoli: Italy. Consiglio Nazionale delle Richerche. Comitato per la Geografia, Geologia e Mineralogia, 1956-1964. 15 vols.

In Italian.

CLIMATOLOGY

174 **Cantu, V.** "The climate of Italy." In *Climates of central and southern Europe*, pp. 127-183. Edited by C. C. Wallen. Amsterdam, The Netherlands: Elsevier Scientific, 1977. (World survey of climatology 6) 248 pp.

A narrative summary with maps, charts, and an unannotated bibliography.

HYDROLOGY

175 **Arone, G., and Faillace, C.** *Bibliografia idrogeologica italiana (1930-1973): acque sotterranee [hydrological bibliography for Italy (1930-1973): ground water]*. Rome: Servizio Geologica Italiana, 1974. (Bollettino 95:2) 230 pp.

In Italian.

BOTANY

176 **Pignatti, Erika, and Pignatti, Sandro.** "Bibliographia phytosociologica: Italia" [bibliography of plant ecology: Italy]. *Excerpta botanica: sectio B: sociologica* 1:4 (1959):265-319; 2:2 (1960):157-159.

Unannotated. Covers the literature from 1900-1959. Contains 593 items. Author and subject indexes.

AGRICULTURE

177 *AGREP: permanent inventory of agricultural research projects in the European Communities*. The Hague: Nijhoff/Junk, 1980. 2 vols.

Coverage includes France, Italy, Greece.

178 **Carrozza, A.** "Appunti di bibliografia sulla riforma fondiaria e agraria in Italia" [bibliographic notes on land and agricultural reform in Italy]. *Rivista Di Diritto Agrario* 49:1 (1970):119- 147.

In Italian.

SOLAR ENERGY

179 **Shea, Carol A.** *Solar energy in Italy: a profile of renewable energy activity in its national context*. Golden CO: U.S. Solar Energy Research Institute, 1980. 72 pp.

A summary and directory of solar energy research activities and organizations in Italy.

ECONOMIC DEVELOPMENT

180 *KOMPASS: repertorio generale dell' economia italiana [register of Italian industry and commerce]*. Milano, Italy: ETAS-KOMPASS Edizioni per l'informazione economica, 1962/63-. Annual.

In Italian. Summaries in Italian, French, English, German, Spanish.

181 *Quarterly economic review of Italy*. London: Economist Intelligence Unit, 1976-. Quarterly.

Includes summary of political and economic news as well as charts and statistics of selected economic indicators. Annual supplement.

MALTA
Repubblika Ta Malta
(formerly a colony of the United Kingdom)

The Maltese Islands are a tiny strategic archipelago located at the center of the Mediterranean Basin. Centuries of invasion and occupation by Phoenicians, Romans, Arabs, Normans, the Spanish, the Knights of St. John, the French, and, up to 1964, the British have created an unusual mixing of culture and language that is distinctly Maltese. The limestone soils increase the natural semiarid dryland so that water conservation is a necessity. Careful cultivation and field terracing have increased the arable lands; potatoes, plant stock, wine, meat, and onions are produced in exportable quantities. With Europe's densest population and few natural resources the Republic has attempted to capitalize on its strategic position, excellent deep water ports, and tourist potential.

Dryland areas of MALTA

DESERTIFICATION RISK	Hyperarid		Arid		Semiarid		Subhumid		Aridity Totals	
	km²	%	km²	%	km²	%	km²	%	km²	%
Very High	By definition, desertification does not exist in hyperarid regions.									
High										
Moderate										
Desertification Totals										
No Desertification							316	100.0	316	100.0
Total Drylands							316	100.0	316	100.0
Non-dryland									0	0.0
Total Area of the Territory									316	100.0

GAZETTEERS

182 *Malta: official standard names approved by the United States Board on Geographic Names.* Washington DC: U.S. Board on Geographic Names, 1971. (Gazetteer 120) 49 pp.

List of 2,350 place-names with latitude/longitude coordinates. Covers the islands of Malta and Gozo.

CLIMATOLOGY

183 **Wallace, J. Allen, Jr.** *Bibliography on climate of Malta and Gibralter.*

See entry 147.

ECONOMIC DEVELOPMENT

184 *Quarterly economic review of Libya, Tunisia, Malta.*

See entry 582.

PORTUGAL
Republica Portuguesa

Portugal, though facing the open Atlantic, can still be considered a Mediterranean nation in its dryland climate, its strong association with Spain, and its Latin language and culture. Roughly half the nation is dryland, primarily subhumid but extending into the semiarid range along the southern Algarve Coast. Relief in the South is generally low and the land provides typical Mediterranean products of wheat, olives, and sheep; pigs graze the mast of extensive oak forests. Some irrigation aids vegetable and rice production. The Algarve coast is a separate entity; mountains separate it from the lands further north and the ancient Moorish influence is strong on the people, their architecture, and their terraced, irrigated gardens. Expanded tourism along the Algarve is a recent development.

The island groups of Madeira and the Azores are considered in separate chapters.

Dryland areas of PORTUGAL

DESERTIFICATION RISK	ARIDITY									
	Hyperarid		Arid		Semiarid		Subhumid		Aridity Totals	
	km²	%	km²	%	km²	%	km²	%	km²	%
Very High	By definition, desertification does not exist in hyperarid regions.									
High										
Moderate					4,956	5.6			4,956	5.6
Desertification Totals					4,956	5.6			4,956	5.6
No Desertification							40,710	46.0	40,710	46.0
Total Drylands					4,956	5.6	40,710	46.0	40,710	51.6
Non-dryland									47,790	48.4
Total Area of the Territory									88,500	100.0

HANDBOOKS

185 Keefe, Eugene K. *Area handbook for Portugal.* Washington DC: U.S. Dept. of Defense, 1977. (American University foreign area studies) 456 pp.

Narrative summary of the nation's history, politics, geography, economy, social life; includes unannotated bibliography.

DIRECTORIES

186 Richards, Robert A. C., ed. *Spain and Portugal.*

See entry 197.

ATLASES

187 Girno, Aristides de Amorim. *Atlas de Portugal/Atlas of Portugal.* Coimbra, Portugal: Instituto de Estudios Geograficos. Faculdade de Letras, 1958. 2nd ed. Unpaged. (40 maps).

In Portuguese and English. Concentrates on European Portugal with some information on overseas provinces and colonies.

GAZETTEERS

188 *Portugal and the Cape Verde Islands: official standard names approved by the United States Board on Geographic Names.* Washington DC: U.S. Board on Geographic Names, 1961. (Gazetteer 50) 321 pp.

List of 25,700 place-names with latitude/longitude coordinates. Covers Portugal, the Azores, the Madeira Islands, and the Cape Verde Islands.

GEOLOGY

189 Acciaivoli, L. de Menezes Correa. *Geologia de Portugal: ensaio bibliografico [geology of Portugal: bibliographic essay].* 2nd ed. Lisboa: Portugal. Direcao Geral de Minas e Servicos Geologicos, 1957-1958. 2 vols.

In Portuguese.

CLIMATOLOGY

190 Lines Escardo, A. "The climate of the Iberian peninsula." In *Climates of northern and western Europe*, pp. 195-239. Edited by C. C. Wallen. Amsterdam, The Netherlands: Elsevier Publishing Co., 1970. (World survey of climatology 5).

Covers Portugal and Spain. A narrative summary of these nations' climates, with maps, charts, and an unannotated bibliography.

BOTANY

191 Pinto da Silva, A. R., and Teles, A. N. "Bibliographia phytosociologica Portugaliae" [bibliography of the plant ecology of Portugal]. *Excerpta botanica: sectio B: sociologica* 4:2 (1962):89-151.

Unannotated. Covers Portugal, the Azores, the Madeira Islands, Cape Verde, Guinea-Bissau, Sao Tome, Angola, Mozambique, and Timor. Contains 623 items.

192 Teles, A. N. "Bibliographie des cartes de vegetation de Portugal" [bibliography of vegetation maps of Portugal]. *Excerpta botanica: sectio B: sociologica* 6:4 (1964): 297-319.

Unannotated. Covers Portugal, the Azores, the Madeira Islands, Cape Verde, Guinea-Bissau, Sao Tome e Principe, Angola, Mozambique, Timor, and Goa. Contains 114 items.

ANTHROPOLOGY

193 Pereira, Benjamim Enes. *Bibliografia analitica de etnografia portuguesa [bibliography of Portuguese ethnography]*. Lisboa: Instituto de Alta Cultura. Centro de Estudos de Etnologia Peninsular, 1965. 670 pp.

In Portuguese. Annotated list of 3,834 references covering the period to 1961.

ECONOMIC DEVELOPMENT

194 *Bibliografia sobre a economia portuguesa [bibliograhy on the Portuguese economy]*. Lisboa: Portugal. Instituto Nacional de Estatistica. Centro de Estudos Economicos, 1948/49-. Irreg.

ECONOMIC DEVELOPMENT

195 *Quarterly economic review of Portugal*. London: Economist Intelligence Unit, 1976-. Quarterly.

Includes summary of political and economic news as well as charts and statistics of selected economic indicators. Annual supplement.

SPAIN
Estado Espanol

Spain is Europe's driest nation in both total area and percentage (save Malta and Cyprus). Only along the Pyrenees, the northern coast, and the northwestern province of Galicia do humid conditions exist.

The Meseta. This interior tableland is composed of a broad plateau divided into a number of semiarid river basins and subhumid mountain ranges. Croplands produce wheat and barley (in the driest area); irrigation is rare, yields are low, and mechanization is fairly recent and uncommon. Sheep are grazed extensively, with goats more common in the south. The river valleys provide water for irrigated vegetables and farming.

The Coast. Here the milder winter climate allows citrus to flourish despite the considerable frost danger compared to other citrus regions of the world. Irrigation allows for very high yields of oranges, of which Spain is a major exporter. Many other crops are cultivated, including grapes, olives, cotton, tobacco, and, unusual in the Mediterranean, rice. Tourism is very important throughout the nation and especially so along the coast.

The Balearic Islands. These four islands possess a semiarid climate both drier and milder than the nearby Spanish coast. As with other Mediterranean islands the local economy is generally at a subsistence level; the rocky landscape supports grazing and some small-scale farming. The primary industry is tourism and a fashionable residential resort for the wealthy.

The Canary Islands are considered in a separate chapter.

Dryland areas of SPAIN

DESERTIFICATION RISK	ARIDITY								Aridity Totals	
	Hyperarid		Arid		Semiarid		Subhumid			
	km²	%	km²	%	km²	%	km²	%	km²	%
Very High	By definition, desertification does not exist in hyperarid regions.		2,488	0.5	63,196	12.7			65,684	13.2
High					70,660	14.2			70,660	14.2
Moderate							7,962	1.6	7,962	1.6
Desertification Totals			2,488	0.5	133,856	26.9	7,962	1.6	144,306	29.0
No Desertification					13,435	2.7	195,062	39.2	208,497	41.9
Total Drylands			2,488	0.5	147,291	29.6	203,023	40.8	352,803	70.9
Non-dryland									144,803	29.1
Total Area of the Territory									497,606	100.0

HANDBOOKS

196 Keefe, Eugene K. *Area handbook for Spain*. Washington DC: U.S. Dept. of Defense, 1976. (American University foreign area studies) 424 pp.

Narrative summary of the nation's history, politics, geography, economy, social life; includes unannotated bibliography.

DIRECTORIES

197 Richards, Robert A. C., ed. *Spain and Portugal*. 2nd ed. Guernsey U.K.: Francis Hodgson, 1974. (Guide to world science 9) 252 pp.

Provides descriptions of scientific research activities and organizations.

STATISTICS

198 *Anuario estadistico de Espana [annual statistics of Spain]*. Madrid: Spain. Instituto Nacional de Estadistica, 1912-. Annual.

In Spanish.

ATLASES

199 *Atlas nacional de Espana [national atlas of Spain]*. Madrid: Spain. Instituto Geografico y Catastral, 1965. 2 vols. and about 50 sheet maps.

In Spanish. Thematic maps appear at the scale of 1:2,000,000. Includes the Canary Islands.

GAZETTEERS

200 *Spain and Andorra: official standard names approved by the United States Board on Geographic Names*. Washington DC: U.S. Board on Geographic Names, 1961. (Gazetteer 51) 651 pp.

List of 50,550 place-names with latitude/longitude coordinates. Covers Spain, the Canary Islands, Spanish North Africa and Andorra.

GEOLOGY

201 "Bibliografia geologica espanola" [bibliography of Spanish geology]. *Acta geologica hispanica* 2:2- (1967-).

Appears annually from 1967 to date. Unannotated. Covers Spain, the Canary Islands, Western Sahara. In Spanish.

CLIMATOLOGY

202 Grimes, Annie E. *An annotated bibliography on the climate of the Balearic Islands*. Washington DC: U.S. Weather Bureau, 1963. (NTIS: AD 660-811) 55 pp.

203 Huerta, Fernando. *Bibliografia meteorologica espanola [bibliography of Spanish meteorology]*. 2nd ed. Madrid: Spain. Instituto Nacional de Meteorologia Biblioteca. Subsecretaria de Aviacion Civil, 1975. 295 pp.

In Spanish.

204 Lines Escardo, A. "The climate of the Iberian peninsula."

See entry 190.

BOTANY

205 Izco, J. "Bibliografia fitosociologica y geobotanica de Espana" [bibliography of plant ecology and geobotany of Spain]. *Excerpta botanica: sectio B: sociologica* 13:2 (1974):134-160; 13:3 (1974):161- 193; 18:2-3 (1979):110-144.

Unannotated. In Spanish.

ECONOMIC DEVELOPMENT

206 *KOMPASS Espana: repertorio general de la economia espanola [general register of the Spanish economy]*. Madrid: KOMPASS Espana, 1960-. Annual.

In Spanish with English summaries.

207 *Quarterly economic review of Spain*. London: Economist Intelligence Unit, 1976-. Quarterly.

Includes summary of political and economic news as well as charts and statistics of selected economic indicators. Annual supplement.

Eastern Europe

Isolated pockets of dryland occur in rainshadow areas of Hungary, Yugoslavia, and Romania. In addition, a marginally Mediterranean-type climate exists along the Black Sea coasts of Bulgaria, Romania, and the Soviet Union. For organizational purposes, the Soviet Union is considered later under Northern and Eastern Asia.

GENERAL BIBLIOGRAPHIES

208 Horecky, Paul Louis, ed. *Southeastern Europe: a guide to basic publications*. Chicago IL: University of Chicago Press, 1969. 755 pp.

Contains 3,018 annotated references. Covers Albania, Bulgaria, Greece, Romania, Yugoslavia. Broad subjects include reference works, general geography, sociology, history and politics, the economy, cultural affairs. Though dated, an excellent guide to the literature.

DIRECTORIES

209 Jones, E. G., ed. *Eastern Europe*. 2nd ed. Guernsey U.K.: Francis Hodgson, 1975. (Guide to world science 10) 234 pp.

Provides descriptions of scientific research activities and organizations. Coverage includes Bulgaria, Hungary, Romania, Yugoslavia.

GEOGRAPHY

210 Kirkpatrick, Meredith. *Environmental problems and policies in Eastern Europe and the U.S.S.R.* Monticello IL: Council of Planning Librarians, 1978. (CPL exchange bibliography 1491) 12 pp.

Unannotated. A brief checklist of sources, mostly in English, very few originating from the nations under consideration. Covers U.S.S.R., Bulgaria, Czechoslovakia, East Germany, Hungary, Poland, Romania, Yugoslavia.

ENERGY

211 Kirkpatrick, Meredith. *Energy resources and energy policies in Eastern Europe and the USSR: a bibliography*. Monticello IL: Vance Bibliographies, 1979. (Public administration series bibliography P-175) 13 pp.

Unannotated. A brief checklist of sources in English.

BULGARIA
Narodna Republika Bulgaria
[The People's Republic of Bulgaria]

Bulgaria's drylands are confined to a subhumid region covering the Rumelian Basin to the Black Sea coast. As the nation's most extensive lowland, the Basin—drained by the Maritsa and Tundzha Rivers—is a major agricultural region and produces lemons (on the coast), cereal crops, and sheep. As irrigation is extended yields have increased.

Dryland areas of BULGARIA

DESERTIFICATION RISK	ARIDITY									
	Hyperarid		Arid		Semiarid		Subhumid		Aridity Totals	
	km²	%	km²	%	km²	%	km²	%	km²	%
Very High	By definition, desertification does not exist in hyperarid regions.									
High										
Moderate										
Desertification Totals										
No Desertification					3,438	3.1	24,068	21.7	27,506	24.8
Total Drylands					3,438	3.1	24,068	21.7	27,506	24.8
Non-dryland									83,405	75.2
Total Area of the Territory									110,911	100.0

GENERAL BIBLIOGRAPHIES

212 **Pundeff, Marin V.** *Bulgaria: a bibliographic guide.* Washington DC: U.S. Library of Congress, 1965. 98 pp.

An excellent narrative bibliography of 1,234 citations, though now somewhat out of date.

HANDBOOKS

213 **Keefe, Eugene K.** *Area handbook for Bulgaria.* Washington DC: U.S. Dept. of Defense, 1974. (American University foreign area studies) 330 pp.

Narrative summary of the nation's history, politics, geography, economy, social life; includes unannotated bibliography.

GEOGRAPHY

214 **Laking, Phyllis W.** *The Black Sea, its geology, chemistry, biology: a bibliography.*

See entry 1613.

ATLASES

215 **Vulkov, Yordan, ed.** *Atlas: Narodna republika Bulgariia [Atlas: Socialist Republic of Bulgaria].* Sofia: Glavno Upravlenie po Geodeziia i Kartografiia, 1973. 168 pp.

In Bulgarian with summary of plate titles in English.

GAZETTEERS

216 *Bulgaria: official standard names approved by the United States Board on Geographic Names.* Washington DC: U.S. Board on Geographic Names, 1959. (Gazetteer 44) 295 pp.

List of 23,700 place-names with latitude/longitude coordinates.

LAND RECLAMATION

217 *Bulgarska literatura po meliioratsii [Bulgarian bibliography on reclamation of land].* Sofia: Tsentur za nauchno—tekhnicheska i ikonomicheska. Informatsiia po selsko i gorsko stopanstvo, 1969-. Frequency unknown.

CLIMATOLOGY

218 **Furlan, D.** "The climates of southeast Europe."

See entry 159.

BOTANY

219 **Kitanov, Boris Pavlov, and Naseva, V.** *Literatura vurkhu visshata flora i rastitelnata geografiia na Bulgariia 1959-1968 [literature on the flora and plant geography in Bulgaria 1959-1968].* Sofia: Bulgarskata akdemiia na naukite. Tsentralna biblioteka, 1975. 270 pp.

In Bulgarian.

ECONOMIC DEVELOPMENT

220 *Quarterly economic review of Rumania, Bulgaria, Albania.*

See entry 241.

HUNGARY
Magyar Nepkoztarsasag
[The Hungarian People's Republic]

Despite its humid continental location, the heart of Hungary is a subhumid dryland. The region between the Danube and Tisza Rivers is called "Mesopotamia" in a reinforcement of locally-accepted Sumerian-Magyar connections. The *puszta* is a salt-saturated waste in the center of the region; surrounding it are more productive steppe-lands that once were the home of semi-nomadic herds of cattle and horses. Now, settled agriculture and increasing irrigation have brought a wide range of cereal and vegetable crops into the region.

Dryland areas of HUNGARY

DESERTIFICATION RISK	ARIDITY									
	Hyperarid		Arid		Semiarid		Subhumid		Aridity Totals	
	km²	%	km²	%	km²	%	km²	%	km²	%
Very High	By definition, desertification does not exist in hyperarid regions.									
High										
Moderate										
Desertification Totals										
No Desertification							18,048	19.4	18,048	19.4
Total Drylands							18,048	19.4	18,048	19.4
Non-dryland									74,984	80.6
Total Area of the Territory									93,032	100.0

GENERAL BIBLIOGRAPHIES

221 Bako, Elemer. *Guide to Hungarian studies.* Stanford CA: Hoover Institution Press, 1973. (Hoover Institution bibliographical series 52) 2 vols.

222 Kabdebo, Thomas. *Hungary.* Oxford, U.K.; Santa Barbara CA: Clio Press, 1980. (World Bibliographical series 15) 280 pp.

A good recent bibliography of the literature; covers most major subjects including physical geography, agriculture, and the environment.

HANDBOOKS

223 Keefe, Eugene K. *Area handbook for Hungary.* Washington DC: U.S. Dept. of Defense, 1973. (American University foreign area studies) 339 pp.

Narrative summary of the nation's history, politics, geography, economy, social life; includes unannotated bibliography.

GEOGRAPHY

224 Balogh, Janos, and Jermy, Tibor. "Hungary." In *Handbook of contemporary developments in world ecology*, pp. 185-204. Edited by Edward John Kormondy and J. Frank McCormick. Westport CT: Greenwood, 1981.

A narrative summary of ecological research in Hungary with an annotated bibliography and list of references.

ATLASES

225 *Magyarorszag Nemzeti Atlasza [the national atlas of Hungary].* Budapest: Kartografiia Vallalat, 1967. 112 pp.

In Hungarian.

Hungary/226-229

GAZETTEERS

226 *Hungary: official standard names approved by the United States Board on Geographic Names*. Washington DC: U.S. Board on Geographic Names, 1961. (Gazetteer 52) 301 pp.

List of 25,000 place-names with latitude/longitude coordinates.

CLIMATOLOGY

227 **Okolowica, W.** "The climate of Poland, Czechoslovakia and Hungary." In *Climates of central and southern Europe*, pp. 75-125. Edited by C. C. Wallen. Amsterdam, The Netherlands: Elsevier Scientific Publishing Co., 1977. (World survey of climatology 6).

A narrative summary with maps, charts, and an unannotated bibliography.

BOTANY

228 **von Soo, Rezso.** "Bibliographia phytosociologica: Hungaria" [bibliography of plant ecology: Hungary]. *Excerpta botanica: sectio B: sociologica* 2:2 (1960):93-156.

Unannotated. Contains 719 items.

ECONOMIC DEVELOPMENT

229 *Quarterly economic review of Hungary*. London: Economist Intelligence Unit, 1982-. Quarterly.

Includes summary of political and economic news as well as charts and statistics of selected economic indicators. Annual supplement. Continues *Quarterly economic review of Czechoslovakia, Hungary*.

ROMANIA
Republica Socialista Romania
[The Socialist Republic of Romania]

Romania's drylands are Eastern Europe's most extensive outside the Soviet Union. The entire Wallachian Plain drained by the lower Danube as well as the Dobruja Plateau along the Black Sea are considered subhumid and continue northeast to join the Ukrainian steppe. Maize and winter wheat are primary crops, followed by rice, soybeans, tobacco, grass, livestock, and fish from both the river and the Black Sea.

Dryland areas of ROMANIA

DESERTIFICATION RISK	Hyperarid km²	Hyperarid %	Arid km²	Arid %	Semiarid km²	Semiarid %	Subhumid km²	Subhumid %	Aridity Totals km²	Aridity Totals %
Very High	By definition, desertification does not exist in hyperarid regions.									
High										
Moderate							2,850	1.2	2,850	1.2
Desertification Totals							2,850	1.2	2,850	1.2
No Desertification							76,475	32.2	76,475	32.2
Total Drylands							79,325	33.4	79,325	33.4
Non-dryland									158,175	66.6
Total Area of the Territory									327,500	100.0

GENERAL BIBLIOGRAPHIES

230 **Fischer-Galati, Stephen A.** *Rumania, a bibliographic guide.* Washington DC: U.S. Library of Congress, 1963. 75 pp.

HANDBOOKS

231 **Keefe, Eugene K.** *Area handbook for Romania.* Washington DC: U.S. Dept. of Defense, 1972. (American University foreign area studies) 319 pp.

Narrative summary of the nation's history, politics, geography, economy, social life; includes unannotated bibliography.

GEOGRAPHY

232 **Laking, Phyllis N.** *The Black Sea, its geology, chemistry, biology: a bibliography.*
See entry 1613.

ATLASES

233 *Atlas: Republica Socialista Romania [atlas of the Socialist Republic of Romania].* Bucharesti: Editura Academies Rupublicii Socialiste Romania, 1974-. Looseleaf.

Text and legends in Romanian, English, French and Russian.

GAZETTEERS

234 *Rumania: official standard names approved by the United States Board on Geographic Names.* Washington DC: U.S. Board on Geographic Names, 1960. (Gazetteer 48) 450 pp.

List of 36,500 place-names with latitude/longitude coordinates.

GEOLOGY

235 **Roman, David, and Codarcea, Alexandru.** *Bibliografia geologica a Romaniei [geologic bibliography of Romania].* Bucharesti: Cultura nationala, 1926. 155 pp.

Romania/236-241

In Romanian. Supplements appeared in 1929, 1939, 1962, 1969, and 1975.

CLIMATOLOGY

236 *Bibliografia meteorologica, 1935-1959 [meteorological bibliography, 1935-1959].* Bucharest: Romania. Institutul de Meteorologie si Hidrologie, 1975. 270 pp.

In English, Romanian, Russian.

237 **Furlan, D.** "The climates of southeast Europe."
See entry 159.

BOTANY

238 **Borza, Alexander.** "Bibliographia phytosociologica: Romania" [bibliography of plant ecology: Romania]. *Excerpta botanica: sectio B: sociologica* 1:2 (1959):134-153.

Unannotated. Covers the literature from 1863-1958.

239 **Borza, Alexander.** "Bibliographie der vegetationskarten Rumaeniens" [bibliography of vegetation maps of Romania]. *Excerpta botanica: sectio B: sociologica* 5:2 (1963):103-107.

Unannotated. Contains 32 items.

AGRICULTURE

240 *Bibliografia agricola romana, 1808-1969 [Romanian agricultural bibliography].* Bucuresti: Academia Republicii Socialiste Romania. Biblioteca. Centrul de Informare si Documentare pentru Agricultura Silvicultura, 1971. 6 vols.

In Romanian.

ECONOMIC DEVELOPMENT

241 *Quarterly economic review of Rumania, Bulgaria, Albania.* London: Economist Intelligence Unit, 1976-. Quarterly.

Includes summary of political and economic news as well as charts and statistics of selected economic indicators. Annual supplement.

YUGOSLAVIA
Socijalisticka Federativna Republika Jugoslavija
[The Socialist Federal Republic of Yugoslavia]

Yugoslavia has a small dryland centering on the city of Skopje in Macedonia where a rainshadow places several narrow river valleys within the subhumid range.

Due to the relatively small importance of these drylands to the nation, few references are provided.

Dryland areas of YUGOSLAVIA

DESERTIFICATION RISK	ARIDITY									
	Hyperarid		Arid		Semiarid		Subhumid		Aridity Totals	
	km²	%	km²	%	km²	%	km²	%	km²	%
Very High	By definition, desertification does not exist in hyperarid regions.									
High										
Moderate										
Desertification Totals										
No Desertification								7,764	3.0	7,764	3.0
Total Drylands								7,764	3.0	7,764	3.0
Non-dryland										248,040	97.0
Total Area of the Territory										255,804	100.0

GENERAL BIBLIOGRAPHIES

242 **Horton, John Joseph.** *Yugoslavia.* Oxford, U.K.: Clio, 1977. (World bibliographical series 1) 194 pp.

A good annotated bibliography covering most major subjects.

243 **Matulic, Rusko.** *Bibliography of sources on Yugoslavia.* Palo Alto CA: Ragusan Press, 1981. 252 pp.

An unannotated listing of some 4,700 books and journal articles on most major subjects.

244 **Petrovich, Michael Boro.** *Yugoslavia, a bibliographic guide.* Washington DC: U.S. Library of Congress, 1974. 270 pp.

An excellent narrative bibliography. The 2,548 citations are provided with Library of Congress call numbers.

245 **Terry, G. M.** *A bibliography of Macedonian studies.* Nottingham, U.K.: Nottingham University Library, 1975. 121 pp.

Concentrates on Macedonian Yugoslavia; more than 1,600 citations in a variety of languages.

HANDBOOKS

246 **McDonald, Gordon C.** *Area handbook for Yugoslavia.* Washington DC: U.S. Dept. of Defense, 1973. (American University foreign area studies) 653 pp.

Narrative summary of the nation's history, politics, geography, economy, social life; includes unannotated bibliography.

GAZETTEERS

247 *Yugoslavia: official standard names approved by the United States Board on Geographic Names.* Washington DC: U.S. Board on Geographic Names, 1961. (Gazetteer 55) 495 pp.

List of 40,000 place-names with latitude/longitude coordinates.

CLIMATOLOGY

248 **Furlan, D.** "The climates of southeast Europe."
See entry 159.

ECONOMIC DEVELOPMENT

249 *Quarterly economic review of Yugoslavia.* London: Economist Intelligence Unit, 1976-. Quarterly.

Includes summary of political and economic news as well as charts and statistics of selected economic indicators. Annual supplement.

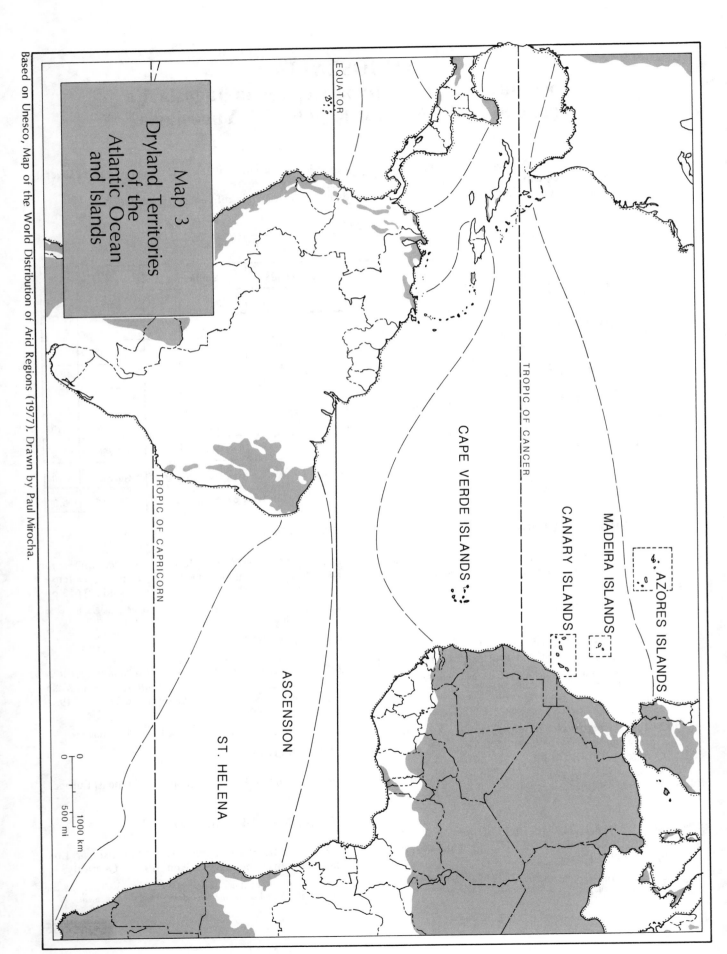

Map 3. Dryland Territories of the Atlantic Ocean and Islands

THE ATLANTIC OCEAN AND ISLANDS

The islands of the Atlantic Ocean, appearing as they do relatively close to the European and African shores, occupy oceanic extensions of the continental climate belts. Moving north to south, one encounters the Azores (humid to subhumid), the Madeiras (humid to semiarid), the Canaries (humid to arid), the Cape Verdes (mostly arid and semiarid), Ascension (subhumid to arid), and St. Helena (humid to arid); Tristan da Cunha is a cool, humid island outside the dryland zone.

Dryland areas of THE ATLANTIC ISLANDS

DESERTIFICATION RISK	ARIDITY									
	Hyperarid		Arid		Semiarid		Subhumid		Aridity Totals	
	km²	%	km²	%	km²	%	km²	%	km²	%
Very High	By definition, desertification does not exist in hyperarid regions.		5,448	15.7	4,295	12.4	1,217	3.5	10,960	31.6
High										
Moderate										
Desertification Totals			5,448	15.7	4,295	12.4	1,217	3.5	10,960	31.6
No Desertification			107	0.3	56	0.2	388	1.1	551	1.6
Total Drylands			5,555	16.0	4,351	12.6	1,605	4.6	11,511	33.2
Non-dryland									23,054	66.8
Total Area of the Territory									34,565	100.0

CLIMATOLOGY

250 **Hastenrath, Stefan, and Lamb, Peter J.** *Climatic atlas of the tropical Atlantic and eastern Pacific Oceans.* Madison WI: University of Wisconsin Press, 1977. 97 pp.

Excludes eastern Africa, the Indian subcontinent, and southeast Asia. Maps compiled from monthly and annual data covering 1911- 1970. Covers area between 30°N to 30°S and 110°W to 20°E.

251 **Hastenrath, Stefan, and Lamb, Peter J.** *Heat budget atlas of the tropical Atlantic and eastern Pacific Oceans.* Madison WI: University of Wisconsin Press, 1978. 90 pp.

Excludes eastern Africa, the Indian subcontinent, and southeast Asia. Maps compiled from monthly and annual data covering 1911- 1970. A sequel to the *Climatic atlas of the tropical Atlantic and eastern Pacific Oceans* (entry 250). Covers area between 30°N to 30°S and 110°W to 20°E.

AZORES ISLANDS
Ilhas dos Acores

(an overseas territory of Portugal comprising three administrative districts;
the dryland island of Santa Maria is in the Ponta Delgado district)

The Azores are a chain of largely volcanic islands lying some 1,500 km (1,000 miles) due west of Lisbon. Though precipitation is relatively low on all the islands, the strong maritime influence brings frequent light rains and fog; based on evapotranspiration, only the southeasternmost island of Santa Maria falls into the subhumid zone. With an area less than 100 square kilometers (39 square miles) Santa Maria supports only vineyards and domestic gardens; the porous limestone bedrock (unlike other islands in the chain) tends to reduce the amount of surface moisture available to vegetation. The Azores' chief importance is its large international airport and subsidiary tourist trade.

See Portugal above for additional entries.

Dryland areas of THE AZORES ISLANDS

DESERTIFICATION RISK	ARIDITY									
	Hyperarid		Arid		Semiarid		Subhumid		Aridity Totals	
	km²	%	km²	%	km²	%	km²	%	km²	%
Very High	By definition, desertification does not exist in hyperarid regions.									
High										
Moderate										
Desertification Totals										
No Desertification								356	4.4	356	4.4
Total Drylands								356	4.4	356	4.4
Non-dryland										7,753	95.6
Total Area of the Territory										8,109	100.0

GAZETTEERS

252 *Portugal and the Cape Verde Islands: official standard names approved by the United States Board on Geographic Names.*

See entry 188.

CANARY ISLANDS
Islas Canarias

(the archipelago comprises two provinces of Spain)

The Canary Islands lie between 100 and 480 km (60 and 300 miles) to the west of Morocco and Western Sahara. Because they are the closest Atlantic island chain to the mainland the Canaries experience less of a maritime climatic influence; the two islands closest to Africa are both fairly low in elevation and mostly arid, while those to the west are higher and better watered on the upper slopes. They all share the winter dominant precipitation regime of the Mediterranean.

Agriculture is well-developed despite limited lowlands and scarce water. Irrigation is used extensively to produce tropical crops along the coasts (bananas, dates, sugarcane), and Mediterranean crops higher up (tomatoes, citrus, figs, potatoes, tobacco, wheat, etc.). Some of the most barren land is given over to grapes, which are carefully tended in pockets hollowed out of volcanic scablands. In addition to agricultural exports the economy depends heavily on fishing, tourism, and port functions.

Dryland areas of THE CANARY ISLANDS

DESERTIFICATION RISK	ARIDITY									
	Hyperarid		Arid		Semiarid		Subhumid		Aridity Totals	
	km²	%	km²	%	km²	%	km²	%	km²	%
Very High	By definition, desertification does not exist in hyperarid regions.		3,956	54.4	2,320	31.9	683	9.4	6,959	95.7
High										
Moderate										
Desertification Totals			3,956	54.4	2,320	31.9	683	9.4	6,959	95.7
No Desertification										
Total Drylands			3,956	54.4	2,320	31.9	683	9.4	6,959	95.7
Non-dryland									314	4.3
Total Area of the Territory									7,273	100.0

ATLASES

253 *Atlas nacional de Espana [national atlas of Spain].*

See entry 199.

GAZETTEERS

254 *Spain and Andorra: official standard names approved by the United States Board on Geographic Names.*

See entry 200.

GEOLOGY

255 "Bibliografia geologica espanola" [bibliography of Spanish geology].

See entry 201.

CLIMATOLOGY

256 Kramer, H. P. "A selective annotated bibliography on the climatology of northwest Africa."

See entry 396.

SOILS

257 *Soils of the Canary Islands.* Slough, U.K.: Commonwealth Agricultural Bureaux, 1978. 15 pp.

Annotated bibliography covering literature published from 1930-1976.

BOTANY

258 Sunding, Per. "Bibliographia phytosociologica: The Canary Islands" [bibliography of plant ecology: The Canary Islands]. *Excerpta botanica: sectio B: sociologica* 10:4 (1970):257-268.

Unannotated. Contains 136 items, chronologically ordered.

CAPE VERDE ISLANDS
Republica de Cabo Verde
(a former colony of Portugal)

The Cape Verde Islands lie between 480 and 760 km (300 and 470 miles) west of Senegal. As a group it is the driest of the Atlantic island chains and its characteristically rugged volcanic surfaces limit agriculture. On the better watered northern slopes and where springs or streams are found, tropical crops such as bananas, coffee, and nuts are grown in exportable quantities. Goats, cattle and pigs are run and fishing is an important source of local food and export income. Mineral products include salt and pozzolana (a volcanic aggregate used in cement).

Dryland areas of THE CAPE VERDE ISLANDS

DESERTIFICATION RISK	ARIDITY									
	Hyperarid		Arid		Semiarid		Subhumid		Aridity Totals	
	km²	%	km²	%	km²	%	km²	%	km²	%
Very High	By definition, desertification does not exist in hyperarid regions.		1,492	37.0	1,960	48.6	443	11.0	3,895	96.6
High										
Moderate										
Desertification Totals			1,492	37.0	1,960	48.6	443	11.0	3,895	96.6
No Desertification										
Total Drylands			1,492	37.0	1,960	48.6	443	11.0	3,895	96.6
Non-dryland										138	3.4
Total Area of the Territory										4,033	100.0

GENERAL BIBLIOGRAPHIES

259 *Contribuicao para una bibliografia sobre Cabo Verde [contribution to a bibliography on Cape Verde].* Lisboa: Portugal. Centro de Documentacao e Informacao, 1977. Various pagings.

Formed by the joining of two specialized bibliographies on Cape Verde. Contains 1,508 entries, with an index of 52 subjects. In Portuguese.

260 **Gibson, Mary Jane.** *Portuguese Africa: a guide to official publications.*

See entry 770.

261 **McCarthy, Joseph M.** *Guinea-Bissau and Cape Verde Islands: a comprehensive bibliography.* New York: Garland, 1977. (Garland reference library of social science 27) 196 pp.

A very comprehensive but unannotated list of references on these two nations; 2,547 citations, mostly in Portuguese.

262 **Zubatsky, David S., comp.** *A guide to resources in United States libraries and archives for the study of Cape Verdes, Guinea (Bissau), Sao Tome-Principe, Angola and Mozambique.* Durham NH: University of New Hampshire. Dept. of History. International Conference Group on Modern Portugal, 1977. (Essays in Portuguese studies 1) 29 pp.

GEOGRAPHY

263 *Draft environmental report on Cape Verde.* Tucson AZ: University of Arizona. Office of Arid Lands Studies. Arid Lands Information Center, 1980.

An excellent narrative summary of the nation's environmental problems with a good unannotated bibliography.

GAZETTEERS

264 *Portugal and the Cape Verde Islands: official standard names approved by the United States Board on Geographic Names.*

See entry 188.

BOTANY

265 **Pinto da Silva, A. R. and Teles, A. N.** Bibliographia phytosociologica Portugaliae [bibliography of the plant ecology of Portugal].

See entry 191.

266 **Sunding, P.** "A botanical bibliography of the Cape Verde Islands." Museo Municipal do Funchal. *Boletim* 31 (1977):100-109.

267 **Teles, A. N.** "Bibliographie des cartes de vegetation du Portugal" [bibliography of vegetation maps of Portugal].

See entry 192.

ECONOMIC DEVELOPMENT

268 *Quarterly economic review of Angola, Guinea-Bissau, Cape Verde, Sao Tome, Principe.*

See entry 782.

HISTORICAL GEOGRAPHY

269 **Lobban, Richard.** *Historical dictionary of the Republics of Guinea-Bissau and Cape Verde.* Metuchen NJ: Scarecrow, 1979. (African historical dictionaries 22) 193 pp.

Entries cover major topics, events, and personalities in the nation's history; includes extensive unannotated bibliography.

MADEIRA ISLANDS
Arquipelago da Madeira

(an overseas territory of Portugal comprising the administrative district of Funchal)

These islands lie some 1,000 km (650 miles) southwest of Lisbon and 860 km (530 miles) due west of Casablanca. Madeira is the name of the largest island. The maritime climate produces high humidity amid frequent fogs so, despite the relatively low precipitation, the islands are quite lush. The driest areas— extending into the semiarid zone—are found along the south coast of Madeira; here is centered the wine industry that produces the famous dessert wine. Slopes are everywhere steep and difficult to cultivate, so laborious terracing and irrigation are used to hold and water the soil.

Dryland areas of THE MADEIRA ISLANDS

| DESERTIFICATION RISK | ARIDITY ||||||||||
| | Hyperarid || Arid || Semiarid || Subhumid || Aridity Totals ||
	km^2	%	km^2	%	km^2	%	km^2	%	km^2	%
Very High	By definition, desertification does not exist in hyperarid regions.				15	1.9	91	11.5	106	13.4
High										
Moderate										
Desertification Totals					15	1.9	91	11.5	106	13.4
No Desertification										
Total Drylands					15	1.9	91	11.5	106	13.4
Non-dryland									692	86.6
Total Area of the Territory									798	100.0

GENERAL BIBLIOGRAPHIES

270 **Rodrigues, Jose Joaquim.** *Catalogo bibliografico do arquipelago da Madeira [bibliographic catalog of the Madeira archipelago]*. Funchal, Madeira: Camara municipal, 1950. 206 pp.

GAZETTEERS

271 *Portugal and the Cape Verde Islands: official standard names approved by the United States Board on Geographic Names.*

See entry 188.

CLIMATOLOGY

272 **Kramer, H. P.** "A selected annotated bibliography on the climatology of northwest Africa."

See entry 396.

BOTANY

273 **Hansen, Alfred.** "A botanical bibliography of the Archipelago of Madeira." Museo Municipal do Funchal. *Boletim* 30 (1976):26-45.

274 **Pinto da Silva, A. R. and Teles, A. N.** "Bibliographia phytosociologica Portugaliae" [bibliography of the plant ecology of Portugal].

See entry 191.

275 **Teles, A. N.** "Bibliographie des cartes de vegetation du Portugal" [bibliography of vegetation maps of Portugal].

See entry 192.

ST. HELENA AND ASCENSION ISLANDS
St. Helena and Dependencies

(a colony of the United Kingdom)

The islands of St. Helena, Ascension, and Tristan da Cunha comprise a single colony; the first two fall into the South Atlantic belt. They are topographically similar, having been formed by volcanic eruption and subsequent erosion of their cones. A roughly concentric pattern of precipitation, aridity, and vegetation exists on both, with zones of arid climate around the coasts, semiarid on the lower slopes, subhumid above, and (on the taller St. Helena only) a peak rising outside the dryland range.

St. Helena. Located some 2,000 km (1,200 miles) due west of Mocamedes, Angola, this island once supported extensive plantations in the early nineteenth century when sailing ships would call for provisions; many exotic plants were introduced, from English oaks to date palms, eucalyptus, bamboo, and tropical fruits. With the passing of the days of sail and the opening of the Suez Canal the economy and population declined. Today much former cropland is now pasture and crops are grown for local consumption only. Some flax fiber and cordage are exported.

Ascension. This island lies 1,300 km (800 miles) northwest of St. Helena (the nearest land) and some 3,200 km (1,980 miles) due west of Luanda, Angola. Poorer than St. Helena in water, soil, and natural vegetation, the island supports introduced grasses and some locally-consumed fruits, vegetables, sheep, and cattle. Its principal importance today is its location along the Atlantic missile range from Cape Canaveral, Florida; tracking and communication facilities are maintained here by the United States.

Dryland areas of ST. HELENA AND ASCENSION ISLANDS

DESERTIFICATION RISK	ARIDITY									
	Hyperarid		Arid		Semiarid		Subhumid		Aridity Totals	
	km²	%	km²	%	km²	%	km²	%	km²	%
Very High	By definition, desertification does not exist in hyperarid regions.									
High										
Moderate										
Desertification Totals										
No Desertification			107	51.3	56	26.7	32	15.2	195	93.2
Total Drylands			107	51.3	56	26.7	32	15.2	195	93.2
Non-dryland									14	6.8
Total Area of the Territory									209	100.0

GAZETTEERS

276 *South Atlantic: official standard names approved by the United States Board on Geographic Names.* Washington DC: U.S. Dept. of the Interior, 1957. (Gazetteer 31) 53 pp.

Contains about 3,750 entries covering St. Helena, Ascension, and other islands in the South Atlantic.

AFRICA

Although Africa does not have the highest percentage of dryland, in total area Africa is the driest continent. Only six of its 45 nations lack any dryland; nine nations are entirely dry, and three are more than 50 percent hyperarid.

Because it straddles the equator the continent presents a general vegetational symmetry; from north to south, Mediterranean sclerophyll gives way to desert, then to savanna grassland, to rainforest, to savanna, to desert, and to the Mediterranean type again at the southern tip.

Two great divisions of dryland territory are recognized: the first, centering on the Sahara, covers northern and western Africa; the second splits off at the Ethiopian Highlands and extends along the eastern slope south to the Cape of Good Hope.

Dryland areas of AFRICA (EXCLUDES MADAGASCAR)

DESERTIFICATION RISK	ARIDITY									Aridity Totals	
	Hyperarid		Arid		Semiarid		Subhumid			
	km²	%	km²	%	km²	%	km²	%	km²	%
Very High	By definition, desertification does not exist in hyperarid regions.		632,982	2.2	812,907	2.8	65,895	0.2	1,511,784	6.2
High	^		4,584,912	15.6	585,991	2.0	144,789	0.5	5,315,692	18.1
Moderate	^		316,575	1.1	2,849,238	9.7	417,135	1.4	3,582,948	12.2
Desertification Totals			5,534,469	18.9	4,248,136	14.5	627,819	2.1	10,410,424	35.5
No Desertification	7,060,325	24.0			237,904	0.8	3,561,603	12.2	10,859,832	37.0
Total Drylands	7,060,325	24.0	5,534,469	18.9	4,486,040	15.3	4,189,422	14.3	21,270,256	72.5
Non-dryland									8,060,000	27.5
Total Area of the Territory									29,330,256	100.0

GENERAL BIBLIOGRAPHIES

277 *Africa, problems and prospects: a bibliographical survey.* Rev. ed. Washington DC: U.S. Dept. of the Army, 1978. 577 pp.

Annotated.

278 Asamani, J. O. *Index Africanus.* Stanford CA: Hoover Institution Press, 1975. (Bibliography 53) 659 pp.

279 *Bibliographie sur les documents officiels africains [bibliography on African government documents].* Tangier, Morocco: African Training and Research Centre in Administration for Development, 1976. 91 pp.

Contains 335 items. Headings in English. Separate subject indexes in French and Arabic.

280 Conover, Helen F. *Africa south of the Sahara, a selected, annotated list of writings.* Washington DC: U.S. Library of Congress, 1964. 354 pp.

281 Duignan, Peter. *Guide to research and reference works on sub-Saharan Africa.* Stanford CA: Hoover Institution Press, 1971. (Bibliographical series 46) 1102 pp.

282 Holdsworth, Mary. *Soviet African studies, 1918-1959: an annotated bibliography.* London: Oxford University Press, 1961. (Chatham House memoranda) 150 pp.

Contains 498 references.

283 Miller, E. Willard, and Miller, Ruby M. *Africa: a bibliography on the Third World.* Monticello IL: Vance Bibliographies, 1981. (Public administration series bibliography P-817) 32 pp.

Unannotated. Some 370 references in English, mainly journal articles appearing in the 1970's.

284 Panofsky, Hans E. *A bibliography of Africana*. Westport CT: Greenwood Press, 1975. (Contributions in librarianship and information science 11) 350 pp.

285 Skurnik, W. A. E. *Sub-Saharan Africa: a guide to information sources*. Detroit MI: Gale Research, 1977. (International relations information guide series 3) 130 pp.

286 South, Aloha, comp. *Guide to Federal archives relating to Africa*. Honolulu HI: Crossroads, 1977. (Archival and bibliographic series) 556 pp.

287 Witherell, Julian W. *The United States and Africa: guide to U.S. official documents and government-sponsored publications on Africa, 1785-1975*. Washington DC: U.S. Library of Congress, 1978. 949 pp.

Covers all Africa, including the Atlantic and Indian Ocean islands, except Egypt. An update for 1976-1980 is in preparation May 1982.

INDEXES AND ABSTRACTS

288 *Abstracta Islamica*. Paris: Geuthner, 1965-. Annual.

In French. A major bibliographic source, arranged by major subjects and regions. Issued as a supplement to the *Revue des Etudes Islamiques*.

289 *Afrika-Bibliographie*. Bonn, Federal Republic of Germany: K. Schroeder, 1960-. Annual.

"A list of scientific literature in German language covering one year."

290 *A current bibliography on African affairs*. Farmingdale NY: Baywood Publishing Co., 1962-. Quarterly.

THESES AND DISSERTATIONS

291 Dinstel, Marion. *List of French doctoral dissertations on Africa, 1884-1961*. Boston MA: G. K. Hall, 1966. 336 pp.

292 Gaignebet, Wanda, comp. *Repertoire des theses africanistes francaises [register of theses of French Africanists]*. Paris: Ecole des Haute Etudes en Sciences Sociales. Centre d'etude Africaines (CARDAN), 1977-. Annual.

In French. Replaced two series of *Bulletin d'information et de liaison, etudes africaines* (-1972), merging *Inventaire de theses africanistes de langue francaise en cours* (-1972), and *Inventaire de theses et memoires africanistes de langue francaise soutenus* (-1972).

293 Koehler, Jochen. *Deutsche Dissertationen uber Afrika: ein Verzeichnis fur die Jahre 1918-1959 [German dissertations on Africa: a list for the years 1918-1959]*. Bonn, Federal Republic of Germany: 1962. Unpaged.

294 Sims, Michael, and Kagan, Alfred. *American and Canadian doctoral dissertations and master's theses on Africa, 1886-1974*. 2nd ed. Waltham MA: African Studies Association. Brandeis University, 1976. 365 pp.

295 *Theses on Africa accepted by universities in the United Kingdom and Ireland*. Cambridge, U.K.: W. Heffer, 1964. 74 pp.

Prepared by Standing Committee on Library Materials on Africa (SCOLMA).

HANDBOOKS

296 *Africa south of the Sahara*. 10th ed. London: Europa Publications, 1980. 1372 pp.

Covers all Africa except Western Sahara, Morocco, Algeria, Tunisia, Libya, and Egypt: includes the Cape Verde Islands, the Comoros, Reunion, Mauritius.

ENCYCLOPEDIAS

297 Oliver, Roland, and Crowder, Michael, eds. *The Cambridge encyclopedia of Africa*. Cambridge, U.K.; New York: Cambridge University Press, 1981. 492 pp.

DICTIONARIES

298 *Abreviations en Afrique/Abkuerzungen in Afrika/Abbreviations in Africa*. Munchen, Federal Republic of Germany: Deutsche Afrika-Gesellschaft, 1969. 260 pp.

A simple list of 4,000 abbreviations used by African organizations, institutions, and governments—primarily in French-speaking Africa. No information regarding the location or function of these organizations is provided. Arranged by abbreviation; not cross-indexed by full name.

DIRECTORIES

299 Chimutengwende, Mary, and Chimutengwende, Chen, eds. *Central Africa*. 2nd ed. Guernsey, U.K.: Francis Hodgson, 1975. (Guide to world science 19) 297 pp.

Provides descriptions of scientific research activities and organizations. Covers all of Africa except the north (Morocco to Egypt and the Sudan) and south (South Africa and Zimbabwe).

300 Duffy, James; Frey, Mitsue; and Sims, Michael. *International directory of scholars and specialists in African studies*. Waltham MA: Crossroads, 1978. 355 pp.

Contains biographical information on Africanists worldwide. Approximately 2,700 entries.

301 *Economic yearbook of member states of the Organization of African Unity/Annuaire economique, des pays membres de l'Organisation de l'Unite Africaine*. Kinshasa, Zaire: EDICA, 1975-. Annual.

In French and English. Although the title promises an economic summary, most of this volume's bulk and research value consists of names and addresses of personnel and offices of member nations. Covers all of Africa except Zimbabwe, Namibia, the Republic of South Africa, and Western Sahara.

302 *Etudes africaines en Europe: bilan et inventoire [African studies in Europe: balance and inventory].* Paris: Editions Karthala, 1981. 2 vols.

In French. An extensive directory of European research on Africa, valuable especially for its lists of organizations and their addresses.

303 **Frey, Mitsue, and Duffy, David.** *Directory of African and Afro-American Studies in the United States.* Waltham MA: African Studies Association, 1971-. Irreg.

STATISTICS

304 **Harvey, Joan M.** *Statistics Africa: sources for market research.* Beckenham, U.K.: CBD Research Ltd., 1970. 175 pp.

GEOGRAPHY

305 **Bederman, Sanford Harold.** *Africa: a bibliography of geography and related disciplines.* 3rd ed. Atlanta GA: Georgia State University. School of Business Administration. Publishing Services Division, 1974. 334 pp.

Unannotated. A selected listing of recent literature published in the English language. Text is a hard-to-read facsimile of a computer print out.

306 **Rogers, Dilwyn J., comp.** *A bibliography of African ecology: a geographically and topically classified list of books and articles.* Westport CT: Greenwood, 1979. (African Bibliographic Center. Special bibliographic series. New series 6) 499 pp.

307 **Sommer, John W.** *Bibliography of African geography, 1940-1964.* Hanover NH: Dartmouth College. Dept. of Geography, 1965. (Geography publications at Dartmouth 3) 139 pp.

ATLASES

308 **Davies, Harold Richard John.** *Tropical Africa: an atlas for rural development.* Cardiff, U.K.: University of Wales Press, 1973. 81 pp.

Covers Senegal, eastward to Ethiopia and Somalia, southward through Zaire and Tanzania. Base maps at scales of 1:20,000,000 and 1:40,000,000.

309 *Grand atlas du continent africain/The atlas of Africa.* Paris: Institut geographique national; New York: Free Press, 1973. 335 pp.

In English or French. French and English editions published simultaneously. Also known as the *Atlas jeune Afrique.* Covers all of Africa, with commentary.

GAZETTEERS

310 *Africa and southwest Asia: official standard names approved by the United States Board on Geographic Names.* Washington DC: U.S. Board on Geographic Names, 1972. (Gazetteer supplement 125) 182 pp.

List of 3,150 place-name corrections with latitude/longitude coordinates.

GEOLOGY

311 *Geological bibliography of Africa.* New York: Unesco, 1963. (UN document E/C no.15/INR/48) 114 pp.

312 **Ridge, John Drew.** *Annotated bibliographies of mineral deposits in Africa, Asia (exclusive of the USSR) and Australasia.* Oxford, U.K.: Pergamon Press, 1976. 545 pp.

CLIMATOLOGY

313 **Gleeson, Thomas Alexander.** *Bibliography of the meteorology of the Mediterranean, Middle East and South Asian areas.*

See entry 122.

314 **Griffiths, J. F., ed.** *Climates of Africa.* Amsterdam, The Netherlands: Elsevier Publishing, 1972. (World survey of climatology 10) 604 pp.

A narrative summary of the continent's climate, with maps, charts, and an unannotated bibliography. Chapters on individual nations are listed separately in this Guide.

315 **Hoeller, Erich, and Stranz, Dietrich.** *Klimahandbuch Afrika/Climatological handbook Africa/Manuel climatique Afrique.* Hamburg, Federal Republic of Germany: Ubersee Verlag, 1975. 96 pp.

In German, English, French.

316 **Jackson, Stanley Percival, comp.** *Climatological atlas of Africa.* Lagos, Nigeria: CCTA/CSA, 1961. (Commission for Technical Cooperation in Africa South of the Sahara. Joint project 1) 8 leaves plus 55 maps.

In English, French, and Portuguese. Scale for most plates 1:15,000,000.

317 **Thompson, B. W.** *The climate of Africa.* Nairobi; New York: Oxford University Press, 1965. 132 pp.

318 **Weiss, Marianne, and Jansen, Anne.** *Duerre in Afrika: klimatische, oekologische, sozio-oekonomische Aspekte: [Bibliographie]/Drought in Africa: climatic, ecological, socio-economic aspects/Secheresse en Afrique: aspects climatologiques, ecologiques et socio- economiques.* Hamburg, Federal Republic of Germany: Institut fuer Afrika-Kunde. Dokumentations-Leitstelle

Afrika, 1976. (Dokumentationsdienst Afrika: Reihe A, 14) 115 pp.

In German, English, French.

HYDROLOGY

319 Rodier, J. *Bibliography of African hydrology.* Paris: Unesco, 1963. (Natural resources research 2) 166 pp.

SOILS

320 *Soil erosion in Africa, southern Asia, Australasia and Oceania (1957-1977).* Harpenden U.K.: Commonwealth Agricultural Bureaux, 1978. (Annotated bibliography) 41 pp.

BIOLOGY

321 Guillarmod, A. J. "Limnological bibliography for Africa south of the Sahara." Limnological Society of Southern Africa. *Journal* 1:1 (1975):37-51.

BOTANY

322 Kuechler, August Wilhelm. *Africa, South America and world maps.* In *International bibliography of vegetation maps*, vol. 4. Lawrence KS: University of Kansas Libraries, 1970. (Library series 36) 561 pp.

An extensive annotated bibliography.

ZOOLOGY

323 Ardizonne, G. D. *A bibliography of African freshwater fish, supplement 1, 1968-1975/Bibliographie des poissons d'eau douce de l'Afrique, supplement 1, 1968-1975.* Rome: FAO. Committee for Inland Fisheries of Africa, 1976. (CIFA occasional paper 5) 40 pp.

In English and French.

324 Mathot, G. "Selective bibliography of the systematic and faunistic zoology of Africa south of the Sahara." *Zooleo (Kinshasa)* 3 (1970):5-289.

325 Matthes, Hubert. *A bibliography of African freshwater fish/Bibliographie des poissons d'eau douce de l'Afrique.* Rome: FAO, 1973.

Maps. In English, French, German.

326 Richard, D. *Bibliographie sur le dromadaire et le chameau [bibliography on the dromedary and the camel].* Maisons-Alfort, France: Institut d'Elevage et de Medicine Veterinaire des Pays Tropicaux, 1980. 137 pp.

In French. Contains 1,465 unannotated citations to the literature in major European languages.

327 Wilmot, B. C., and Wilmot, L. P. "A selected bibliography of literature on odonata from Africa and adjacent islands." Cape Provincial Museums of Natural History. *Annals* 11:10 (1978):195-208.

The Odonata include the dragonflies and damselflies.

AGRICULTURE

328 Lawani, S. M. *Farming systems in Africa: a working bibliography, 1930-1978.* Boston MA: G. K. Hall, 1979. (Bibliographies and guides in African studies) 251 pp.

IRRIGATION

329 Davis, Lenwood G. *Irrigation and water systems in Africa: an introductory survey.* Monticello IL: Council of Planning Librarians, 1977. (CPL exchange bibliography 1206) 24 pp.

Unannotated. Valuable mainly for its coverage of journal articles.

FORESTRY

330 *Bibliografia florestal de interesse africano [forest bibliography regarding Africa].* Lourenco Marques: Instituto de Investigacao Cientifica de Mocambique. Laboratorio de Tecnologia de Produtos Florestais, 1965-.

In Portuguese.

NOMADISM

331 Bullwinkle, Davis A. "Nomadism and pastoralism in Africa: a bibliography." *A current bibliography on African affairs*, n.s. 13:3 (1980-81):303-315.

Unannotated. Limited to English-language publications.

AGRICULTURAL ECONOMICS

332 *Aspects of agricultural policy and rural development in Africa: an annotated bibliography.* Farnham Royal, U.K.: Commonwealth Bureau of Agricultural Economics, 1971. (Annotated bibliographies series B) 5 vols. in 1.

Extensive annotations. Volumes divided by region. Agricultural development considered in the context of national economic development. Covers only material abstracted in *World agricultural economics and rural sociology abstracts* between 1964 and mid-1971.

333 Davis, Lenwood G. *Land usage, reforms and planning in Africa: an introductory survey.* Monticello IL: Council of Planning Librarians, 1977. (CPL exchange bibliography 1372) 13 pp.

Unannotated.

334 Freitag, Ruth S. *Agricultural development schemes in sub-Saharan Africa: a bibliography.* Washington DC: U.S. Government Printing Office, 1963. 189 pp.

335 Neville-Rolfe, Edmund. *Economic aspects of agricultural development in Africa: a selective annotated reading list of reports and studies concerning 40 African countries during the period 1960-1969.* Oxford, U.K.: University of Oxford. Agricultural Economics Research Institute, 1969. 257 pp.

Annotated. Covers works in English, French, and Portuguese.

336 *Rural development in Africa: a bibliography.* Madison WI: University of Wisconsin. Land Tenure Center Library, 1973-. (Training and methods 16/17).

Unannotated. Reflects holdings of the Land Tenure Center Library; includes its call numbers.

337 *Rural development in Africa, 1973- 1980.* Oxford, U.K.: Commonwealth Bureau of Agricultural Economics, 1981. 2 vols.

Annotated.

LAND TENURE

338 *Bibliography on land tenure in Africa.* Rome: FAO. Rural Institutions Division, 1970. 57 pp.

339 *Land tenure and agrarian reform in Africa and the Near East: an annotated bibliography.* Boston MA: G. K. Hall, 1976. 423 pp.

Prepared by the staff of the Land Tenure Center Library of the University of Wisconsin, Madison.

340 Ofori, Patrick. *Land in Africa: its administration, law, tenure and use: a select bibliography.* Nendeln, Liechtenstein: KTO Press, 1978. 200 pp.

MARKETS

341 Smith, Robert H. T. *Periodic markets in Africa, Asia, and Latin America.* Monticello IL: Council of Planning Librarians, 1972. (CPL exchange bibliography 318) 23 pp.

Unannotated.

HUMAN GEOGRAPHY

342 *African abstracts/Bulletin analytique africaniste.* London: International African Institute, 1950-72. 23 vols.

Quarterly review of ethnological, social, and linguistic studies appearing in current periodicals.

DEMOGRAPHY

343 *African demography.* Legon, France: Regional Institute for Population Studies. Research Cooperation and Publications Unit, n.d.

Each issue contains regular bibliographic sections, including on- going projects and recently published documents.

344 *Evolution demographique des capitales des etats francophones d'Afrique noire et de Madagascar [population trends of the capitals of French-speaking states of Black Africa and Madagascar].* Paris: France. Secretariat des missions d'urbanisme et d'habitat. Service documentation- information, 1974. (Bibliographies 15) 51 pp.

Contains 102 items. Provides population tables for approximately 15 years for each city.

345 *Population.* In *Population, education, development in Africa south of the Sahara: a selective annotated bibliography*, vol. 1. Dakar: Unesco Regional Office for Education in Africa, 1978-. 76 pp.

Contains 225 citations, many detailed annotations.

MIGRATION

346 Davis, Lenwood G. *Migration to African cities: an introductory survey.* Monticello IL: Council of Planning Librarians, 1977. (CPL exchange bibliography 1204) 21 pp.

Unannotated.

ANTHROPOLOGY

347 Gibson, G. D. "A bibliography of anthropological bibliographies Africa." *Current anthropology* 10:5 (1969):527-530.

348 Weekes, Richard V., ed. *Muslim peoples: a world ethnographic survey.* Westport CT: Greenwood Press, 1978. 546 pp.

Brief introductory descriptive articles and post-1945 bibliography on contemporary Muslim societies (world-wide). All entries in English. Concentrates on current patterns of living.

349 *Contemporary African women: an introductory bibliographical overview and a guide to women's organizations, 1960-1967.* Washington DC: African Bibliographic Center, 1968. (Special bibliographic series 6:2) 59 pp.

Includes American women's associations interested in Africa. Two- part book: Part I partially annotated bibliographical guide to French and English language sources; Part II directory. Compiled by the Women's African Committee of the African-American Institute.

350 Saulniers, Suzanne Smith, and Rakowski, Cathy A. *Women in the development process: a select bibliography on women in sub-Saharan Africa and Latin America.* Austin TX: University of Texas at Austin. Institute of Latin American Studies, 1977. (Special publication) 287 pp.

Unannotated. Includes only works in English, French, Portuguese, and Spanish.

MEDICINE

351 Patterson, Karl David. *Infectious diseases in twentieth century Africa: a bibliography of their distribution and consequences.* Waltham MA: Crossroads, 1979. 251 pp.

PUBLIC HEALTH

352 Feierman, Steven, comp. *Health and society in Africa: a working bibliography.* Waltham MA: Crossroads, 1979. (Archival and bibliographic series) 210 pp.

Lists studies of witchcraft, sorcery, Islamic healing, Christian Therapeutic cults, use of herbs, etc. Includes works on healers but not the afflictions treated.

353 Jordan, Jeffrey L. *Rural health care and international development in Africa: with additional references to Asia and Latin America.* Monticello IL: Council of Planning Librarians, 1977. (CPL exchange bibliography 1409) 38 pp.

Unannotated. Concentrates on journal articles appearing after 1970.

NUTRITION

354 Leteure, P. C. *Alimentation des populations africaines au Sud du Sahara [nourishment of the African populations south of the Sahara].* Bruxelles: Centre de Documentation Economique et Sociale Africaine, 1965. (Enquetes bibliographiques 13) 221 pp.

Annotated. In French.

355 Leung, Woot-Tsuen (Wu), and Butrum, Ritva Rauanheimo. *A selected bibliography on African foods and nutrition arranged according to subject matter and area.* Washington DC: U.S. Dept. of Health, Education and Welfare. Center for Disease Control. Food Science Information Nutrition Program, 1970. 382 leaves.

Contains 1,400 items. Covers period 1940-1969. Continues *Foods and nutrition* (1966).

356 Newman, Mark D. *Changing patterns of food consumption in tropical Africa: a working bibliography.* East Lansing MI: Michigan State University. Dept. of Agricultural Economics, 1978. (African Rural Economy Program Working Paper 23) 12 pp.

Contains 175 items.

URBAN GEOGRAPHY

357 Ajaegbu, Hyacinth I. *African urbanization: a bibliography.* London: International African Institute, 1972. 78 pp.

Intended as a guide to research into urbanization south of the Sahara. Primarily sociological, economic, and demographic information.

358 Davis, Lenwood G. *Urban growth, development and planning of African towns and cities.* Monticello IL: Council of Planning Librarians, 1977. (CPL exchange bibliography 1277) 27 pp.

Unannotated.

359 *Historique de la planification urbaine des capitales des etats francophones d'Afrique noire et de Madagascar [town planning history of the capitals of French-speaking states of Black Africa and Madagascar].* Paris: France. Secretariat des missions d'urbanisme et d'habitat. Service documentation-information, 1974. (Bibliographie 16) 57 pp.

Contains 141 items.

360 O'Connor, Anthony Michael. *Urbanization in tropical Africa: an annotated bibliography.* Boston MA: G. K. Hall, 1981. (Bibliographies and guides in African studies) 381 pp.

Covers all of Africa except the northern tier (Western Sahara to Egypt) and southern nations of Namibia, Botswana, Republic of South Africa, and Madagascar.

361 Odimuko, C. L., and Bouchard, D. *Urban geography of Africa.* Montreal, Canada: McGill University. Centre for Developing Area Studies, 1973. (Bibliography series 3) 41 pp.

Contains 565 items.

362 *Planification de villages en Afrique au sud du Sahara [village planning in Africa south of the Sahara].* Paris: France. Secretariat des missions d'urbanisme et d'habitat. Service documentation-information, 1975. (Bibliographie 6) 20 pp.

Annotated. Contains 69 items. Omits Southern Africa.

ARCHITECTURE

363 *L'habitat traditionnel en Afrique [traditional housing in Africa].* Paris: France. Secretariat des missions d'urbanisme et d'habitat. Service documentation-information, 1974. (Bibliographie 14) 39 pp.

364 Mekkawi, Mod M. *Bibliography on traditional architecture in Africa.* Rev. ed. Washington DC: Mekkawi, 1979. 117 pp.

Contains some 1,600 citations to works published from 1880-1979. Covers various building types as well as related subjects, such as mural painting, tombs, and environments. Emphasis on folk architectural design. Excludes material on Islamic architecture in North Africa.

HOUSING

365 Davis, Lenwood G. *Housing problems in selected African cities: an introductory bibliography.* Monticello IL: Council of Planning Librarians, 1977. (CPL exchange bibliography 1205) 20 pp.

Unannotated. Covers a number of cities throughout Africa with many more general sources on entire nations.

Africa/366-377

CONSTRUCTION

366 Davis, Lenwood G. *Construction building and planning in selected African cities.* Monticello IL: Council of Planning Librarians, 1977. (CPL exchange bibliography 1202) 23 pp.

Unannotated. Valuable chiefly for its list of journal articles.

ECONOMIC DEVELOPMENT

367 *Africa/Middle East data bank.* Washington DC: Data Resources, 197?.

An online database providing economic data on selected African and Middle Eastern nations. Vendor: Data Resources.

368 Blauvelt, Euan. *Sources of African and Middle-Eastern economic information.* Westport CT: Greenwood, 1982. 2 vols.

369 *Repertoire des projets de recherche en matiere de developpement en Afrique/Register of development research projects in Africa.* Paris: OECD Development Centre, 1979. (Bulletin de liaison entre instituts de recherche et de formation en matiere de developpement, n.s. 1) 106 pp.

Covers 226 then-current projects in 21 countries.

TRANSPORTATION

370 Davis, Lenwood G. *Ports of the African continent: a working bibliography.* Monticello IL: Council of Planning Librarians, 1977. (CPL exchange bibliography 1375) 11 pp.

Unannotated.

TOURISM AND RECREATION

371 Baretje, Rene. *Le tourisme en Afrique: essai bibliographique [tourism in Africa: bibliographic essay].* Aix-en-Provence, France: Centre des hautes etudes touristiques. Faculte de droit, 1976-1978. 2 vol.

372 Houts, Didier van. *International tourism in Africa/Le tourisme international en Afrique.* Antwerp, Belgium: Universiteit Antwerpen-RUCA. College voor de ontwikkelingslanden, 1978. (ALA bibliography 1) 95 pp.

In English and French.

HISTORICAL GEOGRAPHY

373 Hess, Robert L. and Dalvan M. Coger. *Semper ex Africa: a bibliography of primary sources for nineteenth-century tropical Africa as recorded by explorers...* Stanford CA: Hoover Institution, 1972. Hoover bibliographical series: no. 47. 800 pp.

Lists references to over 7,700 books and periodical articles by a variety of early explorers. Limited to that area of the continent exclusive of the Muslim Arab North and of the Republic of South Africa. Geographical organization; author index.

374 Fage, J. D. *An atlas of African history* 2nd ed. New York: Africana Publishing, 1978. 84 pp.

375 Gann, Lewis H., and Duigan, Peter, eds. *A bibliographic guide to colonialism in sub-Saharan Africa.* In Colonialism in Africa, 1870-1960, vol. 5. Edited by Lewis H. Gann and Peter Duigan. Cambridge, U.K.: Cambridge University Press, 1973. (Hoover Institution publications) 552 pp.

Annotated. Geographical arrangement of bibliographies.

PUBLIC ADMINISTRATION

376 Shaw, Robert Baldwin, and Sklar, Richard L. *A bibliography for the study of African politics*, vol. I. Waltham MA: Crossroads, 1977. (University of California at Los Angeles. African Studies Center. Occasional paper 9) 206 pp.

Unannotated. Emphasis on English language works.

377 Solomon, Alan C. *A bibliography for the study of African politics*, vol. II. Waltham MA: Crossroads, 1977. (Archival and bibliographic series) 193 pp.

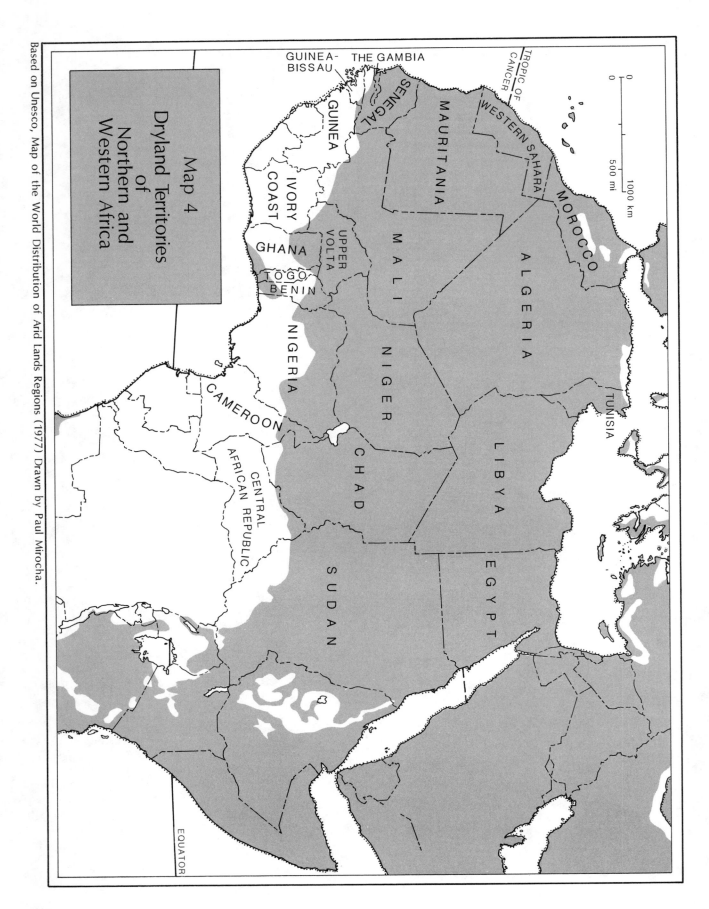

Northern and Western Africa

Nearly all the nations in the northern half of the continent occupy extensive portions of the Sahara, by far the world's largest desert. Its size alone is difficult to grasp for it is almost as large as all fifty United States.

Containing an enormous extent of hyperarid territory, the region presents the most hostile environment to man outside the poles. Nonetheless, the Sahara has provided a home and a living for man for tens of thousands of years at least. Although today the climate is markedly drier and the great herds of cattle, antelope, and elephants have disappeared, rainwater from wetter periods has been stored in underground formations that are now beginning to be tapped. Bringing Saharan lands into cultivation without mining nonrenewable groundwater supplies or otherwise causing long-term environmental destruction can be considered an ultimate goal of dryland research.

More immediate are the problems posed by the Sahel, the wide belt of arid to subhumid savanna that lies to the south of the Sahara and stretches (in its broadest definition) from the Atlantic to the Red Sea. Though not nearly as dry as the Sahara, in good years the region supports a heavy burden of people and grazing animals; in poor years, when drought continues beyond its normal winter season, the land quickly becomes exhausted as all available forage and fuel are consumed. Even massive international relief efforts have not saved millions of people from starvation and disease. Much research into the Sahel grew out of the disastrous 1968-1974 drought so that the region is one of the most intensely studied drylands in the world.

GENERAL BIBLIOGRAPHIES

378 Blaudin de The, Bernard Marie Samuel. *Essai de bibliographie du Sahara francais et des regions avoisinantes [bibliography of French Sahara and neighboring regions].* 2nd ed. Paris: Arts et Metiers Graphiques, 1960. 258 pp.

In French.

379 *Directory of information sources external to the Sahel.* Bamako, Mali: CILSS. Reseau Sahelien d'Information et de Documentation Scientifiques et Techniques, 1979. 95 pp.

380 Kostinko, Gail A. *A selected bibliography of Club du Sahel and CILSS documents.* Washington DC: Koba Associates, 1979. 44 pp.

381 Littlefield, David W. *The Islamic Near East and North Africa: an annotated guide to books in English for non-specialists.* Littleton CO: Libraries Unlimited, 1977. 375 pp.

Extensively annotated. Covers most general subjects; gives Library of Congress and Dewey Decimal call numbers. Covers North Africa (including Sudan and Mauritania), the Middle East, Cyprus.

382 Miller E. Willard, and Miller, Ruby M. *Northern and western Africa: a bibliography on the Third World.* Monticello IL: Vance Bibliographies, 1981. (Public administration series bibliography P-818) 95 pp.

Unannotated. Covers the continent north and west from Sudan.

383 *Sahel bibliographic bulletin/bulletin bibliographique.* East Lansing MI: Sahel Documentation Center, 1977-. Quarterly.

In English and French. A major abstracting journal of documents on the Sahel.

384 *Sahel documentatie: uitgave Koninkliyk Instituut voor de Tropen [Sahel documentation: publication of the Royal Institute for the Tropics].* Amsterdam: Het Instituut, 1977-. Quarterly.

Available in Dutch, English and French. Covers Chad, Mali, Mauritania, Niger, Senegal, Upper Volta, Sudan.

385 Witherell, Julian W. *French-speaking West Africa, a guide to official publications.* Washington DC: U.S. Library of Congress, 1967. 201 pp.

Contains published government documents from mid-19th century to date, of colonial, national, and international bodies.

THESES AND DISSERTATIONS

386 Selim, George Dimitri, comp. *American doctoral dissertations on the Arab world, 1883-1974.*

See entry 1149.

HANDBOOKS

387 *The Middle East and North Africa, 1980-81.*

See entry 1150.

ENCYCLOPEDIAS

388 Gibb, H. A. R.; Kramers, J. H.; and Levi-Provencal, E., eds. *The encyclopaedia of Islam.*

See entry 1128.

DIRECTORIES

389 Wheeler, Stella E. L., and Shah Khawaja T., eds. *Greece, Turkey, and the Arab states.*

See entry 151.

Northern and Western Africa/390-405

GEOGRAPHY

390 *The Sahelian zone: a selected bibliography for the study of its problems.* Rome: FAO Library, 1973. (Occasional bibliography 9) 68 pp.

ATLASES

391 *International atlas of West Africa/Atlas international de l'ouest africain.* Dakar: Institut Fondamental d'Afrique Noire, 1968-. Looseleaf.

Covers Mauritania, Mali, Niger, Senegal, Upper Volta, The Gambia, Ghana, Togo, Benin, Nigeria. In English and French. Scale 1:5,000,000.

GEOLOGY

392 Daveau, S. "Bibliographie pratique pour l'etude du relief en Afrique Occidentale" [bibliography for the study of the relief in western Africa]. *Revue de Geographie de l'Afrique Occidental* 1:2 (1965):229-233.

Unannotated. Most citations are to literature in French.

393 Merabet, Omar. *Bibliographie de l'Algerie du sud (Sahara) et des regions limitrophes [bibliography of southern Algeria (Sahara) and adjacent regions].* Algiers: Algeria. Service de la Carte Geologique de l'Algerie, 1968. (Bulletin 37) 197 pp.

Covers the Saharan portions of Algeria, Morocco, Mauritania, Mali, Niger, Libya, Tunisia.

CLIMATOLOGY

394 Bellot-Courdec, Beatrice. "Bibliographie comentee." In *Secheresse et elevage au Sahel*, pp. 18-55. Bordeaux, France: Universite de Bordeaux, 1978.

In French. Contains 250 annotated citations on drought in the Sahel. Emphasis on the past 10 years.

395 Joyce, Stephen J., and Beudot, Francoise. *Elements de bibliographie sur la secheresse au Sahel [elements for a bibliography of the Sahel drought].* Paris: Organisation de Cooperation et de Developpement Economique, 1976- 1977. 2 vols.

In French. Unannotated. Original edition contains 1,357 items. *Supplement* (1977) contains items 1-401, 88 pp.; *Supplement* 2 (1978) contains items 402-963, 145 pp.

396 Kramer, H. P. "A selective annotated bibliography on the climatology of Africa." *Meteorological abstracts and bibliography* 3:1 (1952):37-79.

Annotated. Covers Algeria, the Canary Islands, Libya, the Madeira Islands, Morocco, Western Sahara, Tunisia, and the Sahara in general.

397 Thran, P., and Broekhuizen, Simon. *Agroclimatic atlas of Europe.*

See entry 114.

HYDROLOGY

398 Bichet, E., and Martin, P. *Collecte stockage utilisation des eaux pluviales dans les pays du Sahel: utilisation des techniques au niveau de village; bibliographie selective et analytique [collecting, storing and using rainwater in the Sahel: using techniques at the village level; a selective annotated bibliography].* Paris: Institut International de Recherche et de Formation, 1976. 99 pp.

In French.

399 Gischler, Christiaan. *Water resources in the Arab Middle East and North Africa.*

See entry 1167.

ZOOLOGY

400 Graber, M. *Bibliographie des parasites internes des animaux domestiques et sauvages du Maghreb, du sahara et de la Mauritanie [bibliography on the internal parasites of domestic and wild animals of Maghreb, Sahara and Mauritania].* Maisons Alfort, France: Institut d'Elevage et de medecine Veterinaire des Pays Tropicaux, 1979. 196 pp.

In French.

401 Harrison, Colin. *An atlas of the birds of the Western Palaearctic*

See entry 116.

AGRICULTURE

402 Shihabi, Mustafa. *Chihabi's dictionary of agricultural and allied terminology: English-Arabic.*

See entry 1171.

PLANT CULTURE

403 Havinden, M. A. "The history of crop cultivation in West Africa: a bibliographical guide." *Economic history review* 23:3 (1970):532-555.

FORESTRY

404 Taylor, George F., III, and Taylor, Beth Ann. "Forestry in the Sahel: a selected bibliography of source materials relating to arid zone forestry and the southern fringe of the Sahara." *A current bibliography on African affairs* 12:1 (1979-80):33-49.

NOMADISM

405 Baumer, Michel, and Bernus, Edmond. "A selective bibliography on nomadism in the Sahelo-Sudanian zones." *Arid lands newsletter* 10 (1976):19-26.

Also printed in *A current bibliography on African affairs* 12:2 (1979-80).

406 Ebolo, Josue E. *Bibliographie retrospective sur le nomadisme pastoral au Sahel, periode 1969-1976 (Haute Volta, Mali, Mauretanie, Niger, Senegal, Tchad) [retrospective bibliography of pastoral nomadism of the Sahel, 1969-1976 (Upper Volta, Mali, Mauritania, Niger, Senegal, Chad)].* Niamey: Commission du Fleuve Niger. Centre de Documentation, 1976. 20 leaves.

Contains 105 items. Includes economy and stock breeding as well as nomadism.

407 *Nomades et nomadisme au Sahara [nomads and nomadism of the Sahara].* Paris: Unesco, 1963. (Arid zone research 19) 195 pp.

In French. Considers traditional and modern nomadism.

AGRICULTURAL ECONOMICS

408 Massing, Andreas. "Bibliography: rural capital formation in West Africa." *Rural africana*, n.s. 2 (1978):95-115.

Contains 240 citations. Includes agriculture, education, development.

409 Walsh, Gretchen. *Access to sources of information on agricultural development in the Sahel.* East Lansing MI: Michigan State University. Dept. of Agricultural Economics, 1976. (African Rural Economy Program working paper 17) 25 pp.

HUMAN GEOGRAPHY

410 Horowitz, Michael M.; Lewis, John van Dusen; and Painter, Tom. *The sociology and political economy of the Sahel: an annotated bibliography.* Binghamton NY: Institute for Development Anthropology, 1979. 35 leaves.

MIGRATION

411 Ebolo, Josue E. *Bibliographie retrospective sur les migrations au Sahel: Haute-Volta, Mali, Mauretanie, Niger, Senegal, Tchad. Periode 1969-1976 [retrospective bibliography of the migrations of the Sahel: Upper Volta, Mali, Mauritania, Niger, Senegal, Chad, 1969-1976].* Niamey: Commission du Fleuve Niger. Centre de Documentation, 1976. 9 pp.

A brief unannotated list of 37 references—mostly in French—covering human migration in the region.

ANTHROPOLOGY

412 Leupen, A. H. A. *Bibliographie des populations Touaregues (Sahara et Soudan Centraux) [bibliography of the Tuareg population (Sahara and Central Sudan].* Leyden, The Netherlands: Afrika-Studiecentrum, 1978. 240 pp.

In French. Contains 1,415 annotated citations to 1976. Includes dissertations. Omits prehistory.

WOMEN'S STUDIES

413 al-Barbar, Aghil M. *The study of Arab women: a bibliography of bibliographies.*

See entry 1181.

414 Al-Qazzaz, Ayad. *Women in the Middle East and North Africa: an annotated bibliography.*

See entry 1180.

415 Meghdessian, Samira Rafidi. *The status of the Arab woman: a select bibliography.*

See entry 1182.

416 Raccagni, Michelle. *The modern Arab woman: a bibliography.*

See entry 1183.

PUBLIC HEALTH

417 Barbar, Aghil M. *Disease, health services and care in the Arab world: an introductory bibliography.*

See entry 1184.

NUTRITION

418 *Bibliography of nutrition in the Sahel.* Silver Spring MD: Intech, 1977. 2 vols.

Contains 2,600 abstracts. Annex to *Nutrition strategy in the Sahel* (1977).

URBAN GEOGRAPHY

419 Barbar, Aghil M. *Bibliography of bibliographies of urbanization in the Arab world.*

See entry 1185.

420 Barbar, Aghil M. *Urbanization in the Arab world: a selected bibliography.*

See entry 1186.

421 Davis, Lenwood G. *Urbanization in the Middle East, with some references to North Africa.*

See entry 1187.

ARCHITECTURE

422 Cigar, Norman. *Architecture of the Sahara.* Monticello IL: Vance Bibliographies, 1981. (Architecture series bibliography A463) 3 pp.

Unannotated. A brief list of 27 sources, mostly French.

423 Prussin, Labelle, and Lee, David. "Architecture in Africa: an annotated bibliography. Part 1: North and West Africa." *Africana library journal* 4:3 2-32.

Contains 820 references to primary sources.

HOUSING

424 Barbar, Aghil M. *Housing in the Arab world: a bibliography.*

See entry 1191.

425 *Habitat rural en Afrique de l'ouest [rural housing in West Africa].* Paris: France. Secretariat des missions d'urbanisme et d'habitat. Service documentation- information, 1974. (Bibliographie 19) 13 pp.

In French. Annotated.

ECONOMIC DEVELOPMENT

426 Bricault, Giselle C., ed. *Major companies of the Arab world—1982.* London: Graham and Trotman Ltd., 1981. 854 pp.

Covers Algeria, Bahrain, Egypt, Iraq, Jordan, Kuwait, Lebanon, Libya, Mauritania, Morocco, Oman, Qatar, Saudi Arabia, Somalia, Sudan, Syria, Tunisia, United Arab Emirates, and the Yemens.

427 *Le developpement economique industriel et commercial en Afrique—bibliographie [industrial and commercial economic development in Africa— bibliography].* Dakar: Chambre de Commerce et d'Industrie de la Region du Cap-Vert, 1980. 207 pp.

In French. Annotated list covering French-speaking West Africa, concentrating on journal articles.

TRANSPORTATION

428 Barbar, Aghil M. *Ports of the Arab world: an annotated bibliography.*

See entry 1198.

429 Barbar, Aghil M. *Transportation in the Arab world.*

See entry 1199.

430 Krummes, Daniel. *Transportation in West Africa: a bibliography.* Monticello IL: Vance Bibliographies, 1979. (Public administration series bibliography P-222) 27 pp.

Unannotated. Covers the Sahelian-Sudanian countries south of the Sahara and west of Cameroon.

ALGERIA
El Djemhouria El Djazairia Eddemokratia Echaabia/ Republique Algerienne Democratique et Populaire

(a former colony of France)

Algeria, Africa's second largest nation, consists of two sharply contrasting regions, both of which offer considerable potential for development.

The Maghreb. The Atlas mountains are to Algeria what the Nile valley is to Egypt: the major source of water and the home for nearly the entire national population. These mountains are, in fact, a series of parallel ranges, foothills, and intervening valleys that become progressively drier toward the south until the Sahara itself is reached. The climate along the coast is typically Mediterranean with moderate temperatures and winter-dominant precipitation. Inland the rise in elevation and the complex topography make generalizations difficult: remnant forests still cover some mountain slopes while the deeper basins are dry enough to hold saline soils. Agriculture is vertically and horizontally zoned, with intensive cultivation in the most favored valleys giving way to poorer subsistence farming and semi-nomadic pastoralism on the slopes and drier inland steppes.

When the French began occupying their African neighbor after 1830, settlers naturally chose the best farmland in the wider plains near the Mediterranean. Native Arabs and Berbers were ruthlessly uprooted and their lands expropriated for Europeans. The destruction of the French vineyards in 1878 by phylloxera greatly encouraged settlement and grape-planting; after 1900 Algeria was producing more wine than France itself, thus becoming the colony's most valuable export. The great differences between wealthy settler and impoverished native led inevitably to revolution; after a bitter three-sided war the new nation in 1962 was left with a shattered economy and the exodus of its (largely European) managers, skilled workers, and plantation farmers. For a nation that once exported great quantities of agricultural products, the necessity now to import grain points to the need for massive rural development.

The Sahara. Algeria commands the largest share of the Great Desert and so faces a monumental challenge for its development. Both the natural and human landscapes are typical for the region: vast expanses of sandy erg, stoney reg, and bedrock hamada are proportioned more or less equally across the surface. Only the Ahaggar and Tasilli Mountains rise high enough to capture a bit more moisture and so create more favorable niches for plants, animals, and human settlement. Outside the mountains only palm-fringed oases—dependent on groundwater— offer shelter and small but self-sustaining economic bases. Outside the oases the Bedouin Arabs have relied on shallow wells and seasonal appearances of pasture for their camels and goats.

A recent development is the discovery of oil and natural gas in the fields toward the Libyan border. These were first opened in the late 1950's and their subsequent exploitation has proved the mainstay of the nation's economy, petroleum replacing agricultural products as the primary export.

Dryland areas of ALGERIA

DESERTIFICATION RISK	ARIDITY								Aridity Totals	
	Hyperarid		Arid		Semiarid		Subhumid			
	km^2	%	km^2	%	km^2	%	km^2	%	km^2	%
Very High	By definition, desertification does not exist in hyperarid regions.		21,436	0.9	59,543	2.5			80,979	3.4
High			273,900	11.5	50,017	2.1			323,917	13.6
Moderate			38,108	1.6	9,527	0.4			47,635	2.0
Desertification Totals			333,444	14.0	119,087	5.0			452,531	19.0
No Desertification	1,900,633	79.8					4,763	0.2	1,905,396	80.0
Total Drylands	1,900,633	79.8	333,444	14.0	119,087	5.0	4,763	0.2	2,357,927	99.0
Non-dryland									23,818	1.0
Total Area of the Territory									2,381,745	100.0

Algeria/431-445

GENERAL BIBLIOGRAPHIES

431 **Lawless, Richard I.** *Algeria.* Oxford, U.K.; Santa Barbara CA: Clio, 1980. (World bibliographical series 19) 215 pp.

Annotated bibliography contains 742 items.

HANDBOOKS

432 **Nelson, Harold D.** *Algeria: a country study.* 3rd ed. Washington DC: U.S. Dept. of Defense, 1979. (American University foreign area studies) 370 pp.

Narrative summary of the nation's history, politics, geography, economy, social life; includes unannotated bibliography.

GEOGRAPHY

433 **Nuttonson, Michael Y.** "Introduction to North Africa and a survey of the agriculture of Morocco, Algeria, and Tunisia, with special reference to their regions containing areas climatically and latitudinally analogous to Israel." In *A survey of North African agro-climatic counterparts of Israel*, vol. 1. Washington DC: American Institute of Crop Ecology, 1961. 608 pp.

GAZETTEERS

434 *Algeria: official standard names approved by the United States Board on Geographic Names.* Washington DC: U.S. Board on Geographic Names, 1972. (Gazetteer 123) 754 pp.

List of 45,200 place-names with latitude/longitude coordinates.

HYDROLOGY

435 **Dazy, Jean.** "Bibliographie hydrogeologique de l'Algeria (1839-1973)" [hydrogeologic bibliography of Algeria, 1839-1973]. Societe d'Histoire Naturelle de l'Afrique Nord. *Bulletin* 65:1-2 (1974):263-370.

In French.

SOILS

436 *Bibliography on desert soils of Morocco, Algeria, Tunisia, Libya (1964-1936).*

See entry 623.

437 *Soils of Algeria (annotated bibliography covering the published literature for 1929-1975).* Harpenden, U.K.: Commonwealth Agricultural Bureaux, 1977. (Annotated bibliography) 7 pp.

BOTANY

438 **Roussine, N., and Sauvage, C.** "Bibliographia phytosociologica: Afrique du Nord" [bibliography of plant ecology: North Africa].

See entry 626.

HUMAN GEOGRAPHY

439 **Vidergar, John J.** *The economic, social and political development of Algeria and Libya.* Monticello IL: Vance Bibliographies, 1978. (Public administration series bibliography P-30) 7 pp.

Unannotated.

MEDICINE

440 **Dedet, J. P.** "The leishmaniasis of North Africa bibliographic list, part 2, Algeria 1860-1974." Institut Pasteur de Tunis. *Archives* 52:1-2 (1975):129-148.

URBAN GEOGRAPHY

441 **Tesdell, Lee S.** *Cities in Algeria.* Monticello IL: Council of Planning Librarians, 1977. (CPL exchange bibliography 1251) 6 pp.

Brief annotations.

ARCHITECTURE

442 **Cigar, Norman.** *North African architecture.*

See entry 629.

ECONOMIC DEVELOPMENT

443 *Quarterly economic review of Algeria.* London: Economist Intelligence Unit, 1952-. Quarterly.

Includes summary of political and economic news as well as charts and statistics of selected economic indicators. Annual supplement.

HISTORICAL GEOGRAPHY

444 **Heggoy, Alf Andrew, and Crout, Robert R.** *Historical dictionary of Algeria.* Metuchen NJ: Scarecrow, 1981. (African historical dictionaries 28) 237 pp.

Entries cover major topics, events, and personalities in the nation's history; includes extensive unannotated bibliography.

PUBLIC ADMINISTRATION

445 **Cigar, Norman.** *Government and politics in North Africa.*

See entry 633.

BENIN
Republique Populaire du Benin
(as Dahomey, a former colony of France)

Benin and Togo, two small nations sandwiched between Ghana and Nigeria, have closely matched environments and economic prospects. Both have two disconnected dryland belts.

The Coastal Lowland. This region falls into the anomalous subhumid zone that stretches along the coast from southeastern Ghana to the Benin-Nigeria frontier. Two rainy seasons—spring and fall—support the oil palm, the region's principal crop and Benin's principal export. Porto-Novo, the capital, and Cotonou, the major seaport, lie just outside this dryland in the nation's southeast corner.

The Savanna. North of the Atakora Range several rivers drain a long savanna slope to the Niger River. The region is a subhumid fringe of the Sahelo-Sudanian Belt that crosses the entire continent. This is cattle, cotton, and groundnut country, neither seriously deficient in moisture nor subject to desertification. Unlike the oil palm coastal lowland, no crop predominates as an export product.

Of the two nations, Benin is the poorer as it lacked the basic capital improvements of roads and railroads that the Germans provided its neighbor (though today a railroad reaches from the coast halfway to the Niger River). Benin's economy is narrowly built upon the oil palm and so suffers the effects of a monocultural agricultural base. Some potential exists for increasing the entrepot functions of the port at Cotonou by extending the railroad north to the Niger River.

Dryland areas of BENIN

DESERTIFICATION RISK	ARIDITY									
	Hyperarid		Arid		Semiarid		Subhumid		Aridity Totals	
	km²	%	km²	%	km²	%	km²	%	km²	%
Very High	By definition, desertification does not exist in hyperarid regions.									
High										
Moderate										
Desertification Totals										
No Desertification					4,167	3.7	56,311	50.0	60,478	53.7
Total Drylands					4,167	3.7	56,311	50.0	60,478	53.7
Non-dryland									52,144	46.3
Total Area of the Territory									112,622	100.0

GENERAL BIBLIOGRAPHIES

446 **Dujarier, M.** *Notes bibliographiques sur le Dahomey [bibliographic notes on Dahomey].* Cotonou, Benin: Institut d'enseignement superieur de Benin, 1971. 94 pp.

GAZETTEERS

447 *Dahomey: official standard names approved by the United States Board on Geographic Names.* Washington DC: U.S. Board on Geographic Names, 1965. (Gazetteer 91) 89 pp.

List of 6,250 place-names with latitude/longitude coordinates.

ECONOMIC DEVELOPMENT

448 *Quarterly economic review of Ivory Coast, Togo, Benin, Niger, Upper Volta.*

See entry 714.

HISTORICAL GEOGRAPHY

449 **Decalo, Samuel.** *Historical dictionary of Dahomey (People's Republic of Benin).* Metuchen NJ: Scarecrow, 1976. (African historical dictionaries 7) 201 pp.

Entries cover major topics, events, and personalities in the nation's history; includes extensive unannotated bibliography.

CAMEROON
Republique Unie du Cameroun

(as Kamerun and Cameroons, a former colony of Germany, then France and the United Kingdom)

Cameroon is a roughly triangular-shaped nation with an arid apex terminating at Lake Chad.

The Sahel. Cameroon's drylands occupy a wedge of the great Sahelo-Sudanian belt between the Benue River and Lake Chad. Agriculture centers on cattle, cotton, and groundnuts. The major difficulty for the region is its distance from the national center; the people and their economy traditionally turn toward adjacent Nigeria.

Cameroon possesses considerable development potential and is beginning to build an industrial sector based on reserves of iron ore and bauxite. The dryland region of the north contributes relatively little to the nation's economy and, moreover, contains Moslem peoples not strongly inclined to turn south for capital or managerial assistance.

Dryland areas of CAMEROON

DESERTIFICATION RISK	ARIDITY								Aridity Totals	
	Hyperarid		Arid		Semiarid		Subhumid			
	km²	%	km²	%	km²	%	km²	%	km²	%
Very High	By definition, desertification does not exist in hyperarid regions.				10,696	2.3			10,696	2.3
High										
Moderate										
Desertification Totals					10,696	2.3			10,696	2.3
No Desertification							41,390	8.9	41,390	8.9
Total Drylands					10,696	2.3	41,390	8.9	52,086	11.2
Non-dryland									412,968	88.8
Total Area of the Territory									465,054	100.0

GENERAL BIBLIOGRAPHIES

450 Bridgman, Jon, and Clarke, David G. *German Africa: a select annotated bibliography.* Stanford CA: Hoover Institution, 1965. (Bibliographical series 19) 120 pp.

Some annotations. Covers present Tanzania, Namibia, Togo, Cameroon.

451 Delancey, Mark W., and Delancey, Virginia H. *A bibliography of Cameroon.* New York: Africana Publishing, 1975. (African bibliography series 4) 673 pp.

Some annotations. Covers the period 1884 to 1972. Most major subjects are included.

452 Dippold, Max F. *Une bibliographie du Cameroun: les ecrits en langue allemande [a bibliography of Cameroon: writings in German].* Freiburg, Federal Republic of Germany: Institut fuer Soziale Zusammenarbeit, 1971. 343 pp.

With 6,266 citations, this is the most complete bibliography of German work on the nation.

453 Witherell, Julian W. *Official publications of French Equitorial Africa, French Cameroons, and Togo, 1946-1958.* Washington DC: U.S. Library of Congress, 1964. 78 pp.

Lists publications of governments concerned with French Equatorial Africa and the trust territories.

HANDBOOKS

454 Nelson, Harold D. *Area handbook for the United Republic of Cameroons.* Washington DC: U.S. Dept. of Defense, 1974. (American University foreign area studies) 335 pp.

Narrative summary of the nation's history, politics, geography, economy, social life; includes unannotated bibliography.

GEOGRAPHY

455 **Beriel, Marie-Magdeleine.** "Contribution a la connaissance de la region du lac Chad (Cameroun, Tchad, Niger): bibliographie analytique" [contribution to the knowledge of the region of Lake Chad (Cameroon, Chad, Niger): analytic bibliography]. Dakar: Institut fondamental d'Afrique noire. *Bulletin*, ser. B, 38 (1976):411-429.

Contains 89 annotations from the period 1948-1974.

456 *Draft environmental profile on United Republic of Cameroon.* Tucson AZ: University of Arizona. Office of Arid Lands Studies. Arid Lands Information Center, 1981. 79 pp.

A narrative summary of the nation's environmental conditions and development problems, with an unannotated bibliography.

ATLASES

457 *Atlas de Cameroun [atlas of Cameroon].* Yaounde: Institut de Recherches Scientifiques du Cameroun, 1960-. 12 maps plus text.

Issued in parts. Scale: 1:1,000,000 and 1:2,000,000.

ATLASES

458 **Laclavere, Georges.** *Atlas de la Republique unie du Cameroun [atlas of the United Republic of Cameroon].* Paris: Editions J. A., 1979. (Les Atlas Jeune Afrique) 72 pp.

GAZETTERS

459 *Cameroon: official standard names approved by the United States Board on Geographic Names.* Washington DC: U.S. Board on Geographic Names, 1962. (Gazetteer 60) 255 pp.

List of 18,000 place-names with latitude/longitude coordinates.

CLIMATOLOGY

460 **Bender, Thomas A., Jr.** *Selected bibliography of climatic maps for Nigeria and the British Cameroons.*
See entry 656.

SOILS

461 *Soils of Cameroon, Central African Republic, Congo, Gabon, and Equatorial Guinea (annotated bibliography covering the published literature for 1942-1975).* Harpenden, U.K.: Commonwealth Agricultural Bureaux, 1977. (Annotated bibliography) 8 pp.

BOTANY

462 **Knapp. R.** "Bibliographia phytosociologica et scientiae vegetationis: Cameroun - Cameroon - Kamerun" [bibliography of plant ecology and vegetation: Cameroon]. *Excerpta botanica: sectio B: sociologica* 17:2 (1978):81-100.

Unannotated. Contains 256 items. Author index.

AGRICULTURAL ECONOMICS

463 **Ferguson, Donald S.** *Selected bibliography of Cameroon agricultural development.* Yaounde: University of Yaounde. Dept. of Rural Economy. National Advanced School of Agriculture, 1973. 13 leaves.

Contains 137 items.

ECONOMIC DEVELOPMENT

464 *Quarterly economic review of Gabon, Congo, Cameroon, C.A.R., Chad, Equatorial Guinea.* London: Economist Intelligence Unit, 1981-. Quarterly.

Includes summary of political and economic news as well as charts and statistics of selected economic indicators. Annual supplement.

HISTORICAL GEOGRAPHY

465 **Le Vire, Victor T., and Nye, Roger P.** *Historical dictionary of Cameroon.* Metuchen NJ: Scarecrow, 1974. (African historical dictionaries 1) 198 pp.

Entries cover major topics, events, and personalities in the nation's history; includes extensive unannotated bibliography.

CENTRAL AFRICAN REPUBLIC

(as Ubangi-Shari, a former colony of France)

The Central African Republic is a landlocked nation straddling the divide between the great rainforests of the Zaire Basin and the dryland belt of the Sahel. The nation itself has one narrow dryland region.

The Chari Valley. The Chari River originates in the uplands of Sudan's Darfur province and flows generally westward in a great curve to enter Lake Chad. It passes through, then forms the border of, the Central African Republic for roughly half its course. The climate in the valley is subhumid and typical of the most southerly fringes of the Sahel; maximum precipitation is in the summer. The land is used for cattle raising, primarily subsistence with little export beyond the region. Roads reach to the regional center of Birao but by and large the Chari Valley has little contact with the national center in the south and so remains of minor economic importance to the nation.

Only selected reference sources are provided.

Dryland areas of CENTRAL AFRICAN REPUBLIC

DESERTIFICATION RISK	ARIDITY									
	Hyperarid		Arid		Semiarid		Subhumid		Aridity Totals	
	km²	%	km²	%	km²	%	km²	%	km²	%
Very High	By definition, desertification does not exist in hyperarid regions.									
High										
Moderate										
Desertification Totals										
No Desertification							30,624	4.9	30,624	4.9
Total Drylands							30,624	4.9	30,624	4.9
Non-dryland									594,353	95.1
Total Area of the Territory									624,977	100.0

GENERAL BIBLIOGRAPHIES

466 Witherell, Julian W. *Official publications of French Equatorial Africa, French Cameroons, and Togo 1946-1958.*

See entry 453.

GAZETTEERS

467 *Central African Republic: official standard names approved by the United States Board on Geographic Names.* Washington DC: U.S. Board on Geographic Names, 1962. (Gazetteer 64) 220 pp.

List of 15,700 place-names with latitude/longitude coordinates.

SOILS

468 *Soils of Cameroon, Central African Republic, Congo, Gabon, and Equatorial Guinea (annotated bibliography covering the published literature for 1942-1975).*

See entry 461.

ECONOMIC DEVELOPMENT

469 *Quarterly economic review of Gabon, Congo, Cameroon, C.A.R., Chad, Equatorial Guinea.*

See entry 464.

HISTORICAL GEOGRAPHY

470 Kalck, Pierre. *Historical dictionary of the Central African Republic.* Metuchen NJ: Scarecrow, 1980. (African historical dictionaries 27) 152 pp.

Entries cover major topics, events, and personalities in the nation's history; includes extensive unannotated bibliography.

CHAD
Republique du Tchad
(a former colony of France)

Chad is a landlocked nation embracing large portions of the Sahara and the Sahel; only a small corner in the extreme south lies outside the drylands. Its two major regions clearly demonstrate the north-south divisions common to all the Sahelian nations.

The Sahel. This region occupies nearly the entire southern half of the nation and ranges from subhumid savanna to the border of the arid Sahara. Cotton is cultivated on the wettest lands, especially along the Chari and Logone Rivers. The region is largely self-sufficient in food and exports meat to Nigeria, Cameroon, and Zaire. With the capital of N'Djamena (formerly Ft. Lamy) near Lake Chad and a small but vital network of roads, the region is the nation's economic core. Perhaps as important, the people here are culturally tied to Black Africa: they are Christian or Animist, sedentary, and both wealthier and better educated than their countrymen to the north.

The Sahara. Chad's portion of the Great Desert takes in large sand seas and several ranges including the extensive Tibesti massif. Except for deposits of salt and deep fields of groundwater, the area lacks known mineral resources. The region is isolated from the south with only nomadic tribes making the regular trek from desert to savanna and back; as Moslems, both poor and poorly educated, their ties are closer to Libya than to the national government.

Dryland areas of CHAD

DESERTIFICATION RISK	ARIDITY									
	Hyperarid		Arid		Semiarid		Subhumid		Aridity Totals	
	km²	%	km²	%	km²	%	km²	%	km²	%
Very High	By definition, desertification does not exist in hyperarid regions.				146,376	11.4	42,372	3.3	188,748	14.7
High			409,596	31.9			62,916	4.9	472,512	36.8
Moderate										
Desertification Totals			409,596	31.9	146,376	11.4	105,288	8.2	661,260	51.5
No Desertification	409,596						164,352	12.8	573,948	44.7
Total Drylands	409,596		409,596	31.9	146,376	11.4	269,640	21.0	1,235,208	96.2
Non-dryland									48,792	3.8
Total Area of the Territory									1,284,000	100.0

HANDBOOKS

471 Nelson, Harold D. *Area handbook for Chad.* Washington DC: U.S. Dept. of Defense, 1972. (American University foreign area studies) 261 pp.

Narrative summary of the nation's history, politics, geography, economy, social life; includes unannotated bibliography.

GEOGRAPHY

472 Beriel, Marie-Magdeleine. Contribution a la connaissance de la region du Lac Tchad (Cameroun, Tchad, Niger): bibliographie analytique. [contribution to the knowledge of the region of Lake Chad (Cameroon, Chad, Niger): analytic bibliography.]

See entry 455.

473 Gabriel, Baldur. "Die Publikationen aus der Forschungsstation Bardai (Tibesti)" [the publications of the Bardai Research Center, Tibesti]. Freie Universitat Berlin. *Pressedienst Wissenschaft* 5 (1974):118-126.

In German with English summary.

ATLASES

474 Cabot, Jean. *Atlas pratique du Tchad [atlas of Chad].* Paris: Institut geographique naturelle, 1972. 76 pp.

In French.

Chad/475-481

GAZETTEERS

475 *Chad: official standard names approved by the United States Board on Geographic Names.* Washington DC: U.S. Board on Geographic Names, 1962. (Gazetteer 65) 232 pp.

List of 16,000 place-names with latitude/longitude coordinates.

BOTANY

476 Knapp, R. "Bibliographia phytosociologica et scientiae vegetationis: Africa Aegyptiaca, Libyca et Tibestica" [bibliography of plant ecology and vegetation: African Egypt, Libya and Tibesti].

Covers northern Chad.
See entry 501.

477 Knapp, R. "Bibliographia phytosociologica et scientiae vegetationis: Mali, Mauretania, Niger, Senegal, Tchad, Volta Superior" [bibliography of plant ecology and vegetation: Mali, Mauritania, Niger, Senegal, Chad, Upper Volta].

See entry 594.

HUMAN GEOGRAPHY

478 "Bibliographie du Tchad: sciences humaines" [bibliography of Chad: social sciences]. 2nd ed. Institut national tchadien pour les sciences humaines. *Etudes et documents tchadiens*, serie A., 5, 1970. 353 pp.

479 *Recherches scientifiques du Tchad: liste des publications relatives aux recherches menees sur le territoire de la Republique du Tchad sous l'egide du C.N.R.S.* [scientific research of Chad: list of publication relating to the secret research on the territory of the Republic of Chad under the direction of the C.N.R.S.]. Paris: France. Centre national de la recherche scientifique, 1974. 57 pp.

Contains 467 references on the human sciences.

ECONOMIC DEVELOPMENT

480 *Quarterly economic review of Gabon, Congo, Cameroon, C.A.R., Chad, Equatorial Guinea.*

See entry 464.

HISTORICAL GEOGRAPHY

481 Decalo, Samuel. *Historical dictionary of Chad.* Metuchen NJ: Scarecrow, 1977. (African historical dictionaries 13) 413 pp.

Entries cover major topics, events, and personalities in the nation's history; includes extensive unannotated bibliography.

EGYPT
The Arab Republic of Egypt

(a former dependency of the United Kingdom)

Ancient geographers recognized that Egypt was the gift of the Nile. This exotic river rises in remote highlands and traverses the world's greatest desert, providing an immensely fertile valley and delta for an extremely dense human population. These superlatives, however, conceal the fact that today even the Nile cannot provide all the people along its banks with an acceptable standard of living. Egypt must expand into its deserts to meet requirements for water, power, food, and living space.

The Nile Delta. This great triangle of fertile land is the principal agricultural area of the nation, brought under extensive cultivation only after engineering works and water control solved inherent problems of flooding and salt buildup. Even now the incursion of Mediterranean sea-water is a major threat. Rice is the staple crop while cotton is grown for export. The ancient city of Alexandria is the chief port and urban center; villages in the thousands dot the delta and are home to nearly half the national population.

The Nile Valley. Enclosed by high escarpments along much of its course, the narrow ribbons of land on either side of the river can be considered immensely attenuated oases. With the completion of the first barrage dams in the nineteenth century, perennial irrigation was made possible. Multiple crops of maize, wheat, cotton, rice, millet, clover, and other foods and fibers are now raised. Cairo is by far the largest city, both of the nation and the continent; its central location places it close to the Delta, the upper valley, the Suez Canal, and the deserts which are always nearby. The region's great engineering work is the Aswan High Dam, which impounds the waters of Lake Nasser and has numerous effects—both positive and negative—on upstream and downstream environments. While the dam has allowed millions of acres to be brought into irrigation, problems with disease, loss of annual silting, and disruption of the people living upstream are still not resolved.

The Eastern Desert and Sinai. This desert fringing the Red Sea is mountainous broken county lacking suitable land for agriculture except in a few favored basins. Although oil, iron, and phosphates are now exploited, the region's importance is largely strategic: the Sinai is a crossroads between continents, holds a major waterway, and weighs heavily on the minds of Egyptian and Israeli military planners.

The Western Desert. This region comprises two-thirds of the national territory and holds the greatest promise for development outside the Nile valley. Though the climate is hyperarid, groundwater is plentiful enough to support several extensive oases. Increasing water supplies from the ground, the Mediterranean, or the Nile could open large areas to cultivation. A great structural depression at Qattarah extends over 100 meters (330 feet) below sea level and thus offers the possibility of creating hydro-power by dropping Mediterranean water into the basin.

Egypt's basic problem is not a lack of resources alone but rather an incredibly expanding human population (doubling every 25 years at the current rate) that has largely negated progress in increasing agricultural yields and arable land. And, although Egypt cannot shirk its leadership in the Arab world, costly military adventures in the Sinai and Yemen have drained essential resources that must be used to ameliorate the lives of the Fellahin upon whom the future of the nation rests.

Dryland areas of EGYPT

DESERTIFICATION RISK	ARIDITY									
	Hyperarid		Arid		Semiarid		Subhumid		Aridity Totals	
	km²	%	km²	%	km²	%	km²	%	km²	%
Very High	By definition, desertification does not exist in hyperarid regions.		10,000	1.0					10,000	1.0
High			41,000	4.1					41,000	4.1
Moderate										
Desertification Totals			42,000	5.1					42,000	5.1
No Desertification	949,000	94.9							949,000	94.9
Total Drylands	949,000	94.9	42,000	5.1					1,000,000	100.0
Non-dryland										
Total Area of the Territory									1,000,000	100.0

Egypt/482-498

GENERAL BIBLIOGRAPHIES

482 *Abstracts of scientific and technical papers published in U.A.R./Resumes analytiques des travaux scientifiques et techniques publies en R.A.U.* Cairo: Egypt. Markaz al-Qawmi lil-Burhath. National Research Centre, Cairo, 1959-. Frequency unknown.

Supersedes *Abstracts of scientific and technical papers published in Egypt* (1955-1959).

483 Geddes, Charles L. *An analytical guide to the bibliographies on modern Egypt and the Sudan.* Denver CO: American Institute of Islamic Studies, 1972. (Bibliographic series 2) 78 pp.

HANDBOOKS

484 Nyrop, Richard F. *Area handbook for Egypt.* 3rd ed. Washington DC: U.S. Dept. of Defense, 1976. (American University foreign area studies) 454 pp.

Narrative summary of the nation's history, politics, geography, economy, social life; includes unannotated bibliography.

STATISTICS

485 *United Arab Republic statistical atlas, 1952-1966.* Cairo: Egypt. Central Agency for Public Mobilisation and Statistics, 1968. 123 pp.

Contains few maps. Valuable for its statistical graphs.

GEOGRAPHY

486 Barth, Hans Karl. *Egypt: geographical bibliography.* Bremen: Universitat Bremen, 1981. 197 pp.

Unannotated.

487 *Draft environmental report of Arab Republic of Egypt.* Tucson AZ: University of Arizona. Office of Arid Lands Studies. Arid Lands Information Center, 1980.

An extensive narrative summary of the nation's environment and economy supported by charts, tables, lists of researchers, and an unannotated bibliography.

488 El-Kassas, Mohamed. "Egypt." In *Handbook of contemporary developments in world ecology*, pp. 447-455. Edited by Edward John Kormondy and J. Frank McCormick. Westport CT: Greenwood, 1981.

A narrative summary of ecological research in Egypt; includes an unannotated list of references.

489 Nuttonson, Michael Y. "Physical environment and agriculture of Libya and Egypt with special reference to their regions containing areas climatically and latitudinally analogous to Israel." In *A survey of North African agro-climatic counterparts of Israel*, vol. 2. Washington DC: American Institute of Crop Ecology, 1961. 452 pp.

Bibliography, pp. 439-452.

GAZETTEER

490 *Egypt and the Gaza Strip: official standard names approved by the United States Board on Geographic Names.* Washington DC: U.S. Dept. of Defense, 1959. (Gazetteer 45) 415 pp.

List of 27,800 place-names with latitude/longitude coordinates. Covers Egypt, the Sinai peninsula and the Gaza Strip.

GEOLOGY

491 Avnimelech, Moshe A. *Bibliography of Levant geology, including Cyprus, Hatay, Israel, Jordania, Lebanon, Sinai, and Syria.*

See entry 1157.

492 *Current bibliography of Middle East geology.*

See entry 1158.

493 Glenn, C. R., and Denman, J. M. *Geologic literature on Egypt, 1933-1978.* Washington DC: U.S. Geological Survey, 1980. (Open file report 80-930) 221 pp.

494 Kaldani, Elias H. *A bibliography of geology and related sciences concerning Egypt up to the end of 1939.* Cairo: Egypt. Government Press, 1941. 428 pp.

495 Said, Rushdi et al. *Bibliography of geology and related sciences concerning Egypt for the period, 1960-1973.* Caire: Egypt. Geological Survey and Mining Authority, 1975. 192 pp.

CLIMATOLOGY

496 Grimes, Annie E. *An annotated bibliography of climatic maps of the United Arab Republic (Egypt).* Washington DC: U.S. Air Weather Service. Climatic Center, 1962. (NTIS: AD 660-866) 23 pp.

Contains 65 references.

497 Hacia, Henry, and Creasi, Vincent J. *An annotated bibliography on the United Arab Republic.* Silver Spring MD: U.S. Dept. of Commerce. Environmental Science Services Administration. Environmental Data Service, 1970. (Technical memorandum EDSTM-BC 104; NTIS: PB 194-692) 41 pp.

498 Kramer, H. P. "A selective annotated bibliography on the climatology of northeast Africa." *Meteorological abstracts and bibliography* 2:10 (1951):831-865.

Covers the period 1862-1951 and the nations of Egypt, Sudan, Ethiopia, Jibuti, and Somalia.

HYDROLOGY

499 Lytle, Elizabeth E. *The Aswan High Dam.* Monticello IL: Council of Planning Librarians, 1977. (CPL exchange bibliography 1334) 14 pp.

Unannotated. Covers social, physical, and natural sciences related to the dam's construction and use.

SOILS

500 *Bibliography on desert soils of Egypt, Eritrea, Somaliland, Aden (1964-1931).* Harpenden U.K.: Commonwealth Bureau of Soils, 196?. 15 pp.

Contains 75 annotated citations.

BOTANY

501 Knapp, R. "Bibliographia phytosociologica et scientiae vegetationis: Africa Aegyptiaca, Libyca et Tibestica" [bibliography of plant ecology and vegetation: African Egypt, Libya and Tibesti]. *Excerpta botanica: sectio B: sociologica* 14:1 (1974):117-134.

Unannotated. Covers all of Libya, all of Egypt except the northeast, and northern Chad.

AGRICULTURE

502 Cigar, Norman. *Agriculture and rural development in Egypt.* Monticello IL: Vance Bibliographies, 1981. (Public administration series bibliography P-851) 7 pp.

Unannotated.

503 *Egyptian agricultural bibliography.* Cairo: Egyptian Documentation and Information Centre for Agriculture, 1978-. Annual.

Unannotated. Introduction in Arabic, French and English. Derived from AGRIS Data Base. Includes author, commodity and participating center indices. Multilingual references.

PLANT CULTURE

504 *Egypt: grassland and fodder crops, 1973- 78 (Bibliography).* Hurley U.K.: Commonwealth Bureau of Pastures and Field Crops, 1978. 13 pp.

Annotated.

AGRICULTURAL ECONOMICS

505 Richards, Alan. *Egypt's agricultural development, 1800-1980: technical and social change.* Boulder CO: Westview Press, 1982. 296 pp.

A narrative summary examining the interactions of technological change, rural social classes, and government policy. Extensive, unannotated bibliography.

ANTHROPOLOGY

506 Coult, Lyman H. *An annotated research bibliography of studies in Arabic, English, and French, of the fellah of the Egyptian Nile, 1798-1955.* Coral Gables FL: University of Miami Press, 1958. 144 pp.

Annotated. Primary emphasis on social and anthropological studies of the fellah (villager) of the Egyptian Nile. Covers period 1798-1955; Arabic, French, and English sources.

URBAN GEOGRAPHY

507 Vidergar, John J. *Urbanization and social welfare programs in Egypt: a bibliography.* Monticello IL: Council of Planning Librarians, 1977. (CPL exchange bibliography 1347) 5 pp.

Unannotated.

ECONOMIC DEVELOPMENT

508 *Quarterly economic review of Egypt.* London: Economist Intelligence Unit, 1976-. Quarterly.

Includes summary of political and economic news as well as charts and statistics of selected economic indicators. Annual supplement.

TRANSPORTATION

509 Blake, Gerald Henry, and Swearingen, W.D., comps. *The Suez Canal: a commemorative bibliography, 1975.* Durham U.K.: University of Durham. Centre for Middle Eastern and Islamic Studies, 1975. (Occasional papers series 4) 49 pp.

TOURISM AND RECREATION

510 al-Barbar, Aghil. *Tourism in the Arab world: an introductory bibliography.*

See entry 1325.

HISTORICAL GEOGRAPHY

511 Baines, John, and Malek, Jaromir. *Atlas of ancient Egypt.* Oxford, U.K.: Phaidon, 1980. 240 pp.

512 King, John Wucher. *Historical dictionary of Egypt.* Metuchen NJ: Scarecrow, 1982. (African historical dictionaries 36).

Entries cover major topics, events, and personalities in the nation's history; includes extensive unannotated bibliography. In press.

PUBLIC ADMINISTRATION

513 Harmon, Robert B. *A selected and annotated guide to the government and politics of Egypt.* Monticello IL: Council of Planning Librarians, 1978. (CPL exchange bibliography 1458) 15 pp.

Brief annotations.

THE GAMBIA
The Republic of The Gambia
(a former colony of the United Kingdom)

The Gambia is Africa's smallest independent nation and a prime example of the continent's political fragmentation based on artificial colonial boundaries. The former British colony is completely surrounded by the former French colony of Senegal and serves as Senegal's natural outlet to the sea. The climate is subhumid, the landscape savanna, and the economy overwhelmingly agricultural; groundnuts and cattle are the only export products. The only diversification is the development of a smuggling trade in black market goods to and from Senegal. The Gambia's greatest development potential lies in unification with its neighbor; with the formalization of a confederation with Senegal in November 1981, this may eventually be achieved.

Dryland areas of THE GAMBIA

DESERTIFICATION RISK	ARIDITY								Aridity Totals	
	Hyperarid		Arid		Semiarid		Subhumid			
	km^2	%	km^2	%	km^2	%	km^2	%	km^2	%
Very High	By definition, desertification does not exist in hyperarid regions.									
High										
Moderate										
Desertification Totals										
No Desertification								9,973	93.3	9,973	93.3
Total Drylands								9,373	93.3	9,973	93.3
Non-dryland										716	6.6
Total Area of the Territory										10,689	100.0

GENERAL BIBLIOGRAPHIES

514 Dorward. D. C., and Butler, A. C. *Government publications relating to The Gambia, 1822-1965*. Wakefield U.K.: E. P. Microform, 1974. 14 pp.

Lists the Blue Books and Government Gazettes available on a 65 reel microfilm collection.

515 Gamble, David P., and Sperling, Louise. *A general bibliography of The Gambia (up to 31 December 1977)*. Boston MA: G. K. Hall, 1979. 266 pp.

Contains over 4,473 unannotated citations, with unusually complete indexes. Most major subjects are covered.

516 Walker, Audrey A. *Official publications of Sierra Leone and Gambia*. Washington DC: U.S. Library of Congress, 1963 92 pp.

Includes materials dating from establishment of colonial governments to date of publication. Almost all reports before 1900 are British.

GEOGRAPHY

517 *Draft environmental report on The Gambia*. Tucson AZ: University of Arizona. Office of Arid Lands Studies. Arid Lands Information Center, 1980.

A narrative summary of the nation's environmental conditions and development problems, supplemented by charts, statistics, and a lengthy unannotated bibliography.

GAZETTEERS

518 *Gambia: official standard names approved by the United States Board on Geographic Names*. Washington DC: U.S. Board on Geographic Names, 1968. (Gazetteer 107) 35 pp.

List of 2,400 place-names with latitude/longitude coordinates.

GEOLOGY

519 **Patterson, D. S.** *Annotated bibliography on the geology, mineral and water resources, mining and mineral industry of The Gambia, 1903-1970.* London: Institute of Geological Sciences, 1970. (Preliminary note 318).

520 **Patterson, D. S.** *Unpublished reports on the geology, mineral and water resources, mining and mineral industry of The Gambia.* London: Institute of Geological Sciences, 1970. (Preliminary note 319).

AGRICULTURAL ECONOMICS

521 **Eicher, Carl K.** *Research on agricultural development in five English-speaking countries in West Africa.*

See entry 664.

HUMAN GEOGRAPHY

522 **Asiedu, Edward Seth.** *Public administration in English-speaking West Africa: an annotated bibliography.*

See entry 666.

ECONOMIC DEVELOPMENT

523 **Langer, Sylvie.** *Selected bibliography on The Gambia, Ghana, Liberia, Nigeria, and Sierra Leone: economic and social aspects with special reference to labour problems.* Geneva: International Institute for Labour Studies, 1972. (International educational materials exchange 7008) 69 pp.

524 *Quarterly economic review of Ghana, Sierra Leone, Gambia, Liberia.*

See entry 546.

HISTORICAL GEOGRAPHY

525 **Gailey, Harry A.** *Historical dictionary of The Gambia.* Metuchen NJ: Scarecrow, 1975. (African historical dictionaries 4) 172 pp.

Entries cover major topics, events, and personalities in the nation's history; includes extensive unannotated bibliography.

GHANA
The Republic of Ghana

(as the Gold Coast, a former colony of the United Kingdom)

Like its neighbors to the east, Ghana extends inland from the coast across several climate-vegetation belts. Two disconnected drylands are identified, both subhumid and marginally dry.

The Coastal Lowland. In Ghana's southeast corner is a portion of the anomalous coastal dryland that extends eastward into Togo and Benin. These Accra Plains are watered by the Volta River and have irrigation potential as part of the huge Volta River Project, West Africa's largest hydroelectric scheme. Accra itself, as the nation's capital, is the chief city. Its port functions have been usurped by new facilities at nearby Tema.

The Savanna. In the northern third of the nation dense forest gives way to the savanna fringe of the Sahel. This is cattle country, supplemented by groundnuts and grain. Rainfall is adequate but variable and heavy storms increase the risks of erosion. Development here is predicated on transport—which is lacking—and cultural ties to the relatively prosperous south, which are weak.

Due to considerable mineral and agricultural wealth, all in the forested areas, Ghana has a reasonably bright economic future even without development of its drylands. The greatest challenge is the maintenance of a viable political state when many of its citizens, with ancient tribal loyalties, fail to share in the prosperity.

Dryland areas of GHANA

DESERTIFICATION RISK	ARIDITY									Aridity Totals	
	Hyperarid		Arid		Semiarid		Subhumid			
	km²	%	km²	%	km²	%	km²	%	km²	%
Very High	By definition, desertification does not exist in hyperarid regions.									
High										
Moderate										
Desertification Totals										
No Desertification							103,663	43.5	103,663	43.5
Total Drylands							103,663	43.5	103,663	43.5
Non-dryland									134,642	56.5
Total Area of the Territory									238,305	100.0

GENERAL BIBLIOGRAPHIES

526 Johnson, A. F. *A bibliography of Ghana, 1930-1961.* Evanston IL: Northwestern University Press, 1964. 210 pp.

Though old and unannotated, still a very complete list of 2,608 citations on most major subjects.

527 Witherell, Julian B., and Lockwood, Sharon B. *Ghana, a guide to official publications, 1872-1968.* Washington DC: U.S. Library of Congress, 1969. 110 pp.

Includes some British documents as well as League of Nations and United Nations documents. Emphasis on documents held by the Library of Congress and other United States libraries.

THESES AND DISSERTATIONS

528 Kafe, Joseph Kofi. *Ghana: an annotated bibliography of academic theses, 1920-1970, in the Commonwealth, the Republic of Ireland, and the United States of America.* Boston MA: G. K. Hall, 1973. 219 pp.

Many extensive annotations.

HANDBOOKS

529 Kaplan, Irving. *Area handbook for Ghana.* Rev. ed. Washington DC: U.S. Dept. of Defense, 1971. (American University foreign area studies) 449 pp.

Narrative summary of the nation's history, politics, geography, economy, social life; includes unannotated bibliography.

GEOGRAPHY

530 **Benneh, George.** *Environment and agricultural development in the Savannah regions of Ghana: an annotated bibliography.* Accra: Ghana. Natural Resources Committee of the Council for Scientific and Industrial Research, 1975. 188 pp.

531 *Draft environmental report on Ghana.* Tucson AZ: University of Arizona. Office of Arid Lands Studies. Arid Lands Information Center, 1980. 172 pp.

A narrative review of the nation's environment and natural resources supplemented by statistics, addresses of organizations, and an unannotated bibliography.

532 **Rogers, Dilwyn J.** "A bibliography on ecology of Africa with special reference to Ghana." *Ghana journal of science* 14 (1974):199-230.

Contains 350 references.

GAZETTEERS

533 *Ghana: official standard names approved by the United States Board on Geographic Names.* Washington DC: U.S. Dept. of Defense, 1967. (Gazetteer 102) 282 pp.

List of 20,000 place-names with latitude/longitude coordinates.

HYDROLOGY

534 **Cochrane, T. W.** *Bibliography of the Volta River project and related matters.* Accra: Volta River Authority, 1971. 83 pp.

BOTANY

535 **Knapp, R.** "Bibliographia phytosociologica et scientiae vegetationis: Ghana" [bibliography of plant ecology and vegetation: Ghana]. *Excerpta botanica: sectio B: sociologica* 11:4 (1971):284-291.

Unannotated. Contains 85 items.

AGRICULTURAL ECONOMICS

536 **Eicher, Carl K.** *Research on agricultural development in five English-speaking countries in West Africa.*

See entry 664.

HUMAN GEOGRAPHY

537 **Aguolu, Christian Chukwunedu.** *Ghana in the humanities and social sciences, 1900-1971: a bibliography.* Metuchen NJ: Scarecrow Press, 1973. 469 pp.

Contains 4,309 annotated citations.

538 **Amedekey, E. Y.** *The culture of Ghana: a bibliography.* Accra: Ghana Universities Press, 1970. 215 pp.

Annotated. Covers general history, economics, anthropology, religion, music, literature.

539 **Asiedu, Edward Seth.** *Public administration in English-speaking West Africa: an annotated bibliography.*

See entry 666.

DEMOGRAPHY

540 **Ewusi, Kodwo.** "Ghana: bibliography on population and development." *African demography* 8 (1978):55-71.

541 *Ghana.* Chapel Hill: Carolina Population Center. Technical Information Service, 1976. (PopScan bibliographies 90) 34 pp.

MEDICINE

542 **Brooks, M.** "A bibliography of hydrobiological work in Ghana including references to onchocerciasis, bilharziases and dracontiasis excluding yellow fever malaria and trypanosomiasis." *Ghana journal of science* 10:1 (1970):49-57.

HOUSING

543 **Simon, Joan C.** *Housing in Ghana.* Monticello IL: Vance Bibliographies, 1979. (Public administration series bibliography P-221) 55 pp.

Some annotation.

ECONOMIC DEVELOPMENT

544 **Langer, Sylvie.** *Selected bibliography on The Gambia, Ghana, Liberia, Nigeria, and Sierra Leone: economic and social aspects with special reference to labour problems.*

See entry 523.

545 **Manu, Comfort Henrietta.** *The economy of Ghana: an annotated bibliography, 1946-1966.* High Wycombe, U.K.: University Microfilms, 1970. 600 leaves.

546 *Quarterly economic review of Ghana, Sierra Leone, Gambia, Liberia.* London: Economist Intelligence Unit, 1976-. Quarterly.

Includes summary of political and economic news as well as charts and statistics of selected economic indicators. Annual supplement.

GUINEA
Republique populaire et revolutionnaire de Guinee

(a former colony of France)

Guinea occupies a physiographically diverse landscape that ranges from the Atlantic coast inland over the Futa Jallon highlands to the upper reaches of the Niger River and thence across the Guinea Highlands. A small subhumid dryland area touches the northern border with Senegal; groundnuts are grown. Otherwise the area is of minimal importance to the national economy.

Only selected reference sources are provided.

Dryland areas of GUINEA

DESERTIFICATION RISK	ARIDITY									Aridity Totals	
	Hyperarid		Arid		Semiarid		Subhumid			
	km²	%	km²	%	km²	%	km²	%	km²	%
Very High	By definition, desertification does not exist in hyperarid regions.									
High											
Moderate											
Desertification Totals											
No Desertification							7,376	3.0	7,376	3.0
Total Drylands							7,376	3.0	7,376	3.0
Non-dryland									238,481	97.0
Total Area of the Territory									245,857	100.0

HANDBOOKS

547 **Nelson, Harold D.** *Area handbook for Guinea.* Washington DC: U.S. Dept. of Defense, 1975. (American University foreign area studies) 385 pp.

Narrative summary of the nation's history, politics, geography, economy, social life; includes unannotated bibliography.

GAZETTEERS

548 *Guinea: official standard names approved by the United States Board on Geographic Names.* Washington DC: U.S. Board on Geographic Names, 1965. (Gazetteer 90) 175 pp.

List of 12,400 place-names with latitude/longitude coordinates.

ECONOMIC DEVELOPMENT

549 *Quarterly economic review of Senegal, Mali, Mauritania, Guinea.*

See entry 596.

HISTORICAL GEOGRAPHY

550 **O'Toole, Thomas E.** *Historical dictionary of Guinea (Republic of Guinea/Conakry).* Metuchen NJ: Scarecrow, 1978. (African historical dictionaries 16) 157 pp.

Entries cover major topics, events, and personalities in the nation's history; includes extensive unannotated bibliography.

GUINEA-BISSAU

(as Portuguese Guinea, a former colony of Portugal)

The tiny nation of Guinea-Bissau occupies a wedge-shaped slice of coastal lowland around the Rio Corubal between Senegal and Guinea. The extreme northeast corner is subhumid dryland and supports groundnut cultivation. Despite the small quantities produced, groundnuts are one of the country's primary exports so the drylands have some economic importance to the nation.

Only selected references are provided.

Dryland areas of GUINEA - BISSAU

DESERTIFICATION RISK	ARIDITY									
	Hyperarid		Arid		Semiarid		Subhumid		Aridity Totals	
	km²	%	km²	%	km²	%	km²	%	km²	%
Very High	By definition, desertification does not exist in hyperarid regions.									
High										
Moderate										
Desertification Totals										
No Desertification								2,204	6.1	2,204	6.1
Total Drylands								2,204	6.1	2,204	6.1
Non-dryland										33,921	93.9
Total Area of the Territory										36,125	100.0

GENERAL BIBLIOGRAPHIES

551 Gibson, Mary Jane. *Portuguese Africa: a guide to official publications.*

See entry 770.

552 McCarthy, Joseph M. *Guinea-Bissau and Cape Verde Islands: a comprehensive bibliography.*

See entry 261.

553 Pyhala, Mikk, and Rylander, Kristina. *Guinea-Bissau: en selektiv litteraturefortecknling [a selected bibliography].* Uppsala, Sweden: Nordiska Afrikainstitutet, 1975. 39 pp.

In English and Swedish. Contains 365 references.

554 Zubatsky, David S., comp. *A guide to resources in United States libraries and archives for the study of Cape Verde, Guinea (Bissau), Sao Tome-Principe, Angola and Mozambique.*

See entry 262.

GAZETTEERS

555 *Portuguese Guinea: official standard names approved by the United States Board on Geographic Names.* Washington DC: U.S. Board on Geographic Names, 1968. (Gazetteer 105) 122 pp.

List of 8,700 place-names with latitude/longitude coordinates.

ECONOMIC DEVELOPMENT

556 *Quarterly economic review of Angola, Guinea-Bissau, Cape Verde, Sao Tome, Principe.*

See entry 782.

HISTORICAL GEOGRAPHY

557 Lobban, Richard. *Historical dictionary of the Republic of Guinea-Bissau and Cape Verde.*

See entry 269.

IVORY COAST
Republique de Cote d'Ivoire
(a former colony of France)

The Ivory Coast contains a sliver of the subhumid savanna fringe of the Sahel in its extreme northeast corner on the border with Ghana. The land is cattle country, served by the town of Bouna which is the terminus of a major road from the capital of Abidjan on the coast. Migrant laborers from Upper Volta pass through the region seeking work in the extensive coffee farms to the south; otherwise the region plays little part in the nation's economy.

Only selected reference sources are provided.

Dryland areas of IVORY COAST

DESERTIFICATION RISK	ARIDITY									
	Hyperarid		Arid		Semiarid		Subhumid		Aridity Totals	
	km²	%	km²	%	km²	%	km²	%	km²	%
Very High	By definition, desertification does not exist in hyperarid regions.									
High										
Moderate										
Desertification Totals										
No Desertification							9,996	3.1	9,996	3.1
Total Drylands							9,996	3.1	9,996	3.1
Non-dryland									312,467	96.9
Total Area of the Territory									322,463	100.0

HANDBOOKS

558 Roberts, T. D. et al. *Area handbook for the Ivory Coast*. 2nd ed. Washington DC: U.S. Dept. of Defense, 1973. (American University foreign area studies) 449 pp.

Narrative summary of the nation's history, politics, geography, economy, social life; includes unannotated bibliography.

ATLASES

559 *Atlas de Cote d'Ivoire [atlas of Ivory Coast]*. Abidjan: ORSTOM, 1971-1979. Unpaged.

In French.

560 *Atlas de la Cote d'Ivoire [atlas of Ivory Coast]*. Paris: Editions J. A., 1978. (Les Atlas Jeune Afrique) 72 pp.

In French. Most plates at a scale of 1:3,600,000.

GAZETTEERS

561 *Ivory Coast: official standard names approved by the United States Board on Geographic Names*. Washington DC: U.S. Board on Geographic Names, 1965. (Gazetteer 89) 250 pp.

List of 17,700 place-names with latitude/longitude coordinates.

ECONOMIC DEVELOPMENT

562 *Quarterly economic review of Ivory Coast, Togo, Benin, Niger, Upper Volta*.

See entry 714.

LIBYA
Al-Jamahiriyah Al-Arabiya Al-Libya Al-Shabiya Al-Ishtirakiya
[The Socialist People's Libyan Arab Jamahiriya]

(as Tripolitania and Cyrenaica, a former colony of Italy)

Libya is the nation on the Mediterranean coast of Africa least favored by its geography. It lacks the high mountain temperate climates of Morocco, Algeria, and Tunisia or the great riverine oasis of Egypt. Libya has only desert—and oil. Two regions are identified, with more similarities than differences.

The Coast. A rising topography and maritime influences give Tripolitania and Cyrenaica the barest touch of the Mediterranean sclerophyll biome: semiarid to subhumid climate and corresponding grassland scrub vegetation. Since ancient times Tripolitania on the west was brought under irrigated cultivation by sedentary farmers; Cyrenaica on the east—with the larger well-watered area—was given over to semi-nomadic pastoralism. Italian colonists after 1911 rebuilt and expanded some of the Roman irrigation structures in an attempt to make the land productive again, but their efforts were cut short by World War II. Now, with capital from oil revenues, development schemes are beginning to bear fruit: one example, in Cyrenaica, will resettle nomads into newly planted olive and almond orchards, with water and power supplied by dams.

The Sahara. Libya's share of the Great Desert is exceeded only by Algeria's; in Libya, in fact, the desert actually reaches to the Mediterranean shore at the Gulf of Sirte. The desert surface offers few resources; nomadic tribes search for rare pasture and scattered oases support orchard crops. Great underground fields of petroleum and water, exploited since the 1950's, promise to transform the nation's economy. The promise will be fulfilled when oil revenues reach the desert and its people: another example of oil-financed development at Al Kufra mines groundwater to irrigate huge circular plots that will supply fodder to sheep owned by the nomads.

Dryland areas of LIBYA

DESERTIFICATION RISK	ARIDITY									
	Hyperarid		Arid		Semiarid		Subhumid		Aridity Totals	
	km²	%	km²	%	km²	%	km²	%	km²	%
Very High	By definition, desertification does not exist in hyperarid regions.		45,748	2.6	22,874	1.3	1,760	0.1	70,382	4.0
High			110,851	6.3	1,760	0.1			112,611	6.4
Moderate										
Desertification Totals			156,599	8.9	24,634	1.4	1,760	0.1	182,993	10.4
No Desertification	1,576,547	89.6							1,576,547	89.6
Total Drylands	1,576,547	89.6	156,599	8.9	24,634	1.4	1,760	0.1	1,759,540	100.0
Non-dryland										
Total Area of the Territory									1,759,540	100.0

GENERAL BIBLIOGRAPHIES

563 **Baiou, M. A.** *Selection of annotated bibliography of Libya.* Tripoli, Libya: Arabic House for Books, 1975. 275 pp.

564 **Baiou, M. A.** *Selection of bibliography of Libya.* Benghazi, Libya: Dar Libya, Al-Taliah, 1967-72. 2 vols.

565 **Hill, R. W.** *A bibliography of Libya.* Durham, U.K.: University of Durham. Dept. of Geography, 1959. (Research paper 1) 100 pp.

566 *Resources on Libya: a critical study of bibliographical resources.* Pittsburgh PA: University of Pittsburgh. Graduate School of Library and Information Sciences, 1972. 27 pp.

Libya/567-582

567 Schlueter, Hans. *Index Libycus: bibliography of Libya, 1957-1969, with supplementary material, 1915-1956.* Boston MA: G. K. Hall, 1972. 305 pp.

Unannotated. Supplements Hill (entry 565). Covers items published in Roman characters only. Subjects range through the natural and social sciences.

568 Schlueter, Hans. *Index Libycus: bibliography of Libya, 1970-1975, with supplementary material.* Boston MA: G. K. Hall, 1979-1981. 2 vols.

Unannotated. Continues entry 567. The second volume provides a complete index to citations appearing in Hill (entry 565), *Index Libycus, 1957-1969* (entry 567), and *Index Libycus, 1970-1975* (this entry).

HANDBOOKS

569 Nelson, Harold D. *Libya: a country study.* 3rd ed. Washington DC: U.S. Dept. of Defense, 1979. (American University foreign area studies) 350 pp.

Narrative summary of the nation's history, politics, geography, economy, social life; includes unannotated bibliography.

GEOGRAPHY

570 Nuttonson, Michael Y. "Physical environment and agriculture of Libya and Egypt."

See entry 489.

GAZETTEERS

571 *Libya: official standard names approved by the United States Board on Geographic Names.* 2nd ed. Washington DC: U.S. Board on Geographic Names, 1973. (Gazetteer 131) 746 pp.

List of 37,500 place-names with latitude/longitude coordinates.

GEOLOGY

572 Ettalhi, J. Azzoyz; Krokovic, D.; and Banerjee, S. *Bibliography of the geology of Libya.* Tarabalus: Libya. Dept. of Geological Researches and Mining, 1978. (Bulletin 11) 135 pp.

Covers through 1976. Updated annually in the Bulletin series.

CLIMATOLOGY

573 Grimes, Annie E. *An annotated bibliography on climatic maps of Libya.* Washington DC: U.S. Weather Bureau, 1961. (NTIS: PB 176-011; AD 664-715) 9 pp.

SOILS

574 *Bibliography on desert soils of Morocco, Algeria, Tunisia, Libya (1964-1936).*

See entry 623.

BOTANY

575 Knapp, R. "Bibliographia phytosociologica et scientiae vegetationis: Africa Aegyptiaca, Libyca et Tibestica [bibliography of plant ecology and vegetation: African Egypt, Libya and Tibesti].

See entry 501.

ZOOLOGY

576 el Awamy, Alad Musa. *A bibliography of the Libyan fauna.* Tripoli, Libya: Dar Markabet Al Fikr, 1971. 77 pp.

HUMAN GEOGRAPHY

577 Vidergar, John J. *The economic, social and political development of Algeria and Libya.*

See entry 439.

DEMOGRAPHY

578 al-Barbar, Aghil M. *The population of Libya.* Monticello IL: Vance Bibliographies, 1980. (Public administration series bibliography P-455) 5 pp.

Annotated. Includes historical sources of demographic information.

PUBLIC HEALTH

579 Barbar, Aghil M. *Health care in Libya: an introductory survey.* Monticello IL: Vance Bibliographies, 1978. (Public administration series bibliography P-142) 6 pp.

Unannotated.

URBAN GEOGRAPHY

580 Barbar, Aghil M. *Urbanization in Libya.* Monticello IL: Council of Planning Librarians, 1977. (CPL exchange bibliography 1241) 5 pp.

Unannotated.

ARCHITECTURE

581 Barbar, Aghil M. *Islamic architecture in Libya: an introductory bibliography.* Monticello IL: Vance Bibliographies, 1979. (Architecture series bibliography A41) 5 pp.

Some annotations.

ECONOMIC DEVELOPMENT

582 *Quarterly economic review of Libya, Tunisia, Malta.* London: Economist Intelligence Unit, 1976-. Quarterly.

Includes summary of political and economic news as well as charts and statistics of selected economic indicators. Annual supplement.

HISTORICAL GEOGRAPHY

583 **Hahn, Lorna.** *Historical dictionary of Libya.* Metuchen NJ: Scarecrow, 1981. (African historical dictionaries 33) 116 pp.

Entries cover major topics, events, and personalities in the nation's history; includes extensive unannotated bibliography.

PUBLIC ADMINISTRATION

584 **al-Barbar, Aghil M.** *Government and politics in Libya, 1969-1978: a bibliography.* Monticello IL: Vance Bibliographies, 1979. (Public administration series bibliography P-388) 139 pp.

Unannotated.

585 **Barbar, Aghil M.** *Public administration in Libya: a bibliography.* Monticello IL: Vance Bibliographies, 1979. (Public administration series bibliography P-192) 5 pp.

Unannotated.

586 **Cigar, Norman.** *Government and politics in North Africa.*

See entry 633.

MALI
Republique du Mali

(as French Sudan, a former colony of France)

Mali joins Niger and Chad as a group of poor, landlocked nations occupying large portions of the Sahara and Sahel.

The Sahel/Savanna. This region is typical of the arid-to-subhumid, summer precipitation, dry savanna of the Sudano-Sahelian belt. The Niger River makes a great northward bend from its source in the Guinea Highlands through this region to pass on through Niger and south to the coast. Along the river cropland is irrigated and supports rice and cotton, while in the wetter savanna lands groundnuts are also grown. A rise of land in the subhumid area along the border with Upper Volta promises some hydropower and irrigation potential; without extensive restructuring of the agricultural system the nation will be unable to reach self-sufficiency in food. The Chinese have helped in both dam- and railroad-building.

The Sahara. Mali's share of the Great Desert is large and largely unproductive, lacking significant oases and mineral deposits.

Dryland areas of MALI

DESERTIFICATION RISK	ARIDITY									
	Hyperarid		Arid		Semiarid		Subhumid		Aridity Totals	
	km²	%	km²	%	km²	%	km²	%	km²	%
Very High	By definition, desertification does not exist in hyperarid regions.		46,957	3.9	15,652	1.3	1,204	0.1	63,813	5.3
High			437,060	36.3					437,060	36.3
Moderate			46,957	3.9	168,563	14.0	36,121	3.0	251,641	20.9
Desertification Totals			530,974	44.1	184,215	15.3	37,325	3.1	752,514	62.5
No Desertification	315,454	26.2			12,040	1.0	80,669	6.7	408,163	33.9
Total Drylands	315,454	26.2	530,974	44.1	196,255	16.3	117,994	9.8	1,160,677	96.4
Non-dryland									43,344	3.6
Total Area of the Territory									1,204,021	100.0

GENERAL BIBLIOGRAPHIES

587 Brasseur, Paule. *Bibliographie generale du Mali (anciens Soudan francais et Haut-Senegal-Niger) [general bibliography of Mali (former French Sudan and Upper Senegal-Niger]*. Dakar: Institut fondamental d'Afrique noire, 1964. (Catalogues et documents 16) 461 pp.

Annotated. An excellent bibliography covering physical and cultural geography, irrigation and agriculture, and economics for Senegal, Mali, Niger, Mauritania, and Upper Volta. Continued by *Bibliographie generale du Mali(1961-1970)*(Dakar: IFAN, 1976).

STATISTICS

588 Imperato, Pascal James, and Imperato, Eleanor M. *Mali: a handbook of historical statistics.* Boston MA: G. K. Hall, 1982. 340 pp.

Contains statistical series on a variety of subjects from climate to social sciences.

GEOGRAPHY

589 Bantje, Han. *A working bibliography of the Western Sahel.* Amsterdam: Royal Tropical Institute, 1975. 46 leaves.

Available in English or French. Annotated. Covers Mali, Mauritania, Upper Volta; focuses on land use, nomadism, social sciences.

GEOGRAPHY

590 Grant, A. Paige, comp. *Draft environmental report on Mali.* Tucson AZ: University of Arizona. Office of Arid Lands Studies. Arid Lands Information Center, 1980. 73 pp.

A narrative summary of the nation's environmental conditions and development problems, supplemented by charts, statistics, and a lengthy unannotated bibliography.

591 Massoni, Colette. *Liste bibliographique des travaux effectues dans le Bassin de Fleuve Niger par les* chercheurs de l'ORSTOM de 1943 a 1968 [bibliographic list of works carried out in the basin of the Niger River by the ORSTOM research workers from 1943-1968]. Bondy, France: O.R.S.T.O.M. Service Central de Documentation, 1971. (Travaux et documents 10) 71 pp.

ATLASES

592　Traore, Issa Baba, and Aubriot, Bernard. *Republique du Mali, mon livret de cartographie [Republic of Mali, my booklet of cartography]*. Paris: Nathan, 1970. 17 pp.

GAZETTEERS

593　*Mali: official standard names approved by the United States Board on Geographic Names*. Washington DC: U.S. Board on Geographic Names, 1965. (Gazetteer 93) 263 pp.

List of 17,800 place-names with latitude/longitude coordinates.

BOTANY

594　Knapp, R. "Bibliographia phytosociologica et scientiae vegetationis: Mali, Mauretania, Niger, Senegal, Tchad, Volta Superior" [bibliography of plant ecology and vegetation: Mali, Mauritania, Niger, Senegal, Chad, Upper Volta]. *Excerpta botanica: sectio B: sociologica* 17:1 (1978):22-32; 19:2 (1979):119-144.

Unannotated. Part I contains items 1-145; Part II contains items 150-499. Author index.

AGRICULTURAL ECONOMICS

595　Hailu, Alem Seged. *Rural development in African countries: a selected bibliography with special reference to Mali and Kenya*. Monticello IL: Vance Bibliographies, 1982. (Public administration series bibliography P-908) 29 pp.

Extensive annotations.

ECONOMIC DEVELOPMENT

596　*Quarterly economic review of Senegal, Mali, Mauritania, Guinea*. London: Economist Intelligence Unit, 1976-. Quarterly.

Includes summary of political and economic news as well as charts and statistics of selected economic indicators. Annual supplement.

HISTORICAL GEOGRAPHY

597　Imperato, Pascal James. *Historical dictionary of Mali*. Metuchen NJ: Scarecrow, 1977. (African historical dictionaries 11) 204 pp.

Entries cover major topics, events, and personalities in the nation's history; includes extensive unannotated bibliography.

MAURITANIA
Republique Islamique de Mauritanie

(a former colony of France)

Mauritania is geographically similar to the nations to the east—Mali, Niger, Chad—but does have the good fortune to possess both a seacoast and sizeable mineral deposits.

The Sahel/Savanna. This region is typical and similar to the nation's neighbors but, in Mali's case, lacks the subhumid fringe found elsewhere. Land cover is as sparse as the precipitation. Some small irrigated croplands exist along the Senegal River but otherwise the country is given over to nomadic pastoralism.

The Sahara. Mauritania's portion of the Great Desert gives it both a seacoast and deposits of copper and iron ore. With capital and technical assistance from Arab states and China these deposits have been developed, rail transportation has been built, and the capacity to produce steel has been achieved.

Dryland areas of MAURITANIA

DESERTIFICATION RISK	ARIDITY								Aridity Totals	
	Hyperarid		Arid		Semiarid		Subhumid			
	km²	%	km²	%	km²	%	km²	%	km²	%
Very High	By definition, desertification does not exist in hyperarid regions.		4,123	0.4					4,123	0.4
High			525,657	51.0	30,921	3.0			556,578	54.0
Moderate			32,982	3.2					32,982	3.2
Desertification Totals			562,762	54.6	30,921	3.0				57.6
No Desertification	437,017	42.4							437,017	42.4
Total Drylands	437,017	42.4	562,762	54.6	30,921	3.0			1,030,700	100.0
Non-dryland										
Total Area of the Territory									1,030,700	100.0

GENERAL BIBLIOGRAPHIES

598 Brasseur, Paule. *Bibliographie generale du Mali (anciens Soudan francais et Haute-Senegal-Niger) [general bibliography of Mali (former French Sudan and Upper Senegal-Niger].*

See entry 587.

599 Toupet, Charles. "Orientation bibliographique sue la Mauritanie" [bibliographic direction on Mauritania]. Institut Francais d'Afrique noire. *Bulletin, serie B* 24 (1962):594-613.

600 Van Maele, Bernard. *Mauritanie: bibliographie [Mauritania: a bibliography].* Nouakchott, Mauritania: Mauritania. Ministere de la planification et de la recherche, 1971. Unpaged.

Contains some 1,400 citations. Preliminary edition.

HANDBOOKS

601 Curran, Brian Dean, and Schrock, Joan. *Area handbook for Mauritania.* Washington DC: U.S. Dept. of Defense, 1972. (American University foreign area studies) 185 pp.

Narrative summary of the nation's history, politics, geography, economy, social life; includes unannotated bibliography.

GEOGRAPHY

602 Bantje, Han. *A working bibliography of the western Sahel.*

See entry 589.

603 Toupet, Charles. "Contribution des geographes a la connaissance de la Mauritanie" [contribution of geographers to the knowledge about Mauritania]. Institut fondamental d'Afrique noire. *Notes africaines* 158 (1978):49-55.

Bibliography preceded by essay which highlights titles by subject.

ATLASES

604 **Toupet, Charles, and Laclavere, Georges.** *Atlas de la Republique Islamique de Mauritanie [atlas of the Islamic Republic of Mauritania]*. Paris: Editions J.A., 1977. (Les Atlas Jeune Afrique) 64 pp.

In French. Most plates at a scale of 1:6,500,000.

GAZETTEERS

605 *Mauritania: official standard names approved by the United States Board on Geographic Names*. Washington DC: U.S. Board on Geographic Names, 1966. (Gazetteer 100) 149 pp.

List of 10,000 place-names with latitude/longitude coordinates.

REMOTE SENSING

606 *[Mauritania]*. Unknown: France. Institut Geographique National, 1954. 146 photos.

A rare collection of aerial photography flown by Mission A.O.F. in 1954. Originals presumably held by "Cliche du Centre de Documentation de Photographies Aeriennes." Black and white positive prints are held by the University of Arizona Library, Tucson, AZ. Scale 1:100,000.

BOTANY

607 **Knapp, R.** "Bibliographia phytosociologica et scientiae vegetationis: Mali, Mauretania, Niger, Senegal, Tchad, Volta Superior" [bibliography of plant ecology and vegetation: Mali, Mauritania, Niger, Senegal, Chad, Upper Volta].

See entry 594.

ECONOMIC DEVELOPMENT

608 *Quarterly economic review of Senegal, Mali, Mauritania, Guinea*.

See entry 596.

HISTORICAL GEOGRAPHY

609 **Gerteiny, Alfred G.** *Historical dictionary of Mauritania*. Metuchen NJ: Scarecrow, 1981. (African historical dictionaries 31) 116 pp.

Entries cover major topics, events, and personalities in the nation's history; includes extensive unannotated bibliography.

MOROCCO
Al-Mamlaka Al-Maghrebia
[the Kingdom of Morocco]
(a former protectorate of France and Spain)

Morocco is an ancient, fiercely independent nation which in some ways is the most favored in northern Africa.

The Coastal Plain. Along the Atlantic Ocean runs a wide semiarid plain served by the ports of Tangier, Rabat, Casablanca, and others. Maritime air masses crossing the cool Canaries Current bring frequent fog and a typical Mediterranean-type climate of winter-dominant precipitation. A number of short rivers flow from the High Atlas across the plain and provide irrigation potential. Crops range from olive trees in the north to increasing reliance on pastoral herding toward the drier south. Of utmost importance as the nation's leading export are some of the world's richest deposits of phosphates, whose market prices have risen along with oil.

The Atlas Highland and Steppe. The core of the nation is a series of roughly parallel mountains that are Morocco's share of the North African Atlas range. The upper slopes and Mediterranean coastal sections are beyond the dryland zone but lower areas and the eastern steppe are subhumid to arid. The esparto grass steppe on the east toward the border with Algeria is called the *Sous*, a land of recurring drought, cold winters, and distinctively fortified villages. Wheat and barley are grown; sheep and goats are herded.

The Sahara. Despite Morocco's historic importance as the chief center of trans-Sahara trade, modern colonialism resulted in the loss of most of the territory through boundary changes. Now the nation occupies only a sliver of the desert beyond the edges of the esparto steppe. A few oases are sites of the only permanent habitations.

The Western Sahara. This territory—the former Spanish Sahara—was annexed by Morocco in 1976. Although a small Algerian-supported independence movement operates out of bases in the desert, Morocco retains full political control. For the purposes of this Guide the area is treated separately, primarily because the published literature generally reflects its former status as a Spanish colony.

Dryland areas of MOROCCO

DESERTIFICATION RISK	ARIDITY									
	Hyperarid		Arid		Semiarid		Subhumid		Aridity Totals	
	km²	%	km²	%	km²	%	km²	%	km²	%
Very High	By definition, desertification does not exist in hyperarid regions.		16,000	3.2	17,000	3.4			33,000	6.6
High			190,500	38.1	100,500	20.1	19,000	3.8	310,000	62.0
Moderate			40,500	8.1	42,000	8.4			82,500	16.5
Desertification Totals			247,000	49.4	159,500	31.9	19,000	3.8		85.1
No Desertification	10,500	2.1					21,500	4.3	32,000	6.4
Total Drylands	10,500	2.1	247,000	49.4	159,500	31.9	40,500	8.1	457,500	91.5
Non-dryland									42,500	8.5
Total Area of the Territory									500,000	100.0

GENERAL BIBLIOGRAPHIES

610 Rishworth, Susan Kroke. *Spanish-speaking Africa: a guide to official publications.*

See entry 751.

HANDBOOKS

611 Nelson, Harold D. *Morocco: a country study.* 4th ed. Washington DC: U.S. Dept. of Defense, 1978. (American University foreign area studies) 410 pp.

Narrative summary of the nation's history, politics, geography, economy, social life; includes unannotated bibliography.

GEOGRAPHY

612 "Bibliographie d'etudes et de recherches sur les zones arides (en particulier au Maroc)" [bibliography of studies and investigations on arid zones (particularly in Morocco)]. *Les cahiers de la recherche agronomique* (Rabat) 19 (1965):71-130.

Contains 571 citations.

613 Nuttonson, Michael Y. *"Introduction to North Africa and a survey of the agriculture of Morocco, Algeria, and Tunisia, with special reference to their regions containing areas climatically and latitudinally analogous to Israel."*

See entry 433.

614 Parker, Susan, comp. *Environmental profile on Morocco (revised draft).* Rev. ed. Tucson AZ: University of Arizona. Office of Arid Lands Studies. Arid Lands Information Center, 1981.

A narrative summary of the nation's environmental conditions and development problems, supplemented by charts, statistics, and a lengthy unannotated bibliography.

ATLASES

615 *Atlas du Maroc* [atlas of Morocco]. Rabat: Comite de geographie du Maroc, 1954-1977(?). 54 maps.

Standard scale 1:2,000,000.

GAZETTEERS

616 *Morocco: official standard names approved by the United States Board on Geographic Names.* Washington DC: U.S. Board on Geographic Names, 1970. (Gazetteer 112) 923 pp.

List of 55,000 place-names with latitude/longitude coordinates.

GEOLOGY

617 Morin, Philippe. *Bibliographie analytique des sciences de la terre: Maroc et regions limitrophes (depuis le debut des recherches geologiques a 1964)* [analytical bibliography of earth sciences: Morocco and adjacent regions (since the beginning of geologic research to 1964)]. Rabat: Editions du Service Geologique du Maroc, 1965. (Morocco. Service Geologique. Notes et memoires 182) 2 vols.

Continued by same title covering period 1965-1969 (Notes et memoires 212, 1970; 407 pp.), and 1970-1976 (Notes et memoires 270, 1979; 755 pp.).

618 Ridge, John Drew. "Morocco." In *Annotated bibliographies of mineral deposits in Africa, Asia (exclusive of the USSR) and Australasia.* Oxford U.K.: Pergamon, 1976. pp. 9-22.

CLIMATOLOGY

619 Grimes, Annie E. *Bibliography of climatic maps of Morocco.* Washington DC: U.S. Weather Bureau, 1960. (NTIS: PB 176-032; AD 664-709) 11 pp.

HYDROLOGY

620 Combe, Michel. "Ressources en eau du Maroc: presentation-evaluation-utilisation: bibliographie essentielle" [water resources in Morocco: occurance, evaluation, utilization: essential bibliography]. France. Bureau de Recherches Geologiques et Minieres. *Bulletin* serie 2, 3:1 (1974):15-20.

621 Combe, Michel, and Thauvin, Jean Pierre. *Bibliographie hydrogeologique analytique du Maroc, 1958-1968* [annotated hydrogeologic bibliography of Morocco, 1958-1968]. Rabat: Morocco. Service Geologique, 1969. (Notes et memoires 220) 178 pp.

In French.

622 Margat, Jean F. *Bibliographie hydrogeologique du Maroc, 1905-1957* [hydrogeologic bibliography of Morocco, 1905-1957]. Rabat: Editions du Service Geologique du Maroc, 1958. (Morocco. Service Geologique. Notes et memoires 142) 63 pp.

Lists 591 references.

SOILS

623 *Bibliography on desert soils of Morocco, Algeria, Tunisia, Libya (1964-1936).* Harpenden, U.K.: Commonwealth Bureau of Soils, 196- (n.d.). 14 pp.

Lists 82 citations, mostly annotated.

624 *Soils of Morocco (annotated bibliography 1962-1975).* Harpenden, U.K.: Commonwealth Agricultural Bureaux, 1977. (Annotated bibliography) 8 pp.

BOTANY

625 Peltier, J. P. "Bibliographie botanique marocaine" [bibliography of Moroccan botany]. Societe des Sciences Naturelles et Physiques du Maroc. *Bulletin* 53:1-2 (1973):247-252.

In French. Covers years 1967-1970 of the main works of Moroccan botany published in Morocco, France, and several other countries.

626 Roussine, N., and Sauvage, C. "Bibliographia phytosociologica: Afrique du Nord" [bibliography of plant ecology: North Africa]. *Excerpta botanica; sectio B: sociologica* 3:1 (1961):34-51.

Unannotated. Covers Morocco, Algeria, Tunisia. Contains over 185 citations; 21 maps. Author index.

URBAN GEOGRAPHY

627 **Cigar, Norman L.** *Modern Moroccan cities.* Monticello IL: Council of Planning Librarians, 1978. (CPL exchange bibliography 1464) 6 pp.

Annotated.

628 **Cigar, Norman L.** *Traditional Moroccan cities.* Monticello IL: Council of Planning Librarians, 1978. (CPL exchange bibliography 1463) 7 pp.

Annotated.

ARCHITECTURE

629 **Cigar, Norman.** *North African architecture.* Monticello IL: Vance Bibliographies, 1978. (Architecture series bibliography A28) 6 pp.

Briefly annotated. Covers Morocco, Algeria, Tunisia; sources are in English and French.

ECONOMIC DEVELOPMENT

630 *Kompas-Maroc: repertoire industriel et commercial/Kompass Maroc: register of Moroccan industry and commerce.* Casablanca: Kompass Maroc-Veto, 1970-. Annual.

In French, Arabic, and English. A directory of commercial firms, products, and trade.

631 *Quarterly economic review of Morocco.* London: Economist Intelligence Unit, 1976-. Quarterly.

Includes summary of political and economic news as well as charts and statistics of selected economic indicators. Annual supplement.

HISTORICAL GEOGRAPHY

632 **Spencer, William.** *Historical dictionary of Morocco.* Metuchen NJ: Scarecrow, 1980. (African historical dictionaries 24) 152 pp.

Entries cover major topics, events, and personalities in the nation's history; includes extensive unannotated bibliography.

PUBLIC ADMINISTRATION

633 **Cigar, Norman.** *Government and politics in North Africa.* Monticello IL: Vance Bibliographies, 1979. (Public administration series bibliography P-171) 14 pp.

Unannotated. Covers Morocco, Algeria, Tunisia, Libya.

NIGER
Republique du Niger
(a former colony of France)

Niger is one of three landlocked nations—between Mali and Chad—that spans large areas of the Sahara and extends southward into the Sahel and Sudanian belt.

The Sahara. Niger's portion of the Great Desert includes the Air Massif and several large oases. The Tuareg keep substantial herds in the Air while a sedentary population of farmers lives in the oases; except for a few mining camps the region is otherwise uninhabited. Uranium, mined at Arlit, has become the nation's principal export.

The Sahel/Savanna. This arid to semiarid region holds the bulk of the nation's population and agricultural potential; it was also the most severely damaged portion of the Sahel during the early 1970's drought. As much as a quarter of the population either emigrated or died. In good years the region produces groundnuts, cotton, and cattle in exportable quantitites.

Dryland areas of NIGER

DESERTIFICATION RISK	ARIDITY								Aridity Totals	
	Hyperarid		Arid		Semiarid		Subhumid			
	km²	%	km²	%	km²	%	km²	%	km²	%
Very High	By definition, desertification does not exist in hyperarid regions.		27,287	2.3	62,880	5.3			90,167	7.6
High			501,851	42.3	55,761	4.7			557,612	47.0
Moderate			8,305	0.7	35,592	3.0			43,897	3.7
Desertification Totals			537,443	45.3	154,233	13.0			691,676	58.3
No Desertification	489,987	41.3			4,746	0.4			494,733	41.7
Total Drylands	489,987	41.3	537,443	45.3	158,979	13.4			1,186,408	100.0
Non-dryland										
Total Area of the Territory									1,186,408	100.0

GENERAL BIBLIOGRAPHIES

634 Brasseur, Paule. *Bibliographie generale du Mali (anciens Soudan francaise et Haute-Senegal-Niger)* [general bibliography of Mali (former French Sudan and Upper Senegal-Niger)].

See entry 587.

GEOGRAPHY

635 Beriel, Marie-Magdeleine. "Contribution a la connaissance de la region du Lac Tchad (Cameroon, Tchad, Niger): bibliographie analytique" [contribution to the knowledge of the region of Lake Chad (Cameroon, Chad, Niger): analytic bibliography].

See entry 455.

636 *Draft environmental report on Niger.* Tucson AZ: University of Arizona. Office of Arid Lands Studies. Arid Lands Information Center, 1980. 160 pp.

A narrative summary of the nation's environmental and developmental problems supported by maps, charts, statistics, and an extensive unannotated bibliography.

637 Massoni, Colette. *Liste bibliographique des travaux effectues dan le Bassin du Fleuve Niger par les chercheurs de l'ORSTOM de 1943 a 1968* [bibliographic list of works carried out in the basin of the Niger River by the ORSTOM research workers from 1943- 1968].

See entry 591.

ATLASES

638 **Bernus, Edmond, and Hamidon, Sidikon A.** *Atlas du Niger [atlas of Niger].* Paris: Editions J.A., 1980. (Les Atlas Jeune Afrique) 64 pp.

In French. Most plates at scale of 1:7,000,000.

GAZETTEERS

639 *Niger: official standard names approved by the United States Board on Geographic Names.* Washington DC: U.S. Board on Geographic Names, 1966. (Gazetteer 99) 207 pp.

List of 14,700 place-names with latitude/longitude coordinates.

BOTANY

640 **Knapp. R.** "Bibliographia phytosociologica et scientiae vegetationis: Mali, Mauretania, Niger, Senegal, Tchad, Volta Superior" [bibliography of plant ecology and vegetation: Mali, Mauritania, Niger, Senegal, Chad, Upper Volta].

See entry 594.

ECONOMIC DEVELOPMENT

641 *Quarterly economic review of Ivory Coast, Togo, Benin, Niger, Upper Volta.*

See entry 714.

HISTORICAL GEOGRAPHY

642 **Decalo, Samuel.** *Historical dictionary of Niger.* Metuchen NJ: Scarecrow, 1979. (African historical dictionaries 20) 358 pp.

Entries cover major topics, events, and personalities in the nation's history; includes extensive unannotated bibliography.

NIGERIA
The Federal Republic of Nigeria
(a former colony of the United Kingdom)

Nigeria, Africa's most populous nation, is roughly one-third dryland. This northern area lies at the edge of the Sahelo-Sudanian belt and ranges from subhumid to semiarid.

The Hausa Plains. This is a higher elevation, short-grass savanna region between the basins of the Sokoto River and Lake Chad. It is the North's economic heart, producing cattle (foremost), groundnuts, cotton, maize, millet, and other cereal crops. Groundnuts are the principal cash crop; Nigeria is usually the world's largest exporter when drought doesn't interfere.

The Sokoto Basin. In the northwest corner of the nation this lowland region supports cattle but lacks sufficient water supplies for denser settlement and more diversified cropping. Transport into the area is poorly developed; no railroads reach the town of Sokoto.

The Chad Basin. In the northeast corner, along Lake Chad and the border with Niger, is the driest region of the nation, extending into the arid range. Cattle and groundnuts are raised in most favored areas, though, like the Sokoto Basin, low and extremely variable rainfall limits productivity. Transport is better developed but the region is still far removed from national economic centers.

Dryland areas of NIGERIA

DESERTIFICATION RISK	ARIDITY								Aridity Totals	
	Hyperarid		Arid		Semiarid		Subhumid			
	km²	%	km²	%	km²	%	km²	%	km²	%
Very High	By definition, desertification does not exist in hyperarid regions.				63,740	6.9			63,740	6.9
High			10,162	1.1	36,027	3.9			46,189	5.0
Moderate										
Desertification Totals			10,162	1.1	99,767	10.8			109,929	11.9
No Desertification					161,660	17.5	144,109	15.6	305,769	33.1
Total Drylands			10,162	1.1	261,427	28.3	144,109	15.6	415,698	45.0
Non-dryland									508,075	55.0
Total Area of the Territory									923,773	100.0

GENERAL BIBLIOGRAPHIES

643 **Lockwood, Sharon Burdge, comp.** *Nigeria: a guide to official publications.* Rev. ed. Washington DC: U.S. Library of Congress, 1966. 166 pp.

Lists 2,451 publications issued by Nigerian governments from 1841-1965. Includes bibliography of additional sources.

HANDBOOKS

644 **Nelson, Harold D.** *Area handbook for Nigeria.* 2nd ed. Washington DC: U.S. Dept. of Defense, 1972. (American University foreign area studies) 485 pp.

Narrative summary of the nation's history, politics, geography, economy, social life; includes unannotated bibliography.

THESES AND DISSERTATIONS

645 "Theses and dissertations in the Faculty of the Social Sciences of Nigerian Universities." *Nigerian journal of economic and social studies* 16 (1974):381-384.

Contains 57 items.

GEOGRAPHY

646 **Aiyepeku, Wilson O.** *Geographical literature on Nigeria, 1901-1970 and annotated bibliography.* Boston MA: G. K. Hall, 1974. 214 pp.

647 Egunjobi, James K. "Nigeria." In *Handbook of contemporary developments in world ecology*, pp 469-484. Edited by Edward John Kormondy and J. Frank McCormick. Westport CT: Greenwood, 1981.

A good narrative summary of ecological research in Nigeria, with an annotated list of major research papers and an unannotated list of references.

648 Massoni, Colette. *Liste bibliographique des travaux effectues dan le Bassin du Fleuve Niger par les chercheurs de l'ORSTOM de 1943 a 1968 [bibliographic list of works carried out in the basin of the Niger River by the ORSTOM research workers from 1943-1968].*

See entry 591.

649 Posnett, N. W.; Reilly, P. M.; and Whitfield, P., comps. *Nigeria*. Surbiton, U.K.: United Kingdom. Overseas Development Administration. Land Resources Division, 1971. (Land resource bibliography 2) 3 vols.

650 Udo, Reuben K. *Geographical regions of Nigeria*. Berkeley CA: University of California Press, 1970. 212 pp.

651 Visser, S. A., ed. "Bibliography on the River Niger with special reference to the Kainji Dam." In *Ecology*, pp. 113-126. Ibadan: Published for the Nigerian Institute of Social and Economic Research, 1970.

652 Williams, Geoffrey J., and Achema, John A., comps. "A current bibliography of the Savanna states of Nigeria." *Savanna* 4 (1975).

Includes post-graduate dissertations and some government publications. Omits newspapers and fiction. Special emphasis on periodical literature. Primarily Nigerian publications.

ATLASES

653 Barbour, K. Michael et al. *Nigeria in maps*. London: Hodder and Stoughton, 1982. 148 pp.

Contains recent small-scale maps and extensive text covering a variety of topics.

GAZETTEERS

654 *Nigeria: official standard names approved by the United States Board on Geographic Names*. Washington DC: U.S. Board on Geographic Names, 1971. (Gazetteer 117) 641 pp.

List of 42,000 place-names with latitude/longitude coordinates.

GEOLOGY

655 Ridge, John Drew. "Nigeria." In *Annotated bibliographies of mineral deposits in Africa, Asia (exclusive of the USSR) and Australasia*, pp. 22-27. Oxford, U.K.: Pergamon, 1976.

CLIMATOLOGY

656 Bender, Thomas A., Jr. *Selected bibliography of climatic maps for Nigeria and the British Cameroons*. Washington DC: U.S. Weather Bureau, 1960. 15 pp.

657 Ene, Ngozi. "A bibliography of the climate of Nigeria." *Nigerian geographical journal* 5:1 (1962):53-60.

Annotated.

658 Griffiths, J. F. "Nigeria." In *Climates of Africa*, pp. 167-192. Edited by J. F. Griffiths. Amsterdam, The Netherlands: Elsevier Publishing Co., 1972. (World survey of climatology 10) 604 pp.

A narrative summary of the nation's climate, with maps, charts, and an unannotated bibliography.

659 Kowal, Jan M., and Knabe, Danota T. *An agroclimatological atlas of the northern states of Nigeria, with explanatory note*. Zaria: Ahmadu Bello University Press, 1972. 111 pp. plus 16 maps.

Most plates at a scale of 1:2,000,000.

BOTANY

660 Knapp, R. "Bibliographia phytosociologica et scientiae vegetationis: Nigeria" [bibliography of plant ecology and vegetation: Nigeria]. *Excerpta botanica: sectio B; sociologica* 10:3 (1969):231-242; 17:2 (1978):101-117.

Unannotated.

ZOOLOGY

661 Medler, John T. *Insects of Nigeria: check list and bibliography*. Ann Arbor MI: American Entomological Institute, 1980. (Memoirs 30) 919 pp.

662 Toye, Beatrice Olukemi, comp. *Bibliography of entomological research in Nigeria: 1900-1973: a bibliography*. Ibadan: Entomological Society of Nigeria, 1974. (Occasional publication 16) 133 pp.

AGRICULTURE

663 Agboola, S. *An agricultural atlas of Nigeria*. Oxford, U.K.: Oxford University Press, 1979. 248 pp.

Bibliography: pp. 211-230.

AGRICULTURAL ECONOMICS

664 Eicher, Carl K. *Research on agricultural development in five English-speaking countries in West Africa*. New York: The Agricultural Development Council, 1970. 152 pp.

Discusses agricultural development strategies and research (1950-1969) in The Gambia, Ghana, Liberia, Nigeria, and Sierra Leone. Extensive unannotated bibliographies.

HUMAN GEOGRAPHY

665 **Aguolu, Christian Chukwunedu.** *Nigeria: a comprehensive bibliography in the humanities and social sciences, 1900-1971.* Boston MA: G. K. Hall, 1973. 620 pp.

666 **Asiedu, Edward Seth.** *Public administration in English-speaking West Africa: an annotated bibliography.* Boston MA: G. K. Hall, 1977. 365 pp.

Covers various social aspects of Nigeria, Ghana, The Gambia, Liberia, and Sierra Leone, 1945-1975. Subjects include public administration, political development, agricultural development, economic development, agricultural development, economic development, mineral resources, housing, public health, education, demography. Contains 4,377 citations.

DEMOGRAPHY

667 **Emezi, Herbert O.** "Nigerian population studies: a partial bibliography of periodical literature, 1950-1970." *A current bibliography on African affairs* 6:3 (1973):333-344.

Unannotated.

668 **Lucas, David and McWilliam, John.** *Population in Nigeria: a select bibliography.* Chapel Hill NC: Carolina Population Center, 1974. (Technical information service, bibliography series 8) 15 pp.

669 **Lucas, David and McWilliam, John.** *A survey of Nigerian population literature.* Lagos: Lagos University. Human Resources Research Unit, 1976. (Population series monograph 4) 143 pp.

ANTHROPOLOGY

670 **Ita, Nduntuei O.** *Bibliography of Nigeria: a survey of anthropological and linguistic writings from the earliest times to 1966.* London: Frank Cass, 1971. 271 pp.

ECONOMIC DEVELOPMENT

671 **Langer, Sylvie.** *Selected bibliography on The Gambia, Ghana, Liberia, Nigeria, and Sierra Leone: economic and social aspects with special reference to labour problems.*

See entry 523.

672 *Quarterly economic review of Nigeria.* London: Economist Intelligence Unit, 1976-. Quarterly.

Includes summary of political and economic news as well as charts and statistics of selected economic indicators. Annual supplement.

673 **Stanley, Janet L.** *Bibliography on labour force participation in Nigeria, 1970-1976.* Montreal: McGill University. Centre for Developing-Area Studies, 1978. (Bibliography series 9) 28 pp.

An unannotated list of 181 citations.

SENEGAL
Republique du Senegal
(as part of French West Africa, a former colony of France)

Senegal is a small nation at the western edge of the Sahel that nonetheless manages to include a range of dryland belts, from subhumid to arid. The nation is topographically uniform: low savannas are crossed by westward flowing rivers, the largest being the Gambia (which enters the nation of The Gambia on its lower course), the Senegal, and the Casamance. Its position in the Sahelo-Sudanian belt means erratic precipitation but the rivers and the sea give Senegal access to sufficient water supplies for a diversified agricultural economy. Groundnuts (foremost), cotton, rice, sugar, vegetables, and other crops are grown, with emphasis recently being placed on increasing non- groundnut acreage along the rivers.

Senegal still shows the legacy of colonial favoritism under the French; Dakar—the capital, chief port, and former capital of all French West Africa—is a major city with an important manufacturing economy. The nation is, in fact, one of the most industrialized in Africa; food and ore-processing are dominant.

A confederation with The Gambia is being negotiated.

Dryland areas of SENEGAL

DESERTIFICATION RISK	Hyperarid km²	Hyperarid %	Arid km²	Arid %	Semiarid km²	Semiarid %	Subhumid km²	Subhumid %	Aridity Totals km²	Aridity Totals %
Very High					46,820	23.8			46,820	23.8
High	By definition, desertification does not exist in hyperarid regions.		7,475	3.8					7,475	3.8
Moderate					63,345	32.2	9,639	4.9	72,984	37.1
Desertification Totals			7,475	3.8	110,165	56.0	9,639	4.9	127,279	64.7
No Desertification					6,885	3.5	52,328	26.6	59,213	30.1
Total Drylands			7,475	3.8	117,050	59.5	61,967	31.5	186,492	94.8

	km²	%
Non-dryland	10,230	5.2
Total Area of the Territory	196,722	100.0

GENERAL BIBLIOGRAPHIES

674 **Brasseur, Paule.** *Bibliographie generale du Mali (anciens Soudan francais et Haut-Senegal-Niger) [general bibliography of Mali (former French Sudan and Upper Senegal- Niger)].*

See entry 587.

675 **Porges, Laurence.** *Bibliographie des regions du Senegal: complement pour la periode des origines a 1965 et mise a jour 1966-1973 [bibliography of the regions of Senegal: covering the period 1965-1973].* Paris: Mouton, 1977. 637 pp.

HANDBOOKS

676 **Nelson, Harold D.** *Area handbook for Senegal.* 2nd ed. Washington DC: U.S. Dept. of Defense, 1974. (American University foreign area studies) 410 pp.

Narrative summary of the nation's history, politics, geography, economy, social life; includes unannotated bibliography.

GEOGRAPHY

677 *Draft environmental report on Senegal.* Tucson AZ: University of Arizona. Office of Arid Lands Studies. Arid Lands Information Center, 1980. 105 pp.

A narrative summary of the nation's environment and development problems, with maps, charts, and an unannotated bibliography.

ATLASES

678 *Atlas national du Senegal [national atlas of Senegal]*. Dakar: Institut fondamental d'Afrique noire, 1977. 147 pp.

In French.

GAZETTEERS

679 *Senegal: official standard names approved by the United States Board on Geographic Names*. Washington DC: U.S. Board on Geographic Names, 1965. (Gazetteer 88) 194 pp.

List of 13,600 place-names with latitude/longitude coordinates.

BOTANY

680 Knapp, R. "Bibliographia phytosociologica et scientiae vegetationis: Mali, Mauretania, Niger, Senegal, Tchad, Volta Superior" [bibliography of plant ecology and vegetation: Mali, Mauritania, Niger, Senegal, Chad, Upper Volta].

See entry 594.

681 Lebrun, J. P. "Nouvelle contribution a la connaissance de la flore de la Republique du Senegal et bibliographie botanique senegalaise 1941-1965" [new contribution to the knowledge of the flora in the Republic of Senegal and botanical bibliography of Senegal 1941-1965]. Societe Botanique de France. *Bulletin* 116: 5/6 (1969):249-277.

In French with English summary.

AGRICULTURAL ECONOMICS

682 Kostinko, Gail, and Dione, Josue. *An annotated bibliography of rural development in Senegal, 1975-1980*. East Lansing MI: Michigan State University. Dept. of Agricultural Economics, 1980. (African rural economy paper 23) 73 pp.

Covers the period 1975 to February 1980.

WOMEN'S STUDIES

683 Kane, F. *Bibliographie, femmes senegalaises [bibliography, women of Senegal]*. Dakar: United Nations. Institut africain de developpement economique et de planification, 1976. 23 pp.

Contains 163 citations, some annotated.

ECONOMIC DEVELOPMENT

684 Busch, Lawrence. *Guinea, Ivory Coast, and Senegal: a bibliography on development*. Monticello IL: Council of Planning Librarians, 1973. (CPL exchange bibliography 427) 48 pp.

Some annotation.

685 *Quarterly economic review of Senegal, Mali, Mauritania, Guinea*.

See entry 596.

HISTORICAL GEOGRAPHY

686 Colvin, Lucie Gallisted. *Historical dictionary of Senegal*. Metuchen NJ: Scarecrow, 1981. (African historical dictionaries 23) 339 pp.

Entries cover major topics, events, and personalities in the nation's history; includes extensive unannotated bibliography.

THE SUDAN
Jamhuryat es-Sudan Al Democratia
[The Democratic Republic of Sudan]

(as the Anglo-Egyptian Sudan, a former protectorate of Egypt and the United Kingdom)

The Sudan is Africa's largest nation and spans environments ranging from tropical rain forest to swampland to savanna grassland to hyperarid desert. Two dryland belts parallel those of nations to the west; a third is unique to the Sudan.

The Sahara. Sudan's portion of the Great Desert extends across the northern third of the nation. Most of the region is bare of vegetation and totally unproductive except in the valley of the Nile itself; here both subsistence crops and cotton are grown. The most important settlement is Port Sudan on the Red Sea coast; it serves as the nation's only outlet to the sea and is connected to inland centers via road and rail. The Nile is not continuously navigable through the region so land transport is essential.

The Sahelo-Sudanian Savanna. The great middle third of the nation covers the eastern-most extension of the arid to semiarid Sahelian savanna. The presence of the White and Blue Niles (which come together at the capital of Khartoum) makes large-scale irrigation feasible; new schemes between the two rivers promise to greatly increase Sudan's output of cotton, sorghum, and groundnuts. On the Kordofan Plateau west of the White Nile gum arabic, harvested from acacia trees, is exported in large quantities. Transportation is developed but distance is still a negative factor in outlying areas.

The Sudd. This region is a wetland within a dryland: although the climate is primarily semiarid vast swamps spread far beyond the banks of the White Nile. Seasonal flooding and a tortuous drainage pattern make the region virtually impassible except within the main channel of the river; explorers were repeatedly stopped by these swamps in their efforts to locate the source of the Nile. Today, the region offers enormous development potential with, unfortunately, unknown effects on the unique ecosystem. A great canal, called the Jonglei, will cut through the Sudd to offer a more direct channel for the White Nile and its traffic. Large areas of swamp have been planted in kenaf, an agave plant that produces a sturdy fiber; coffee, tea, rice, tobacco, papyrus, and cattle are all being raised.

Dryland areas of SUDAN

DESERTIFICATION RISK	ARIDITY								Aridity Totals	
	Hyperarid		Arid		Semiarid		Subhumid			
	km^2	%	km^2	%	km^2	%	km^2	%	km^2	%
Very High	By definition, desertification does not exist in hyperarid regions.		125,000	5.0	150,000	6.0			275,000	11.0
High			435,000	17.4			12,500	0.5	447,500	17.9
Moderate			55,000	2.2	452,500	18.1	102,500	4.1	610,000	24.4
Desertification Totals			615,000	24.6	602,500	24.1	115,000	4.6	1,332,500	53.3
No Desertification	687,500	27.5					177,500	7.1	865,000	34.6
Total Drylands	687,000	27.5	615,000	24.6	602,500	24.1	292,500	11.7	2,197,500	87.9
								Non-dryland	302,500	12.1
								Total Area of the Territory	2,500,000	100.0

GENERAL BIBLIOGRAPHIES

687 Geddes, Charles L. *An analytical guide to the bibliographies on modern Egypt and the Sudan.*
See entry 483.

688 Hill, Richard Leslie. *A bibliography of the Anglo-Egyptian Sudan from the earliest times to 1937.* London: Oxford University Press, 1939. 213 pp.
Continued by Nasri (1962) and Ibrahim and Nasri (1965) below.

689 **Ibrahim, A., and el Nasri, Abdel Rahman.** "Sudan bibliography, 1959-1963." *Sudan notes and records* 46 (1965):130-166.

Continues Hill (1939) above and Nasri (1962) below.

690 **el Nasri, Abdel Rahman.** *A bibliography of the Sudan, 1938-1958.* London; New York: Oxford University Press, 1962. 171 pp.

Contains over 2,800 citations; continues Hill (1939) and is continued by Ibrahim and Nasri (1965) above.

691 **Vidergar, John J.** *Bibliography on Afghanistan, the Sudan and Tunisia.*

See entry 1205.

HANDBOOKS

692 **Nelson, Harold D.** *Area handbook for the Democratic Republic of Sudan.* Rev. ed. Washington DC: U.S. Dept. of Defense, 1973. (American University foreign area studies) 351 pp.

Narrative summary of the nation's history, politics, geography, economy, social life; includes unannotated bibliography.

THESES AND DISSERTATIONS

693 *Dissertation abstracts: University of Khartoum, 1958-1975.* Khartoum: Khartoum University Press, 1977. 192 pp.

Contains 290 items. Lengthy abstracts.

GEOGRAPHY

694 **Beshir, M. E., and Obeid, M.** "Sudan." In *Handbook of contemporary developments in world ecology*, pp. 511-524. Edited by Edward John Kormondy and J. Frank McCormick. Westport CT: Greenwood, 1981.

A good narrative summary of ecological research in the Sudan, with an unannotated list of references.

695 **Speece, Mark.** *Draft environmental profile of The Democratic Republic of Sudan.* Tucson AZ: University of Arizona. Office of Arid Lands Studies. Arid Lands Information Center, 1982. 302 pp.

A narrative summary of the nation's environmental conditions and development problems, with charts, maps, and several unannotated bibliographies.

GAZETTEERS

696 *Sudan: official standard names approved by the United States Board on Geographic Names.* Washington DC: U.S. Board on Geographic Names, 1962. (Gazetteer 68) 358 pp.

List of 25,000 place-names with latitude/longitude coordinates.

CLIMATOLOGY

697 **Kramer, H. P.** "A selective annotated bibliography on the climatology of northeast Africa."

See entry 498.

698 **Wallace, J. Allen, Jr.** *An annotated bibliography on climatic maps of Sudan.* Washington DC: U.S. Weather Bureau, 1962. (NTIS: AD 660-868) 9 pp.

699 **Wallace, J. Allen, Jr.** *An annotated bibliography on the climate of Sudan.* Washington DC: U.S. Weather Bureau, 1964. (NTIS: AD 660-804) 41 pp.

Contains 103 citations to 1964.

HYDROLOGY

700 **Lytle, Elizabeth E.** *The Aswan High Dam.*

See entry 499.

701 **Nash, H. G.** *Review of hydrology and groundwater resource studies in the Sudan (with bibliography).* London: Institute of Geological Sciences. Geophysical and Hydrological Division, 1979. 100 pp.

BOTANY

702 **Knapp, R.** "Bibliographia phytosociologica et scientiae vegetationis: Aethiopia, Somalia, Sudan, Afar et Issa, Socotra" [bibliography of plant ecology and vegetation: Ethiopia, Somalia, Sudan, Afars and Issas, Socotra]. *Excerpta botanica: sectio B: sociologica* 13:2 (1974):91-106.

Unannotated. Contains 193 items. Author index.

AGRICULTURE

703 **Cigar, Norman.** *Agriculture and rural development in the Sudan.* Monticello IL: Vance Bibliographies, 1981. (Public administration series bibliography P-700) 4 pp.

Unannotated.

704 **Mamoun, Izz Eldin.** *Bibliography of agriculture and veterinary sciences in the Sudan, up to 1974.* 2nd ed. Khartoum: National Council for Research. Agricultural Research Council, 1978. 325 pp.

Contains 4,578 items. Published literature only. Covers period to 1974.

FORESTRY

705 **Bayoumi, A. A., comp.** *A forest bibliography of the Sudan to 1973.* Khartoum: Sudan. National Council for Research. Agricultural Research Council, 1974. 159 pp.

Unannotated list of 2,111 references covering forests, forestry, fire, and more general topics such as climate and vegetation.

NOMADISM

706 **Ahmed, Abdel Ghaffar M.** "A bibliography of nomadic studies in the Sudan." In *Some aspects of pastoral nomadism in the Sudan*, pp. 211-218. Edited by Abdel Ghaffar M. Ahmed. Khartoum: Sudan National Population Committee, 1976.

COOPERATIVES

707 **Saghayroun, Atif A. Rahman.** *Bibliography on cooperative movement in the Sudan*. Khartoum: Sudan. National Council for Research, 1977. 28 pp.

ANTHROPOLOGY

708 **van Garsse, Yvan.** *Ethnological and anthropological literature on the three southern Sudan provinces: Upper Nile, Bahr el Ghazel, Equatoria*. Wien, Austria: Institut fur Volkerkunde der Universitat Wien, 1972. (Acta Ethnologica et Linguistica 29/Series Africana 7) 88 pp.

Unannotated listing of 1,072 citations.

HISTORICAL GEOGRAPHY

709 **Voll, John Obert.** *Historical dictionary of the Sudan*. Metuchen NJ: Scarecrow, 1978. (African historical dictionaries 17) 175 pp.

Entries cover major topics, events, and personalities in the nation's history; includes extensive unannotated bibliography.

ECONOMIC DEVELOPMENT

710 *Quarterly economic review of Sudan*. London: Economist Intelligence Unit, 1976-. Quarterly.

Includes summary of political and economic news as well as charts and statistics of selected economic indicators. Annual supplement.

TOGO
Republique Togolaise
(a former colony of Germany, then France)

Togo and Benin, two small nations sandwiched between Ghana and Nigeria, have closely matched environments and economic prospects. Both have two disconnected dryland belts.

The Coastal Lowland. This region is the middle portion of the anomalous subhumid zone that stretches along the coast from southeastern Ghana to the Benin-Nigeria frontier. Two rainy seasons—spring and fall—support the oil palm, although this plays a much smaller role in Togo's economy than in neighboring Benin. With the capital and chief port at Lome this region is the nation's economic and administrative center.

The Savanna. Again like Benin, Togo occupies a subhumid fringe of the Sahelo-Sudanian belt; it is drained by the Oti River and is both remote and poorly integrated into the national economy.

Despite its very small size, even when compared to Benin, Togo has a number of economic advantages. The Germans had early turned the land into a model colony by the standards of the time, building roads and railways and establishing an administrative structure. Even when English-speaking Togo joined Ghana in 1956 the French-speaking remainder was left with the old German capital at Lome, several rail lines, and deposits of phosphate and iron ore that have proved crucial to the nation after its independence.

Dryland areas of TOGO

DESERTIFICATION RISK	ARIDITY									Aridity Totals	
	Hyperarid		Arid		Semiarid		Subhumid			
	km²	%	km²	%	km²	%	km²	%	km²	%
Very High	By definition, desertification does not exist in hyperarid regions.									
High										
Moderate										
Desertification Totals										
No Desertification							17,136	30.6	17,136	30.6
Total Drylands							17,136	30.6	17,136	30.6
Non-dryland									38,864	69.4
Total Area of the Territory									56,000	100.0

GENERAL BIBLIOGRAPHIES

711   Bridgman, Jon. *German Africa: a select annotated bibliography.*

See entry 450.

712   Witherell, Julian W. *Official publications of French Equatorial Africa, French Cameroons, and Togo 1946-1958.*

See entry 453.

GAZETTEERS

713   *Togo: official standard names approved by the United States Board on Geographic Names.* Washington DC: U.S. Board on Geographic Names, 1966. (Gazetteer 98) 100 pp.

List of 7,000 place-names with latitude/longitude coordinates.

ECONOMIC DEVELOPMENT

714   *Quarterly economic review of Ivory Coast, Togo, Benin, Niger, Upper Volta.* London: Economist Intelligence Unit, 1976-. Quarterly.

Includes summary of political and economic news as well as charts and statistics of selected economic indicators. Annual supplement.

HISTORICAL GEOGRAPHY

715   Decalo, Samuel. *Historical dictionary of Togo.* Metuchen NJ: Scarecrow, 1976. (African historical dictionaries 9) 243 pp.

Entries cover major topics, events, and personalities in the nation's history; includes extensive unannotated bibliography.

TUNISIA
Al-Djoumhouria Attunusia
[The Republic of Tunisia]

(a former colony of France)

Tunisia—the smallest of the three Maghreb states— nonetheless possesses the same range of climate belts as Morocco and Algeria. A humid north coast rises to the Atlas highland which, in turn, gives way to steppe and desert to the south. Tunisia has the additional influence of an eastern coast so that nearly the entire nation is less than 250 km (150 miles) from the sea.

The Majardah Valley and Coastal Plain. This semiarid region follows the drainage of the Majardah River as it parallels the north coast and crosses the low Bizerte Plain; this is the nation's heartland where most of its rural and urban population lives. The Majardah Valley holds great potential for irrigated agriculture and has received the strongest development support to solve its problems of flooding, deforestation, soil salination, and land tenure. Durum wheat, citrus, and vegetables are favored; extensive vineyards planted by the French are being allowed to decline. Along the eastern coast especially are extensive olive orchards as well as a developing fishing industry.

The Atlas Mountains and Steppe. These mountains are not nearly as high in Morocco as in neighboring Algeria; because of this the barriers to transport into the steppe and desert beyond have never been great. Traditional subsistence agriculture is still the norm. Sheep are grazed extensively, pockets of forest are exploited, and great stretches of esparto grass are harvested. Some deposits of lead, iron, phosphate, and zinc are also important.

The Sahara. Tunisia lacks Algeria's great desert hinterland but its position along a low coastline provides easy access around the Atlas range; for example, pipelines from Algerian and Tunisian oilfields are routed to the eastern Tunisian coast. Scattered oases support traditional date palm plantations, while citrus and olives are grown along a narrow coastal belt.

Dryland areas of TUNISIA

DESERTIFICATION RISK	ARIDITY									
	Hyperarid		Arid		Semiarid		Subhumid		Aridity Totals	
	km²	%	km²	%	km²	%	km²	%	km²	%
Very High	By definition, desertification does not exist in hyperarid regions.		54,005	32.9	3,776	2.3			57,781	35.2
High			33,323	20.3	35,456	21.6			68,779	41.9
Moderate										
Desertification Totals			87,328	53.2	39,232	23.9			126,560	77.1
No Desertification	30,368	18.5							30,368	18.5
Total Drylands	30,368	18.5	87,328	53.2	39,232	23.9			156,928	95.6
Non-dryland									7,222	4.4
Total Area of the Territory									164,150	100.0

GENERAL BIBLIOGRAPHIES

716   Bennett, Norman Robert. *A study guide for Tunisia.* Boston MA: Boston University. African Studies Center, 1968. 50 pp.

717   Findlay, Allan M.; Findlay, Anne M.; and Lawless, Richard I. *Tunisia.* Oxford, U.K.; Santa Barbara CA: Clio Press, 1982. (World bibliographical series 33) 251 pp.

An annotated bibliography of 895 items covering most major subject areas; limited to English- and some French-language titles.

718 **Vidergar, John J.** *Bibliography on Afghanistan, the Sudan and Tunisia.*

See entry 1205.

HANDBOOKS

719 **Nelson, Harold D.** *Tunisia: a country study.* 2nd ed. Washington DC: U.S. Dept. of Defense, 1979. (American University foreign area studies) 326 pp.

Narrative summary of the nation's history, politics, geography, economy, social life; includes unannotated bibliography.

GEOGRAPHY

720 *Environmental report of Tunisia (revised draft).* Tucson AZ: University of Arizona. Office of Arid Lands Studies, 1981. 81 pp.

A summary of geography, resources, and environmental problems. Especially valuable for its lists of environmental legislation and organizations. Includes unannotated lists of references.

721 **Nuttonson, Michael Y.** "Introduction to North African and a survey of the agriculture of Morocco, Algeria, and Tunisia, with special reference to their regions containing areas climatically and latudinally analagous to Israel."

See entry 433.

GAZETTEERS

722 *Tunisia: official standard names approved by the United States Board on Geographic Names.* Washington DC: U.S. Board on Geographic Names, 1964. (Gazetteer 81) 399 pp.

List of 23,000 place-names with latitude/longitude coordinates.

GEOLOGY

723 **Memmi, L.** *Bibliographie geologique de la Tunisie [Tunisian bibliography of geology].* Tunis: Tunisia. Service Geologique, 1977. (Bibliographique geologique de Tunisie 7) 83 pp.

724 **Morin, Philippe.** *Bibliographie analytique des sciences de la terre: Tunisie et regions limitrophes (depuis de debut des recherches geologiques a 1971) [analytical bibliography of the earth sciences: Tunisia and surrounding regions: from the beginning of geologic research to 1971].* Paris: Centre national de la recherche scientifique, 1972. (Publication serie geologie 13) 644 pp.

In French. Updated annually by the *Bibliographie geologique de la Tunisie [geologic bibliography of Tunisia].*

725 **Ridge, John Drew.** "Tunisia." In *Annotated bibliographies of mineral deposits in Africa, Asia (exclusive of the USSR) and Australasia,* pp 170-176. Oxford U.K.: Pergamon, 1976.

CLIMATOLOGY

726 **Grimes, Annie E.** *Bibliography of climatic maps of Tunisia.* Washington DC: U.S. Weather Bureau, 1960. (NTIS: PB 176-033; AD 664-710) 11 pp.

SOILS

727 *Bibliography on desert soils of Morocco, Algeria, Tunisia, Libya (1964-1936).*

See entry 623.

728 *Soils of Tunisia (annotated bibliography covering the published literature for 1932- 1975).* Harpenden, U.K.: Commonwealth Agricultural Bureaux, 1977. (Annotated bibliography) 4 pp.

BOTANY

729 **Roussine, N., and Sauvage, C.** "Bibliographia phytosociologica: Afrique du Nord" [bibliography of plant ecology: North Africa].

See entry 626.

AGRICULTURAL ECONOMICS

730 **Cigar, Norman.** *Agriculture and rural development in Tunisia.* Monticello IL: Vance Bibliographies, 1981. (Public administration series bibliography P-850) 7 pp.

Unannotated.

ANTHROPOLOGY

731 **Louis, Andre.** *Bibliographie ethno-sociologique de la Tunisie [ethno-sociological bibliography of Tunisia].* Tunis: Institute Belles Lettres Arabes, 1977. 393 pp.

MEDICINE

732 **Dedet, J-P.** "Les leishmanioses en Afrique du Nord: list bibliographique des travaux: 1. Tunisie (1884-1972)" [leishmaniasis in North Africa: bibliographical list of works, part 1, Tunisia 1884-1972]. Institut Pasteur de Tunis. *Archives* 49:3 (1972):207-226.

In French.

PUBLIC HEALTH

733 **Indesha, N.** *Public health in Tunisia.* Monticello IL: Vance Bibliographies, 1981. (Public administration series bibliography P-702) 6 pp.

Unannotated.

Tunisia/734-739

734 **Montague, Joel.** "Disease and public health in Tunisia: 1882-1970: an overview of the literature and its sources: part 1." *A current bibliography on African affairs*, new series 4:4 (1971):250-260.

A narrative bibliography; the promised part 2 apparently never appeared.

URBAN GEOGRAPHY

735 **Barbar, Aghil M.** *Urbanization in Tunisia.* Monticello IL: Council of Planning Librarians, 1978. (CPL exchange bibliography 1521) 5 pp.

Unannotated.

ARCHITECTURE

736 **Cigar, Norman.** *North African architecture.*

See entry 629.

ECONOMIC DEVELOPMENT

737 *Quarterly economic review of Libya, Tunisia, Malta.*

See entry 582.

TOURISM AND RECREATION

738 **al-Barbar, Aghil.** *Tourism in the Arab world: an introductory bibliography.*

See entry 1325.

PUBLIC ADMINISTRATION

739 **Cigar, Norman.** *Government and politics in North Africa.*

See entry 633.

UPPER VOLTA
Republique de Haute-Volta
(a former colony of France)

Upper Volta is the smallest and most populous of the landlocked Sahelian nations; it may also be the poorest country in Africa. Ironically, water is somewhat more plentiful here than in Niger or Chad. Most of Upper Volta falls into the semiarid to subhumid categories with only a tiny extension into the arid zone in the far north. Two highland areas—the Mossi Plateau around the capital and the uplands around Bobo Dioulasso to the west— are relatively free of disease and so hold the bulk of the population. Soils, however, are thin and badly degraded from cattle herding. Groundnuts and cotton are grown. The best land exists in the river villages below the highlands but river blindness, sleeping sickness, and the simulium fly all inhibit exploitation of the fertile, well-watered bottomlands. Manganese and iron deposits in the north hold some promise but until the transportation network is extended one of the nation's principal exports will remain its labor supply.

Dryland areas of UPPER VOLTA

DESERTIFICATION RISK	ARIDITY									
	Hyperarid		Arid		Semiarid		Subhumid		Aridity Totals	
	km²	%	km²	%	km²	%	km²	%	km²	%
Very High	By definition, desertification does not exist in hyperarid regions.		9,320	3.4	108,279	39.5	20,559	7.5	138,158	50.4
High										
Moderate					63,322	23.1			63,322	23.1
Desertification Totals			9,320	3.4	171,601	62.6	20,559	7.5	201,480	73.5
No Desertification					16,721	6.1	38,103	13.9	54,824	20.0
Total Drylands			9,320	3.4	188,322	68.7	58,662	21.4	256,304	93.5
Non-dryland									17,818	6.5
Total Area of the Territory									274,122	100.0

GENERAL BIBLIOGRAPHIES

740 Brasseur, Paule. *Bibliographie generale du Mali (anciens Soudan francais et Haute-Senegal-Niger) [general bibliography of Mali (former French Sudan and Upper Senegal- Niger)].*

See entry 587.

741 Gagner, Lorraine. *Upper Volta: a selected and partially annotated bibliography.* New York: U.N. Institute for Training and Research, 1972. 13 leaves.

GEOGRAPHY

742 Bantje, Han. *A working bibliography of the western Sahel.*

See entry 589.

743 *Draft environmental report on Upper Volta.* Tucson AZ: University of Arizona. Office of Arid Lands Studies. Arid Lands Information Center, 1980. 138 pp.

Narrative on the nation's environment and development problems, with maps, charts, and unannotated bibliography.

744 Pallier, Ginette. *Geographie generale de la Haute-Volta [general geography of Upper Volta].* Limoges: U.E.R. des lettres et sciences humaines, 1978. 241 pp.

In French. A monograph on the nation's physical, social, and economic geography, with illustrations, maps, and charts. Includes 8-page unannotated bibliography.

Upper Volta/745-750

ATLASES

745 **Peron, Yves, and Zalacain, Victoire.** *Atlas de la Haute-Volta [atlas of Upper Volta].* Paris: Editions J. A., 1975. (Les Atlas Jeune Afrique) 47 pp.

In French. Most plates at a scale of 1:5,000,000.

GAZETTEERS

746 *Upper Volta: official standard names approved by the United States Board on Geographic Names.* Washington DC: U.S. Board on Geographic Names, 1965. (Gazetteer 87) 168 pp.

List of 11,900 place-names with latitude/longitude coordinates.

REMOTE SENSING

747 *Remote sensing newsletter.* Ouagadougou: Regional Remote Sensing Center, 1980-. Quarterly.

BOTANY

748 **Knapp, R.** "Bibliographia phytosociologica et scientiae vegetationis: Mali, Mauretania, Niger, Senegal, Tchad, Volta Superior" [bibliography of plant ecology and vegetation: Mali, Mauritania, Niger, Senegal, Chad, Upper Volta].

See entry 594.

ECONOMIC DEVELOPMENT

749 *Quarterly economic review of Ivory Coast, Togo, Benin, Niger, Upper Volta.*

See entry 714.

HISTORICAL GEOGRAPHY

750 **McFarland, Daniel Miles.** *Historical dictionary of Upper Volta (Haute Volta).* Metuchen NJ: Scarecrow, 1978. (African historical dictionaries 14) 217 pp.

Entries cover major topics, events, and personalities in the nation's history; includes extensive unannotated bibliography.

WESTERN SAHARA

(as Rio de Oro or Spanish Sahara, a former colony of Spain; now organized into Moroccan provinces)

This former Spanish colony—called simply Sahara by the local inhabitants—was partially occupied by Morocco since 1975 and completely so by 1979. It borders a long arid Atlantic coast and extends inland to an undemarcated frontier with Mauritania. The population is sparse and consists primarily of nomadic tribes who cross the frontiers at will. A small fishing community exports its catch to the Canary Islands; otherwise, the territory has no other known resources except the world's richest deposits of phosphate. This treasure makes Western Sahara a land worth fighting for: Morocco, Algeria, Mauritania, and the native *Saharouis* have yet to resolve their claims of sovereignty.

Dryland areas of WESTERN SAHARA

DESERTIFICATION RISK	ARIDITY									
	Hyperarid		Arid		Semiarid		Subhumid		Aridity Totals	
	km²	%	km²	%	km²	%	km²	%	km²	%
Very High	By definition, desertification does not exist in hyperarid regions.									
High			140,980	53.0					140,980	53.0
Moderate										
Desertification Totals			140,980	53.0					140,980	53.0
No Desertification	125,020	47.0							125,020	47.0
Total Drylands	125,020	47.0	140,980	53.0					266,000	100.0
Non-dryland										
Total Area of the Territory									266,000	100.0

GENERAL BIBLIOGRAPHIES

751 Rishworth, Susan Knoke. *Spanish-speaking Africa: a guide to official publications.* Washington DC: U.S. Library of Congress, 1973. (Series of bibliographies on African administrations) 68 pp.

Contains 640 unannotated citations from 19th century to 1972, including Spanish government publications. Excludes Spanish islands off Africa, but covers Equitorial Guinea, Spanish Sahara (now Western Sahara), Ifni (now part of Morocco), and the Spanish zones of Morocco (now part of Morocco).

GAZETTEERS

752 *Spanish Sahara: official standard names approved by the United States Board on Geographic Names.* Washington DC: U.S. Board on Geographic Names, 1969. (Gazetteer 108) 52 pp.

List of 3,000 place-names with latitude/longitude coordinates.

GEOLOGY

753 "Bibliografia geologica espanola."

See entry 201.

HISTORICAL GEOGRAPHY

754 Hodges, Tony. *Historical dictionary of Western Sahara.* Metuchen NJ: Scarecrow, 1982. (African historical dictionaries 35) 431 pp.

Entries cover major topics, events, and personalities in the nation's history; includes extensive unannotated bibliography.

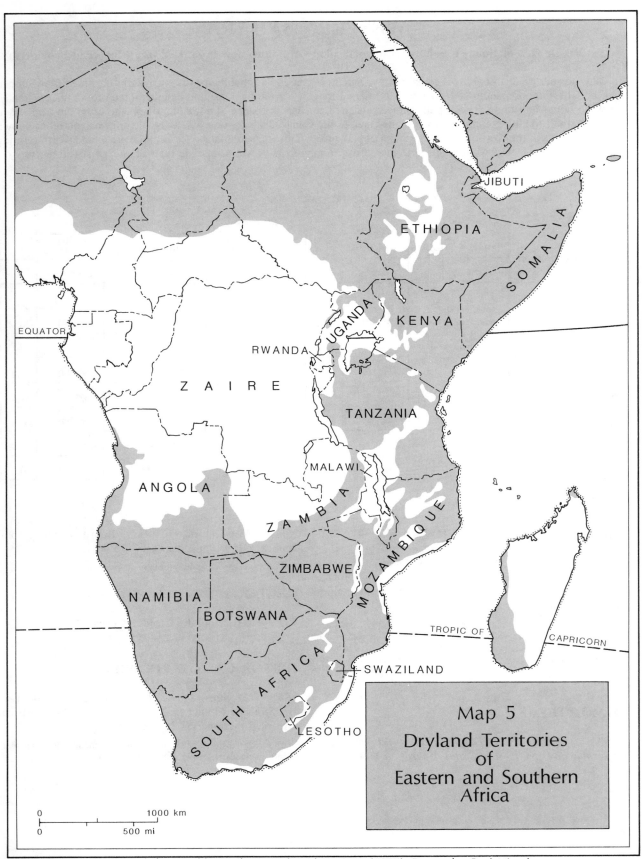

Map 5. Dryland Territories of Eastern and Southern Africa

Based on Unesco, Map of the World Distribution of Arid Regions (1977). Drawn by Paul Mirocha.

Eastern and Southern Africa

This portion of the continent presents a more complex and less extreme picture of aridity. Hyperarid regions are confined to the fog desert of Namibia, the chemical hell of Ethiopia's Danakil Depression, and a narrow coastal strip of Somalia. Though extensive arid areas exist, much of the high tableland of eastern and southern Africa lies in the semiarid to subhumid range and supports extensive grassland savannas and scrub forest. Though wetter than true desert, these regions resemble the Sahel in their sensitivity to overgrazing and other processes of desertification.

For organizational purposes, Madagascar is considered below under the Indian Ocean Islands.

GENERAL BIBLIOGRAPHIES

755 Miller, E. Willard, and Miller, Ruby M. *Tropical Eastern and Southern Africa: a bibliography on the Third World.* Monticello IL: Vance Bibliographies, 1981. (Public administration series bibliography P-819) 82 pp.

Unannotated. Covers the continent from Cameroon, Central African Republic, and Ethiopia south. The Republic of South Africa is omitted; Lesotho, Swaziland, and Namibia are included.

ENCYCLOPEDIAS

756 Rosenthal, Eric. *Encyclopaedia of southern Africa.* 7th ed. Cape Town: Juta, 1978. 577 pp.

A one-volume work covering Namibia, Botswana, and nations to the south.

757 *Standard encyclopaedia of Southern Africa.* Cape Town: Nasou Ltd., 1970-1976. 12 vols.

A regional encyclopedia covering the Republic of South Africa principally but ranging beyond to touch most of sub-Saharan Africa and the surrounding islands. "...[A] work which is already widely accepted as authoritative and definitive." (Musiker, *South Africa*)

THESES AND DISSERTATIONS

758 Pollak, Oliver B., and Pollak, Karen. *Theses and dissertations on Southern Africa: an international bibliography.* Boston MA: G. K. Hall, 1976. 236 pp.

Covers the nations of Angola, Botswana, Lesotho, Malawi, Mozambique, Zimbabwe, South Africa, Namibia, Swaziland, and Zambia with master's theses and doctoral dissertations produced in these nations and in Europe, North America, Israel, India, and Australia.

CLIMATOLOGY

759 *Bibliography of regional meteorological literature.* Pretoria: South Africa. Weather Bureau, 1950-.

Volume 1 (1950) covers period to 1947; volume 2 (1972) covers period 1948-1960; volume 3 (1966) covers Antarctica; volume 4 (1975) covers period 1960-1972.

BOTANY

760 Bullock, A. A. *Flora of southern Africa: bibliography of South African botany (up to 1951).* Pretoria: South Africa. Dept. of Agricultural Technical Services, 1978. 194 pp.

ZOOLOGY

761 Petersen, J. C. Briand, and Casebeer, R. L. "A bibliography relating to the ecology and energetics of East African large mammals." *East African wildlife journal* 9 (1971):1-23.

Covers Kenya, Zimbabwe, South Africa, Tanzania, Uganda, Zaire, Zambia, and Zanzibar.

ANIMAL CULTURE

762 Little, Peter D. *The socio-economic aspects of pastoralism and livestock development in Eastern and Southern Africa: an annotated bibliography.* Madison WI: University of Wisconsin. Land Tenure Center, 1980.

ENERGY

763 Cavan, Ann. *Energy resources in southern Africa: a select bibliography.* Washington DC: African Bibliographic Center, 1981. 65 pp.

ANTHROPOLOGY

764 van Warmelo, Nicolaas Jacobus, comp. *Anthropology of Southern Africa in periodicals to 1950: an analysis and index.* Johannesburg: Witwatersrand University Press, 1977. 1484 pp.

"A voluminous bibliography which is encyclopaedic in scope. Gives an in-depth analysis of over 400 worldwide periodicals covering 155 years of anthropological material relevant to Southern Africa. It is a virtual concordance in its coverage of place-names, tribes, etc." (Musiker, *South Africa*)

Eastern and Southern Africa/765-769

HOUSING

765 **Barr, Charles W.** *Eastern Africa: references related to developmental planning and housing.* Monticello IL: Council of Planning Librarians, 1972. (CPL exchange bibliography 330) 75 pp.

Some annotated entries. Covers Ethiopia, Kenya, Tanzania, Uganda, Zambia, Zimbabwe, Malawi, South Africa, Lesotho.

766 *Habitat rural en Afrique centrale et en Afrique de l'est [rural housing in Central and East Africa].* Paris: France. Secretariat des missions d'urbanisme et d'habitat. Service documentation-information, 1975. (Bibliographie 20) 11 pp.

Contains 31 annotated entries. In French.

ECONOMIC DEVELOPMENT

767 **Ketso, L. Victor.** "A provisional bibliography on labour migrancy (with special reference to southern Africa)." In *Towards a strategy of labour migration*, pp. 31-164. Roma, Lesotho: I.L.O.-N.U.L. Research Project, 1977.

"A bibliography of 1,169 entries compiled with the assistance of C. A. Perrings and E. M. Sebatane. Some 750 of the items relate directly to Southern Africa, of which approximately 100 refer specifically to Lesotho." (Willet and Ambrose, *Lesotho*, p. 418)

768 *SADEX (Southern Africa Development Information/Documentation Exchange).* Washington DC: African Bibliographic Center, 1979-. 2 per month.

Contains an annotated bibliography in each issue covering economic development in Angola, Botswana, Lesotho, Malawi, Mozambique, Namibia, Swaziland, Tanzania, Zambia, Zimbabwe. Excludes the Republic of South Africa.

TRANSPORTATION

769 **Krummes, Daniel C.** *Transportation in East Africa: a bibliography.* Monticello IL: Vance Bibliographies, 1980. (Public administration series bibliography P-595) 27 pp.

Unannotated. Covers Ethiopia, Jibuti, Kenya, Uganda, Tanzania, Somalia, Zambia.

ANGOLA
Republica Popular de Angola
(a former colony of Portugal)

The dryland regions of Angola are extensions and peripheries of regions with arid cores outside the nation.

The Namib. This coastal desert, centered south of Angola in neighboring Namibia, extends north of the border (the Cunene River) to embrace most of the Angolan coast, although the hyperarid conditions of the Namib proper barely reach the port of Mocamedes. The greatest length of the coastal plain is arid to semiarid and shares the cool, foggy conditions common to the Namib proper. The Cunene River, flowing south out of the central plateau and then west to the sea, is the major stream and holds promise for both hydroelectric development and irrigated agriculture. Dams on this river and on the other streams that cut across the arid plain to the sea have begun to be built. The region is economically important for its seaports (including the capital at Luanda), its mineral resources (salt, diamonds, oil), and its potential for irrigated agriculture (at present limited to sugar cane and palm products around Benguela).

The Southeastern Savanna. This region is a dry savanna extension of the arid Kalahari core of nearby Botswana. It is a large, amorphous, subhumid area, the least developed region in the nation. Its greatest potential is probably for extensive cattle ranching, though few roads and no railroads traverse the area.

Though a large share of the national territory is dry, Angola's economy rests primarily on the exploitation of extensive mineral and agricultural resources in the wetter regions. Capital improvements in irrigation works (for the coast and Cunene River basin) and transportation (for the southeast) are required before the drylands can be more fully utilized. Though retarded by Portugal's neglect and damaged by civil war, the nation has a sufficiently diverse resource base for a relatively prosperous future.

Dryland areas of ANGOLA

DESERTIFICATION RISK	ARIDITY									
	Hyperarid		Arid		Semiarid		Subhumid		Aridity Totals	
	km²	%	km²	%	km²	%	km²	%	km²	%
Very High	By definition, desertification does not exist in hyperarid regions.				2,492	0.2			2,492	0.2
High			21,194	1.7	22,441	1.8			43,635	3.5
Moderate			13,714	1.1	71,062	5.7	53,608	4.3	138,384	11.1
Desertification Totals			34,908	2.8	95,995	7.7	53,608	4.3	184,511	14.8
No Desertification	6,234	0.5			6,234	0.5	291,728	23.4	304,196	24.4
Total Drylands	6,234	0.5	34,908	2.8	102,229	8.2	345,336	27.7	488,707	39.2
Non-dryland									757,993	60.8
Total Area of the Territory									1,246,700	100.0

GENERAL BIBLIOGRAPHIES

770 **Gibson, Mary Jane.** *Portuguese Africa: a guide to official publications.* Washington DC: U.S. Library of Congress, 1967. 217 pp.

Contains 2,831 entries. Lists publications from 1850-1964 of Portuguese Africa and Portugal. Covers Cape Verde Islands, Angola, Mozambique, and Guinea-Bissau.

771 **Zubatsky, David S.** *A guide to resources in United States libraries and archives for the study of Cape Verde, Guinea (Bissau), Sao Tome-Principe, Angola and Mozambique.*

See entry 262.

Angola/772-783

HANDBOOKS

772 **Kaplan, Irving.** *Angola: a country study.* 2nd ed. Washington DC: U.S. Dept. of Defense, 1979. (American University foreign area studies) 286 pp.

Narrative summary of the nation's history, politics, geography, economy, social life; includes unannotated bibliography.

GAZETTEERS

773 *Angola: official standard names approved by the United States Board on Geographic Names.* Washington DC: U.S. Board on Geographic Names, 1956. (Gazetteer 20) 234 pp.

List of 19,200 place-names with latitude/longitude coordinates.

GEOLOGY

774 **Ribeiro, Marilia da Cunha Ferro, and Silva, A. T. S. Ferreira da.** *Bibliografia geologica de Angola, 1970- 1972 [Angolan bibliography of geology, 1970-1972].* Luanda: Angola. Servicos de Geologia e Minas, 1973. 29 pp.

In Portuguese.

775 **Silva, A. T. S. Ferreira da.** *Bibliografia geologica de Angola [Angolan bibliography of geology].* Luanda: Angola. Servicos de Geologia e Minas, 1971. (Memoria 10) 130 pp.

In Portuguese.

CLIMATOLOGY

776 **Carraway, D. M.** *Bibliography on the climate of Angola.* Washington DC: U.S. Weather Bureau, 1956. (NTIS: AD 669-410) 26 pp.

777 **Ferreira, H. Amorim.** *Bibliografia meteorologica e geofisica de Angola [Angolan bibliography of meteorology and geophysics].* Luanda: Angola. Servico Meteorologico, 1957. 14 pp.

An unannotated list of around 110 citations.

778 **Wallace, J. Allen, Jr.** *An annotated bibliography of climatic maps of Angola.* Washington DC: U.S. Weather Bureau, 1962. (NTIS: AD 660-851) 14 pp.

BOTANY

779 **Pinto da Silva, A. R., and Teles, A. N.** "Bibliographia phytosociologica Portugaliae" [bibliography of the plant ecology of Portugal].

See entry 191.

780 **Teles, A. N.** "Bibliographie des cartes de vegetation du Portugal" [bibliography of vegetation maps of Portugal].

See entry 192.

ANTHROPOLOGY

781 **Strohmeyer, Eckhard, and Moritz, Walter.** *Umfassende bibliographie der Volker Namibiens (Sudwestafrikas) und Sudwestangolas [comprehensive bibliography of the people of Namibia (Southwest Africa) and southwest Angola].*

See entry 919.

ECONOMIC DEVELOPMENT

782 *Quarterly economic review of Angola, Guinea-Bissau, Cape Verde, Sao Tome, Principe.* London: Economist Intelligence Unit, 1978-. Quarterly.

Includes summary of political and economic news as well as charts and statistics of selected economic indicators. Annual supplement.

HISTORICAL GEOGRAPHY

783 **Martin, Phyllis M.** *Historical dictionary of Angola.* Metuchen NJ: Scarecrow, 1980. (African historical dictionaries 26) 174 pp.

Entries cover major topics, events, and personalities in the nation's history; includes extensive unannotated bibliography.

BOTSWANA
The Republic of Botswana

(as the Bechuanaland Protectorate, a former colony of the United Kingdom)

Botswana occupies the center of southern Africa's great dryland belt, roughly overlapping the area of the Kalahari Desert. Although the landscape is relatively uniform in climate and topography, at least three regions can be distinguished.

The Eastern Frontier. The heart of the nation in fact hugs the eastern border with South Africa and Zimbabwe. The topography is slightly higher and more broken, rainfall slightly greater, and good cropland more abundant. The nation's best roads and only rail line unify the region and link it with neighboring nations. The primary agricultural product is cattle, which are brought to Africa's largest slaughterhouse and cannery at Lobatse near the capital. Chilled carcasses are then shipped to South Africa, Zambia, and the United Kingdom. Extensive mining industries have arisen in the region to exploit diamonds, copper, nickel, and coal.

The Kalahari. This desert region is characterized by a gently undulating plateau ranging between 1,000 and 1,300 meters (3,300 and 4,200 feet) above sea level. As deserts go, the Kalahari is not terribly dry, falling in fact in the semiarid category except for the extreme southwest. Rainfall is, however, erratic and the product of localized summer thunderstorms. Surface drainage is poor and largely confined to internal basins. Groundwater brought up from boreholes provides most of the water for humans and livestock. Vegetation cover consists of grasses and acacias, more a savanna than true desert. Overall the land has some potential as a cattle range but is easily abused.

There is a people—the Bushmen—who deserve special mention as one of the premier human adaptors to the limitations of dryland environments. Small bands roam the veld in search of antelope, giraffe, ostrich, and smaller game. They possess a remarkable knowledge of the food resources of their land and, though confined to an extremely low energy level in their economy, their culture and their peculiar physical traits (short stature, tough skin, provision for fat storage) have allowed them to survive.

The Okavango. This region is geographically a part of the Kalahari; nonetheless it possesses some peculiar characteristics of its own. The Okavango River flows from the Angolan highlands across the desert into a vast swampland in northern Botswana. This inland delta harbors tsetse fly, malaria, and a huge concentration of wildlife; it also offers the potential for a sizeable irrigation project that could produce rice, millet, maize, or many other crops. A wildlife-based tourist industry is beginning to grow. At some future date, hard decisions will have to be made regarding the region's future and its contribution to the nation's development.

Botswana is one of the world's poorest nations, with an economy geared almost totally toward export of raw materials to South Africa. The primary industry—cattle production—suffers from concentrated ownership, diseased stock, poor transport, and uncertain markets. Recently-developed mineral resources promise quick infusions of capital but offer little long-term enrichment for the people.

Dryland areas of BOTSWANA

DESERTIFICATION RISK	ARIDITY									
	Hyperarid		Arid		Semiarid		Subhumid		Aridity Totals	
	km²	%	km²	%	km²	%	km²	%	km²	%
Very High	By definition, desertification does not exist in hyperarid regions.									
High										
Moderate			32,775	5.7	526,125	91.5	16,100	2.8	575,000	100.0
Desertification Totals			32,775	5.7	526,125	91.5	16,100	2.8	575,000	100.0
No Desertification										
Total Drylands			32,775	5.7	526,125	91.5	16,100	2.8	575,000	100.0
Non-dryland										
Total Area of the Territory									575,000	100.0

Botswana/784-802

GENERAL BIBLIOGRAPHIES

784 **Balima, Mildred Grimes.** *Botswana, Lesotho, and Swaziland: a guide to official publications, 1868-1968.* Washington DC: U.S. Library of Congress, 1972. 84 pp.

Contains 791 entries. Includes some British and South African documents. Excludes United Nations reports.

785 *Botswana: a bibliography.* Uppsala, Sweden: Lantbrukshoegskolan. Rural Development Section, 1973. (Ru-Develop documentation. Serie land) 41 leaves.

Contains 284 items from 1967-1972. Includes library location.

786 **Mohome, Paulus, and Webster, John B.** *A bibliography on Bechuanaland.* Syracuse NY: Syracuse University. Program of Eastern African Studies. Bibliographic Section, 1966. (Occasional bibliography 5) 58 pp.

787 **Stevens, Pamela.** *Bechuanaland: bibliography.* 4th ed. Cape Town: University of Cape Town Libraries, 1969. (Bibliographical series) 27 pp.

GEOGRAPHY

788 **Henderson, F. I., and Opschoor, J. B.** *Botswana's environment: an annotated bibliography.* Gaborone: University College of Botswana. National Institute of Development and Cultural Research, 1981. (Working bibliography 5) 81 pp.

Includes 375 citations emphasizing resource exploitation.

GAZETTEERS

789 *Guide to the villages of Botswana.* Gaborone: Botswana. Central Statistics Office, 1973. 221 pp.

A detailed guide to rural settlement based on the 1971 census. Includes information for each village on population, physical facilities, dwellings, commercial establishments, and access to government services.

790 **Leistner, O. A., and Morris, J. W.** *Southern African place names.*

See entry 963.

791 *South Africa: official standard names approved by the United States Board on Geographic Names.*

See entry 965.

GEOLOGY

792 **Bennett, J. D.** *Annotated bibliography and index of geology of Botswana, 1967-1970.* Gaborone: Botswana. Geological Survey and Mines Dept., 1971. 70 pp.

793 *Bibliography and subject index of South African geology.*

See entry 966.

794 **Laughton, C. A., comp.** *Annotated bibliography and index to the geology of Botswana to 1966.* Gaborone: Botswana. Geological Survey Dept., 1967? 172 pp.

CLIMATOLOGY

795 **Hitchcock, R. K.** "Bibliography." In Symposium on drought in Botswana, Gaborone, Botswana, 1978. *Proceedings*, pp. 294-305. Edited by Madalon T. Hinchey. Worcester, MA: Botswana Society with Clark University Press, 1979.

796 **Schulze, B. R.** "South Africa."

See entry 975.

HUMAN GEOGRAPHY

797 **Crush, Jonathan S.** "Diffuse development: Botswana, Lesotho and Swaziland since independence." *A current bibliography on African affairs*, new series, 13:4 (1981):393-423.

A narrative summary and unannotated bibliography on these nation's social, political, and economic development.

MIGRATION

798 **Kerven, Carol.** *National migration study: bibliography.* Gaborone: Botswana. Central Statistics Office. Ministry of Agriculture. Rural Sociology Unit. National Migration Study, 1979. 17 leaves.

ANTHROPOLOGY

799 **Tobias, Phillip V., ed.** *The Bushmen: San hunters and herders of Southern Africa.* Cape Town: Human and Rousseau, 1978. 206 pp.

Emphasis on physical anthropology.

ECONOMIC DEVELOPMENT

800 **Crush, J. S.** *The post-colonial development of Botswana, Lesotho and Swaziland.* Monticello IL: Vance Bibliographies, 1978. (Public administration series bibliography P-138) 12 pp.

Unannotated.

801 *Quarterly economic review of Namibia, Botswana, Lesotho, Swaziland.*

See entry 921.

HISTORICAL GEOGRAPHY

802 **Stevens, Richard P.** *Historical dictionary of the Republic of Botswana.* Metuchen NJ: Scarecrow, 1975. (African historical dictionaries 5) 189 pp.

Entries cover major topics, events, and personalities in the nation's history; includes extensive unannotated bibliography.

ETHIOPIA

(as Abyssinia, a former colony of Italy)

Occupying the highest ground on the continent, Ethiopia has a long history of isolation punctuated by sporadic—and sometimes disastrous—contacts with the outside world. Three regions are recognized.

The Highland Plateau. The core of the nation is an elevated and deeply bisected tableland that in many places rises above the drylands surrounding it. The highest belt of land is called the *dega* and, at above 2,500 meters (8,200 feet), it supports wheat, barley, and grazing pasture. The lower slopes of the highlands form a temperate belt, the *woina dega*, between 1,500 and 2,500 meters (4,900 and 8,200 feet); here are most of the nation's scarce forests as well as croplands that provide grapes, citrus, cereals, and fodder for cattle and sheep. Despite sufficient water and good soils, the strong natural agricultural potential is undermined by poor roads and inefficient cultivation practices. Soil erosion is severe in many locations.

Eritrea and the Danakil. The Red Sea coast and the rift valleys immediately inland are among the most inhospitable lands on the earth. Most of the region is given over to nomadic cattle grazing with small pockets of cotton cultivation. Desertification risk is high and the effects of drought in 1969-1975 were severe enough to kill hundreds of thousands of people and drive even more into relief camps; as a source of food the land became totally exhausted.

The Danakil is a hyperarid depression formed by the slow spreading of two tectonic plates. Because this process usually occurs beneath the sea the region has provoked intense scientific interest as a natural mineral factory.

The Ogaden Desert. A huge triangular area forms the eastern quarter of the nation, an arid land both physiographically and culturally tied to neighboring Somalia. Again, as with Eritrea, nomadic cattle herders comprise the bulk of the population; also like Eritrea, overgrazing is severe and famine stalks the land. Sporadic warfare also erupts as the two nations vie for the region's control.

Dryland areas of ETHIOPIA

DESERTIFICATION RISK	ARIDITY									
	Hyperarid		Arid		Semiarid		Subhumid		Aridity Totals	
	km²	%	km²	%	km²	%	km²	%	km²	%
Very High	By definition, desertification does not exist in hyperarid regions.		72,000	7.2	25,000	2.5			97,000	9.7
High			325,000	32.5					325,000	32.5
Moderate					145,000	14.5	18,000	1.8	163,000	16.3
Desertification Totals			397,000	39.7	170,000	17.0	18,000	1.8	585,000	58.5
No Desertification	14,000	1.4			6,000	0.6	216,000	21.6	236,000	23.6
Total Drylands	14,000	1.4	397,000	39.7	176,000	17.6	234,000	23.4	821,000	82.1
Non-dryland									179,000	17.9
Total Area of the Territory									1,000,000	100.0

GENERAL BIBLIOGRAPHIES

803 Brown, Clifton F., comp. *Ethiopian perspectives: a bibliographical guide to the history of Ethiopia.* Westport CT: Greenwood, 1978. (African Bibliographic Center. Special bibliographic series; new series 5) 264 pp.

Not limited to history.

804 Darch, Colin. *A Soviet view of Africa: an annotated bibliography on Ethiopia, Somalia and Djibouti.* Boston MA: G. K. Hall, 1980. (Bibliographies and guides in African studies) 200 pp.

Annotated. Citations cover most subjects; the introduction is unusually informative.

Ethiopia/805-821

805 Hidaru, Alula, and Rahmato, Dessalegn. *A short guide to the study of Ethiopia: a general bibliography*. Westport CT: Greenwood, 1976. (African Bibliographic Center. Special bibliographic series; new series 2) 176 pp.

806 Kassahun, Checole. "Eritrea: a preliminary bibliography." *Africana journal* 6 (1975):303-314.

HANDBOOKS

807 Kaplan, Irving. *Area handbook for Ethiopia*. Rev. ed. Washington DC: U.S. Dept. of Defense, 1971. (American University foreign area studies) 543 pp.

Narrative summary of the nation's history, politics, geography, economy, social life; includes unannotated bibliography.

GEOGRAPHY

808 Reilly, P. M. *Ethiopia*. Surrey, U.K.: United Kingdom. Ministry of Overseas Development. Land Resources Division, 1978. (Land resource bibliography 10) 280 pp.

Unannotated. Covers agriculture, climatology, cultural studies, economics, geoscience, natural resources, soil science, water resources.

ATLASES

809 Mariam, Mesfin Wolde. *An atlas of Ethiopia*. Rev. ed. Asmare, Ethiopia: Il Poligrafico, 1970. 84 pp.

Scale of most maps at 1:2,000,000.

810 *National atlas of Ethiopia*. Addis Ababa: Ethiopian Mapping Agency, 1981. 92 leaves.

"Preliminary edition." Covers physical and cultural geography generally. Most maps at a base scale of 1:4,000,000. Narrative text includes statistical data.

GAZETTEERS

811 *Ethiopia, Eritrea and the Somalilands: official standard names approved by the United States Board on Geographic Names*. Washington DC: U.S. Board on Geographic Names, 1978. (Gazetteer 165) 521 pp.

List of 9,800 place-names with latitude/longitude coordinates. Covers Ethiopia, Jibuti and Somalia.

CLIMATOLOGY

812 Griffiths, J. F. "Ethiopian highlands." In *Climates of Africa*, pp. 369-388. Edited by J. F. Griffiths. Amsterdam, The Netherlands: Elsevier Publishing Co., 1972. (World survey of climatology 10).

A narrative summary of the nation's climate, with maps, charts, and an unannotated bibliography. For the Eritrean coast along the Red Sea, see entry 813.

813 Griffith, J. F. "The Horn of Africa."
See entry 929.

814 Grimes, Annie E. *Bibliography of climatic maps for Ethiopia, Eritrea, and the Somalilands*. Washington DC: U.S. Weather Bureau, 1959. (NTIS: PB 176-030; AD 665-179) 13 pp.

815 Kramer, H. P. "A selective annotated bibliography on the climatology of northeast Africa."
See entry 498.

BOTANY

816 Knapp, R. "Bibliographia phytosociologica et scientiae vegetationis: Aethiopia, Somalia, Sudan, Afar et Issa, Socotra" [bibliography of plant ecology and vegetation: Ethiopia, Somalia, Sudan, Afars and Issas, Socotra].
See entry 702.

SOILS

817 *Bibliography on desert soils of Egypt, Eritrea, Somaliland, Aden (1964-1931)*.
See entry 500.

RURAL GEOGRAPHY

818 Cohen, John M. *A select bibliography on rural Ethiopia*. Addis Ababa: Haile Sellassie I University Library, 1971. (Ethiopian bibliographical series 4) 82 pp.

ECONOMIC DEVELOPMENT

819 *Quarterly economic review of Uganda, Ethiopia, Somalia, Djibouti*.
See entry 1052.

HISTORICAL GEOGRAPHY

820 Marcus, Harold G. *The modern history of Ethiopia and the Horn of Africa: a select and annotated bibliography*. Stanford CA: Hoover Institution Press, 1972. 641 pp.

Extensively annotated. Covers Ethiopia, Jibuti, Somalia. The 2,042 citations are derived primarily from geographical journals from Europe and North America; emphasis is on historical geography, exploration and travel, military campaigns.

821 Prouty, Chris, and Rosenfeld, Eugene. *Historical dictionary of Ethiopia*. Metuchen NJ: Scarecrow, 1981. (African historical dictionaries 32) 436 pp.

Entries cover major topics, events, and personalities in the nation's history; includes extensive unannotated bibliography.

JIBUTI
The Republic of Jibuti

(as French Somaliland, the French Territory of Afars and Issas, and as Djibouti, a former colony of France)

Jibuti is a tiny arid nation surrounding an excellent harbor just beyond the southern end of the Red Sea. Because it serves as the natural port and railroad terminus for Ethiopia, both that nation and Somalia have designs on Jibuti. Its economy is based almost solely on its port functions with an additional small trade in leather and skins destined for France. The population is evenly divided between the pastoral Afar (with ties to Ethiopia) and more urbanized Issa (with ties to Somalia); most trade is in the hands of a community of Arabs and Asians.

The nation's prospects lie with French aid and technical assistance but also with reaching some compromise with Ethiopia and Somalia before warfare settles its fate.

Dryland areas of JIBUTI

DESERTIFICATION RISK	ARIDITY									
	Hyperarid		Arid		Semiarid		Subhumid		Aridity Totals	
	km²	%	km²	%	km²	%	km²	%	km²	%
Very High	By definition, desertification does not exist in hyperarid regions.									
High			23,000	100.0					23,000	100.0
Moderate										
Desertification Totals			23,000	100.0					23,000	100.0
No Desertification										
Total Drylands			23,000	100.0					23,000	100.0
Non-dryland										
Total Area of the Territory									23,000	100.0

GENERAL BIBLIOGRAPHIES

822 Darch, Colin. *A Soviet view of Africa: an annotated bibliography on Ethiopia, Somalia and Djibouti.*
See entry 804.

GAZETTEERS

823 *Ethiopia, Eritrea and the Somalilands: official standard names approved by the United States Board on Geographic Names.*
See entry 811.

GEOLOGY

824 *Bibliographie geologique et miniere (1930-1959) [geologic and mining bibliography, 1930-1959].* Djibuti: French Somaliland. Service des Travaux Publics, 1960.

CLIMATOLOGY

825 Griffith, J. F. "The Horn of Africa."
See entry 929.

826 Grimes, Annie E. *Bibliography of climatic maps for Ethiopia, Eritrea, and the Somalilands.*
See entry 814.

827 Kramer, H. P. "A selective annotated bibliography on the climatology of northeast Africa."
See entry 498.

828 Vitale, Charles S. *Annotated bibliography on the climate of French Somaliland.* Silver Spring MD: U.S. Weather Bureau, 1968. (NTIS: AD 680-446) 34 pp.
Contains 47 annotated entries.

Jibuti/829-832

BOTANY

829 **Knapp, R.** "Bibliographia phytosociologica et scientiae vegetationis: Aethiopia, Somalia, Sudan, Afar et Issa, Socotra" [bibliography of plant ecology and vegetation: Ethiopia, Somalia, Sudan, Afars and Issas, Socotra].

See entry 702.

ECONOMIC DEVELOPMENT

830 **Clarke, Walter Sheldon.** *A developmental bibliography for the Republic of Djibouti.* Djibouti: Clarke, 1979. 250 pp.

831 *Quarterly economic review of Uganda, Ethiopia, Somalia, Djibouti.*

See entry 1052.

HISTORICAL GEOGRAPHY

832 **Marcus, Harold G.** *The modern history of Ethiopia and the Horn of Africa: a select and annotated bibliography.*

See entry 820.

KENYA
Jamhuri ya Kenya/The Republic of Kenya

(as British East Africa, a former colony of the United Kingdom)

Kenya is the leading economic power in East Africa, blessed by some of the best agricultural land, a diversified economy, and a stable, conservative political climate.

The Highlands. This region is a well-watered volcanic plateau extending above 1,500 meters (4,900 feet) in elevation; it is subhumid or semiarid only on its lower slopes, escarpments, and rift valleys. Cattle are run on the largest portion of the Highlands but profitable crops of coffee, tea, pyrethrum, sisal, and various cereals, vegetables, and fruits make the region a remarkable African garden. Much of the land was brought into its present state of cultivation by white settlers from Britain who entered the area after a series of natural disasters (smallpox, rinderpest, drought, locust invasion) temporarily depopulated the eastern highlands after 1898. With the return of the native tribesmen conflicts between black and white were inevitable and increasingly violent. After the Mau Mau uprising of the early 1950's and the granting of independence in 1963 reforms in land tenure were implemented with some success.

The Lowland Savanna. Below the highland escarpments stretches a large, amorphous, semiarid savanna southeast to the Indian Ocean coast. Rainfall is erratic and population is sparse but the region offers some development potential. From the nation's seaport of Mombasa runs the main railway inland to Nairobi and beyond, so much of the savanna has access to transport routes and export markets. Mombasa itself is Kenya's second and East Africa's third largest city and has important manufacturing and entrepot functions.

The Northern Desert. Another ill-defined region extends across the northern tier of the nation from the vicinity of Lake Rudolf to the Somalia border. This arid land is poor in resources and periodically devastated by drought. Its peoples have little affinity with the tribes in the Highlands and share the same subsistence level of economy with those in neighboring Ethiopia and Somalia. Attempts to develop the area—for instance, encouraging a fishing industry on Lake Rudolf—have met with very limited success. The presence of nomadic Somali herdsmen along the border has encouraged Somalia to press for annexation of their territory.

Dryland areas of KENYA

DESERTIFICATION RISK	ARIDITY									
	Hyperarid		Arid		Semiarid		Subhumid		Aridity Totals	
	km²	%	km²	%	km²	%	km²	%	km²	%
Very High	By definition, desertification does not exist in hyperarid regions.		7,574	1.3					7,574	1.3
High			196,919	33.8			2,330	0.4	199,249	34.2
Moderate					217,892	37.4	29,713	5.1	247,605	42.5
Desertification Totals			204,493	35.1	217,892	37.4	32,043	5.5	454,428	78.0
No Desertification							52,434	9.0	52,434	9.0
Total Drylands			204,493	35.1	217,892	37.4	84,477	14.5	506,862	87.0
Non-dryland									75,738	13.0
Total Area of the Territory									582,600	100.0

GENERAL BIBLIOGRAPHIES

833 **Collison, Robert Lewis.** *Kenya.* Oxford, U.K.; Santa Barbara CA: Clio, 1982. (World bibliographical series 25) 157 pp.

Provides 596 annotated references covering most major subjects; especially good for geography, agriculture, sociology.

834 **Howell, John Bruce.** *East African community: subject guide to official publications.* Washington DC: U.S. Library of Congress, 1976. 272 pp.

Contains 1,812 semi-annotated entries for the period 1926 to 1974 covering the nations of Kenya, Tanzania, and Uganda.

835 Howell, John Bruce. *Kenya: subject guide to official publications.* Washington DC: U.S. Library of Congress, 1978. 423 pp.

Contains 3,048 semi-annotated entries for the period 1886 to 1975.

836 Webster, John B. et al, comps. *A bibliography on Kenya.* Syracuse NY: Syracuse University. Program of Eastern African Studies, 1967. (Eastern Africa bibliographical series 2) 461 pp.

Unannotated. Contains 7,210 items.

HANDBOOKS

837 Kaplan, Irving D. *Area handbook for Kenya* 2nd ed. Washington DC: U.S. Dept. of Defense, 1976. (American University foreign area studies) 472 pp.

Narrative summary of the nation's history, politics, geography, economy, social life; includes unannotated bibliography.

GEOGRAPHY

838 "Bibliography of Serengeti, Tanzania scientific publications."

See entry 1011.

839 Isaac, G. "Researches in the area formerly known as 'East Rudolf,' a commentary and classified bibliography." *Paleoecology of Africa and the surrounding islands and Antarctica* 9 (1976):109-123.

840 Lundgren, Bjorn, and Samuelson, Ann-Marie. *Land use in Kenya and Tanzania: a bibliography.* Stockholm: Royal College of Forestry. International Rural Development Division, 1975. 152 leaves.

841 Ulfstrand, Stattan. *Ecology in semi- arid east Africa: a selection of recent ecological references.* Stockholm: Swedish Natural Science Research Council, 1971. (Ecological Research Committee bulletin 11) 62 pp.

Covers Kenya, Tanzania, and Uganda. Subjects covered include soils, vegetation, animal life, land use, wildlife conservation.

ATLASES

842 *Atlas of Kenya.* Nairobi: Survey of Kenya, 1961. 46 leaves.

Base map scale 1:3,000,000.

843 *National atlas of Kenya.* 3rd ed. Nairobi: Survey of Kenya, 1970. 103 pp.

Base map scale 1:3,000,000.

844 Taylor, David Ruxton Fraser. *A computer atlas of Kenya, with a bibliography of computer mapping.* Ottawa: Carleton University. Dept. of Geography, 1971. 121 leaves.

Contains 64 computer maps based on 1962-1968 data.

GAZETTEERS

845 *Kenya: official standard names approved by the United States Board on Geographic Names.* 2nd ed. Washington DC: U.S. Board on Geographic Names, 1978. (Gazetteer 169) 470 pp.

List of 30,000 place-names with latitude/longitude coordinates.

846 Dosaj, N. P., and Walsh, J. *Bibliography of the geology of Kenya, 1859-1968.* Nairobi: Kenya. Geological Survey, 1970. (Bulletin 10) 65 pp.

CLIMATOLOGY

847 Griffiths, J. F. "Eastern Africa." In *Climates of Africa.* pp. 313-347. Edited by J. F. Griffiths. Amsterdam, The Netherlands: Elsevier Publishing Co., 1972. (World survey of climatology 10).

Covers Kenya, Uganda, Tanzania. A narrative summary with maps, charts and an unannotated bibliography.

BOTANY

848 Knapp, R. "Comentarii et bibliographia phytosociologica et scientiae vegetationis: Kenya, Tanzania, Uganda" [commentary and bibliography of plant ecology and vegetation: Kenya, Tanzania, Uganda]. *Excerpta botanica: sectio B: sociologica* 10:3 (1969):204-230; 14:1 (1974):1-16; 19:1 (1979):45-61.

Unannotated.

AGRICULTURAL ECONOMICS

849 Hailu, Alem Seged. *Rural development in African countries: a selected bibliography with special reference to Mali and Kenya.*

See entry 595.

850 McLoughlin, Peter F. M. *Research on agricultural development in East Africa.* New York: Agricultural Development Council, 1967. 112 pp.

Unannotated. Though old, includes much material originating in Africa. Covers Kenya, Tanzania, Uganda.

HUMAN GEOGRAPHY

851 Norgaard, Ole. *Kenya in the social sciences: an annotated bibliography, 1967-1979.* Nairobi: Kenya Literature Bureau, 1980.

PUBLIC HEALTH

852 Hartnig, Charles W. *Health and the social sciences: a selected bibliography with special reference to Kenya and East Africa.* Monticello IL: Council of Planning Librarians, 1976. (CPL exchange bibliography 996) 8 pp.

Unannotated.

NUTRITION

853 **Blankhart, D. M.** *Analytical documentation on human nutrition in Kenya: author's list, publications, and list of key indexing words.* Nairobi: University of Nairobi. Medical Research Centre. Nutrition Department, 1973. 28 leaves.

URBAN GEOGRAPHY

854 **Obudho, Robert A., and Obudho, Constance E.** *Urbanization, city, and regional planning of metropolitan Kisumu, Kenya: bibliographical survey of an East African city.* Monticello IL: Council of Planning librarians, 1972. (CPL exchange bibliography 278) 26 pp.

Unannotated. The city of Kisumu is located on the humid shore of eastern Lake Victoria, but the majority of citations refer to Kenya or East Africa generally.

ECONOMIC DEVELOPMENT

855 **Killick,. Tony.** *The economies of East Africa.* Boston MA: G. K. Hall, 1976. (Bibliographies and guides in African Studies) 150 pp.

English language publications from 1963 to 1974. Does not include unpublished theses or routine government documents, such as annual reports. Primarily the work of non-Africans. Covers Kenya, Tanzania, Uganda.

856 **Molnos, Angela** *Development in Africa: planning and implementation: a bibliography (1964-1969) and outline with some emphasis on Kenya, Tanzania, and Uganda.* Nairobi: East African Academy. Research Information Centre, 1970. (Information circular 3) 120 pp.

857 *Quarterly economic review of Kenya.* London: Economist Intelligence Unit, 1976-. Quarterly.

Includes summary of political and economic news as well as charts and statistics of selected economic indicators. Annual supplement.

TOURISM AND RECREATION

858 **Mascarenhas, Ophelia C.** "Tourism in East Africa: a bibliographical essay." *A current bibliography on African affairs*, new series 4:5 (1971):315-326.

Covers Kenya, Uganda, Tanzania. Considers the tourist industry, its organization, and present trends.

HISTORICAL GEOGRAPHY

859 **Ogot, Bethwell A.** *Historical dictionary of Kenya.* Metuchen NJ: Scarecrow, 1981. (African historical dictionaries 29) 279 pp.

Entries cover major topics, events, and personalities in the nation's history; includes extensive unannotated bibliography.

LESOTHO
The Kingdom of Lesotho
(as Basutoland, a former colony of the United Kingdom)

The tiny nation of Lesotho is an independent native enclave surrounded by South Africa. It sits high in the Drakensberg Mountains and, although its drylands are entirely subhumid, it has two distinct dryland regions.

The "Lowlands". Only low relative to the rest of the nation, this region is a rolling plateau of mesas, sandy soils, and severely degraded vegetation cover along the western border. The bulk of the population lives here, primarily as cultivators of maize and millet.

The Foothills. As the land rises eastward to the peaks of the Drakensberg it becomes steeper and more barren but also less exploited and hence less abused than the western area. Cattle raising predominates.

Lesotho's development rests on improving routine agricultural practices to halt the severe erosion experienced in the most productive parts of the nation. Crop yields have declined steadily, though winter wheat is exported to South Africa in exchange for maize. In addition to all its manufactured goods, Lesotho must import significant supplies of food as well.

Dryland areas of LESOTHO

DESERTIFICATION RISK	ARIDITY									
	Hyperarid		Arid		Semiarid		Subhumid		Aridity Totals	
	km²	%	km²	%	km²	%	km²	%	km²	%
Very High	By definition, desertification does not exist in hyperarid regions.									
High										
Moderate										
Desertification Totals										
No Desertification							19,842	65.4	19,842	65.4
Total Drylands							19,842	65.4	19,842	65.4
Non-dryland									10,498	34.6
Total Area of the Territory									30,340	100.0

GENERAL BIBLIOGRAPHIES

860   Balima, Mildred Grimes. *Botswana, Lesotho and Swaziland: a guide to official publications 1868-1968.*

See entry 784.

861   Willet, Shelagh M., and Ambrose, David P. *Lesotho: a comprehensive bibliography.* Santa Barbara CA: Clio, 1980. (World bibliographical series 3) 499 pp.

An impressive compilation of 2,562 annotated references covering most subjects.

HANDBOOKS

862   Ambrose, David P. *The guide to Lesotho.* 2nd ed. Johannesburg: Winchester, 1976. 370 pp.

A book-length narrative providing detailed information on geography, history, travel, and government affairs.

GEOGRAPHY

863   Hilty, Steven L., comp. *Draft environmental profile of The Kingdom of Lesotho.* Tucson AZ: University of Arizona. Office of Arid Lands Studies. Arid Lands Information Center, 1982. 121 pp.

A narrative summary of the nation's environment and development problems, with maps, charts, and an unannotated bibliography.

ATLASES

864   Ambrose, David P., and Perry, J. W. B. *Atlas for Lesotho.* 3rd ed. Johannesburg: Collins Longman Atlases, 1978. 75 pp.

Contains 3 pages of maps specifically covering Lesotho.

865   Talbot, A. M., and Talbot, W. J. *Atlas of the Union of South Africa/Atlas van die unie van Suid-Afrika.*

See entry 962.

GAZETTEERS

866 Leistner, O. A., and Morris, J. W. *Southern African place names.*

See entry 963.

867 *South Africa: official standard names approved by the United States Board on Geographic Names.*

See entry 965.

CLIMATOLOGY

868 Schulze, B. R. "South Africa."

See entry 975.

LAND TENURE

869 Eckert, Jerry Bruce. *Lesotho's land tenure: an analysis and annotated bibliography.* Madison WI: University of Wisconsin. Land Tenure Center, 1980.

HUMAN GEOGRAPHY

870 Crush, Jonathan S. "Diffuse development: Botswana, Lesotho and Swaziland since independence."

See entry 797.

ECONOMIC DEVELOPMENT

871 Crush, J. S. *The post-colonial development of Botswana, Lesotho and Swaziland.*

See entry 800.

872 *Quarterly economic review of Namibia, Botswana, Lesotho, Swaziland.*

See entry 921.

873 Wilken, Gene C., and Amiet, Carolyn F. *Bibliography for planning and development in Lesotho: a partially annotated bibliography of holdings and references developed for the Lesotho Agricultural Sector Analysis Project.* Ft. Collins CO: Colorado State University. Dept. of Economics, 1977. (Research report 1) 244 pp.

Also published in *African research and documentation* 18 (1978) and *International African bibliography* 9:4 (1979).

HISTORICAL GEOGRAPHY

874 Haliburton, Gordon MacKay. *Historical dictionary of Lesotho.* Metuchen NJ: Scarecrow, 1977. (African historical dictionaries 10) 223 pp.

Entries cover major topics, events, and personalities in the nation's history; includes extensive unannotated bibliography.

MALAWI
The Republic of Malawi
(as Nyassaland, a former colony of the United Kingdom)

Malawi is a narrow landlocked nation running along the length of the western shore of Lake Malawi and beyond to stop within 30 km (18 miles) of the Zambezi River. Its subhumid drylands are concentrated in the southern third of the country along the Shire River valley. Cotton and groundnuts are grown and contribute to a diverse agricultural economy. Ties with South Africa are strong, in part because a major Malawian export is manpower for the South African mining industry.

Dryland areas of MALAWI

DESERTIFICATION RISK	ARIDITY									
	Hyperarid		Arid		Semiarid		Subhumid		Aridity Totals	
	km²	%	km²	%	km²	%	km²	%	km²	%
Very High	By definition, desertification does not exist in hyperarid regions.									
High										
Moderate										
Desertification Totals										
No Desertification							25,496	27.1	25,496	27.1
Total Drylands							25,496	27.1	25,496	27.1
Non-dryland									68,586	72.9
Total Area of the Territory									94,082	100.0

GENERAL BIBLIOGRAPHIES

875 **Brown, Edward E. et al.** *A bibliography of Malawi.* Syracuse NY: Syracuse University. Program of Eastern African Studies, 1965. (East Africa bibliographical series 1) 161 pp.

Unannotated. Does not include government publications.

876 *National register of research publications, 1965-1975.* Lilongwe: Malawi. National Research Council, 1976. 43 leaves.

877 **Walker, Audrey A.** *The Rhodesias and Nyasaland: a guide to official publications.* Washington DC: U.S. Library of Congress, 1965. 285 pp.

Contains 1,889 entries from 1889 to 1963. Includes publications of African and British governments as well as inter-territorial agencies. Covers Malawi, Zambia, Zimbabwe.

HANDBOOKS

878 **Nelson, Harold D.** *Area handbook for Malawi.* Washington DC: U.S. Dept. of Defense, 1975. (American University foreign area studies) 353 pp.

Narrative summary of the nation's history, politics, geography, economy, social life; includes unannotated bibliography.

GEOGRAPHY

879 **Varady, Robert G., comp.** *Draft environmental profile of Malawi.* Tucson AZ: University of Arizona. Office of Arid Lands Studies. Arid Lands Information Center, 1982. 196 pp.

A narrative summary of the nation's environment and development problems, with maps, charts, and an unannotated bibliography.

ATLASES

880 Agnew, Swanzie, and Stubbs, Michael, eds. *Malawi in maps.* London: University of London Press, 1972. 143 pp.

Standard scale 1:50,000.

GAZETTEERS

881 *Malawi: official standard names approved by the United States Board on Geographic Names.* Washington DC: U.S. Board on Geographic Names, 1970. (Gazetteer 113) 161 pp.

List of 10,200 place-names with latitude/longitude coordinates.

CLIMATOLOGY

882 Torrance, J. D. "Malawi, Rhodesia and Zambia." In *Climates of Africa*, pp. 409-460. Edited by J. F. Griffiths. Amsterdam, The Netherlands: Elsevier Publishing Co., 1972. (World survey of climatology 10).

A narrative summary of these nations' climates, with maps, charts, and an unannotated bibliography.

SOILS

883 *Soils of Malawi, Rhodesia and Zambia (annotated bibliography covering the published literature for 1930-1975).* Harpenden, U.K.: Commonwealth Agricultural Bureaux, 1977. (Annotated bibliography) 13 pp.

RURAL GEOGRAPHY

884 Turner, P. V. "Rural Malawi: an approach to a current bibliography." *Rural africana* 21 (1973):59-66.

Bibliographical essay.

ECONOMIC DEVELOPMENT

885 *An interim bibliography of development in Malawi.* Limbe, Malawi: University of Malawi. Chancellor College. Library, 1972? 26 pp.

886 *Quarterly economic review of Zimbabwe, Malawi.* London: Economist Intelligence Unit, 1980-. Quarterly.

Includes summary of political and economic news as well as charts and statistics of selected economic indicators. Annual supplement.

HISTORICAL GEOGRAPHY

887 Crosby, Cynthia A. *Historical dictionary of Malawi.* Metuchen NJ: Scarecrow, 1980. (African historical dictionaries 25) 169 pp.

Entries cover major topics, events, and personalities in the nation's history; includes extensive unannotated bibliography.

MOZAMBIQUE
The People's Republic of Mozambique
(as Moçambique, a former colony of Portugal)

Mozambique is a former Portuguese colony on the southeastern coast of Africa opposite the island of Madagascar. Most of the nation lies below the great escarpment that defines the plateau of the region; Mozambique occupies the widest section of coastal plain in southern Africa and serves as a remarkable natural outlet for five other nations (Malawi, Zambia, Zimbabwe, South Africa's Transvaal, and Swaziland).

The South. Here the coastal plain is widest. The land is subhumid, becoming semiarid toward the Zimbabwe border, with summer precipitation and high humidity all year. The vegetation is scrub savanna, often sparse due to drought and overgrazing. Soils are best developed along the rivers that cross the plain (the Limpopo, the Sabi) and between dunes near the coast. The Portuguese succeeded Arab traders as the major outside influence, establishing a capital at Maputo (formerly Lorenco Marques). This site serves as a port and principal rail-link between Johannesburg, South Africa, and the sea. A similar function is performed at Beira for the rail-line into Zimbabwe. These relationships with the powerful inland economies have given Mozambique strong and prosperous urban centers with large markets for local agricultural products. Much of this prosperity, however, was based on a strict and often cruel forced labor program in cooperation with South African mining companies. Alternative—and less drastic—development prospects rest with the lower Limpopo valley irrigation scheme and hydropower exploitation on the Zambezi.

The Zambezi Valley. Mozambique occupies a long stretch of the lower Zambezi, the legacy of intrepid Portuguese explorers. The upper portion of the nation's share of the river is semiarid, becoming wetter downstream to exceed the subhumid zone at the river's delta. The region holds great promise for development; in addition to the Cabara Bassa Dam (whose hydropower will be sold to South Africa) deposits of coal and iron ore have been located.

The North. This region is large, populous, less dry, and much less developed than the Zambezi Valley and the South. The Portuguese wrecked havoc on the local peasant village economy by forcing farmers to plant cotton in place of traditional subsistence crops. Even when famine occurred the authorities used force to maintain production. Not surprisingly, this region became the center of the Frelimo revolt that eventually brought Portuguese rule to an end. The war forced thousands off the land into temporary exile in Malawi and Tanzania. Development efforts must concentrate here, especially in rebuilding subsistence agriculture. There is great potential in locating mineral reserves, improving the transportation system, and constructing a modern port at Nacala.

Dryland areas of MOZAMBIQUE

DESERTIFICATION RISK	Hyperarid		Arid		Semiarid		Subhumid		Aridity Totals	
	km²	%	km²	%	km²	%	km²	%	km²	%
Very High	By definition, desertification does not exist in hyperarid regions.									
High										
Moderate					109,895	14.0			109,895	14.0
Desertification Totals					109,895	14.0			109,895	14.0
No Desertification							524,354	66.8	524,354	66.8
Total Drylands					109,895	14.0	524,354	66.8	634,249	80.8
Non-dryland									150,712	19.2
Total Area of the Territory									784,961	100.0

GENERAL BIBLIOGRAPHIES

888 **Chonchol, Maria-Edy.** *Guide bibliographique du Mozambique: environnement naturel, developpement et organisation villageoise [bibliographic guide for Mozambique: natural environment, development and village organization].* Paris: L'Harmattan, 1979. 135 pp.

In French. A good narrative bibliography covering sources in several languages on the nation's natural environment, economic development, and social organization.

889 **Gibson, Mary Jane.** *Portuguese Africa: a guide to official publications.*

See entry 770.

890 **Zubatsky, David S.** *A guide to resources in United States libraries and archives for the study of Cape Verde, Guinea (Bissau), Sao Tome-Principe, Angola and Mozambique.*

See entry 262.

HANDBOOKS

891 **Kaplan, Irving.** *Area handbook for Mozambique.* 2nd ed. Washington DC: U.S. Dept. of Defense, 1977. (American University foreign area studies) 240 pp.

Narrative summary of the nation's history, politics, geography, economy, social life; includes unannotated bibliography.

ATLASES

892 *Atlas de Mocambique [atlas of Mozambique].* Lourenco Marques: Mozambique. Direcao dos Servicos de Agrimensura, 1962. 43 pp.

Scale 1:1,000,000 or 1:6,000,000.

GAZETTEERS

893 *Mozambique: official standard names approved by the United States Board on Geographic Names.* Washington DC: U.S. Board on Geographic Names, 1969. (Gazetteer 109) 505 pp.

List of 32,500 place-names with latitude/longitude coordinates.

BOTANY

894 **Pinto da Silva, A. R., and Teles, A. N.** "Bibliographia phytosociologica Portugaliae" [bibliography of the plant ecology of Portugal].

See entry 191.

895 **Teles, A. N.** "Bibliographie des cartes de vegetation du Portugal" [bibliography of vegetation maps of Portugal].

See entry 192.

CLIMATOLOGY

896 **Griffiths, J. F.** "Mozambique." In *Climates of Africa*, pp. 389-408. Edited by J. F. Griffiths. Amsterdam, The Netherlands: Elsevier Publishing Co., 1972. (World survey of climatology 10).

A narrative summary of the nation's climate, with maps, charts, and an unannotated bibliography.

AGRICULTURE

897 *Boletim bibliografico/Bibliographic bulletin.* Lourenco Marques: Instituto de Investigacao Agronomica de Mocambique. Centro de Documentacao Agraria, 1973-. Freq. unknown.

898 *Mocambique: food and agriculture sector: preliminary edition.* Uppsala, Sweden: Swedish University of Agriculture, Forestry and Veterinary Medicine. International Rural Development Division, 1977. 65 pp.

HUMAN GEOGRAPHY

899 **Enevoldsen, Thyge, and Johnsen, Vibe.** *A political, economic and social bibliography on Mocambique with main emphasis on the period 1965-1978.* Copenhagen: Centre for Development Research, 1978. 60 pp.

Annotated list stressing political and economic development.

ECONOMIC DEVELOPMENT

900 *Quarterly economic review of Tanzania, Mozambique.*

See entry 1032.

NAMIBIA

(as German Southwest Africa, then Southwest Africa/Suidwes Afrika, a former colony of Germany and then a mandated territory of South Africa)

Namibia is a huge, sparsely-populated, dryland territory that continues to lack a clearly defined political administration after over 60 years of South African occupation.

The Namib Desert. Along the so-called Skeleton Coast the cold Benguela current produces a classic fog desert with almost no precipitation but frequent invasions of cool, moist air. Plant and animal life show unusual adaptations to this environment. Resources are rich along the coast: sardines and pilchards are the base for an extensive fishing industry, while incredible—and still unexplained—deposits of diamonds are excavated from ancient dune sands. Labor for the mines is handled through strict contract arrangements with the native tribes.

The Central Plateau. This region of arid to semiarid savanna holds the bulk of the agricultural population, including extensive white-owned plantation farms. Cattle and Karakul sheep are the principal livestock; meat and dairy products are exported. The land, however, is harsh enough to require importation of subsistence cereals in dry years. Mineral resources here are remarkable even by South African standards: copper, lead, tin, uranium, zinc, and many other precious metals are produced with the help of cheap contract labor.

The Kalahari Desert. The eastern slopes of the plateau gradually give way to the lower, semiarid, thornscrub plain of the Kalahari basin. Plantation farming is limited to a few irrigated valleys; most of the land is given over to cattle raising, often at the bare subsistence level in the black homelands. The region lacks the mineral wealth ubiquitous to the west.

A note on the black homelands. Together with South Africa, Namibia has a system of social, political, and economic division based on the assumption that native blacks are best treated by segregation into landlocked enclaves roughly corresponding to their traditional tribal territories. These homelands are carefully defined to exclude the most desirable mineral and agricultural lands, leaving the homeland population vulnerable to economic exploitation by neighboring white settlers and commercial interests. In Namibia the ten homelands are concentrated along the border with Angola, in the Caprivi Strip and the Botswana border, and in disconnected areas of the Central Plateau.

Dryland areas of NAMIBIA

DESERTIFICATION RISK	ARIDITY								Aridity Totals	
	Hyperarid		Arid		Semiarid		Subhumid			
	km²	%	km²	%	km²	%	km²	%	km²	%
Very High	By definition, desertification does not exist in hyperarid regions.		113,439	13.8	44,389	5.4			157,828	19.2
High			218,657	26.6	69,872	8.5			288,529	35.1
Moderate					247,428	30.1	29,593	3.6	277,021	33.7
Desertification Totals			332,096	40.4	361,689	44.0	29,593	3.6	723,378	88.0
No Desertification	89,600	10.9					9,043	1.1	98,643	12.0
Total Drylands	89,600	10.9	332,096	40.4	361,689	44.0	38,636	4.7	822,021	100.0
Non-dryland										
Total Area of the Territory									822,021	100.0

GENERAL BIBLIOGRAPHIES

901 Bridgman, Jon, and Clarke, David G. *German Africa: a select annotated bibliography.*

See entry 450.

902 Roukens de Lande, E. J. *South West Africa, 1946-1960: a selective bibliography.* Cape Town: University of Cape Town. School of Librarianship, 1961. 51 pp.

Contains 322 references. Continues Welch, *South West Africa: a bibliography, 1919-1946* (entry 904).

903 Vogt, Martin. "Bibliographical aids for studies on South West Africa." *Basler Afrika Bibliographien Mitteilungen* 13 (1975):21-32.

Contains 68 titles with annotations.

904 Welch, Florette Jean. *South-West Africa: a bibliography, 1919-1946.* 2nd ed. Cape Town: University of Cape Town. School of Librarianship, 1967. (Bibliographical series) 41 pp.

Contains 343 references. Continued by Roukens de Lange, *South West Africa, 1946-1960: a selective bibliography* (entry 902).

GEOGRAPHY

905 Botha, Laurette Isabella. *The Namib desert: a bibliography.* Cape Town: University of Cape Town Libraries, 1970. 26 pp.

906 Logan, Richard F. *Bibliography of South West Africa: geography and related fields.* Windhoek: South West Africa Scientific Society, 1969. (Scientific research in South West Africa 8) 152 pp.

Contains 2,000 annotated entries, coverage through 1966.

GAZETTEERS

907 Leistner, O. A., and Morris, J. W. *Southern African place names.*

See entry 963.

908 *South Africa: official standard names approved by the United States Board on Geographic Names.*

See entry 965.

GEOLOGY

909 *Bibliography and subject index of South African geology.*

See entry 966.

910 *Bibliography of geology of South West Africa.* Windhoek: South West Africa. Geological Survey, n.d.

911 Martin, Henno A., comp. *A bibliography of geological and allied subjects, South West Africa.* Capetown: University of Capetown. Chamber of Mines. Precambrian Research Unit, 1965. (Bulletin 1) 87 pp.

Covers works to 1961.

912 Ridge, John Drew. "South West Africa." In *Annotated bibliographies of mineral deposits in Africa, Asia (exclusive of the USSR) and Australasia,* pp 163-170. Oxford U.K.: Pergamon, 1976.

CLIMATOLOGY

913 Schulze, B. R. "South Africa."

See entry 975.

HYDROLOGY

914 Stengel, H. W. von. *Bibliographie Wasserwirtschaft in Suedwestafrika/Bibliography of water management in South West Africa.* Basel, Switzerland: Basler Afrika Bibliographien, 1974. (Mitteilungen 10) 65 pp.

SOILS

915 Ganssen, Robert. *Suedwest-Afrika, Boden und Bodenkulture [South West Africa, soils and soil cultivation].* Berlin, Federal Republic of Germany: Verlag von Dietrich Reimer, 1963. 160 pp.

In German. Bibliography: pp. 125-127, 46 references.

ZOOLOGY

916 Haacke, W. D. "Selected publications on the herpetology of the Namib desert." *Namib bulletin* 1977. pp. 12-13.

Transvaal Museum Bulletin, supplement 2.

HUMAN GEOGRAPHY

917 Fontolliet, Micheline. *Bibliography on Namibia.* Geneva, Switzerland: Lutheran World Federation. Documentation Office, 1976-77. 4 vols.

Includes 1,032 annotated references emphasizing politics and economics.

918 Schoeman, Elna. *The Namibian issue, 1920-1980: a select and annotated bibliography.* Boston MA: G. K. Hall, 1982. 247 pp.

Includes 1,489 annotated references emphasizing politics and international law; economics, sociology, military affairs are also covered.

ANTHROPOLOGY

919 Strohmeyer, Eckhard, and Moritz, Walter. *Umfassende bibliographie der Volker Namibiens (Sudwestafrikas) und Sudwestangolas [comprehensive bibliography of the people of Namibia (South West Africa) and south west Angola].* Starnberg, Federal Republic of Germany: Max-Planck Institute, 1975- .

To be completed in several volumes. Covers ethnography, language, prehistory. Chapter titles are in German, English, and French.

920 Tobias, Phillip V., ed. *The Bushmen: San hunters and herders of Southern Africa.*

See entry 799.

ECONOMIC DEVELOPMENT

921 *Quarterly economic review of Namibia, Botswana, Lesotho, Swaziland.* London: Economist Intelligence Unit, 1982- . Quarterly.

Includes summary of political and economic news as well as charts and statistics of selected economic indicators. Annual supplement.

RWANDA
The Rwandese Republic
(a former colony of Germany, then Belgium)

Rwanda—a tiny nation between Zaire, Uganda, Tanzania, and Burundi—has only a sliver of subhumid dryland along the Kagera River on its eastern edge. The region is swampy and sparsely populated (in contrast to areas to the west which rank among Africa's highest population densities); the Kagera National Park is a significant nature preserve.

Only selected reference sources are provided.

Dryland areas of RWANDA

DESERTIFICATION RISK	ARIDITY								Aridity Totals	
	Hyperarid		Arid		Semiarid		Subhumid			
	km²	%	km²	%	km²	%	km²	%	km²	%
Very High	By definition, desertification does not exist in hyperarid regions.									
High										
Moderate										
Desertification Totals										
No Desertification								5,766	21.9	5,766	21.9
Total Drylands								5,766	21.9	5,766	21.9
Non-dryland										20,564	78.1
Total Area of the Territory										26,330	100.0

GEOGRAPHY

922 **Arnould, Eric, comp.** *Draft environmental profile of Rwanda*. Tucson AZ: University of Arizona. Office of Arid Lands Studies. Arid lands Information Center, 1981. 180 pp.

A narrative summary of the nation's environment and development problems, with maps, charts, and an unannotated bibliography.

GAZETTEERS

923 *Rwanda: official standard names approved by the United States Board on Geographic Names*. Washington DC: U.S. Board on Geographic Names, 1964. (Gazetteer 85) 44 pp.

List of 3,000 place-names with latitude/longitude coordinates.

SOMALIA
Al-Jumhouriya As-Somaliya Al-Domocradia
[the Somali Democratic Republic]

(as British and Italian Somaliland, former colonies of the United Kingdom and Italy)

Somalia forms the Horn of Africa, the great eastward projection of the continent into the Indian Ocean. It is topographically diverse with a full range of dryland zones and a long seacoast. Of international importance are Somali claims on lands in nearby Jibuti, Ethiopia, and Kenya, all of which are economically important to Somali pastoralists.

The Northern Mountains. This region was formerly occupied by the British as British Somaliland. Rugged mountains rise high enough above the hyperarid north coast to capture moisture and create small pockets of land with subhumid climates and perennial water supplies. The area is given over entirely to nomadic pastoralism, with sheep, goats, and camels finding an export market in nearby Arabia. The ancient port of Berbera is the region's primary trading center and has military importance as a strategic site near the entrance to the Red Sea.

The Central and Southern Plateau. South of the mountain zone a long monotonous low plateau of thorn-scrub desert stretches to the Kenya border. Most of the land supports traditional Somali herds but along the nation's two perennial streams—the Giuba and Shebelle Rivers—irrigated agriculture flourishes. Sorghum, maize, and cassava are the staple crops, while sugar and bananas are exported. This favored area forms the hinterland of the national capital and port of Mogadishu.

Dryland areas of SOMALIA

DESERTIFICATION RISK	ARIDITY									
	Hyperarid		Arid		Semiarid		Subhumid		Aridity Totals	
	km²	%	km²	%	km²	%	km²	%	km²	%
Very High	By definition, desertification does not exist in hyperarid regions.		45,360	7.2	33,390	5.3			78,750	12.5
High			379,260	60.2			9,450	1.5	388,710	61.7
Moderate					146,160	23.2	3,780	0.6	149,940	23.8
Desertification Totals			424,620	67.4	179,550	28.5	13,230	2.1	617,400	98.0
No Desertification	12,600	2.0							12,600	2.0
Total Drylands	12,600	2.0	424,620	67.4	179,550	28.5	13,230	2.1	630,000	100.0
Non-dryland										
Total Area of the Territory									630,000	100.0

GENERAL BIBLIOGRAPHIES

924 Darch, Colin. *A Soviet view of Africa: an annotated bibliography on Ethiopia, Somalia, and Djibouti.*

See entry 804.

925 Salad, Mohamed Khalief, comp. *Somalia: a bibliographical survey.* Westport CT: Greenwood Press, 1977. (African Bibliographic Center. Special bibliographic series: n.s. 4) 468 pp.

HANDBOOKS

926 Kaplan, Irving. *Area handbook for Somalia* 2nd ed. Washington DC: U.S. Dept. of Defense, 1977. (American University foreign area studies) 392 pp.

Narrative summary of the nation's history, politics, geography, economy, social life; includes unannotated bibliography.

GAZETTEERS

927 *Ethiopia, Eritrea and the Somalilands: official standard names approved by the United States Board on Geographic Names.*

See entry 811.

Somalia/928-937

GEOLOGY

928 Puri, R. K. *Bibliography relating to geology, mineral resources, paleontology, etc., of Somali Republic.* Hargeisa: Somali Republic, 1961. (Survey report RKP/1) 13 leaves.

Unannotated.

CLIMATOLOGY

929 Griffiths, J. F. "The Horn of Africa." In *Climates of Africa*, pp 133-165. Edited by J. F. Griffiths. Amsterdam, The Netherlands: Elsevier Publishing Co., 1972. (World survey of climatology 10).

Covers Somalia, Jibuti, and the Eritrean coast of Ethiopia. A narrative summary with maps, charts, and an unannotated bibliography.

930 Grimes, Annie E. *Bibliography of climatic maps for Ethiopia, Eritrea, and the Somalilands.*

See entry 814.

931 Kramer, H. P. "A selective annotated bibliography on the climatology of northeast Africa."

See entry 498.

932 Vitale, Charles S. *Bibliography of the climate of the Somali Republic.* Washington DC: U.S. Weather Bureau, 1967. (NTIS: AD 670-048; PB 176-237) 60 pp.

SOILS

933 *Bibliography on desert soils of Egypt, Eritrea, Somaliland, Aden (1964-1931).*

See entry 500.

BOTANY

934 Knapp, R. "Bibliographia phytosociologica et scientiae vegetationis: Aethiopia, Somalia, Sudan, Afar et Issa, Socotra" [bibliography of plant ecology and vegetation: Ethiopia, Somalia, Sudan, Afars and Issas, Socotra].

See entry 702.

ECONOMIC DEVELOPMENT

935 *Quarterly economic review of Uganda, Ethiopia, Somalia, Djibuti.*

See entry 1052.

HISTORICAL GEOGRAPHY

936 Castagno, Margaret. *Historical dictionary of Somalia.* Metuchen NJ: Scarecrow, 1975. (African historical dictionaries 6) 213 pp.

Entries cover major topics, events, and personalities in the nation's history; includes extensive unannotated bibliography.

937 Marcus, Harold G. *The modern history of Ethiopia and the Horn of Africa: a select and annotated bibliography.*

See entry 820.

SOUTH AFRICA
Republiek van Suid-Afrika
(as the Union of South Africa, a former colony of the United Kingdom)

The Republic of South Africa is Africa's major economic and industrial power. However, despite its immense natural wealth and the extraordinary standard of living it has provided white citizens, inequality of opportunity for the black majority still places large sections of the country into the category of a developing nation. The physical, ethnic, and political geography of the nation is complex; three dryland regions are identified here.

The High Veld. This semiarid to subhumid upland plateau is the geographical and economic heartland, centered largely on the basin of the Vaal River. The most productive lands are given over to cattle, sheep, dairying, and maize; irrigated crops include tobacco, cotton, and citrus. Although the white-owned farms are extremely productive, the real wealth of the region lies beneath the soil: deposits of gold, diamonds, copper, coal, chromium, and other ores make South Africa an extraordinary treasurehouse. Upon this powerful base rest the major manufacturing industries, dominated by steel, foods, textiles, and chemicals. Labor for these enterprises comes from native blacks and from contract workers from Malawi and Mozambique, while white managers and professionals have brought great prosperity to the cities of Johannesburg and Pretoria.

The Northwest Desert. The Orange River, once it passes beyond the Drakensburg Mountains, flows generally westward through an increasingly arid land. Although this is the natural hinterland of the great port of Cape Town, development here has not kept pace with that of the High Veld. Agriculture is limited to extensive sheep and goat-raising, and to stretches of irrigation along the Orange.

The Kalahari Desert. Along the northern border with Botswana the Kalahari extends into South African territory. Agriculture here alternates between white-owned cattle ranching and black-homeland subsistence farming.

A note on the black homelands. As with Namibia, South Africa has a system of social, political, and economic division based on the assumption that native blacks are best treated by segregation into enclaves roughly corresponding to their traditional tribal territories. These homelands are carefully defined to exclude the most desirable mineral and agricultural lands, leaving the homeland populations vulnerable to economic exploitation by neighboring white settlers and commercial interests. In South Africa four homelands (Transkei, Venda, BophuthaTswana, and Ciskei) have been granted "independence," meaning, in practice, internal self-government only. An economic development plan that designates border areas for favored industrial growth holds some promise in extending South Africa's prosperity to its black populations; whether such a scheme will succeed before protest and violence destroy the homeland plan remains to be seen.

Dryland areas of SOUTH AFRICA

DESERTIFICATION RISK	ARIDITY									
	Hyperarid		Arid		Semiarid		Subhumid		Aridity Totals	
	km²	%	km²	%	km²	%	km²	%	km²	%
Very High	By definition, desertification does not exist in hyperarid regions.		43,733	3.6					43,733	3.6
High			303,527	24.8	183,236	15.0	38,593	3.2	525,356	43.0
Moderate			48,234	3.9	118,205	9.7	6,454	0.5	172,893	14.1
Desertification Totals			395,494	32.3	303,441	24.7	45,047	3.7	741,982	60.7
No Desertification	6,269	0.5			7,409	0.6	431,642	35.3	445,320	36.4
Total Drylands	6,269	0.5	395,494	32.3	308,850	25.3	476,689	39.0	1,187,302	97.1
Non-dryland									34,864	2.9
Total Area of the Territory									1,222,166	100.0

South Africa/938-955

GENERAL BIBLIOGRAPHIES

938 **Kalley, Jacqueline Audrey.** *The Transkei region of South Africa, 1877-1978: an annotated bibliography.* Boston MA: G. K. Hall, 1980. (Bibliographies and guides in African studies) 218 pp.

Contains 1,439 items. Includes government publications.

939 **Kotze, D. A.** *Bibliography of official publications of the Black South African homelands.* Pretoria: University of South Africa, 1979. (Documenta 19) 80 pp.

940 *KWIC index of research bulletins.* Pretoria: Human Sciences Research Council, 1971-.

"Gives an alphabetical list of keywords of projects and of research workers and research organizations." (Musiker, *South Africa*)

941 **Musiker, Reuben.** *South Africa.* Oxford, U.K.; Santa Barbara CA: Clio, 1979. (World bibliographical series 7) 194 pp.

An excellent guide by a respected South African bibliographer; 577 major references with many more noted in the annotations.

942 *National register of research projects: natural sciences.* Pretoria: South Africa. Office of the Scientific Adviser to the Prime Minister, 1971-1973. 3 vols.

"Covers current research projects being conducted at research institutions, universities, government departments, and the private sector, in the fields of physical and biological sciences, engineering, medicine, and agriculture. For each project the book gives field of study, title of project, names of leaders, co-investigators and technicians, year of commencement and expected duration." (Musiker, *South Africa*)

THESES AND DISSERTATIONS

943 **Malan, Stephanus Immelman.** *Union catalogue of theses and dissertations of the South African universities, 1942-1958.* Potchefstroom, South Africa: Potchefstroom University for Christian Higher Education, 1959.

944 **Robinson, Anthony, and Lewin, Meredith.** *Catalogue of theses and dissertations accepted for degrees by the South African universities, 1918-1941.* Capetown: The Author, 1943. 155 pp.

945 *Union catalogue of theses and dissertations of the South African universities.* Potchefstroom, South Africa: Potchefstroom University for Christian Higher Education. Ferdinand Postma Library, 1959-. Annual.

HANDBOOKS

946 **Kaplan, Irving.** *Area handbook for the Republic of South Africa.* Washington DC: U.S. Dept. of Defense, 1971. (American University foreign area studies) 845 pp.

Narrative summary of the nation's history, politics, geography, economy, social life; includes unannotated bibliography.

947 *South Africa, 1980-1981: official yearbook of the Republic of South Africa.* 7th ed. Johannesburg: Chris van Rensburg, 1981. 975 pp.

"A statistical yearbook and general work of reference on the Republic of South Africa..." (Musiker, *South Africa*)

DIRECTORIES

948 *Directory of scientific research organizations in South Africa.* Pretoria: South African Council for Scientific and Industrial Research, 1967-. Annual.

949 **Herbst, J. F., and Milligan, G. A., eds.** *South Africa and Rhodesia.* 2nd ed. Guernsey, U.K.: Francis Hodgson, 1975. (Guide to world science 20) 204 pp.

Provides descriptions of scientific research activities and organizations.

950 *Scientific and technical societies in South Africa/Wetenskaplike en tegniese verenigings in Suid-Afrika.* Pretoria: South African Council for Scientific and Industrial Research, 1967-. Annual.

In English and Afrikaans.

STATISTICS

951 **Naude, M. H.** *A guide to statistical sources in the Republic of South Africa.* Pretoria: South Africa. Bureau of Market Research, 1972. 214 pp.

952 *Quarterly bulletin of statistics - South African Reserve Bank/Statistiese Kwartaalblad - Suid-Afrikaanse Reserwebank.* Pretoria: South African Reserve Bank, 194_-1965. 78 nos.

In Afrikaans and English. "The current official source of statistical information..." (Musiker, *South Africa*)

953 *South Africa statistics/Suid-Afrikanse statistieke.* Pretoria: South Africa. Dept. of Statistics, 1968-. Every 2 years.

"This work contains statistical tables for the following subjects: population, migration, vital statistics, health, education, social security, judicial statistics, labour, prices, agriculture, fisheries, mining, industry, trade, transport, communication, public finance, towns, currency, banking, and general finance. The work does not give descriptive information." (Musiker, *South Africa*). In English and Afrikaans.

954 *Southern Africa data.* Pretoria: Africa Institute, 1969-. Looseleaf.

"Gives statistics and facts on health, education, population, transport, etc." (Musiker, *South Africa*)

955 *Statistiese corsig ven swart ontwikkeling/Statistical survey of black development.* Pretorio: Benso, 19?-. Annual.

In Afrikaans and English. Benso = Buro vir Ekonomiese Navorsing, Samewerking en Ontwikkeling. (Bureau for Economic Research, Cooperation and Development).

GEOGRAPHY

956 de Graff, Gerrit. "Republic of South Africa." In *Handbook of contemporary developments in world ecology*, pp. 485-510. Westport CT: Greenwood, 1981.

A narrative summary of ecological research in South Africa, with an unannotated list of references.

957 *The independent Venda*. Pretoria: Benso, 1979. 197 pp.

A general narrative guide to the territory and its government prepared by the staff of the Bureau for Economic Research: Cooperation and Development (Benso) and the Institute for Development Studies, Rand Afrikaans University. Venda is a black homeland located in the northeast corner of the Transvaal Province of South Africa, near the border with Zimbabwe.

958 Nuttonson, Michael Y. *The physical environment and agriculture of the Union of South Africa with special reference to its winter-rainfall regions containing areas climatically and latitudinally analogous to Israel*. Washington DC: American Institute of Crop Ecology, 1961. 459 pp.

Bibliography: pp. 452-459.

DESERTIFICATION

959 Wilcocks, Julia Ruth Nadene. *Desert encroachment in South Africa*. Johannesburg: University of the Witwatersrand. Dept. of Geography and Environmental Studies, 1977. 349 pp.

Includes annotated bibliography of about 500 items. Continued by: J. R. N. Wilcocks and J. A. Nijlard, *Bibliography to update dissertation on desert encroachment in South Africa* (Johannesburg: University of the Witwatersrand, 1978), 18 pp.

MAP COLLECTIONS

960 Merrett, Christopher Edward. *A selected bibliography of Natal maps, 1800-1977*. Boston MA: G. K. Hall, 1979. (Bibliographies and guides in African studies) 226 pp.

ATLASES

961 *Ontwikkelingsatlas/Development atlas*. Pretoria: South Africa. Dept. of Planning, 1966-. Looseleaf.

962 Talbot, A. M., and Talbot, W. J. *Atlas of the Union of South Africa/Atlas van die unie van Suid-Afrika*. Pretoria: South Africa. Government Printer, 1960. 178 pp.

Includes Swaziland and Lesotho. In English and Afrikaans. Standard scale 1:4,000,000.

GAZETTEERS

963 Leistner, O. A., and Morris, J. W. *Southern African place names*. Grahamstown, South Africa: Cape Provincial Museums, 1976. (Annals 12) 565 pp.

A gazetteer of some 42,000 names with latitude/longitude coordinates, covering South Africa, Namibia, Botswana, Lesotho, and Swaziland.

964 Raper, P. E. *Source guide for toponymy and topology/Bronnengids vir toponimie en topologie*. Pretoria: South African Centre of Onomastic Sciences, 1975. (Onomastics series 5) 478 pp.

"A comprehensive source guide for the study of places and place-names. Includes bibliographical references to books, pamphlets, theses, periodical articles and newspaper articles." (Musiker, *South Africa*)

965 *South Africa: official standard names approved by the United States Board on Geographic Names*. Washington DC: U.S. Board on Geographic Names, 1954. (Gazetteer 166) 2 vols.

List of 44,000 place-names with latitude/longitude coordinates. Covers the Republic of South Africa, Botswana, Lesotho, Namibia and Swaziland.

GEOLOGY

966 *Bibliography and subject index of South African geology*. Pretoria: South Africa. Geological Survey, 1957-. Annual.

Covers South Africa, Namibia, Botswana.

967 *Bibliography of South African Geology, 1936-1956: index of authors*. Pretoria: South Africa. Geological Survey, 1972. 72 pp.

968 Haughton, S. H. *Bibliography of South African Geology, 1936-1956: subject index*. Pretoria: South Africa. Geological Survey, 1973. 72 pp.

969 Hall, Arthur Lewis. *Bibliography of South African geology*. Pretoria: South Africa. Geological Survey, 1922-1939. 6 volumes. (Memoirs 18, 22, 25, 27, 30, 37).

Memoir 18 covers the period to the end of 1920; Memoir 22 is the subject index to #18; Memoir 25 covers the period 1921-1925; Memoir 27 covers the period 1926-1930; Memoir 30 is the author index covering 1931-1935; Memoir 37 is the subject index covering 1921-1935.

970 *The mineral potential and mining development in the Black Homelands of South Africa*. Johannesburg: Chris van Rensburg, 1977. 127 pp.

971 *Mineral resources of the Republic of South Africa*. 5th ed. Pretoria: South Africa. Geological Survey. Dept. of Mines, 1976. (Handbook 7) 462 pp.

South Africa/972-987

972 *Minerals; a report for the Republic of South Africa/Minerale;'n verslag ten opsigte van die Republiek van Suid Afrika.* Pretoria: Government Printer, 195_- (Dept. of Mines. Quarterly information circular). Quarterly.

"Gives data on production and exports of gold, silver, diamonds and other minerals." (Musiker, *South Africa*)

973 Ridge, John Drew. "South Africa." In *Annotated bibliographies of mineral deposits in Africa, Asia (exclusive of the USSR) and Australasia*, pp. 46-163. Oxford, U.K.: Pergamon, 1976.

CLIMATOLOGY

974 Schulze, B. R. *Climate of South Africa: general survey.* Pretoria: South Africa. Geological Survey, 1965. (Climate of South Africa part 8) 330 pp.

975 Schulze, B. R. "South Africa." In *Climates of Africa*, pp. 501-586. Edited by J. F. Griffiths. Amsterdam, The Netherlands: Elsevier Publishing Co., 1972. (World survey of climatology 10).

Covers Namibia, Botswana, Lesotho, Swaziland, and South Africa. A narrative summary of these nations' climate with maps, charts and an unannotated bibliography.

976 Wallace, J. Allen, Jr. *An annotated bibliography of climatic maps of Republic of South Africa.* Washington DC: U.S. Weather Bureau, 1962. (NTIS: AD 660-863) 24 pp.

HYDROLOGY

977 *Water 75.* Johannesburg: Erudita, 1975. 227 pp.

"Wide-ranging survey of South African water problems and resources. Includes useful information on major irrigation projects and water schemes." (Musiker, *South Africa*)

SOILS

978 *Soils of South Africa (1962-1977).* Harpenden, U.K.: Commonwealth Agricultural Bureaux, 1978. (Annotated bibliography) 14 pp.

ZOOLOGY

979 Goodwin, Scheila McMillan. *South African animal life.* Johannesburg: University of the Witwatersrand. Dept. of Bibliography, Librarianship and Topography, 1972. 66 pp.

"Annotated list of 352 monographs and journal articles on marine and terrestrial life." (Musiker, *South Africa*)

AGRICULTURE

980 *Agriculture in South Africa.* 2nd ed. Johannesburg: Chris van Rensburg, 1978. 175 pp.

"A compendium of South African farming theory, practice, facts and figures." (Musiker, *South Africa*)

981 *Bibliography of South African government publications, part 2: Dept. of Agricultural Technical Services and Dept. of Agricultural Economics and Marketing, 1910- 1972.* Pretoria: South Africa. Division of Library Services, 1978. 610 pp.

"Comprehensive bibliography of official agricultural publications for South Africa. Also published in Afrikaans." (Musiker, *South Africa*)

982 Hattingh, Phillipus Stefanus, and Moody, Elize. "Agriculture in the Black homelands and the Republic of Transkei." *South African journal of African affairs* 7:1 (1977):73-86.

Contains 265 post-1950 citations. Appendices list regular publications on the homelands and published maps.

PLANT CULTURE

983 Gorter, G. J. M. A. "A new guide to South African literature on plant diseases." South Africa. Dept. of Agricultural Technical Services. *Technical communication* 111 (1973):1-61.

HUMAN GEOGRAPHY

984 Kalley, Jaqueline Audrey. *BophuthaTswana politics and the economy: a select and annotated bibliography.* Johannesburg: South Africa. Institute of International Affairs, 1978. (South African Institute of International Affairs 4) 39 pp.

Includes citations to many government documents.

985 *Register of current research in the humanities at the universities.* Pretoria: South Africa. Human Sciences Research Council, 1946-. Annual.

In English and Afrikaans.

986 Schapera, Isaac et al., eds. *Select bibliography of South African native life and problems.* London: Oxford University Press, 1941. 244 pp.

Original volume covers period to 1938; supplement 1 covers 1939-1949, supplement 2 1950-1958, supplement 3 1958-1963, supplement 4 1964-1970. "This basic work is a critically annotated bibliography of books, periodicals, articles, and government reports. It covers such topics as law, economics, education, religion, health, and social services." (Musiker, *South Africa*)

987 Weiss, Marianne. *Sudafrika: Wirtschaft und Politik: Auswahlbibliographie/South Africa: economy and politics: a selected bibliography.* Hamburg, Federal Republic of Germany: Institut fur Afrika-Kunde, 1977. (Reihe A, 15) 169 pp.

In German and English. Includes over 1,000 annotated citations covering political, economic, and social conditions of the black population within the Republic of South Africa.

ANTHROPOLOGY

988 **Neser, L., comp.** *Zulu ethnography: a classified bibliography.* Kwa-Dlangezwa, South Africa: University of Zululand for the KwaZulu Documentation Centre, 1976. (Publications series 3. Specialized publication 18) 92 leaves.

Contains 504 citations to published and unpublished material to mid-1975.

ARCHITECTURE

989 **Greig, Doreen E.** *A guide to architecture in South Africa.* Cape Town: Timmins, 1971. (Bibliography) 237 pp.

ECONOMIC DEVELOPMENT

990 **de Kock, C. I.** *A guide to directories, year books and buyers' guides in the Republic of South Africa.* Pretoria: University of South Africa. Bureau of Market Research, 1978. (Research report 63) 74 pp.

"An annotated list of published sources, principally geared to commerce and industry." (Musiker, *South Africa*)

991 *Quarterly economic review of South Africa.* London: Economist Intelligence Unit, 1982-. Quarterly.

Includes summary of political and economic news as well as charts and statistics of selected economic indicators. Annual supplement.

HISTORICAL GEOGRAPHY

992 **Boeseken, A. J. et al.** *Geskiedenis atlas vir Suid-Afrika [history atlas for South Africa].* 2nd ed. Cape Town: Nasou, 1971. 92 pp.

"An important work for establishing the correct spelling and form of geographical and historical names and terms." (Musiker, *South Africa*)

993 **Muller, C. F. J. et al.** *A select bibliography of South African history: a guide for historical research.* Pretoria: University of South Africa, 1966. (Communication D3) 215 pp.

"The original work and its supplement cover 4,000 entries, wide-ranging in scope and forming a useful checklist on the subject; unfortunately not annotated." (Musiker, *South Africa*) See also Supplement (1974); 166 pages (Communication D13).

994 **Wilson, Monica, and Thompson, Leonard M.** *The Oxford history of South Africa.* New York: Oxford University Press, 1969-71. 2 vols.

"Specialist contributors known for their scholarship make this one of the standard histories of the country. An excellent bibliography completes each volume." (Musiker, *South Africa*)

PUBLIC ADMINISTRATION

995 **Geyser, O. et al., eds.** *Bibliography on South African political history, 1902-1974.* Boston MA: G. K. Hall, 1979. 3 volumes. (Bibliographies and guides in African studies).

Compiled at the Institute for Contemporary History, University of the Orange Free State.

996 **Horrell, Muriel, comp.** *Survey of race relations in South Africa.* Johannesburg, South Africa: South African Institute of Race Relations, 1951/52-. Annual.

"A valuable comprehensive review of developments and trends in legislation, government action and opposition...Themes covered include: political developments, black political activity, churches, defence, justice, police, prisons, security legislation, employment, homelands, etc." (Musiker, *South Africa*)

SWAZILAND
The Kingdom of Swaziland
(a former colony of the United Kingdom)

Swaziland is a tiny landlocked nation surrounded on three sides by South Africa and on a fourth by Mozambique. It covers a portion of the subhumid dryland belt that lies to the southeast of the Drakensberg highlands. A number of rivers cross the land and provide water for irrigation; rice, sugar cane, citrus, pineapple, sisal, and cotton are all produced. Roads and rail lines link these agricultural areas with markets in Transvaal, Natal, and Mozambique. Timber products, sugar, iron ore, asbestos, cattle, and foodstuffs are exported and give the nation a relatively healthy and diversified economic base, even though ties with South Africa's manufacturing facilities are very strong.

Dryland areas of SWAZILAND

DESERTIFICATION RISK	ARIDITY								Aridity Totals	
	Hyperarid		Arid		Semiarid		Subhumid			
	km²	%	km²	%	km²	%	km²	%	km²	%
Very High	By definition, desertification does not exist in hyperarid regions.									
High										
Moderate										
Desertification Totals										
No Desertification								16,165	92.9	16,165	92.9
Total Drylands								16,165	92.9	16,165	92.9
Non-dryland										1,235	7.1
Total Area of the Territory										17,400	100.0

GENERAL BIBLIOGRAPHIES

997 Balima, Mildred Grimes. *Botswana, Lesotho and Swaziland: a guide to official publications 1868-1968.*

See entry 784.

998 *Swaziland official publications, 1880-1972: a bibliography of the original and microfiche edition.* Pretoria: South Africa. State Library, 1975. (Bibliographies 18) 190 pp.

ATLASES

999 Talbot, A. M. and Talbot, W. S. *Atlas of the Union of South Africa/Atlas van die unie van Suid-Afrika.*

See entry 962.

GAZETTEERS

1000 Leistner, O. A., and Morris, J. W. *Southern African place names.*

See entry 963.

1001 *South Africa: official standard names approved by the United States Board on Geographic Names.*

See entry 965.

CLIMATOLOGY

1002 Schulze, B. R. "South Africa."

See entry 975.

HUMAN GEOGRAPHY

1003 Crush, Jonathan S. "Diffuse development: Botswana, Lesotho and Swaziland since independence."

See entry 797.

ECONOMIC DEVELOPMENT

1004 Crush, J. S. *The post-colonial development of Botswana, Lesotho and Swaziland.*

See entry 800.

1005 *Quarterly economic review of Namibia, Botswana, Lesotho, Swaziland.*

See entry 921.

HISTORICAL GEOGRAPHY

1006 Grotpeter, John J. *Historical dictionary of Swaziland.* Metuchen NJ: Scarecrow, 1975. (African historical dictionaries 3) 25 pp.

Entries cover major topics, events, and personalities in the nation's history; includes extensive unannotated bibliography.

TANZANIA
The United Republic of Tanzania

(as Tanganyika and Zanzibar, former colonies of Germany, then the United Kingdom)

Tanzania occupies the center of eastern Africa's dryland belt; its high undulating plateaus are in fact the drainage divide for three of Africa's great rivers, the Nile, the Zaire, and the Zambezi. The nation's dryland core is the semiarid Masai Steppe along the border with Kenya. Most of the rest of the country is only slightly better watered and falls into the subhumid zone.

The Masai Steppe. This high semiarid savanna is scored by the deep trenches of the Great Rift Valley and punctured by the volcanoes of Mounts Kilimanjaro and Meru. Most of the land can support little more than poor herds of cattle, though the slopes of the volcanoes above the dry zone are good for coffee, maize, peas, beans, pyrethrum, and sisal. In some ways the most productive lands are those given over to the great parks and game preserves; tourism in these areas is an important component of the national economy.

The Southeastern Lowlands. A hilly, humid belt separates the Masai Steppe from the coastal lowlands that extend farthest inland in the southeast corner. Settlement here is sparse, due as much to the tsetse fly as to the subhumid climate. The giant Selous Game Reserve occupies part of the region while, toward the coast, agriculture becomes more important. In the southeast corner near the border with Mozambique can be found the vestiges of the monumentally-mismanaged East African Groundnut Scheme, a British attempt in the late 1940's to establish the crop on land ill-suited for the plant. Sisal, cashews, and even some remnant groundnuts continue to be grown.

The Western Savanna. West of the Masai Steppe stretches another subhumid region to the shores of Lake Tanganyika. It is similar to the southeastern area in size, population, and tsetse infestation. The most favored land supports tobacco and cotton. Scattered deposits of salt, gold, lead, and diamonds contribute to the nation's small mining industry.

Dryland areas of TANZANIA

DESERTIFICATION RISK	ARIDITY									Aridity Totals	
	Hyperarid		Arid		Semiarid		Subhumid			
	km²	%	km²	%	km²	%	km²	%	km²	%
Very High	By definition, desertification does not exist in hyperarid regions.									
High										
Moderate				223,391	23.7	39,588	4.2	262,979	27.9
Desertification Totals				223,391	23.7	39,588	4.2	262,979	27.9
No Desertification							514,648	54.6	514,648	54.6
Total Drylands				223,391	23.7	554,236	58.8	777,627	82.5
Non-dryland									164,951	17.5
Total Area of the Territory									942,578	100.0

GENERAL BIBLIOGRAPHIES

1007 Bates, Margaret L. *A study guide for Tanzania.* Brookline MA: Boston University. African Studies Center, 1969. 83 pp.

1008 Bridgman, Jon, and Clarke, David G. *German Africa: a select annotated bibliography.*

See entry 450.

1009 Howell, John Bruce. *East African community: subject guide to official publications.*

See entry 834.

HANDBOOKS

1010 Kaplan, Irving. *Tanzania: a country study.* 2nd ed. Washington DC: U.S. Dept. of Defense, 1978. (American University foreign area studies) 344 pp.

Narrative summary of the nation's history, politics, geography, economy, social life; includes unannotated bibliography.

GEOGRAPHY

1011 "Bibliography of Serengeti, Tanzania scientific publication." In *Serengeti, dynamics of an ecosystem*, pp. 362-382. Edited by A. R. E. Sinclair and M. Norton-Griffiths. Chicago IL: University of Chicago Press, 1979.

1012 Lundgren, Bjorn, and Samuelson, Ann-Marie. *Land use in Kenya and Tanzania: a bibliograpy.*
See entry 840.

1013 Townshend, J., and Wisner, B. *Bibliography of Dodoma region.* Dar es Salaam: University of Dar es Salaam. Bureau of Resource Assessment and Land Use Planning, 1971. (Research report 43) 21 pp.
Contains 245 citations with some annotations. The Dodoma region is semiarid savanna inland from Dar es Salaam.

1014 Ulfstrand, Stattan. *Ecology in semi-arid East Africa: a selection of recent ecological references.*
See entry 841.

ATLASES

1015 *Atlas of Tanzania.* Dar es Salaam: Tanzania. Surveys and Mapping Division, 1976. 39 pp.
Scale of most maps is 1:3,000,000.

1016 Berry, Leonard, ed. *Tanzania in maps.* London: University of London Press, 1971. 172 pp.
Scale of most maps is 1:2,700,000.

GAZETTEERS

1017 *Tanzania: official standard names approved by the United States Board on Geographic Names.* Washington DC: U.S. Board on Geographic Names, 1965. (Gazetteer 92) 236 pp.
List of 16,500 place-names with latitude/longitude coordinates.

GEOLOGY

1018 *Bibliography of the geology and mineral resources of Tanzania to December, 1967.* Dar es Salaam: University of Dar es Salaam. Bureau of Resource Assessment and Land Use Planning, 1969. (Research notes 5c) 249 pp.

1019 Nilsen, Odd. *A bibliography of the mineral resources of Tanzania.* Uppsala, Sweden: Scandinavian Institute of African Studies, 1980. 92 pp.
Unannotated. Covers published and some unpublished literature to 1979.

CLIMATOLOGY

1020 Griffiths, J. F. "Eastern Africa."
See entry 847.

HYDROLOGY

1021 Heijnen, J. E. *The river basins in Tanzania: a bibliography.* Dar es Salaam: University of Dar es Salaam. Bureau of Resource Assessment and Land Use Planning, 1970. (Research note 5e) 34 pp.

SOILS

1022 Cook, Alison. *A soils bibliography of Tanzania.* Dar es Salaam: University of Dar es Salaam. Bureau of Resource Assessment and Land Use Planning, 1975. (Research paper 39) 49 pp.

BOTANY

1023 Knapp, R. "Commentarii et bibliographia phytosociologica et scientiae vegetationis: Kenya, Tanzania, Uganda" [commentary and bibliography of plant ecology and vegetation: Kenya, Tanzania, Uganda].
See entry 848.

AGRICULTURE

1024 De Vries, James. *Selected bibliography on agricultural extension in Tanzania.* Morogoro, Tanzania: University of Dar es Salaam. Faculty of Agriculture, Forestry and Veterinary Science, 1978. (Technical paper 3) 44 pp.

AGRICULTURAL ECONOMICS

1025 Kai-Samba, Ibrahim B.; Mbwana, Salum S.; and Mchomba, Vallery G. *Development for self-reliance: a bibliography of contributions from the Faculty of Agriculture, Forestry and Veterinary Science, University of Dar es Salaam, 1969-1977.* Morogoro, Tanzania: University of Dar es Salaam. Faculty of Agriculture, 1977. 154 pp.

1026 Kocher, James E., and Fleischer, Beverly. *A bibliography on rural development in Tanzania.* East Lansing MI: Michigan State University. Dept. of Agricultural Economics, 1979. (Rural development paper 3) 77 pp.

1027 McHenry, Dean E., Jr. *Ujamaa villages in Tanzania: a bibliography.* Uppsala, Sweden: Scandinavian Institute of African Studies, 1981. 69 pp.
Covers period to 1980. The *Ujamaa* villages are communal agricultural settlements established by the national government.

1028 McLoughlin, Peter F. M. *Research on agricultural development in East Africa.*

See entry 850.

HUMAN GEOGRAPHY

1029 Hundsdorfer, Volkhard, and Kuper, Wolfgang. *Bibliographie zur sozialwissenschaftlichen Erforschung Tanzanias [bibliography of the sociological development of Tanzania].* Munchen, Federal Republic of Germany: Weltforum Verlag, 1974. (Materialien zu Entwicklung und Politik 6) 231 pp.

ECONOMIC DEVELOPMENT

1030 Killick, Tony. *The economies of East Africa.*

See entry 855.

1031 Molnos, Angela. *Development in Africa: planning and implementation: a bibliography (1964-1969) and outline with some emphasis on Kenya, Tanzania, and Uganda.*

See entry 856.

1032 *Quarterly economic review of Tanzania, Mozambique.* London: Economist Intelligence Unit, 1978-. Quarterly.

Includes summary of political and economic news as well as charts and statistics of selected economic indicators. Annual supplement.

TOURISM AND RECREATION

1033 Mascarenhas, Ophelia C. "Tourism in East Africa: a bibliographical essay."

See entry 858.

HISTORICAL GEOGRAPHY

1034 Kurtz, Laura S. *Historical dictionary of Tanzania.* Metuchen NJ: Scarecrow, 1978. (African historical dictionaries 15) 331 pp.

Entries cover major topics, events, and personalities in the nation's history; includes extensive unannotated bibliography.

PUBLIC ADMINISTRATION

1035 Rweyemamu, A. H. *Government and politics in Tanzania: a bibliography.* Nairobi: East African Academy. Research Information and Publication Services, 1972. (Information circular 6) 39 pp.

UGANDA
The Republic of Uganda
(as Buganda, a former colony of the United Kingdom)

Uganda is a landlocked, populous, and generally well watered nation straddling the Nile as it leaves Lake Victoria. Despite a crippling civil war brought on by the excesses of Idi Amin the nation has considerable agricultural and economic potential. Sizeable dryland areas exist in the north, northeast, and southwest corners of the country.

The North. A subhumid to semiarid belt extends from the shores of Lake Mobutu (formerly Albert) eastwards to the border with Kenya. Cattle and sorghum are the principal subsistence products. The region is poorly served by the transportation network so it lacks the necessary connections with the primary city of Kampala to encourage development.

The Ankole Region. This former kingdom occupies the subhumid divide between Lake Victoria and Lake Idi Amin (formerly Edward) in the nation's southwest corner. Cattle and bananas are the principal subsistence products while tea, coffee, and tobacco are grown for cash. The road network to Kampala is more developed than in the North.

Dryland areas of UGANDA

DESERTIFICATION RISK	ARIDITY									
	Hyperarid		Arid		Semiarid		Subhumid		Aridity Totals	
	km²	%	km²	%	km²	%	km²	%	km²	%
Very High	By definition, desertification does not exist in hyperarid regions.									
High										
Moderate					15,870	6.7			15,870	6.7
Desertification Totals					15,870	6.7			15,870	6.7
No Desertification							78,401	33.1	78,401	33.1
Total Drylands					15,870	6.7	78,401	33.1	94,271	39.8
Non-dryland									142,589	60.2
Total Area of the Territory									236,860	100.0

GENERAL BIBLIOGRAPHIES

1036 Collison, Robert L., comp. *Uganda.* Oxford, U.K.; Santa Barbara CA: Clio, 1981. (World bibliographical series 11) 159 pp.

Annotated bibliograpahy contains 521 items. Covers most major topics.

1037 Gray, Beverly Ann. *Uganda: subject guide to official publications.* Washington DC: Library of Congress, 1977. 271 pp.

2,442 unannotated entries for period 1893-1974. Includes publications issued by Uganda, Great Britain and international organizations.

1038 Howell, John Bruce. *East African Community: subject guide to official publications.*

See entry 834.

HANDBOOKS

1039 Herrick, Allison Butler. *Area handbook for Uganda.* Washington DC: U.S. Dept. of Defense, 1969. (American University foreign area studies) 456 pp.

Narrative summary of the nation's history, politics, geography, economy, social life; includes unannotated bibliography.

GEOGRAPHY

1040 Langlands, B. W. *Inventory of geographical research at Makerere University, 1947-1972.* Kampala: Makerere University. Dept. of Geography, 1972. (Occasional paper 50) 191 pp.

Uganda/1041-1053

1041 Rossi, Georges. "Etat actual des travaux de geographie physique sur l'Ouganda" [review of recent work on the physical geography of Uganda]. *Madagascar: revue de geographie* (1979):159-165.

In French. References range from 1960-1972.

1042 Ulfstrand, Stattan. *Ecology in semi-arid East Africa: a selection of recent ecological references.*

See entry 841.

1043 Varady, Robert G., comp. *Draft environmental profile of Uganda.* Tucson AZ: University of Arizona. Office of Arid Lands Studies, 1982. 258 pp.

A narrative summary of the nation's environment and development problems, with maps, charts, and an extensive unannotated bibliography.

ATLASES

1044 *Atlas of Uganda. Entebbe: Uganda. Lands and Surveys Dept., 1962. 83 pp.*

Base map scale 1:1,500,000

GAZETTEERS

1045 *Uganda: official standard names approved by the United States Board on Geographic Names.* Washington DC: U.S. Board on Geographic Names, 1964. (Gazetteer 82) 167 pp.

List of 11,900 place-names with latitude/longitude coordinates.

CLIMATOLOGY

1046 Griffiths, J. F. "Eastern Africa."

See entry 847.

BOTANY

1047 Knapp, R. "Commentarii et bibliographia phytosociologica et scientiae vegetationis: Kenya, Tanzania, Uganda" [commentary and bibliography of plant ecology: Kenya, Tanzania, Uganda].

See entry 848.

AGRICULTURAL ECONOMICS

1048 McLoughlin, Peter F. M. *Research on agricultural development in East Africa.*

See entry 850.

MEDICINE

1049 Hall, S. A., and Langlands, B. W., eds. *Uganda atlas of disease distribution.* Nairobi: East African Publishing House, 1975. 165 pp.

Although the data on which this atlas is based are from the mid-1960's this is nonetheless an excellent summary of the nation's medical geography.

ECONOMIC DEVELOPMENT

1050 Killick, Tony. *The economies of East Africa.*

See entry 855.

1051 Molnos, Angela. *Development in Africa: planning and implementation: a bibliography (1964-1969) and outline with some emphasis on Kenya, Tanzania, and Uganda.*

See entry 856.

1052 *Quarterly economic review of Uganda, Ethiopia, Somalia, Djibouti.* London: Economist Intelligence Unit, 1978-. Quarterly.

Includes summary of political and economic news as well as charts and statistics of selected economic indicators. Annual supplement.

TOURISM AND RECREATION

1053 Mascarenhas, Ophelia C. "Tourism in East Africa: a bibliographical essay."

See entry 858.

ZAIRE
Republique de Zaire

(as the Belgian Congo or Congo Free State, a former colony of Belgium)

Zaire possesses a tiny area of subhumid dryland along the shore of Lake Mobutu (formerly Albert) at the extreme eastern edge of the nation. The region supports some coffee plantings; otherwise its significance to Zaire is minimal. No reference sources are provided.

Dryland areas of ZAIRE

DESERTIFICATION RISK	Hyperarid		Arid		Semiarid		Subhumid		Aridity Totals	
	km²	%	km²	%	km²	%	km²	%	km²	%
Very High	By definition, desertification does not exist in hyperarid regions.									
High										
Moderate										
Desertification Totals										
No Desertification								4,691	0.2	4,691	0.2
Total Drylands								4,691	0.2	4,691	0.2
Non-dryland										2,340,718	99.8
Total Area of the Territory										2,345,409	100.0

ZAMBIA
The Republic of Zambia
(as Northern Rhodesia, a former colony of the United Kingdom)

Zambia lies across a high undulating plateau at the center of southern Africa. Its economic heartland—the copperbelt—borders Zaire's Shaba Province in the humid climatic zone. Zambia's drylands consist of a single belt of semiarid/subhumid lowland formed by the Zambezi and Luangwa Rivers. Maize, tobacco, and groundnuts are the principal crops of the region. Irrigation and hydroelectric potential is high; the Kariba Dam on the Zambezi impounds the world's largest artificial lake and generates most of the electricity consumed by both Zambia and Zimbabwe. Now that majority rule has finally come to Zimbabwe, Zambia should benefit from renewed economic contact with its wealthier neighbor.

Dryland areas of ZAMBIA

DESERTIFICATION RISK	ARIDITY									
	Hyperarid		Arid		Semiarid		Subhumid		Aridity Totals	
	km²	%	km²	%	km²	%	km²	%	km²	%
Very High	By definition, desertification does not exist in hyperarid regions.									
High										
Moderate					58,704	7.8	4,516	0.6	63,220	8.4
Desertification Totals					58,704	7.8	4,516	0.6	63,220	8.4
No Desertification					12,042	1.6	231,807	30.8	243,849	32.4
Total Drylands					70,746	9.4	236,323	31.4	307,069	40.8
Non-dryland									445,551	59.2
Total Area of the Territory									752,620	100.0

GENERAL BIBLIOGRAPHIES

1054 Walker, Audrey A. *The Rhodesias and Nyasaland: a guide to official publication.*

See entry 877.

HANDBOOKS

1055 Kaplan, Irving. *Zambia: a country study.* 3rd ed. Washington DC: U.S. Dept. of Defense, 1979. (American University foreign area studies) 308 pp.

Narrative summary of the nation's history, politics, geography, economy, social life; includes unannotated bibliography.

GEOGRAPHY

1056 Posnett, N. W., and Reilly, P. M., comps. *Zambia.* Tolworth Tower, U.K.: U.K. Ministry of Overseas Development. Land Resources Division, 1977. (Land resource bibliography 9) 200 pp.

1057 Speece, Mark W., comp. *Draft environmental profile of Zambia.* Tucson AZ: University of Arizona. Office of Arid Lands Studies. Arid Lands Information Center, 1982. 288 pp.

A narrative summary of the nation's environment and development problems, with maps, charts, and an extensive unannotated bibliography.

ATLASES

1058 Adika, G. H., ed. *Atlas of the population of Zambia.* Lusaka, Zambia: Zambia. National Council for Scientific Research, 1977. Unpaged.

Covers physical environment and demography, with detailed census maps at 1:1,250,000 based on 1963 and 1969 data.

1059 Davies, D. Hywel, ed. *Zambia in maps.* London: University of London Press, 1971. 128 pp.

GAZETTEERS

1060 *Zambia: official standard names approved by the United States Board on Geographic Names.* Washington DC: U.S. Board on Geographic Names, 1972. (Gazetteer 122) 585 pp.

List of 38,000 place-names with latitude/longitude coordinates.

GEOLOGY

1061 **Ridge, John Drew.** "Zambia." In *Annotated bibliographies of mineral deposits in Africa, Asia (exclusive of the USSR) and Australasia*, pp. 195-209. Oxford U.K.: Pergamon, 1976.

1062 **Snowball, George J.** *Annotated bibliography and index of the geology of Zambia, 1931-1959.* Lusaka: Zambia. Geological Survey, 1965. 121 pp.

Updated by biennial bibliographies.

CLIMATOLOGY

1063 **Torrance, J. D.** "Malawi, Rhodesia and Zambia."

See entry 882.

SOILS

1064 *Soils of Malawi, Rhodesia and Zambia (annotated bibliography covering the published literature for 1930-1975).*

See entry 883.

BOTANY

1065 **Knapp, R.** "Commentarii et bibliographia phytosociologiae et scientiae vegetationis: Zambia" [commentary and bibliography of plant ecology and vegetation: Zambia]. *Excerpta botanica: sectio B: sociologica* 19:3 (1979):221-256.

Contains unannotated bibliography and narrative summary (in German and English) of research on vegetation in Zambia.

ANIMAL CULTURE

1066 **Chimwano, A. M. P.** *A bibliography of livestock production literature in Zambia.* Lusaka: University of Zambia. School of Agricultural Sciences, 1981. 38 pp.

Unannotated; 475 citations.

HUMAN GEOGRAPHY

1067 **Rau, William E.** *A bibliography of pre-independence Zambia: the social sciences.* Boston MA: G. K. Hall, 1978. (Bibliographies and guides in African studies) 357 pp.

ECONOMIC DEVELOPMENT

1068 *Quarterly economic review of Zambia.* London: Economist Intelligence Unit, 1976-. Quarterly.

Includes summary of political and economic news as well as charts and statistics of selected economic indicators. Annual supplement.

HISTORICAL GEOGRAPHY

1069 **Grotpeter, John J.** *Historical dictionary of Zambia.* Metuchen NJ: Scarecrow, 1979. (African historical dictionaries 19) 410 pp.

Entries cover major topics, events, and personalities in the nation's history; includes extensive unannotated bibliography.

ZIMBABWE
The Republic of Zimbabwe

(as Southern Rhodesia, then Rhodesia, a former colony of the United Kingdom)

The coming of majority rule to Zimbabwe in 1980 ended that nation's status as a political (if not entirely economic) pariah among African states. Although landlocked and bordered by mountains, desert, and disease-ridden river valleys Zimbabwe has diverse resources and a basically strong economy. Nearly all the land is dry to some extent.

The Central Plateau. The heartland of Zimbabwe is a subhumid tableland savanna that supports the bulk of the population and agriculture. Before majority rule, black Africans maintained a precarious subsistence economy based on maize, cattle, millet, groundnuts, and poorer grades of tobacco. Most of the best lands were reserved for the white plantation farmers; tobacco was by far the biggest crop, followed by cotton, cattle, tea, and others. The greatest challenge to the new government will be the reorganization of this extremely inequitable system of land tenure.

Of even greater importance to the economy than agriculture is Zimbabwe's considerable mineral wealth. A long narrow geological formation known as the Great Dyke cuts across the central plateau and holds gold, chromium, asbestos, nickel, and iron.

The Zambezi Valley. This semiarid valley bordering Zambia is the chief provider of energy for the nation. The Kariba Dam impounds the world's largest artificial lake and generates most of Zimbabwe's electricity. Coal deposits at Wankie near Victoria Falls are an additional energy source. Agriculture in the Zambezi Valley is limited; large areas of scrub forest remain unexploited due to the prevalence of the tsetse fly. However, irrigation potential below the Kariba Dam remains high.

The Southwest Lowveld. This semiarid region stretches below the Central Plateau to the Limpopo River and to the borders with Mozambique and South Africa. Large irrigation projects have been undertaken to open up these lands to agriculture.

Dryland areas of ZIMBABWE

DESERTIFICATION RISK	ARIDITY									Aridity Totals	
	Hyperarid		Arid		Semiarid		Subhumid			
	km²	%	km²	%	km²	%	km²	%	km²	%
Very High	By definition, desertification does not exist in hyperarid regions.									
High										
Moderate					134,657	34.5	67,523	17.3	202,180	51.8
Desertification Totals					134,657	34.5	67,523	17.3	202,180	51.8
No Desertification							177,590	45.5	177,590	45.5
Total Drylands					134,657	34.5	245,113	62.8	379,770	97.3
Non-dryland									10,532	2.7
Total Area of the Territory									390,308	100.0

GENERAL BIBLIOGRAPHIES

1070 Pollak, Oliver B., and Pollak, Karen. *Rhodesia/Zimbabwe.* Oxford U.K.; Santa Barbara CA: Clio, 1979. (World bibliographical series 4) 195 pp.
Annotated bibliography of 496 references covering most major subjects.

1071 Pollak, Oliver B., and Pollak, Karen. *Rhodesia/Zimbabwe: an international bibliograpy.* Boston MA: G. K. Hall, 1977. (Bibliographies and guides in African studies) 620 pp.

1072 Walker, Audrey A. *The Rhodesias and Nyasaland: a guide to official publications.*

See entry 877.

INDEXES AND ABSTRACTS

1073 *Zimbabwe research index.* Salisbury: Zimbabwe. Scientific Liaison Office, 1979-. Annual.

Continues: *Rhodesia research index.* Salisbury: Rhodesia. Scientific Liaison Office, 1970-. Annual.

HANDBOOKS

1074 Nelson, Harold D. *Area handbook for Southern Rhodesia.* Washington DC: U.S. Dept. of Defense, 1975. (American University foreign area studies) 393 pp.

Narrative summary of the nation's history, politics, geography, economy, social life; includes unannotated bibliography.

DIRECTORIES

1075 Herbst, J. F., and Milligan, G. A., eds. *South Africa and Rhodesia.*

See entry 949.

ATLASES

1076 Collins, Michael Owen. *Rhodesia: its natural resources and economic development.* Salisbury: M. O. Collins, 1965. 52 pp.

Scale of principal maps: 1:1,000,000 and 1:2,500,000.

GAZETTEERS

1077 *Southern Rhodesia: official standard names approved by the United States Board on Geographic Names.* Washington DC: U.S. Board on Geographic Names, 1973. (Gazetteer 132) 362 pp.

List of 22,500 place-names with latitude/longitude coordinates.

GEOLOGY

1078 Ridge, John Drew. "Rhodesia." In *Annotated bibliographies of mineral deposits in Africa, Asia (exclusive of the USSR) and Australasia*, pp. 27-46. Oxford U.K.: Pergamon, 1976.

1079 Smith, Craig C., and Van der Heyde, H. E. "Rhodesian geology: a bibliography and brief index to 1968." National Museum of Rhodesia. *Occasional papers* 4:31B (1971):323-575.

CLIMATOLOGY

1080 Torrance, J. D. "Malawi, Rhodesia and Zambia."

See entry 882.

SOILS

1081 *Soils of Malawi, Rhodesia and Zambia (annotated bibliography covering the published literature for 1930-1975).*

See entry 883.

PLANT CULTURE

1082 McClymont, D. S., comp. *A selected bibliography of tobacco production information, 1.1.1903 to 1.1.1978.* Salisbury, Zimbabwe: s.n., 1980. 161 pp.

Covers Zimbabwe.

ECONOMIC DEVELOPMENT

1083 Clarke, D. G. "The economics of underdevelopment in Rhodesia: an essay on selected bibliography." *A current bibliography on African affairs*, new series 6:3 (1973):293-332.

1084 *Quarterly economic review of Zimbabwe, Malawi.*

See entry 886.

HISTORICAL GEOGRAPHY

1085 Rasmussen, R. Kent. *Historical dictionary of Rhodesia/Zimbabwe.* Metuchen NJ: Scarecrow, 1979. (African historical dictionaries 18) 445 pp.

Entries cover major topics, events, and personalities in the nation's history; includes extensive unannotated bibliography.

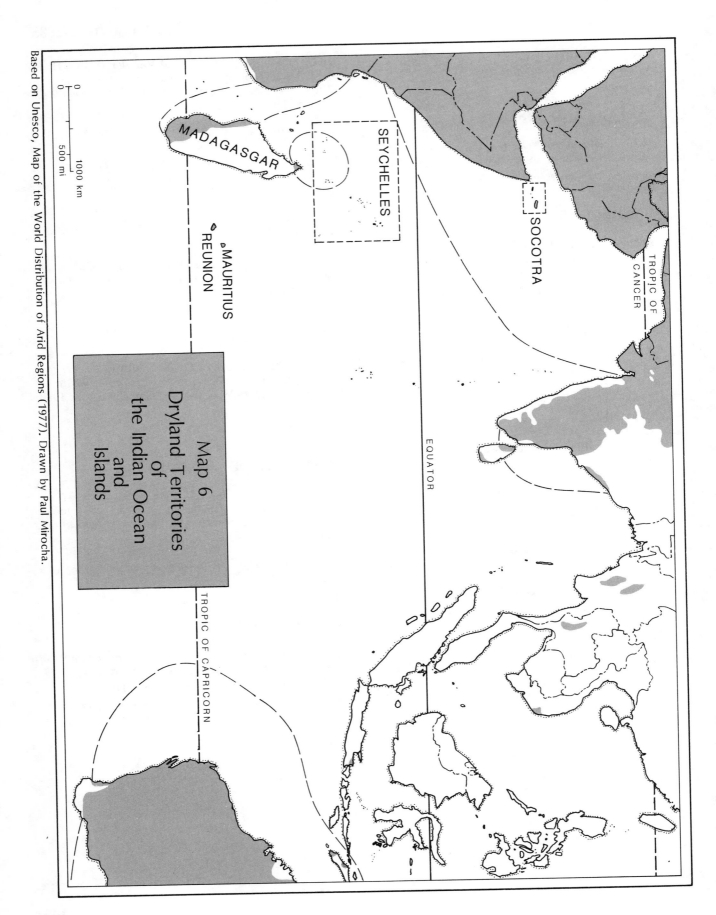

Map 6. Dryland Territories of the Indian Ocean and Islands

THE INDIAN OCEAN AND ISLANDS

Several islands in the Indian Ocean exhibit dryland environments to varying extents. Those closest to the African mainland are most closely affected by mainland climate belts (Socotra, Seychelles, Madagascar) while those further from shore have drylands due to rainshadow effects (Mauritius, Reunion).

For organizational purposes, Sri Lanka is considered under South and Southeastern Asia.

Dryland areas of THE INDIAN OCEAN ISLANDS

DESERTIFICATION RISK	ARIDITY									
	Hyperarid		Arid		Semiarid		Subhumid		Aridity Totals	
	km²	%	km²	%	km²	%	km²	%	km²	%
Very High	By definition, desertification does not exist in hyperarid regions.		2,970	0.4	30,897	4.9			33,867	5.3
High			16,831	2.7					16,831	2.7
Moderate					19,607	3.1	81,996	12.9	101,603	16.0
Desertification Totals			19,801	3.1	50,504	8.0	81,996	12.9	152,301	24.0
No Desertification					306	0.1	6,917	1.1	7,277	1.1
Total Drylands			19,801	3.1	50,864	8.0	88,913	14.0	159,578	25.1
Non-dryland									474,796	74.9
Total Area of the Territory									634,374	100.0

GEOGRAPHY

1086 Hadley, D. G., and Schmidt, D. L. *Bibliography of the Red Sea, its margins and its Rift extension (Red Sea research, 1970-1975)*. Jiddah, Saudi Arabia: Saudi Arabia. Ministry of Petroleum and Mineral Resources, 1975. (Mineral resources bulletin 22-R) 46 pp.

1087 Sukhwal, B. L. *A bibliography of theses and dissertations in geography on South Asia.*

See entry 1401.

1088 Yentsch, Anne. *A partial bibliography of the Indian Ocean*. Woods Hole MA: Woods Hole Oceanographic Institution, 1962. (Contribution 1286) 398 pp.

Unannotated. Covers expeditions, oceanography, geology, meteorology and, especially, biology.

ATLASES

1089 *Indian Ocean atlas*. Washington DC: U.S. Central Intelligence Agency, 1976. 80 pp.

Narrative text and maps of the ocean and its islands.

CLIMATOLOGY

1090 Hastenrath, S., and Lamb, Peter J. *Climatic atlas of the Indian Ocean*. Madison WI: University of Wisconsin Press, 1979. 2 volumes.

1091 Rao, K. N. "Tropical cyclones of the Indian seas." In *Climates of southern and western Asia*, pp. 257-324. Edited by K. Takahashi and H. Arakawa. Amsterdam, The Netherlands: Elsevier Scientific Publishing Co., 1981. (World survey of climatology 9).

Covers a meteorological phenomenon with important effects on the climate of surrounding East Africa, Arabia, and India. A narrative summary with maps, charts, and an unannotated bibliography.

TOURISM AND RECREATION

1092 Baretje, Rene. "Bibliographie sur le tourisme dans l'Ocean Indien" [bibliography of tourism in the Indian Ocean]. *Annuaire des pays de l'Ocean Indien* 1 (1976):555-567.

MADAGASCAR
The Democratic Republic of Madagascar
(a former colony of France)

One of the world's largest islands, Madagascar lies off the coast of Mozambique and physiographically resembles the African mainland. Two disconnected dryland regions are recognized.

The Southwest. The major dryland belt centers on an arid sliver of the extreme southwest coast and becomes gradually wetter toward the central plateau. Summer rain predominates. The land is given over to cotton and cattle, though paddy rice—the nation's staple crop—is grown along several of the short rivers that cross the coastal plain. The entire region is subject to desertification in varying degrees.

Diego Suarez. The land surrounding this deep water port formed by Cap d'Ambre and the Bay of Diego Suarez is a tiny subhumid dryland that supports beef cattle. In a nation lacking good seaports Diego Suarez would be a major asset but its location at the extreme northern tip of the island places it out of practical reach of the most economically productive area.

Note: Sources covering Africa in general often include Madagascar as well.

Dryland areas of MADAGASCAR

DESERTIFICATION RISK	Hyperarid km²	Hyperarid %	Arid km²	Arid %	Semiarid km²	Semiarid %	Subhumid km²	Subhumid %	Aridity Totals km²	Aridity Totals %
Very High			2,970	0.5	30,897	5.2			33,897	5.7
High	By definition, desertification does not exist in hyperarid regions.		13,071	2.2					13,071	2.2
Moderate					19,607	3.3	81,996	13.8	101,603	17.1
Desertification Totals			16,041	2.7	50,504	8.5	81,996	13.8	148,541	25.0
No Desertification							5,347	0.9	5,347	0.9
Total Drylands			16,041	2.7	50,504	8.5	87,343	14.7	153,888	25.9
Non-dryland									440,292	74.1
Total Area of the Territory									594,180	100.0

GENERAL BIBLIOGRAPHIES

1093 Grandidier, Guillaume. *Bibliographie de Madagascar (1500-1955)* [bibliography of Madagascar]. Paris: Comite de Madagascar, 1905; Paris: Societe d'editions geographiques maritimes et coloniales, 1935; Tananarive: Institut de recherche scientifique de Madagascar, 1957; reprint ed., Nedeln, Liechtenstein: Kraus, 1978. 3 vols.

In French. Unannotated. Fully indexed.

1094 Witherell, Julian W. *Madagascar and adjacent islands: a guide to official publications.* Washington DC: Library of Congress, 1965. 58 pp.

Contains 927 items on Madagascar (1896-1958), the Comoro Islands (1896 to date), Reunion, Mauritius and Seychelles.

HANDBOOKS

1095 Nelson, Harold D. *Area handbook for the Malagasy Republic.* Washington DC: U.S. Dept. of Defense, 1973. (American University foreign area studies) 327 pp.

Narrative summary of the nation's history, politics, geography, economy, social life; includes unannotated bibliography.

GEOGRAPHY

1096 Donque, G. "Bilan de dix-sept annees de recherches au Laboratoire de Geographie de l'Universite de Madagascar" [schedule of 17 years of research at the Geography Laboratory of the University of Madagascar]. *Madagascar: revue de geographie* 32 (1978):99-114.

1097 Donque, G. "Index analytique des articles, notes et compte-rendus parus dans les trente premiers numeros de *Madagascar: revue de geographie*" [analytic index of articles, notes and comments for the first 30 numbers of *Madagascar: revue de geographie*]. *Madagascar: revue de geographie* 30 (1977):145-62.

ATLASES

1098 Le Bourdiee, Francoise; Battistini, Rene; and Le Bourdiee, Paul. *Atlas de Madagascar [atlas of Madagascar]*. Tananarive: Madagascar. Bureau pour le developpement de la production agricole, 1969.

Most plates at scale 1:2,000,000.

GAZETTEERS

1099 *Madagascar, Reunion, and the Comoro Islands: official standard names approved by the United States Board on Geographic Names*. Washington DC: U.S. Board on Geographic Names, 1955. (Gazetteer 2) 498 pp.

List of 20,000 place-names with latitude/longitude coordinates.

GEOLOGY

1100 Besairie, Henri, and Hottin, G. *Bibliographie geologique de Madagascar 1940-1973*. Tananarive: Madagascar. Bureau de Geologie, 1974. (Document 189) 215 pp.

CLIMATOLOGY

1101 Griffiths, J. F., and Ranaivoson, R. "Madagascar." In *Climates of Africa*, pp. 461-499. Edited by J. F. Griffiths. Amsterdam, The Netherlands: Elsevier Publishing Co., 1972. (World survey of climatology 10).

A narrative summary of the nation's climate, with maps, charts, and an unannotated bibliography.

DEMOGRAPHY

1102 *Evolution demographique des capitales des etats francophones d'Afrique noire et de Madagascar [population trends of the capitals of French-speaking states of Black Africa and Madagascar]*.

See entry 344.

1103 Lacombe, Bernard. *Bibliographie commentee des etudes de population a Madagascar [bibliography of population studies of Madagascar]*. Paris: ORSTOM, 1975 (Initiations, documents techniques 27) 67 pp.

MEDICINE

1104 Brygoo, Edouard R. [medical bibliography of Madagascar]. Academie Malgache (Tananarive). *Memoires* 42 (1968):23-187.

In French. Contains 4,800 references. Updated in 1975. See entry 1105.

1105 Brygoo, Edouard R. [medical bibliography of Madagascar previous to 1968. 1st addition]. Institut Pasteur de Madagascar. *Archives* 44:1 (1975):217-244.

In French. Supplement to 1968 bibliography. Corrects or completes 27 items; adds 240 new references. See entry 1104.

1106 Coulanges, Pierre, and Coulanges, M. [annotated bibliography of the works of the Pasteur Institute of Madagascar from 1898-1978]. Institut Pasteur de Madagascar. *Archives* 1979:1-307. (Numero special).

In French.

URBAN GEOGRAPHY

1107 *Historique de la planification urbaine des capitales des etats francophones d'Afrique noire et de Madagascar [town planning history of the capitals of French-speaking states of Black Africa and Madagascar]*.

See entry 359.

ECONOMIC DEVELOPMENT

1108 *Quarterly economic review of Madagascar, Mauritius, Seychelles, Comoros*. London: Economist Intelligence Unit, 1978. Quarterly.

Includes summary of political and economic news as well as charts and statistics of selected economic indicators. Annual supplement.

HISTORICAL GEOGRAPHY

1109 Rajemisa-Raolison, Regis. *Dictionnaire historique et geographique de Madagascar [historical and geographical dictionary of Madagascar]*. Fianarantsoa, Madagascar: Librairie Ambozontany, 1966. 383 pp.

In French.

MAURITIUS

(a former colony of the United Kingdom)

Located about 240 km (150 miles) northeast of Reunion and about 920 km (570 miles) due east of Madagascar, the island of Mauritius is generally very wet but does possess a subhumid dryland along its northwestern coast. Fertile volcanic soils and plentiful water support extensive sugar cane plantations that dominate the economy. The dryland area, however, is used primarily for grazing though, with irrigation, a wide variety of tropical crops could be grown. The island's economic problems stem from a large unemployed population and almost total reliance on sugar as an export; inequitable land tenure has placed more than half the sugar acreage in the hands of a score of large estates controlled by French-descended whites.

Dryland areas of MAURITIUS

DESERTIFICATION RISK	ARIDITY									
	Hyperarid		Arid		Semiarid		Subhumid		Aridity Totals	
	km²	%	km²	%	km²	%	km²	%	km²	%
Very High	By definition, desertification does not exist in hyperarid regions.									
High										
Moderate										
Desertification Totals										
No Desertification								204	10.0	204	10.0
Total Drylands								204	10.0	204	10.0
Non-dryland										1,836	90.0
Total Area of the Territory										2,040	100.0

GENERAL BIBLIOGRAPHIES

1110 Witherell, Julian W. *Madagascar and adjacent islands: a guide to official publications.*

See entry 1094.

HANDBOOKS

1111 Stoddard, Theodore L. *Area handbook for the Indian Ocean territories.*

See entry 1116.

GAZETTEERS

1112 *Indian Ocean: official standard names approved by the United States Board on Geographic Names.* Washington DC: U.S. Dept. of the Interior, 1957. (Gazetteer 32) 54 pp.

List of 4,000 place-names with latitude/longitude coordinates. Covers Mauritius, Seychelles, and other islands.

SOILS

1113 *Soils of Mauritius, Reunion and the Seychelles (annotated bibliography covering the published literature for 1931-1975).*

See entry 1119.

HISTORICAL GEOGRAPHY

1114 Riviere, Lindsay. *Historical dictionary of Mauritius.* Metuchen NJ: Scarecrow, 1982. (African historical dictionaries 34) 172 pp.

Entries cover major topics, events, and personalities in the nation's history; includes extensive unannotated bibliography.

REUNION

(an overseas department of France)

Reunion lies 240 km (150 miles) southwest of Mauritius and some 645 km (400 miles) east of Madagascar. It is a virtual twin of Mauritius in its rugged volcanic topography, the existence of a leeward dryland, its reliance on sugar cane as a plantation export, overpopulation, and the attendant economic difficulties of expanding a limited resource base. Reunion is considerably drier overall, with a semiarid coastal strip on the west and a subhumid zone embracing the tableland and upper slopes of over half the island. Although a variety of tropical fruits and other crops can be grown sugar cane, molasses, and rum are virtually the only exports.

Dryland areas of REUNION

DESERTIFICATION RISK	ARIDITY									
	Hyperarid		Arid		Semiarid		Subhumid		Aridity Totals	
	km²	%	km²	%	km²	%	km²	%	km²	%
Very High	By definition, desertification does not exist in hyperarid regions.									
High										
Moderate										
Desertification Totals										
No Desertification					360	14.2	1,179	46.4	1,539	60.6
Total Drylands					360	14.2	1,179	46.4	1,539	60.6
Non-dryland									1,003	39.4
Total Area of the Territory									2,542	100.0

GENERAL BIBLIOGRAPHIES

1115 **Witherell, Julian W.** *Madagascar and adjacent islands: a guide to official publications.*

See entry 1094.

HANDBOOKS

1116 **Stoddard, Theodore L.** *Area handbook for the Indian Ocean territories.* Washington DC: U.S. Dept. of Defense, 1971. (American University foreign area studies) 160 pp.

Narrative summary of the nation's history, politics, geography, economy, social life; includes unannotated bibliography. Covers Maldives, Mauritius, Seychelles, Reunion.

ATLASES

1117 *Atlas des departements Francais d'Outre-Mer: I. La Reunion [atlas of French overseas departments: I.*

Reunion]. Paris: France. Centre d'Etudes de Geographie Tropicale (C.N.R.S.)/Institute Geographique National, 1975. Unpaged; 37 maps.

In French.

GAZETTEERS

1118 *Madagascar, Reunion and the Comoro Islands: official standard names approved by the United States Board on Geographic Names.*

See entry 1099.

SOILS

1119 *Soils of Mauritius, Reunion and the Seychelles (annotated bibliography covering the published literature for 1931-1975).* Harpenden, U.K.: Commonwealth Agricultural Bureaux, 1977. (Annotated bibliography) 4 pp.

SEYCHELLES
The Republic of Seychelles
(a former colony of the United Kingdom)

The Seychelles chain stretches north of Madagascar toward the equator; Mahe, the largest and most populated island, lies some 1,800 km (1,100 miles) due east of Mombasa, Kenya. Mahe, Preslin, and other islands in the northeast corner of the chain are rugged, granitic, and wet. To the southwest lie the Amirante, Farquhar, Cosmoledo, and Aldabra islands, all of which are low, coraline, and subhumid. Although they define much of the areal extent of the nation, the drier islands are virtually uninhabited and remain unspoiled by development. Aldabra—as large as Mahe—is an especially valuable example of the pristine coral atoll with ten percent of its biota unique to the island. This and several others in the chain have potential as natural preserves, tourist resorts, and military bases.

Dryland areas of SEYCHELLES

DESERTIFICATION RISK	ARIDITY									
	Hyperarid		Arid		Semiarid		Subhumid		Aridity Totals	
	km²	%	km²	%	km²	%	km²	%	km²	%
Very High	By definition, desertification does not exist in hyperarid regions.										
High											
Moderate											
Desertification Totals											
No Desertification							187	42.0	187	42.0
Total Drylands							187	42.0	187	42.0
Non-dryland									258	58.0
Total Area of the Territory									445	100.0

GENERAL BIBLIOGRAPHIES

1120 Witherell, Julian W. *Madagascar and adjacent islands: a guide to official publications.*
See entry 1094.

HANDBOOKS

1121 Stoddard, Theodore L. *Area handbook for the Indian Ocean territories.*
See entry 1116.

GAZETTEERS

1122 *Indian Ocean: official standard names approved by the United States Board on Geographic Names.*
See entry 1112.

SOILS

1123 *Soils of Mauritius, Reunion and the Seychelles (annotated bibliography covering the published literature for 1931-1975).*
See entry 1119.

SOCOTRA

(a part of the People's Democratic Republic of Yemen)

The arid island of Socotra lies 400 km (250 miles) from the South Yemen coast; though it is physiographically a part of Africa—and is 100 km (80 miles) closer to Somalia—it is administered from the capital at Aden. Socotra and neighboring islets rise from coral reefs; the main island extends almost to 1,500 meters (5,000 feet) and intercepts some moisture from winter monsoons. Surface water is scarce and there is little cultivation except for small vegetable plots and date palm oases in the wadis. Bedouin herdsmen roam the hills and occasionally send out clarified butter, aloe juice, and civit musk for export; the Negroes and Arabs along the coast produce dried fish, pearls, and ambergis. Due to reefs and strong monsoonal winds coastal navigation is hazardous. Nonetheless, the island's location near the entrance to the Red Sea makes it a strategic base.

Note: See sources covering Africa and the Middle East generally for additional entries.

Dryland areas of SOCOTRA

DESERTIFICATION RISK	ARIDITY								Aridity Totals	
	Hyperarid		Arid		Semiarid		Subhumid			
	km²	%	km²	%	km²	%	km²	%	km²	%
Very High	By definition, desertification does not exist in hyperarid regions.									
High			3,760	100.0					3,760	100.0
Moderate										
Desertification Totals			3,760	100.0					3,760	100.0
No Desertification										
Total Drylands			3,760	100.0					3,760	100.0
Non-dryland										
Total Area of the Territory									3,760	100.0

BOTANY

1124 **Knapp, R.** "Bibliographia phytosociologica et scientiae vegetationis: Aethiopia, Somalia, Sudan, Afar et Issa, Socotra" [bibliography of plant ecology and vegetation: Ethiopia, Somalia, Sudan, Afars and Issas, Socotra].

See entry 702.

SRI LANKA

Note: The Democratic Republic of Sri Lanka (Ceylon) is included in the section South and Southeast Asia below.

ASIA

The extent of Asia's drylands is nearly equal to that of Africa, and the presence of hyperarid regions like the Rub al-Khali, the Nafud, and the Takla Makan give the continent a degree of aridity exceeded only by the Sahara. Asia's peculiarity is the extent to which its drylands fall into cold winter belts. With a landmass so large and fringing mountains so high, the dry steppes of Central Asia are removed from the moderating influences of maritime air.

The northern landmass and the ocean to the south also set up strong periodic inrushes of air called monsoons. Although local effects vary considerably, in general the winter monsoon is made up of dry, cold air blowing out of Central Asia while the summer monsoon brings warm, wet air in from the Indian Ocean. Monsoonal effects increase across southern Asia from west to east so that the Middle East feels little of the regime while Southeast Asia experiences the full dramatic change from dry to wet when the winds abruptly shift. The agricultural value of the monsoons is immense, especially when their failure to arrive at the expected date exposes crops to the full fury of a summer tropical sun at a time when evapotranspiration is highest.

Two great Asian dryland areas are recognized. One extends across the southern portion of the continent to embrace the Middle East and South to Southeast Asia. The other is defined by the cold winters prevalent across the Soviet Union, Mongolia, and northern China.

Dryland areas of ASIA (INCLUDES ENTIRE USSR)

DESERTIFICATION RISK	ARIDITY									
	Hyperarid		Arid		Semiarid		Subhumid		Aridity Totals	
	km^2	%	km^2	%	km^2	%	km^2	%	km^2	%
Very High	By definition, desertification does not exist in hyperarid regions.		471,818	0.9	290,451	0.6	60,102	0.1	822,371	1.6
High			6,487,522	12.9	615,203	1.2	243,524	0.5	7,346,249	14.6
Moderate			206,980	0.4	4,358,465	8.7	1,452,936	2.9	6,018,381	12.0
Desertification Totals			7,166,320	14.2	5,264,119	10.5	1,756,562	3.5	14,187,001	28.2
No Desertification	1,891,694	3.8	844,571	1.7	700,147	1.4	3,758,328	7.5	7,194,740	14.4
Total Drylands	1,891,694	3.8	8,010,891	15.9	5,964,266	11.9	5,514,890	11.0	21,381,741	42.6
Non-dryland									28,771,412	57.4
Total Area of the Territory									50,153,153	100.0

INDEXES AND ABSTRACTS

1125 *Abstracta Islamica.*
See entry 288.

THESES AND DISSERTATIONS

1126 *Doctoral dissertations on Asia: an annotated bibliographical journal of current international research.* Ann Arbor MI: Xerox University Microfilms, 1975-. 2 per year.

HANDBOOKS

1127 *The Far East and Australia: a survey and directory of Asia and the Pacific.* London: Europa, 1969-. Annual.

Countries covered include: Afghanistan, Australia, Burma, China, India, Mongolia, Pakistan, and U.S.S.R. in Asia.

ENCYCLOPEDIAS

1128 **Gibb, H. A. R.; Kramers, J. H.; and Levi-Provencal, E., eds.** *The encyclopaedia of Islam.* New ed. Leiden, The Netherlands: E. J. Brill, 1954-.

A major work of reference under painfully slow production for some 25 years; as of this writing the alphabet through the letter K has appeared in a series of separate fascicles. Full, signed articles with bibliographies provide extensive coverage of Islamic peoples and nations, their history, culture, and civilization. Good historical maps appear; a full atlas is promised.

DIRECTORIES

1129 *International directory of centers for Asian studies.* Hong Kong: Asian Research Service, 1975/6-. Annual.

STATISTICS

1130 Harvey, Joan M. *Statistics Asia and Australasia: sources for market research.* Bechenham, U.K.: CBD Research, 1974. 238 pp.

In English, French, and German.

1131 *Statistical yearbook for Asia and the Pacific.* Bangkok: United Nations. Economic and Social Commission for Asia and the Pacific, 1973-. Annual.

In English and French. Continues: Statistical yearbook for Asia and the Far East (1968-1972).

GAZETTEERS

1132 *Asia: official standard names approved by the United States Board on Geographic Names.* Washington DC: U.S. Board on Geographic Names, 1972. (Gazetteer supplement 127) 137 pp.

List of 2,075 place-name corrections with latitude/longitude coordinates. Does not include Cambodia, Laos, Republic of China, Hong Kong or Macao.

GEOLOGY

1133 Ridge, John Drew. *Annotated bibliographies of mineral deposits in Africa, Asia (exclusive of the U.S.S.R.) and Australasia.*

See entry 312.

CLIMATOLOGY

1134 Kramer, H. P. "Climatology of the Middle East and Central Asia: a selected annotated bibliography." *Meteorological abstracts and bibliography* 2:6 (1951):453-480.

1135 Yao, Augustine Y. M. *Precipitation probability for eastern Asia.* Silver Spring MD: U.S. Environmental Data Service, 1971. (NOAA atlas 1) 71 pp.

Covers Pakistan, India, southeast Asia, Indonesia, eastern China, Korea, and Japan.

BOTANY

1136 Kuechler, August Wilhelm. *International bibliography of vegetation maps: volume 3: U.S.S.R., Asia, and Australia.*

See entry 1651.

ZOOLOGY

1137 Richard, D. *Bibliographie sur le dromadaire et le chameau [bibliography on the dromedary and the camel].*

See entry 326.

MARKETS

1138 Smith, Robert H. T. *Periodic markets in Africa, Asia, and Latin America.*

See entry 341.

ENERGY

1139 *Energy atlas of Asia and the Far East.* New York: United Nations. Economic Commission for Asia and the Far East, 1970. 25 pp.

Covers Afghanistan, Iran, India, Pakistan, Sri Lanka, Indonesia, Australia, Japan, etc. Omits China.

ANTHROPOLOGY

1140 Weekes, Richard V., ed. *Muslim peoples: a world ethnographic survey.*

See entry 348.

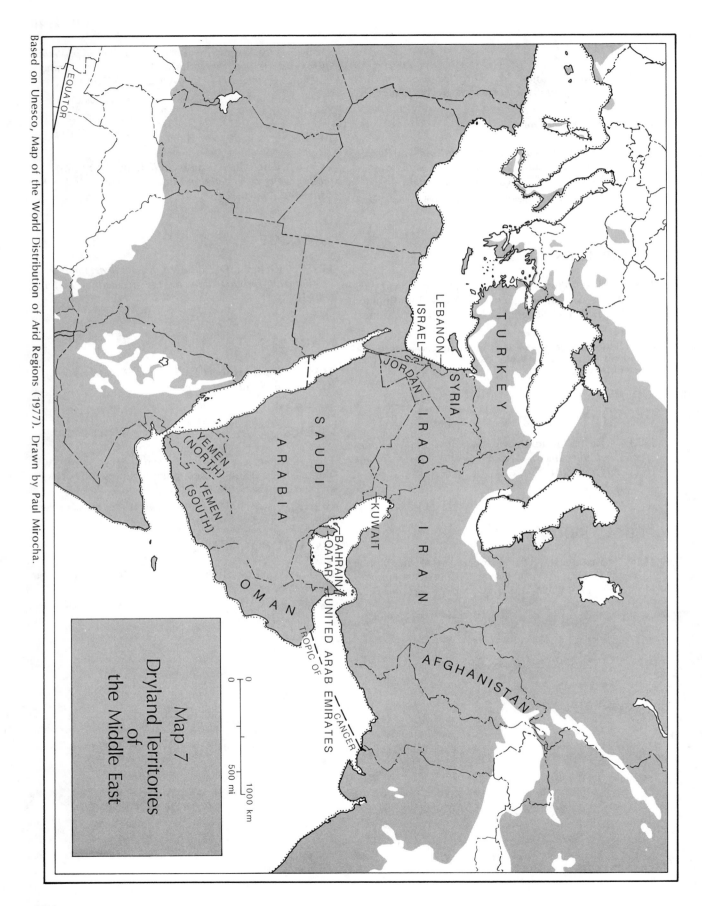

The Middle East

The nations of the Middle East are defined here as all those from Turkey to Afghanistan south to the tip of the Arabian Peninsula. Most of the area lies outside the path of the South Asian monsoons and exhibits a pronounced summer drought that intensifies the effects of aridity. Despite the proximity of ocean bodies around the area and the resulting high coastal humidity, precipitation is extremely meager and unreliable through large areas.

Two broad physiographic belts exist. One is a mountainous upland extending from Turkey into Afghanistan with an extension along the Mediterranean into Israel; here the mountains produce cool to cold winters and isolated peaks outside the drylands hold snow for spring runoff. South of the mountain belt, from the Tigris-Euphrates Valley to the tip of the Arabian Peninsula, most of the land is a low alternation of alluvial plains, lava scablands, stoney plateaus, and sand seas. Only the highlands of Yemen and Oman break into the semiarid to subhumid range.

Socially and economically the region is going through a period of extraordinary change. A strategic geopolitical position, conflict over Israel, and immense petroleum wealth have thrust the people of the region into positions of power and influence they have not enjoyed since the heyday of Islamic expansion over five centuries ago.

GENERAL BIBLIOGRAPHIES

1141 Anderson, Margaret. *Arabic materials in English translation: a bibliography of works from the pre-Islamic period to 1977.* Boston MA: G. K. Hall, 1980. 249 pp.

1142 Anthony, John Duke. *The states of the Arabian Peninsula and the Gulf littoral: a selected bibliography.* Washington DC: Middle East Institute, 1973. 21 pp.

1143 Atiyeh, George Nicholas, comp. *The contemporary Middle East, 1948-1973: a selective and annotated bibliography.* Boston MA: G. K. Hall, 1974. 664 pp.

1144 Geddes, Charles L. *Analytical guide to bibliographies on the Arab Fertile Crescent.* Denver CO: American Institute of Islamic Studies, 1975. (Bibliographic series 8) 131 pp.

1145 Geddes, Charles L. *Analytical guide to the bibliographies on the Arabian peninsula.* Denver CO: American Institute of Islamic Studies, 1974. (Bibliographic series 4) 50 pp.

1146 Hopwood, Derek, and Grimwood-Jones, Diana, eds. *Middle East and Islam: a bibliographical introduction.* Zug, Switzerland: Inter Documentation Co., 1972. (Bibliotheca Asiatica 9) 368 pp.

1147 Littlefield, David W. *The Islamic Near East and North Africa: an annotated guide to books in English for non-specialists.*
See entry 381.

ONLINE DATABASES

1148 *MEDAB (Middle East database).* Parsippany NJ: New York Times Information Service, 1980-. Updated daily.

Available online only. Indexes Middle Eastern, British, and U.S. publications and covers information on current events, economics, business, etc. Includes Arabic-language items. Vendor: New York Times Information Service.

THESES AND DISSERTATIONS

1149 Selim, George Dimitri, comp. *American doctoral dissertations on the Arab world, 1883-1974.* Washington DC: Library of Congress, 1976. 2nd ed. 173 pp.

Alphabetical listing of 1,825 dissertations from the U.S. and Canada. Titles relating to Islam and Arab minorities are listed regardless of geographical location.

HANDBOOKS

1150 *The Middle East and North Africa 1980-81.* London: Europa, 1980-81. 27th edition. 985 pp.

A standard source for recent information on the nations of the region stressing political and economic trends, events, statistics, names and addresses. Includes information on regional organizations, a who's who of important people, and a list of research institutions.

1151 *Middle East annual review.* Saffron Walden, U.K.: Middle East Review, 1974- Annual.

1152 Wazaife, Rashid, ed. *The Arabian year book 1978.* Kuwait: Dar Al Sayassah Press, 1978. 964 pp.

Covers the Persian Gulf states of Kuwait, Saudi Arabia, Bahrain, Qatar, United Arab Emirates, and Oman.

DIRECTORIES

1153 Wheeler, Stella E. L., and Shah, Khawaja T., eds. *Greece, Turkey, and the Arab states.*
See entry 151.

STATISTICS

1154 McCarthy, Justin, Jr. *The Arab world, Turkey, and the Balkans (1878-1914): a handbook of historical statistics.* Boston: G. K. Hall, 1982. 309 pp.

The Middle East/1155-1171

A wide-ranging compendium of Ottoman imperial statistics covering modern Greece, Turkey, Iraq, Syria, Lebanon, Jordan, Israel, and northwestern Saudi Arabia. Subjects include climate, population, medical demography, finance, mining, agriculture.

GAZETTEERS

1155 *Africa and southwest Asia: official standard names approved by the United States Board on Geographic Names.*

See entry 310.

1156 Scoville, Sheila A., ed. *Gazetteer of Arabia: a geographical and tribal history of the Arabian peninsula.* Graz, Austria: Akademische Druck und Verlagsanstalt, 1979-. 4 vols.

Based on a 1917 British gazetteer and brought up to date in the 1970's. As of November 1981, volumes 2, 3, 4 not yet published.

GEOLOGY

1157 Avnimelech, Moshe A. *Bibliography of Levant geology, including Cyprus, Hatay, Israel, Jordania, Lebanon, Sinai and Syria.* Jerusalem: Israel Program for Scientific Translations, 1965-1969. 2 volumes.

Covers Cyprus, Syria, Turkish Hatay, Lebanon, Jordan, Israel, and the Sinai peninsula of Egypt.

1158 *Current bibliography of Middle East geology.* Jerusalem: Geological Survey of Israel. Library, 1976-. Monthly.

Covers the Arabian peninsula, Cyprus, Egypt, Iran, Iraq, Israel, Jordan, Libya, the Mediterranean Sea, the Red Sea, Syria, and Turkey.

CLIMATOLOGY

1159 "Bibliography." In *Handbook of the weather in the Gulf: between Iran and the Arabian Peninsula*, vol. 2. London: Imcos Marine Ltd., 1974. (IM 103) 19 pp.

1160 *Climatic atlas of Europe.*

See entry 113.

1161 Gleeson, Thomas Alexander. *Bibliography of the meteorology of the Mediterranean, Middle East and South Asian areas.*

See entry 122.

1162 Grimes, Annie E. *An annotated bibliography on the climate of the Arabian Peninsula.* Washington DC: U.S. Weather Bureau, 1960. (NTIS: AD 664-696) 42 pp.

Contains 124 items.

1163 Kramer, H. P. "Selective annotated bibliography on the climatology of the Near East." *Meteorological abstracts and bibliography* 2:5 (1951):373-404.

Covers 240 items for the period 1856-1951.

1164 "Surface wind data." In *Handbook of the weather in the Gulf: between Iran and the Arabian Peninsula*, vol. 1. London: Imcos Marine Ltd., 1974. (IM 102) 103 pp.

Consists almost entirely of data and charts.

1165 Taha, M. F. et al. "The climate of the Near East." In *Climates of southern and western Asia*, pp. 183-255. Edited by K. Takahashi and H. Arakawa. Amsterdam, The Netherlands: Elsevier Scientific Publishing Co., 1981. (World survey of climatology 9).

Covers the region from Turkey into Iran south through the Arabian Peninsula; also Cyprus. A narrative summary with maps, charts, and an unannotated bibliography.

HYDROLOGY

1166 Dost, H. *Bibliography on land and water utilization in the Middle East.* Wageningen, The Netherlands: Wageningen Agricultural University College, 1953. 115 pp.

1167 Gischler, Christiaan. *Water resources in the Arab Middle East and North Africa.* Cambridge, U.K.: Middle East and North African Studies Press, 1979. 132 pp.

A review of previously published sources and unpublished reports.

ZOOLOGY

1168 Allouse, Bashir E., comp. *A bibliography on the invertebrate fauna of Iraq and neighboring countries. I. Molluscs.* Baghdad: Iraq Natural History Museum, 1956. (Publication 8) 38 pp.

Lightly annotated. Covers the Middle East (Turkey to Yemen, Syria to Iran.)

1169 Allouse, Bashir E. *A bibliography on the vertebrate fauna of Iraq and neighboring countries.* Baghdad: Iraq Natural History Museum, 1954-55. (Publications 4, 5, 6, 7) 4 vols.

Lightly annotated. Covers the Middle East (Turkey to Yemen, Syria to Iran). Divided as follows: I. Mammals (Publication 4; 1954) II. Birds (Publication 5; 1954) III. Reptiles and amphibians (Publication 6; 1955) IV. Fishes (Publication 7; 1955)

1170 Harrison, Colin. *An atlas of the birds of the Western Palaearctic.*

See entry 116.

AGRICULTURE

1171 Shihabi, Mustafa. *Chihabi's dictionary of agricultural and allied terminology: English-Arabic.* Beirut: Libraire de Liban, 1978. 907 pp.

AGRICULTURAL ECONOMICS

1172 Taylor, Donald C. *Research on agricultural development in selected Middle Eastern countries.* New York: Agricultural Development Council, 1968. 166 pp.

LAND TENURE

1173 *Land tenure and agrarian reform in Africa and the Near East: an annotated bibliography.*

See entry 339.

1174 *Near East and South Asia: a bibliography.* Madison WI: The University of Wisconsin. Land Tenure Center, 1971. (Training and methods 13) 74 pp.

Unannotated. Reflects the holdings of the Land Tenure Center library; includes its call numbers. Continued by Supplements 2 (1976) and 3 (1976).

HUMAN GEOGRAPHY

1175 Hansen, Gerda et al. *Wirtschaft, Gesellschaft und Politik der Staaten der Arabischen Halbinsel: eine bibliographische Einfuhrung [economy, society and politics of the Arabian peninsula: a bibliographic introduction].* Hamburg: Deutsches Orient-Institut, 1976. (Dokumentationsdienst Moderner Orient. Reihe A:7) 271 pp.

Annotated. In German. Social and political development of Arabian Peninsula during the past 30 years. Includes journals and monographs in English, French, German and Arabic.

1176 Simon, Reeva S. *The modern Middle East: a guide to research tools in the social sciences.* Boulder CO: Westview, 1978. (Westview special studies on the Middle East) 283 pp.

1177 Vidergar, John J. *The southern Arabian Peninsula: social and economic development.* Monticello IL: Council of Planning Librarians, 1978. (CPL exchange bibliography 1540) 6 pp.

Unannotated. Covers Saudi Arabia, Kuwait, and other states on the peninsula. Includes works in Arabic.

ANTHROPOLOGY

1178 Lytle, Elizabeth E. *A bibliography of the Kurds, Kurdistan, and the Kurdish question.* Monticello IL: Council of Planning Librarians, 1977. (CPL exchange bibliography 1301) 15 pp.

Unannotated. The Kurds occupy parts of five nations: Turkey, Syria, Iraq, Iran, and Transcaucasian U.S.S.R.

1179 Lytle, Elizabeth E. *The Palestinian refugees: a selected bibliography.* Monticello IL: Council of Planning Librarians, 1977. (CPL exchange bibliography 1403) 32 pp.

Unannotated. Includes 471 English and French sources in most subject areas appearing between 1948 and 1976.

WOMEN'S STUDIES

1180 Al-Qazzaz, Ayad. *Women in the Middle East and North Africa: an annotated bibliography.* Austin TX: University of Texas at Austin. Center for Middle Eastern Studies, 1977. (Middle East monographs 2) 178 pp.

Extensive annotations. English language items only. Includes unpublished papers. Emphasis on women and Islam.

1181 al-Barbar, Aghil M. *The study of Arab women: a bibliography of bibliographies.* Monticello IL: Vance Bibliographies, 1980. (Public administration series bibliography P-436) 4 pp.

Annotated.

1182 Meghdessian, Samira Rafidi, comp. *The status of the Arab woman: a select bibliography.* London: Mansell, 1980. 176 pp.

Unannotated. Contains mostly post-1950 literature in English or other Western European languages. Covers books, articles (including unpublished ones), theses, and bibliographies.

1183 Raccagni, Michelle. *The modern Arab woman: a bibliography.* Metuchen NJ: Scarecrow, 1978. 262 pp.

Some brief annotations. Includes Arabic sources. Lists modern works including dissertations in progress in 1976.

PUBLIC HEALTH

1184 Barbar, Aghil M. *Disease, health services and care in the Arab world: an introductory bibliography.* Monticello IL: Vance Bibliographies, 1979. (Public administration series bibliography P-160) 11 pp.

Unannotated.

URBAN GEOGRAPHY

1185 Barbar, Aghil M. *Bibliography of bibliographies of urbanization in the Arab world.* Monticello IL: Vance Bibliographies, 1978. (Public administration series bibliography P-114) 4 pp.

Unannotated. Covers English-language sources only.

1186 Barbar, Aghil M. *Urbanization in the Arab world: a selected bibliography.* Monticello IL: Council of Planning Librarians, 1977. (CPL exchange bibliography 1198) 18 pp.

Unannotated. Includes English and French citations; no Arabic.

1187 Davis, Lenwood G. *Urbanization in the Middle East, with some references to North Africa.* Monticello IL: Council of Planning Librarians, 1978. (CPL exchange bibliography 1539) 20 pp.

Unannotated.

1188 Mahayni, Riad G. *Urbanization in the Middle East.* Monticello IL: Vance Bibliographies, 1978. (Public administration series bibliography P-23) 18 pp.

Annotated.

1189 Samaan, A. G. *Urbanization in the Middle East*. Monticello IL: Council of Planning Librarians, 1976. (CPL exchange bibliography 1037) 18 pp.

Unannotated.

ARCHITECTURE

1190 Barbar, Aghil M. *The architecture of the Arabian house*. Monticello IL: Vance Bibliographies, 1979. (Architecture series bibliography A85) 9 pp.

Some entries annotated.

HOUSING

1191 Barbar, Aghil M. *Housing in the Arab world: a bibliography*. Monticello IL: Council of Planning Librarians, 1978. (CPL exchange bibliography 1520) 7 pp.

Unannotated.

ECONOMIC DEVELOPMENT

1192 *Africa/Middle East data bank*.

See entry 367.

1193 Blauvelt, Evan. *Sources of African and Middle-Eastern economic information*.

See entry 368.

1194 Bricault, Giselle C., ed. *Major companies of the Arab world - 1982*.

See entry 426.

1195 *A cumulation of a selected and annotated bibliography of economic literature on the Arabic-speaking countries of the Middle East, 1938-1960*. Boston MA: G. K. Hall, 1967. 358 pp.

Includes articles, reports, monographs, official documents in English, French, and Arabic. Cumulated at the School of Oriental and African Studies, University of London, from the bibliography prepared by the Economic Research Institute, American University of Beirut.

1196 Fleming, Quentin W. *A guide to doing business on the Arabian Peninsula*. New York: AMACOM, 1981. 150 pp.

1197 Indesha, N. *Oil and economy in the Middle East: a selected bibliography*. Monticello IL: Vance Bibliographies, 1981. (Public administration series bibliography P-708) 21 pp.

Some annotations.

TRANSPORTATION

1198 Barbar, Aghil M. *Ports of the Arab world: an annotated bibliography*. Monticello IL: Council of Planning Librarians, 1977. (CPL exchange bibliography 1243) 6 pp.

Briefly annotated.

1199 Barbar, Aghil M. *Transportation in the Arab world*. Monticello IL: Vance Bibliographies, 1978. (Public administration series bibliography P-143) 7 pp.

Unannotated.

HISTORICAL GEOGRAPHY

1200 Lorimer, John Gordon. *Gazetteer of the Persian Gulf, Oman, and Central Arabia*. Calcutta: Superintendent of Government Printing, 1900-1915; reprint ed., Farnborough, U.K.: Gregg International, 1970. 2 vols. in 6.

The primary source of historical data for the region, long classified as "secret" by the British government.

PUBLIC ADMINISTRATION

1201 Cigar, Norman. *Government and politics in the Arabian Peninsula*. Monticello IL: Vance Bibliographies, 1979. (Public administration series bibliography P-170) 10 pp.

Annotated. Covers materials appearing in the 1960's and 1970's.

1202 Cigar, Norman. *Government and politics of the Middle East*. Monticello IL: Vance Bibliographies, 1979. (Public administration series bibliography P-196) 16 pp.

Unannotated. Covers materials appearing since 1945.

AFGHANISTAN
The Democratic Republic of Afghanistan
(since 1979 occupied by the Soviet Union)

Except for the higher slopes of the Hindu Kush in the northeast, the landlocked nation of Afghanistan is entirely dry. Its complex mountain structure and deep river valleys create several arid cores and a variety of local climates.

Registan. In the southwest the perennial Helmand and Arghandab Rivers flow through a hot summer/cool winter desert that supports only scattered nomads and their herds of sheep and camels. The rivers themselves, however, give life to oases and offer the nation's greatest potential for extensive agricultural development and hydroelectric power. The Helmand supported irrigated agriculture 2,000 years ago and still forms the focus of modern development projects; the Japanese began work here in 1937 and the Americans continued the effort with the formation of the Helmand Valley Authority, now an Afghan enterprise. Kandahar is the region's leading city and serves as a major entrepot for trade with nearby Pakistan.

Bactria. In the north the land slopes away from the Hindu Kush and forms another broad valley, this containing the Amu Darya which borders the Soviet Union. Geographically and historically both banks of the river and its native Turkmen, Uzbek, and Tadjik peoples are one land and culture; the ancient kingdom of Bactria and its capital at Balkh was a prosperous center of central Asian trade and a remarkable crossroads for the distant cultures of the Mediterranean, Indian, and Chinese civilizations. Today the river itself is underutilized and the scattered settlements must rely on local water supplies to support their irrigated oases. Wheat, barley, corn, rice, sugar beets, cotton, and various fruits are grown; along the dry steppe slopes Karakul sheep—whose skins provide the nation's leading export—are raised. Mazar-i-Sharik is the largest city in what is Afghanistan's major agricultural region.

Hazarajat. This region, the land of the Hazaras, is the central highland core of the nation. It is mostly a semiarid warm summer/cold winter steppe covered, at best, with grasses and low shrubs; stripping these slopes for fuel has denuded large areas. The sedentary Hazaras, separated from their fellow Afghans by a different language and religious sect, occupy the most favorable agricultural belt between 2,500 and 3,700 meters (8,000 and 12,000 feet). Few roads penetrate this vast region so it forms the nation's virtually unconquerable core.

The East. Eastern Afghanistan is the best-watered region with high mountains extending well outside the dryland zone; from here flow the Kunduz, the Helmand, and the Kabul rivers, each of great importance to the surrounding deserts. Kabul, Jalalabad, and other large towns occupy fertile basins that require irrigation to supplement the spring rains. Some dam-building for water storage and power generation has occurred. Although not the nation's agricultural center, the eastern region contains the capital city and a strategic position that makes it the most developed part of the country.

Dryland areas of AFGHANISTAN

DESERTIFICATION RISK	ARIDITY									
	Hyperarid		Arid		Semiarid		Subhumid		Aridity Totals	
	km²	%	km²	%	km²	%	km²	%	km²	%
Very High	By definition, desertification does not exist in hyperarid regions.		73,171	11.5					73,171	11.5
High			94,804	14.9					94,804	14.9
Moderate					301,590	47.4	108,801	17.1	410,391	64.5
Desertification Totals			167,975	26.4	301,590	47.4	108,801	17.1	578,366	90.9
No Desertification			11,453	1.8					11,453	1.8
Total Drylands			179,428	28.2	301,590	47.4	108,801	17.1	589,819	92.9
Non-dryland									46,447	7.1
Total Area of the Territory									636,266	100.0

Afghanistan/1203-1218

GENERAL BIBLIOGRAPHIES

1203 *Bibliography of Russian works on Afghanistan.* London: Central Asian Research Centre, 1956. 12 pp.

1204 Pearson, J. D., ed. *South Asian bibliography: a handbook and guide.*

See entry 1398.

1205 Vidergar, John J. *Bibliography on Afghanistan, The Sudan and Tunisia.* Monticello IL: Vance Bibliographies, 1978. (Public administration series bibliography P- 141) 7 pp.

Unannotated.

1206 Wilber, Donald N. *Annotated bibliography of Afghanistan.* 3rd ed. New Haven CT: HRAF, 1968. (Behaviour science bibliographies) 252 pp.

HANDBOOKS

1207 Smith, Harvey H. *Area handbook for Afghanistan.* 4th ed. Washington DC: U.S. Dept. of Defense, 1973. (American University foreign area studies) 453 pp.

Narrative summary of the nation's history, politics, geography, economy, social life; includes unannotated bibliography.

DIRECTORIES

1208 Winter, Alan; Kamm, Antony; and Khan, M. N. G. A., eds. *China, India, and Central Asia.*

See entry 1548.

GEOGRAPHY

1209 Lytle, Elizabeth Edith. *A bibliography of the geography of Afghanistan: background for planning.* Monticello IL: Council of Planning Librarians, 1976. (CPL exchange bibliography 1133) 40 pp.

Unannotated.

ATLASES

1210 Sahab, Abbas. *General atlas of Afghanistan.* Tehran: Sahab Geographic and Drafting Institute, 1973(?). 201 pp.

GAZETTEERS

1211 Adamec, Ludwig W., ed. *Historical and political gazetteer of Afghanistan.* Graz, Austria: Akademische Druck und Verlagsanstalt, 1972-.

Based on a secret 1871 British gazetteer and brought up to date in the 1970's. As of November 1981, five volumes have appeared.

1212 *Afghanistan: official standard names approved by the United States Board on Geographic Names.* Washington DC: U.S. Board on Geographic Names, 1971. (Gazetteer 119) 170 pp.

List of 10,000 place-names with latitude/longitude coordinates.

GEOLOGY

1213 *Bibliography of the geology of Afghanistan up to 1964.* Kabul: German Geological Mission to Afghanistan, 1964. 18 pp.

1214 Kaestner, Hermann. *Bibliographie zur Geologie Afghanistans und unmittelbar angrenzender Gebiete (stand ende 1970) [bibliography of the geology of Afghanistan and immediately adjacent areas, through 1970].* Hannover, Federal Republic of Germany: Geologischen Landesamtern der Bundesrepublik Deutschland, 1971. (Geologischen Jahrbuch. Beihefte 114) 43 pp.

In German, with English, French, and Russian summaries.

SOILS

1215 *Bibliography on desert soils of India, Pakistan and Afghanistan (1962-1945).*

See entry 1443.

BOTANY

1216 Meher-homji, V. M.; Gupta, Raj Kumar; and Freitag, H. "Bibliography on 'plant ecology' in Afghanistan." *Excerpta botanica: sectio B: sociologica* 12:4 (1973):310-315.

Unannotated. Contains 80 items. Author index.

URBAN GEOGRAPHY

1217 Bonine, Michael E. *Urbanization and city structures in contemporary Iran and Afghanistan: a selected annotated bibliography.*

See entry 1245.

ARCHITECTURE

1218 Harmon, Robert B. *Architectural trends in Afghanistan: a selected bibliography.* Monticello IL: Vance Bibliographies, 1982. (Architecture series A-768) 12 pp.

Annotated.

ECONOMIC DEVELOPMENT

1219 *Quarterly economic review of Pakistan, Bangladesh, Afghanistan.*

See entry 1513.

HISTORICAL GEOGRAPHY

1220 **Hanifi, M. Jamil.** *Historical and cultural dictionary of Afghanistan.* Metuchen NJ: Scarecrow, 1976. (Historical and cultural dictionaries of Asia 5) 191 pp.

1221 **Schwartzberg, Joseph E., ed.** *A historical atlas of South Asia.*

See entry 1410.

PUBLIC ADMINISTRATION

1222 **Cigar, Norman.** *Government and politics of the "Northern Tier": Turkey, Iran and Afghanistan.*

See entry 1380.

BAHRAIN
The State of Bahrain
(a former dependency of the United Kingdom)

Bahrain is a small low-lying island located in the Gulf of Bahrain between the Qatar peninsula and the mainland of Saudi Arabia. Fresh-water springs have encouraged trading settlements here for thousands of years and traces remain of the ancient civilization of Dilmun. Pearl fishing was an important activity as late as the nineteenth century; oil and natural gas have now taken their place as the leading resources and exports.

Dates, citrus, alfalfa, and vegetables are grown and there is some livestock and poultry as well as a fishing industry. Major manufacturing industries are being developed and include a large aluminum smelter, a ship-building and repair yard, and a petrochemical complex.

Dryland areas of BAHRAIN

DESERTIFICATION RISK	Hyperarid		Arid		Semiarid		Subhumid		Aridity Totals	
	km²	%	km²	%	km²	%	km²	%	km²	%
Very High	By definition, desertification does not exist in hyperarid regions.									
High			660	100.0					660	100.0
Moderate										
Desertification Totals			660	100.0					660	100.0
No Desertification										
Total Drylands			660	100.0					660	100.0
Non-dryland										
Total Area of the Territory									660	100.0

GENERAL BIBLIOGRAPHIES

1223 **Kabeel, Soraya M.** *Source book on Arabian Gulf states, Arabian Gulf in general, Kuwait, Bahrain, Qatar, and Oman.*

See entry 1313.

HANDBOOKS

1224 **Nyrop, Richard F.** *Area handbook for the Persian Gulf states.*

See entry 1314.

GAZETTEERS

1225 *Bahrain, Kuwait, Qatar, and United Arab Emirates: official standard names approved by the United States Board on Geographic Names.* Washington DC: U.S. Board on Geographic Names, 1976. (Gazetteer 143) 145 pp.

List of 7,650 place-names with latitude/longitude coordinates.

ECONOMIC DEVELOPMENT

1226 *Quarterly economic review of Bahrain, Qatar, Oman, the Yemens.* London: Economist Intelligence Unit, 1980-. Quarterly.

Includes summary of political and economic news as well as charts and statistics of selected economic indicators. Annual supplement.

IRAN
The Islamic Republic of Iran
(known as Persia before 1935)

Iran is a large, topographically diverse nation with great natural wealth, a long history of civilization, and a tendency for its people to erupt in extreme religious and political fanaticism. Nearly all the land is dry to some extent.

The Northwest. Facing the border with Turkey and the Soviet Union, the northwest corner is a broken mountain/plateau complex with a semiarid cold winter climate and stunted steppe-scrub vegetation. Agriculture flourishes where irrigation works bring wellwater to the volcanic soils, especially around the salty Lake Urmia; wheat is the principal crop, while vegetables, grapes, cotton, rice, tobacco, melons, and opium poppies are also grown. The uplands support nomadic herds of sheep and goats. The people of this region are a diverse mixture with strong ties to neighboring groups in Turkey, Iraq, and the Soviet Union; these include Kurds, Turks, Armenians, and Azerbaijanis.

The Zagros-Makran Mountains. Along Iran's western and southern borders the Zagros and Makran ranges rise from deserts on either side to form a wall that isolates the nation from the Mesopotamian lowlands and the sea. The higher slopes carry snow through the winter months and provide runoff for pastures and limited agriculture. Deforestation due to goat grazing and charcoal harvesting has been extensive, contributing greatly to the degraded aspect of the landscape.

The Interior Basins. This region is the geographical heart of Iran and also the most inhospitable. The climate is arid to hyperarid with cool winters, very hot summers, and extreme summer drought. Drainage cannot reach the sea so vast salt flats have developed in the lowlands. Agriculture and settlement are limited to the basins immediately adjacent to the Zagros Mountains where ancient qanat irrigation tunnels carry water from the highland pediment.

The Eastern Highlands. Iran's northeastern corner is a semiarid highland facing the Soviet Union and Afghanistan. The region is similar to Iran's northwest with the presence of scrub and grassland on the slopes and agricultural oases in the most favored valleys.

Dryland areas of IRAN

DESERTIFICATION RISK	ARIDITY									
	Hyperarid		Arid		Semiarid		Subhumid		Aridity Totals	
	km²	%	km²	%	km²	%	km²	%	km²	%
Very High	By definition, desertification does not exist in hyperarid regions.		47,792	2.9					47,792	2.9
High			960,784	58.3	6,592	0.4			967,376	58.7
Moderate			47,792	2.9	341,136	20.7	141,728	8.6	530,656	32.2
Desertification Totals			1,056,368	64.1	347,728	21.1	141,728	8.6	1,545,824	93.8
No Desertification	28,016	1.7					13,184	0.8	41,200	2.5
Total Drylands	28,016	1.7	1,056,368	64.1	347,728	21.1	154,912	9.4	1,587,024	96.3
Non-dryland									60,976	3.7
Total Area of the Territory									1,648,000	100.0

GENERAL BIBLIOGRAPHIES

1227 Ehlers, Eckart. *Iran: ein bibliographischer Forschungsbericht mit Kommentaren und Annotionen [Iran: a bibliographic research survey with comments and annotations].* Munchen, Federal Republic of Germany: K. G. Saur, 1980. (Bibliographien zur regionalen Geographie und Landeskunde 2) 441 pp.

In English and German.

HANDBOOKS

1228 Nyrop, Richard F. *Iran: a country study.* 3rd ed. Washington DC: U.S. Dept. of Defense, 1978. (American University foreign area studies) 492 pp.

Narrative summary of the nation's history, politics, geography, economy, social life; includes unannotated bibliography.

DIRECTORIES

1229 Winter, Alan; Kamm, Antony; and Khan, M. N. G. A., eds. *China, India, and Central Asia.*

See entry 1548.

GEOGRAPHY

1230 Burgess, R. L.; Mokhtarzadeh, A.; and Cornwallis, L. *A preliminary bibliography of the natural history of Iran.* Shiraz, Iran: Pahlavi University. College of Arts and Sciences, 1966. (Science bulletin 1:220) 143 pp.

In English and Farsi. Contains 1,719 items.

1231 Giul, K. K.; Lappalainen, T. N.; and Polushkin, V. A. *Kaspiiskoe More [Caspian Sea].*

See entry 1609.

1232 Snead, Rodman E. *Physical geography of the Makran coastal plain of Iran: final report, reconnaissance phase.* Albuquerque NM: University of New Mexico for Office of Naval Research, 1970. (NTIS: AD 707-745) 715 pp.

Includes a 779-item bibliography.

ATLASES

1233 *Atlas of Iran: white revolution proceeds and progresses.* Tehran: Sahab Geographic and Drafting Institute, 1971. 190 pp.

In English, French, Farsi.

GAZETTEERS

1234 Adamec, Ludwig W., ed. *Historical gazetteer of Iran.* Graz, Austria: Akademische Druck und Verlagsanstalt, 1976-.

Based on early twentieth century British sources and brought up to date in the 1970's.

1235 *Iran: official standard names approved by the United States Board on Geographic Names.* Washington DC: U.S. Board on Geographic Names, 1956. (Gazetteer 19) 578 pp.

List of 46,000 place-names with latitude/longitude coordinates.

GEOLOGY

1236 Berberian, Manuel, and Tchalenko, J. S. *Bibliography with abstracts on the seismicity and tectonics of Iran.* Tehran: Iran. Geological Survey, 1976.

1237 Ridge, John Drew. "Iran." In *Annotated bibliographies of mineral deposits in Africa, Asia (exclusive of the USSR) and Australasia*, pp. 253-258. Oxford U.K.: Pergamon, 1976.

1238 Rosen, Norman C. *Bibliography of geology of Iran.* Tehran: Iran. Geological Survey, 1969. (Special publication 2) 77 pp.

CLIMATOLOGY

1239 *Climatic atlas of Iran.* Tehran: University of Tehran. Institute of Geography, 1965. 12 pp., 117 plates.

In English and Farsi. Scale 1:5,000,000.

1240 Peterson, A. D. *Bibliography on the climate of Iran.* Washington DC: U.S. Weather Bureau, 1957. (NTIS: PB 176-024; AD 664-694) 26 pp.

1241 Weight, M. L., and Gold, H. *An annotated bibliography of climatic maps of Iran.* Washington DC: U.S. Weather Bureau, 1962. (NTIS: AD 660-857) 17 pp.

SOILS

1242 *Soils of Iran and Iraq (annotated bibliography 1944-1975).* Harpenden, U.K.: Commonwealth Agricultural Bureaux, 1977. (Annotated bibliography) 10 pp.

HUMAN GEOGRAPHY

1243 Mahdi, Ali-Akbar. *A selected bibliography of academic theses dealing with socio-politico-economic aspects of Iranian society.* Monticello IL: Vance Bibliographies, 1980. (Public administration series bibliography P-599) 14 pp.

Unannotated.

1244 Vidergar, John J. *The economic and social development of Iran.* Monticello IL: Council of Planning Librarians, 1977. (CPL exchange bibliography 1380) 6 pp.

Unannotated.

URBAN GEOGRAPHY

1245 Bonine, Michael E. *Urbanization and city structures in contemporary Iran and Afghanistan: a selected annotated bibliography.* Monticello IL: Council of Planning Librarians, 1975. (CPL exchange bibliography 875) 31 pp.

Extensive annotations.

ECONOMIC DEVELOPMENT

1246 Bartsch, William H., and Bharier, Julian. *The economy of Iran, 1940-1970.* Durham U.K.: University of Durham, 1971. (Publication 2) 114 pp.

Unannotated.

1247 **Indesha, N.** *The Iranian economy: a selected bibliography.* Monticello IL: Vance Bibliographies, 1981. (Public administration series bibliography P-703) 6 pp.

Unannotated.

1248 **Mahdi, Ali-Akbar.** *A selected bibliography on political economy of Iran.* Monticello IL: Vance Bibliographies, 1980. (Public administration series bibliography P-598) 104 pp.

Unannotated.

1249 *Quarterly economic review of Iran.* London: Economist Intelligence Unit, 1976-. Quarterly.

Includes summary of political and economic news as well as charts and statistics of selected economic indicators. Annual supplement.

HISTORICAL GEOGRAPHY

1250 **Bastani-Parizi, Mohammad Ebrahim, et al.** *Historical atlas of Iran.* Tehran: Tehran University Press, 1971. Approximately 250 pp.

In English, French, Farsi. Contains 26 maps; .historical-anthropological. Scale 1:7,500,000.

PUBLIC ADMINISTRATION

1251 **Cigar, Norman.** *Government and politics of the "Northern Tier": Turkey, Iran and Afghanistan.*

See entry 1380.

IRAQ
Al Jumhouriya al 'Iraqia
[The Republic of Iraq]
(a former dependency of the United Kingdom)

Iraq occupies the basin of the Tigris and Euphrates Rivers, the location of one of man's earliest and most extensive agricultural civilizations. The presence of these two rivers gives climatically arid Iraq a vital resource base, similar to what the Nile provides Egypt, the Indus provides Pakistan, or, indeed, the Colorado provides the Southwestern United States.

The Tigris-Euphrates Lowland. As rich as this land is, the twin exotic rivers and their floodplain present difficult management problems. The alluvial soils have a high calcium content due to limestone formations in their place of origin; to this is added heavy doses of salts leached from the high water table that prevails in vast areas where poor drainage and centuries of irrigation have done their worst. Although development projects in the region have usually concentrated on bringing more water into agricultural lands, without equal attention being paid to ridding the soil of excess water, efforts to increase agricultural production are doomed.

Wheat is the principal crop, followed by barley (more salt tolerant and therefore planted closest to the Gulf), rice (planted in the delta marshlands), dates (in which Iraq leads the world in production), maize, sorghum, cotton, vegetables and other crops.

A peculiar subregion is the delta of the two rivers, formed where they meet and create a main channel (called the Shatt al Arab) and many distributaries. Access into the area is virtually impossible except by small boat and the isolation has allowed the native Marsh Arabs to maintain considerable cultural and economic independence.

The Northern Highlands. At the upper reaches of the Tigris Iraq occupies an extensive upland portion of the highland where Iran, Iraq, and Turkey meet. This is the home of the Kurds and other non-Arab peoples who have historically been in conflict with those from the lowlands. With a semiarid to subhumid climate rain-fed winter wheat is cultivated in favored valleys while the slopes support herds of sheep and goats.

The Western Desert. Beyond the Euphrates to the west the fertile lowlands give way to an extensive arid gravel desert that merges imperceptibly into neighboring Syria, Jordan, and Saudi Arabia. Save for scattered nomads the region is uninhabited.

Dryland areas of IRAQ

DESERTIFICATION RISK	Hyperarid km²	Hyperarid %	Arid km²	Arid %	Semiarid km²	Semiarid %	Subhumid km²	Subhumid %	Aridity Totals km²	Aridity Totals %
Very High	By definition, desertification does not exist in hyperarid regions.		90,320	20.6	2,630	0.6			92,950	21.2
High			288,060	65.7	24,114	5.5			312,174	71.2
Moderate							33,322	7.6	33,322	7.6
Desertification Totals			297,080	86.3	26,744	6.1	33,322	7.6	438,446	100.0
No Desertification										
Total Drylands			297,080	86.3	26,744	6.1	33,322	7.6	438,446	100.0
Non-dryland										
Total Area of the Territory									438,466	100.0

HANDBOOKS

1252 **Nyrop, Richard F.** *Iraq: a country study.* 3rd ed. Washington DC: U.S. Dept. of Defense, 1979. (American University foreign area studies) 320 pp.

Narrative summary of the nation's history, politics, geography, economy, social life; includes unannotated bibliography.

GEOGRAPHY

1253 **Lytle, Elizabeth E.** *The geography of Iraq.* Monticello IL: Council of Planning Librarians, 1977. (CPL exchange bibliography 1294) 36 pp.

Unannotated.

GAZETTEERS

1254 *Iraq: official standard names approved by the United States Board on Geographic Names.* Washington DC: U.S. Board on Geographic Names, 1957. (Gazetteer 37) 175 pp.

List of 13,200 place-names with latitude/longitude coordinates.

GEOLOGY

1255 **Alsinawi, S. A., and Naqash, A. B.** "Bibliography on the geology of Iraq." University of Baghdad. College of Science. *Bulletin* 17:2 (1976):517-569.

CLIMATOLOGY

1256 *Bibliography on the climate of Iraq.* Washington DC: U.S. Weather Bureau, 1958. (NTIS: PB 176-035; AD 665-182) 14 pp.

Contains 50 annotated citations.

1257 *Climatological atlas for Iraq.* Baghdad: Iraq. Meteorological Dept. Climatological Section, 1962. (Publication 13) 217 leaves.

Scale 1:5,000,000.

1258 **Roman, S. J.** *Bibliography of climatic maps for Iraq.* Washington DC: U.S. Weather Bureau, 1958. (NTIS: PB 176-038; AD 665-176) 19 pp.

1259 *Secondary bibliography on the climate of Iraq.* Washington DC: U.S. Weather Bureau, 1958. (NTIS: PB 176-023; AD 665-185) 14 pp.

SOILS

1260 *Soils of Iran and Iraq (annotated bibliography 1944-1975).*

See entry 1242.

BOTANY

1261 **Hadac, E., and Kreeb, Karlheinz.** "Bibliographia phytosociologica: Iraq" [bibliography of plant ecology: Iraq]. *Excerpta botanica: sectio B: sociologica* 3:1 (1961):78; 7:2 (1966):102-104.

Unannotated. Part I contains items 1-12; Part II contains items 13-34. Author indexes.

ZOOLOGY

1262 **Allouse, B. E.** "A bibliography of the vertebrate fauna of Iraq and neighboring countries' mammals, 2nd compilation." Iraq Natural History Research Centre and Museum. *Publication* 25 (1968):1, 336.

AGRICULTURE

1263 **Cigar, Norman.** *Agriculture and rural development in the Fertile Crescent: Lebanon, Syria, Jordan and Iraq.*

See entry 1323.

MIGRATION

1264 **Vidergar, John J.** *Migration studies of Iraq.* Monticello IL: Council of Planning Librarians, 1977. (CPL exchange bibliography 1379) 7 pp.

Unannotated.

MEDICINE

1265 **Abul-Hab, J., and Abdul-Rasoul, M. S.** "Bibliography of medical entomology in Iraq." *Bulletin of endemic diseases* 15:1/2 (1974):71-83.

HISTORICAL GEOGRAPHY

1266 **Beck, Martinus Adrianus.** *Atlas of Mesopotamia: a survey of the history and civilization of Mesopotamia from the Stone Age to the fall of Babylon.* New York: Nelson, 1962. 164 pp.

ECONOMIC DEVELOPMENT

1267 *Quarterly economic review of Iraq.* London: Economist Intelligence Unit, 1976-. Quarterly.

Includes summary of political and economic news as well as charts and statistics of selected economic indicators. Annual supplement.

ISRAEL
Medinat Israel
[The State of Israel]

(established from portions of Palestine, which was a mandated territory of the United Kingdom)

Israel is a tiny nation—so small that one can view the sea from beyond its eastern border—but nonetheless it possesses immense historical, political, and strategic importance. Because it occupies the most fertile land along the route from Asia to Africa it has long served as a channel of cultural and military invasion where European, Middle Eastern, and North African peoples have crossed and clashed.

The land itself ranges from a narrow semiarid coastal strip with a typically Mediterranean climate and vegetation to a hyperarid desert immediately inland and to the south. As in so much of the Middle East, centuries of cultivation and grazing have depleted the natural stands of timber and denuded the soil cover; however, capital improvements and Israeli innovation in dryland agricultural techniques have restored large areas to a high productivity to the point where the nation serves as a model illustrating the possibilities of arid lands development. Unfortunately, Israeli technological expertise has not been matched by the social and political expertise needed to solve the nation's increasingly unmanageable problems concerning disenfranchised Palestinians and Israel's hostile Arab neighbors.

The North. This region—corresponding to ancient Galilee, Samaria, and Judaea—is subhumid and holds most of the nation's naturally arable lands. Upland areas receive considerable precipitation (to 1,000 mm or 40 inches or so) and are being reforested to their original vegetation cover. On gentler slopes and lowlands agriculture is intensive and highly productive, most so on the coastal Plain of Sharon which is the center of Israel's important citrus industry. Other crops follow the typical Mediterranean pattern: wheat, barley, olives, figs, grapes, cotton, tobacco, sugar beets, etc. The biggest hurdles to dryland development in the region include the social and political questions of Palestinian land tenure, new Israeli settlements, and assimilation of the West Bank's land and people into a peaceful political state.

The Negev Desert. This region stands in sharp contrast to the agriculturally prosperous north, though here the development of dryland agricultural technology is as dramatic as anywhere on earth. Although the Negev is naturally a hyperarid desert of rocky tablelands and dry wadis it supported intensive irrigated agriculture since ancient times. Modern Israelis have utilized trickle irrigation, hydroponics, aquaculture, genetic manipulation, and other unconventional methods to supplement equally promising ancient techniques of runoff farming to reclaim lands uncultivated for thousands of years.

Dryland areas of ISRAEL
(INCLUDES THE WEST BANK, GAZA, GOLAN HEIGHTS)

DESERTIFICATION RISK	ARIDITY								Aridity Totals	
	Hyperarid		Arid		Semiarid		Subhumid			
	km²	%	km²	%	km²	%	km²	%	km²	%
Very High	By definition, desertification does not exist in hyperarid regions.									
High			17,173	18.2					17,173	18.2
Moderate					40,008	42.4	11,418	12.1	51,426	54.5
Desertification Totals			17,173	18.2	40,008	42.4	11,418	12.1	68,599	72.7
No Desertification	25,760	27.3							25,760	27.3
Total Drylands	25,760	27.3	17,713	18.2	40,008	42.4	11,418	12.1	94,359	100.0
Non-dryland										
Total Area of the Territory									94,359	100.0

GENERAL BIBLIOGRAPHIES

1268 Emanuel, Muriel. *Israel: a survey and bibliography.* New York: St. Martin's Press, 1971. 309 pp.

INDEXES AND ABSTRACTS

1269 *Index to Jewish periodicals.* Columbia Heights OH: College of Jewish Studies Press, June/Aug 1963-. 2 per year.

Indexes English-language journals covering Jewish history, culture, and political concerns. Coverage is worldwide but of obvious importance to research on Israel.

HANDBOOKS

1270 *Israel government yearbook 5740 (1979).* Jerusalem: Israel. Government Printer, 1980. Annual.

In Hebrew with some English translations.

1271 Nyrop, Richard F. *Israel: a country study.* 2nd ed. Washington DC: U.S. Dept. of Defense, 1979. (American University foreign area studies) 414 pp.

Narrative summary of the nation's history, politics, geography, economy, social life; includes unannotated bibliography.

ENCYCLOPEDIAS

1272 *Encyclopaedia Judaica.* New York: Macmillan, 1972. 16 volumes.

A major work of reference covering the land of Israel and the Jewish people. Full articles are signed and include bibliographies.

DIRECTORIES

1273 Dudai, Yadin, ed. *Israel.* 2nd ed. Guernsey, U.K.: Francis Hodgson, 1975. (Guide to world science 13) 184 pp.

Provides descriptions of scientific research activities and organizations.

STATISTICS

1274 *Statistical abstract of Israel.* Jerusalem: Israel. Central Bureau of Statistics, 1949/50-. Annual.

In English and Hebrew.

GEOGRAPHY

1275 Creasi, Vincent J. et al. *A selected annotated bibliography of environmental studies of Israel (1960-1969).* Washington DC: U.S. Air Force, 1970. (Report ETAC- TN-70-4; NTIS: AD 705-199) 59 pp.

Contains 119 citations.

1276 Orni, E., and Efrat, E. *Geography of Israel.* 3rd ed. Jerusalem: Israel Universities Press, 1973. 551 pp.

1277 Slobodkin, Lawrence B., and Loya, Yossef. "Israel." In *Handbook of contemporary developments in world ecology*, pp. 549-559. Edited by Edward John Kormondy and J. Frank McCormick. Westport CT: Greenwood, 1981.

A brief narrative on ecological research in Israel. Unlike the other chapters in the book, this one lacks references, bibliographies and "hard" information on research organizations and activities.

1278 Speece, Mark, comp. *Draft environmental report on West Bank and Gaza.* Tucson AZ: University of Arizona. Office of Arid Lands Studies. Arid Lands Information Center, 1980. 52 pp.

A narrative summary of the two region's environment and development problems.

ATLASES

1279 Amiran, David H. K.; Shackar, Arie; and Kimhi, Israel. *Atlas of Jerusalem.* Berlin, Federal Republic of Germany: W. de Gruyter, 1973. 8 pp. 53 col. maps.

In English and Hebrew. Most maps at a scale of 1:30,000.

1280 *Atlas of Israel; cartography, physical geography, human and economic geography, history.* 2nd ed. Jerusalem: Israel. Ministry of Labour. Survey of Israel, 1970. Various pagings.

Includes areas under Israeli administration since the June 1967 war. Historic maps before 1948 cover Palestine. Emphasis on historical geography. Commentaries accompany each map. First edition published in Hebrew as *Atlas Yisrael* (1956-1964).

GAZETTEERS

1281 *Israel: official standard names approved by the United States Board on Geographic Names.* Washington DC: U.S. Board on Geographic Names, 1970. (Gazetteer 114) 123 pp.

List of 7,400 place-names with latitude/longitude coordinates.

1282 *Place names in Israel.* Jerusalem: Israel Program for Scientific Translations, 1962. 402 pp.

Provides locations, history, population; focuses on "new" settlements (as of 1962) established by immigrants. All Hebrew names are transliterated into the Roman alphabet.

GEOLOGY

1283 Bartov, Yosef, comp. *Geological survey of Israel research in Sinai, 1967-1980: collected reprints.* Jerusalem: Israel. Ministry of Energy and Infrastructure. Geological Survey, 1980. (Special publication 1) 2 vols.

Includes bibliographies.

CLIMATOLOGY

1284 Gold, H. K. *An annotated bibliography on the climate of Israel.* Washington DC: Air Weather Service Climatic Center, 1962. (NTIS: AD 660-872) 50 pp.

Israel/1285-1300

SOILS

1285 *Bibliography on desert soils of Israel and Jordan (1964-1954).* Harpenden, U.K.: Commonwealth Bureau of Soils, 196- (n.d.). 9 pp.

Contains 41 citations.

BIOLOGY

1286 Sarig, S. "Bibliography on fish ponds and inland waters in Israel except Lake Kinnereth and the Dead Sea bibliography." *Bamidgeh* 20:2 (1968):35-71.

1287 Steinitz, H., and Oren, O. H. "Bibliography on Lake Kinnereth/Lake Tiberias." Haife. Tahanah le- heker ha- dayig ha-yami./Haifa. Sea Fisheries Research Station. *Bulletin* 53 (1968):3-48.

BOTANY

1288 Gruenberg-Fertig, I., and Zohary, M. "Bibliographia phytosociologica: Israel" [bibliography of plant ecology: Israel]. *Excerpta botanica: sectio B: sociologica* 1:3 (1959):202-212; 5:2 (1963):157-160; 15:4 (1976):266-276.

Unannotated. Part I contains items 1-119; Part II contains items 120-155; Part III contains items 156-297.

1289 Tadmor, R. *Bibliography: vegetation of Israel.* Jerusalem: Hebrew University. Botany Dept. 1972. 73 pp.

In English and Hebrew. Contains about 450 items.

PLANT CULTURE

1290 Shalhevet, J., ed. *Irrigation of field and orchard crops under semi-arid conditions.* Rev. ed. Bet Dagan, Israel: International Irrigation Information Center, 1979. (I.I.I.C. publication 1) 124 pp.

Reports on studies in Israel, 1954-1974, to establish optimum irrigation regime for various field crops. Provides data from Israel. Bibliography contains Hebrew and English sources.

AGRICULTURAL ECONOMICS

1291 Gat, Z., and Marton, Shraga T. *Rural development in Israel: a list of publications in languages other than Hebrew.* Tel Aviv: Ministry of Agriculture. Dept. for Agricultural Cooperation with Developing Countries. Centre for Comparative Studies on Agricultural Development, 1966. 69 pp.

HUMAN GEOGRAPHY

1292 Vidergar, John J. *Israel: social and economic development.* Monticello IL: Vance Bibliographies, 1978. (Public administration series bibliography P-31) 4 pp.

Unannotated.

WOMEN'S STUDIES

1293 Lytle, Elizabeth E. *Women in Israel: a selected bibliography.* Monticello IL: Vance Bibliographies, 1979. (Public administration series bibliography P-194) 19 pp.

Unannotated.

KIBBUTZ

1294 Erickson, Judith B. *The Israeli kibbutz.* Monticello IL: Council of Planning Librarians, 1978. (CPL exchange bibliography 1454) 67 pp.

Unannotated.

1295 Rabin, Albert I. *Kibbutz studies: a digest of books and articles on the kibbutz by social scientists, educators, and others.* East Lansing MI: Michigan State University Press, 1971. 124 pp.

Extensive annotations.

1296 Shur, Shimon. *Kibbutz bibliography.* Tel Aviv, Israel: Council for Higher Education of the Federation of Kibbutz Movements, 1971. 110 pp.

Unannotated. Based on E. Cohen's "Bibliography of the Kibbutz" (1964).

1297 Shur, Shimon et al. *The kibbutz: a bibliography of scientific and professional publications in English.* Darby PA: Norwood, 1981. 103 pp.

Unannotated list of 951 references.

URBAN GEOGRAPHY

1298 Golany, Gideon. *City and regional planning and development in Israel.* Monticello IL: Council of Planning Librarians, 1968. (CPL exchange bibliography 56) 30 pp.

Unannotated.

ECONOMIC DEVELOPMENT

1299 *Quarterly economic review of Israel.* London: Economist Intelligence Unit, 1976-. Quarterly.

Includes summary of political and economic news as well as charts and statistics of selected economic indicators. Annual supplement.

PUBLIC ADMINISTRATION

1300 Cigar, Norman. *Government and politics in Israel.* Monticello IL: Vance Bibliographies, 1979. (Public administration series bibliography P-169) 7 pp.

Unannotated.

JORDAN
Al Mamlaka al Urduniya al Hashemiyah
[The Hashemite Kingdom of Jordan]

(as Trans-Jordan, a former mandated territory of the United Kingdom)

Jordan is a small resource-poor nation whose most valuable lands—the West Bank of the Jordan River—were lost to Israel in 1967. The remaining territory is almost entirely arid except for the semiarid hills just east of the Jordan River.

The Northern Hills. Although lying in the rainshadow of Israel's Judean Hills, this semiarid region is Jordan's most fertile land. Amman—the capital and chief settlement—receives about 280 mm (11 inches) of rainfall annually, which supports a heavily-grazed scrub vegetation. Lowlands along the Jordan River and other tributary wadis support irrigated and rain-fed crops; wheat and barley are the predominant cereals followed by smaller acreages of olives, sorghum, fruits and vegetables, and others. The limits of arable land have been reached unless large-scale water diversion schemes can deliver water to the arid lands to the east and south.

The Eastern Desert and Dead Sea Coast. This arid to hyperarid region holds little promise for development without a great increase in available water supplies. Two resources deserve mention. One is the existence of quantities of dissolved minerals in the Dead Sea; Jordan operates an evaporating plant within sight of a similar Israeli plant at the Sea's southern end. The second resource is Jordan's small frontage on the Gulf of Aqaba where the port of Aqaba gives the nation access to the sea.

Dryland areas of JORDAN (EXCLUDES THE WEST BANK)

DESERTIFICATION RISK	ARIDITY									
	Hyperarid		Arid		Semiarid		Subhumid		Aridity Totals	
	km²	%	km²	%	km²	%	km²	%	km²	%
Very High	By definition, desertification does not exist in hyperarid regions.									
High			62,699	63.9					62,699	68.9
Moderate			2,002	2.2	8,099	8.9			10,101	11.1
Desertification Totals			64,701	71.1	8,099	8.9			72,800	80.0
No Desertification	18,200	20.0							18,200	20.0
Total Drylands	18,200	20.0	64,701	71.1	8,099	8.9			91,000	100.0
Non-dryland										
Total Area of the Territory									91,000	100.0

HANDBOOKS

1301 **Nyrop, Richard F.** *Jordan: a country study.* 3rd ed. Washington DC: U.S. Dept. of Defense, 1980. (American University foreign area studies) 310 pp.

Narrative summary of the nation's history, politics, geography, economy, social life; includes unannotated bibliography.

GAZETTEERS

1302 *Jordan: official standard names approved by the United States Board on Geographic Names.* Washington DC: U.S. Board on Geographic Names, 1971. (Gazetteer 3) 419 pp.

List of 22,000 place-names with latitude/longitude coordinates. Covers Jordan, the West Bank of the Jordan River currently under Israeli occupation, and the so-called No Man's Lands between Jordan and Israel.

CLIMATOLOGY

1303 *al-Atlas al-Marakhi lil-Urdun/Climatic atlas of Jordan.* Amman: Jordan. Meteorological Dept., 1971. 127 leaves.

In English and Arabic. Based on data contained in the *Jordan climatological data handbook, 1980.*

1304 Carraway, D. M. *Annotated bibliography of climatic maps of Jordan.* Washington DC: U.S. Weather Bureau, 1961. (NTIS: PB 176-027; AD 664-718) 13 pp.

1305 Wallace, J. Allen, Jr. *An annotated bibliography on the climate of Jordan.* Washington DC: U.S. Weather Bureau, 1961. (NTIS: AD 662-589) 33 pp.

Contains 95 citations.

SOILS

1306 *Bibliography on desert soils of Israel and Jordan (1964-1954).*

See entry 1285.

BOTANY

1307 Kasapligil, Baki. *A bibliography on the botany and forestry of the Hashemite Kingdom of Jordan.* Amman: FAO. Expanded Technical Assistance Program, 1955. 17 pp.

Includes 154 unannotated references.

AGRICULTURE

1308 Cigar, Norman. *Agriculture and rural development in the Fertile Crescent: Lebanon, Syria, Jordan and Iraq.*

See entry 1323.

URBAN GEOGRAPHY

1309 Tesdell, Lee S. *Urbanization and economic growth in Jordan and Syria.* Monticello IL: Council of Planning Librarians, 1978. (CPL exchange bibliography 1449) 7 pp.

Some annotations.

1310 Tesdell, Lee S. *A working bibliography for use in urban planning: Amman, Jordan—the history of its social and economic development (1878-1948).* Monticello IL: Vance Bibliographies, 1980. (Public administration series bibliography P-528) 14 pp.

Unannotated.

ARCHITECTURE

1311 Tesdell, Lee S. *History of architecture in Jordan.* Monticello IL: Vance Bibliographies, 1980. (Architecture series bibliography A393) 10 pp.

Unannotated.

ECONOMIC DEVELOPMENT

1312 *Quarterly economic review of Syria, Jordan.* London: Economist Intelligence Unit, 1978-. Quarterly.

Includes summary of political and economic news as well as charts and statistics of selected economic indicators. Annual supplement.

KUWAIT
Dowlat al Kuwait
[The State of Kuwait]
(a former protectorate of the United Kingdom)

Before the discovery of oil, Kuwait was one of several tiny Arabian coastal settlements with a small pearl fishery, negligable agriculture, and some local maritime trade. Now, with the development of incredibly rich oil reserves, Kuwait, like its neighbors, has been transformed into an immensely prosperous state with a per capita income perhaps the highest in the world. Lacking significant surface resources and even a dependable source of fresh water, Kuwait's challenge is to transform its wealth into a stable economic base that will allow the nation to survive once the oil is gone. Several strategies are being pursued: technical and administrative education for all qualified citizens, development of service industries (especially banking) for the region, and expansion of non-oil industries such as boat-building, shrimp fishing, and fruit and vegetable farming.

Dryland areas of KUWAIT

DESERTIFICATION RISK	ARIDITY								Aridity Totals	
	Hyperarid		Arid		Semiarid		Subhumid			
	km²	%	km²	%	km²	%	km²	%	km²	%
Very High	By definition, desertification does not exist in hyperarid regions.									
High			24,280	100.0					24,280	100.0
Moderate										
Desertification Totals			24,280	100.0					24,280	100.0
No Desertification										
Total Drylands			24,280	100.0					24,280	100.0
Non-dryland										
Total Area of the Territory									24,280	100.0

GENERAL BIBLIOGRAPHIES

1313 **Kabeel, Soraya M.** *Source book on Arabian Gulf states, Arabian Gulf in general, Kuwait, Bahrain, Qatar, and Oman.* Kuwait: Kuwait University. Libraries Dept., 1975. 427 pp.

An extensive, though unannotated, list of 3,377 books and journal articles.

HANDBOOKS

1314 **Nyrop, Richard F.** *Area handbook for the Persian Gulf states.* Washington DC: U.S. Dept. of Defense, 1977. (American University foreign area studies) 448 pp.

Narrative summary of the nation's history, politics, geography, economy, social life; includes unannotated bibliography. Covers Kuwait, Bahrain, Qatar, U. A. E., and Oman.

GAZETTEERS

1315 *Bahrain, Kuwait, Qatar, and United Arab Emirates: official standard names approved by the United States Board on Geographic Names.*

See entry 1225.

ECONOMIC DEVELOPMENT

1316 *Quarterly economic review of Kuwait.* London: Economist Intelligence Unit, 1978-. Quarterly.

Includes summary of political and economic news as well as charts and statistics of selected economic indicators. Annual supplement.

LEBANON
Al-Jumhouriya al-Lubnaniya
[The Republic of Lebanon]
(a former mandated territory of France)

Lebanon is a classic example of a tiny nation with considerable natural and human resources that nonetheless suffers terribly because of its strategic position between powerful warring factions both within and beyond the nation's borders. Like Israel, it occupies the fertile routeway from Asia to Africa but, unlike Israel, Lebanon lacks a desert interior and so historically has turned to the sea.

The nation is so small that at the scale of other countries there is only a single dryland region; however, even Lebanon has important differences in local physiography and land use. The coastal plain is a very fertile subhumid region that supports citrus, olives, bananas, sugar cane, and vegetable crops. Inland, the Lebanon Mountains rise quite high (to 3,000 meters or 10,000 feet) and provide important water supplies from its winter snowpack; the famous cedars grow in isolated stands, and deep canyons support ancient monasteries and other havens for minority peoples. The Bekaa Valley is Lebanon's most extensive agricultural area; it lies east of the mountains and is drained by the Litani River, which itself provides considerable hydroelectric potential. Beyond the Bekaa Valley the Anti-Lebanon Mountains are drier, support primarily herdsmen, and form the border with Syria.

Dryland areas of LEBANON

DESERTIFICATION RISK	ARIDITY								Aridity Totals	
	Hyperarid		Arid		Semiarid		Subhumid			
	km²	%	km²	%	km²	%	km²	%	km²	%
Very High	By definition, desertification does not exist in hyperarid regions.									
High										
Moderate					4,337	41.7			4,337	41.7
Desertification Totals					4,337	41.7			4,337	41.7
No Desertification							6,063	58.3	6,063	58.3
Total Drylands					4,337	41.7	6,063	58.3	10,400	100.0
Non-dryland										
Total Area of the Territory									10,400	100.0

HANDBOOKS

1317 **Smith, Harvey H.** *Area handbook for Lebanon.* 2nd ed. Washington DC: U.S. Dept. of Defense, 1974. (American University foreign area studies) 354 pp.

Narrative summary of the nation's history, politics, geography, economy, social life; includes unannotated bibliography.

GAZETTEERS

1318 *Lebanon: official standard names approved by the United States Board on Geographic Names.* Washington DC: U.S. Board on Geographic Names, 1970. (Gazetteer 115) 676 pp.

List of 37,000 place-names with latitude/longitude coordinates.

CLIMATOLOGY

1319 **Grimes, Annie E.** *An annotated bibliography on climate maps of Lebanon.* Washington DC: U.S. Weather Bureau, 1960. (NTIS: PB 176-005 and AD 664-708) 14 pp.

1320 **Grimes, Annie E.** *An annotated bibliography on climate maps of Lebanon.* Washington DC: U.S. Weather Bureau, 1961. (NTIS: PB 176-004 and AD 664-698) 42 pp.

SOILS

1321 **Butters, B.** *Soils of Syria and Lebanon (1973-1939).*

See entry 1359.

ZOOLOGY

1322 **Kumerloeve, Hans.** "Die Saugetiere (Mammalia) Syriens und des Libanon" [the mammals (mammalia) of Syria and Lebanon: a preliminary review (as of 1974)].

See entry 1360.

AGRICULTURE

1323 **Cigar, Norman.** *Agriculture and rural development in the Fertile Crescent: Lebanon, Syria, Jordan and Iraq.* Monticello IL: Vance Bibliographies, 1981. (Public administration series bibliography P-698) 5 pp.

Unannotated.

ECONOMIC DEVELOPMENT

1324 *Quarterly economic review of Lebanon, Cyprus.* London: Economist Intelligence Unit, 1978-. Quarterly.

Includes summary of political and economic news as well as charts and statistics of selected economic indicators. Annual supplement.

TOURISM AND RECREATION

1325 **al-Barbar, Aghil.** *Tourism in the Arab world: an introductory bibliography.* Monticello IL: Vance Bibliographies, 1980. (Public administration series bibliography P-601) 5 pp.

Unannotated. Covers Egypt, Lebanon, Syria, and Tunisia.

OMAN
Saltanat Oman
[The Sultanate of Oman]

Oman occupies a long stretch of southeastern Arabian coastline from the strategic Strait of Hormuz to the border with South Yemen; inland there exists an ill-defined border with Saudi Arabia.

The Coastal Plain. This region is a hot, arid, and—due to the humidity—extremely uncomfortable place to live. Agriculture is generally limited to scattered plots of sugar cane, dates, and fruit; fishing is at least as important. Most of the land, if used at all, is given over to cattle and camel grazing. Northwest of Muscat are extensive date plantations which recently have expanded with the rebuilding of the ancient *aflaj* irrigation system as well as with increased pumping of groundwater.

The Green Mountains (al Jabal al Akhdar). These highlands inland of the northern coast actually extend into the semiarid zone and form, with the Yemen Highlands, the only naturally favored agricultural land on the Arabian Peninsula. Hillsides are carefully terraced with rock walls and the annual 300 mm (12 inch) precipitation supports onions, garlic, and fodder crops. The frankincense tree (*Boswellia sacra*) produces most of the world's supply of the fragrance.

The Rub' al-Khali. Oman's portion of this vast desert is small but includes important oil fields at the perimeter. Oil revenue—although not as great as that of other nations in the area—has made possible considerable development and modernization.

Dryland areas of OMAN

DESERTIFICATION RISK	ARIDITY								Aridity Totals	
	Hyperarid		Arid		Semiarid		Subhumid			
	km²	%	km²	%	km²	%	km²	%	km²	%
Very High	By definition, desertification does not exist in hyperarid regions.		4,895	1.8					4,895	1.8
High			156,099	57.4					156,099	57.4
Moderate					23,932	8.8	11,966	4.4	35,898	13.2
Desertification Totals			160,994	59.2	23,932	8.8	11,966	4.4	196,892	72.4
No Desertification	75,058	27.6							75,058	27.6
Total Drylands	75,058	27.6	160,994	59.2	23,932	8.8	11,966	4.4	271,950	100.0
Non-dryland										
Total Area of the Territory									271,950	100.0

GENERAL BIBLIOGRAPHIES

1326 Clements, Frank A. *Oman.* Oxford U.K.; Santa Barbara CA: Clio, 1981. (World bibliographical series 29) 217 pp.

Annotated; 890 citations. Covers most major subjects.

1327 Duster, Joachin, and Scholz, Fred. *Bibliographie uber das Sultanat Oman [bibliography of the Sultanate of Oman].* Hamburg, Federal Republic of Germany: Dokumentations-Leitstelle Moderner Orient, 1980. 141 pp.

Unannotated. Introduction in German and English.

1328 Kabeel, Soraya M. *Source book on Arabian Gulf states, Arabian Gulf in general, Kuwait, Bahrain, Qatar, and Oman.*

See entry 1313.

1329 King, Russell, and Stevens, J. H., comp. *A bibliography of Oman, 1900-1970.* Durham U.K.: University of Durham. Centre for Middle Eastern and Islamic Studies, 1973. (Occasional Papers 2) 14 leaves.

1330 Shannon, Michael Owen. *Oman and Southeastern Arabia: a bibliographic survey.* Boston MA: G. K. Hall, 1978. 165 pp.

A partially annotated bibliography of 988 citations, compiled in an attempt to cover most of the primary and secondary works on Oman and surrounding territories through 1976.

INDEXES AND ABSTRACTS

1331 *Oman studies bibliographic info.* Tubingen, Federal Republic of Germany: Oman Studies, 1982-. 3 per year.

HANDBOOKS

1332 Nyrop, Richard F. *Area handbook for the Persian Gulf states.*

See entry 1314.

1333 Whelan, John, ed. *Oman; a MEED practical guide.* London: Middle East Economic Digest, 1981. 198 pp.

A detailed handbook covering various aspects of the nation of greatest use to businessmen, travellers, and those interested in modern cultural life. Includes maps, directories of addresses, statistics, bibliography.

STATISTICS

1334 *Statistical yearbook.* Muscat: Oman. Directorate General of National Statistics, 1976-. Annual.

The most extensive collection of statistics on the Sultanate.

GEOGRAPHY

1335 Speece, Mark, comp. *Environmental profile of the Sultanate of Oman.* Tucson AZ: University of Arizona. Office of Arid Lands Studies. Arid Lands Information Center, 1981.

A narrative summary of the nation's environment and development problems, with charts, maps, and unannotated bibliography.

GAZETTEERS

1336 *Oman: official standard names approved by the United States Board on Geographic Names.* Washington DC: U.S. Board on Geographic Names, 1976. (Gazetteer 139) 97 pp.

List of 5,600 place-names with latitude/longitude coordinates.

ECONOMIC DEVELOPMENT

1337 *Quarterly economic review of Bahrain, Qatar, Oman, the Yemens.*

See entry 1226.

HISTORICAL GEOGRAPHY

1338 Anthony, John Duke et al. *Historical and cultural dictionary of the Sultanate of Oman and the Emirates of eastern Arabia.* Metuchen NJ: Scarecrow, 1976. (Historical and cultural dictionaries of Asia 9) 136 pp.

Includes unannotated bibliography.

QATAR
The State of Qatar
(a former dependency of the United Kingdom)

Qatar occupies an arid peninsula extending from the Arabian mainland into the Gulf; it is hot, humid, nearly barren of vegetation, yet rich in oil and natural gas. Traditional subsistence for the small population was based on fishing and pearling, with some grazing away from the coast. Today a desalination plant supplies fresh water, and vegetables are grown in sufficient quantity to supply the nation for half the year.

Dryland areas of QATAR

DESERTIFICATION RISK	ARIDITY									
	Hyperarid		Arid		Semiarid		Subhumid		Aridity Totals	
	km²	%	km²	%	km²	%	km²	%	km²	%
Very High	By definition, desertification does not exist in hyperarid regions.									
High			8,800	80.0					8,800	80.0
Moderate										
Desertification Totals			8,800	80.0					8,800	80.0
No Desertification	2,200	20.0							2,200	20.0
Total Drylands	2,200	20.0	8,800	80.0					11,000	100.0
Non-dryland										
Total Area of the Territory									11,000	100.0

GENERAL BIBLIOGRAPHIES

1339 **Kabeel, Soraya M.** *Source book on Arabian Gulf states, Arabian Gulf in general, Kuwait, Bahrain, Qatar, and Oman.*

See entry 1313.

HANDBOOKS

1340 **Nyrop, Richard F.** *Area handbook for the Persian Gulf states.*

See entry 1314.

GAZETTEERS

1341 *Bahrain, Kuwait, Qatar, and United Arab Emirates: official standard names approved by the United States Board on Geographic Names.*

See entry 1225.

ECONOMIC DEVELOPMENT

1342 *Quarterly economic review of Bahrain, Qatar, Oman, the Yemens.*

See entry 1226.

SAUDI ARABIA
Al-Mamlaka al-'Arabiya as-Sa'udiya
[The Kingdom of Saudi Arabia]

Saudi Arabia has emerged from a forgotten corner of the world to become a major economic power whose control of a large share of the West's petroleum reserves raises new and perplexing questions for the industrialized nations. Equally important to the Saudis will be their handling of this unparalleled redistribution of wealth and how it can be used to enhance the long-term economic future of the nation. Save for a sliver of semiarid land near the border with North Yemen, all of Saudi Arabia is arid to hyperarid.

The Nejd. This is the cultural heart of the nation, unconquered by foreign powers and the home of Bedouin tribesmen from whom the House of Saud has emerged. The land is arid—with a scanty winter rainfall—and broken by low escarpments into shallow basins with interior drainage. Occasional pockets of sand dunes and lava flows reduce the grazing lands still further. Where groundwater emerges from springs or can be reached with wells oases are found and date cultivation flourishes. The city of Riyadh, once a semipermanent desert encampment, now serves as the administrative center of the nation and holds a population of 700,000, nearly 10 percent of the country's total.

The Northern Frontier. This region of sand seas, rocky desert, and volcanic scabland is the nation's exposed northern border with Jordan and Iraq. Its importance is more strategic than economic, although a pipeline crosses the region bringing oil from the Gulf fields to the Mediterranean.

The Hasa Coast. Saudi Arabia's Gulf coastline and lands immediately inland are the nation's economic heart, the center of oil production and refining as well as of secondary industries. This is also an agricultural center, due to the presence of artesian water formations and large oasis settlements. The date palm is the predominant crop, followed by figs, citrus, vegetables, cotton, and cereal grains.

The Hejaz. At the opposite side of the nation, along the Red Sea coast and inland, lies the spiritual heart of Arabia and, indeed, of all Islam. The mountainous Hejaz contains the holy cities of Mecca and Medina as well as the cosmopolitan port city of Jiddah through which millions pass on pilgrimage to Mecca. The land ranges from hyperarid along the northern coast to a slightly less dry coast in the south (called Tihama and extending into North Yemen) backed by semiarid slopes that are Saudi Arabia's wettest lands. This favored upland (roughly corresponding to Asir Province) supports some rain-fed agriculture and captures enough moisture to aid in irrigating the lower arid slopes to either side.

The Rub' al Khali. The southwestern quadrant of Saudi Arabia consists of a vast hyperarid region known as the Empty Quarter, considered so desolate that even the Bedouin had failed to penetrate it extensively. Not until oil exploration parties crossed the region in motor tractor caravans did details of the land emerge. Much of the region is covered by formidable mountains of sand that are fixed in position and separated by salt flats and desert pavement; ironically, what little rain falls is sucked into the sand so fast that the accumulated supply of groundwater lies within a few feet of the surface of the interdune flats. The economic importance of the region is minimal except for oil fields along its margins; the sand seas also harbor the larvae of the desert locust whose periodic eruptions devastate huge areas of the Middle East and neighboring Africa.

Dryland areas of SAUDI ARABIA

DESERTIFICATION RISK	ARIDITY									
	Hyperarid		Arid		Semiarid		Subhumid		Aridity Totals	
	km²	%	km²	%	km²	%	km²	%	km²	%
Very High	By definition, desertification does not exist in hyperarid regions.									
High			1,087,200	45.3	36,000	1.5			1,123,200	46.8
Moderate			12,000	0.5					12,000	0.5
Desertification Totals			1,099,200	45.8	36,000	1.5			1,135,200	47.3
No Desertification	1,264,800	52.7							1,264,800	52.7
Total Drylands	1,264,800	52.7	1,099,200	45.8	36,000	1.5			2,400,000	100.0
Non-dryland										
Total Area of the Territory									2,400,000	100.0

Saudi Arabia/1343-1354

GENERAL BIBLIOGRAPHIES

1343 Clements, Frank A. *Saudi Arabia.* Oxford U.K.; Santa Barbara CA: Clio, 1979. (World bibliographical series 5) 197 pp.

An annotated bibliographic guide of 789 references. Emphasis on current works; strongest on history, economics, oil. Much weaker on physical geography, anthropology.

HANDBOOKS

1344 Mostyn, Trevor, ed. *Saudi Arabia: a MEED practical guide.* London: Middle East Economic Digest, 1981. 280 pp.

A detailed handbook covering various aspects of the nation of greatest use to businessmen, travellers, and those interested in modern cultural life. Includes maps, directories of addresses, statistics, bibliography.

1345 Nyrop, Richard F. *Area handbook for Saudi Arabia.* 3rd ed. Washington DC: U.S. Dept. of Defense, 1977. (American University foreign area studies) 389 pp.

Narrative summary of the nation's history, politics, geography, economy, social life; includes unannotated bibliography.

GEOGRAPHY

1346 Batanouny, K. H. *Natural history of Saudi Arabia: a bibliography.* Jiddah, Saudi Arabia: King Abdulaziz University, 1978. 121 pp.

Subjects covered include physical geography, land use, and geographical exploration.

ATLASES

1347 Bindagji, Hussein Hamza. *Atlas of Saudi Arabia.* 3rd ed. Oxford U.K.: Oxford University Press, 1978. 61 pp.

GAZETTEERS

1348 *Saudi Arabia: official standard names approved by the United States Board on Geographic Names.* Washington DC: U.S. Board on Geographic Names, 1978. (Gazetteer 164) 374 pp.

List of 20,800 place-names with latitude/longitude coordinates. Covers Saudi Arabia and the so-called Neutral Zone between Saudi Arabia and Iraq.

GEOLOGY

1349 *Annotated bibliography pre-1970 A.D..* Jiddah, Saudi Arabia: Saudi Arabia. Directorate General of Mineral Resources, 1980. (Mineral resource bulletin 19) 126 pp.

1350 *Annotated bibliography: 1970-1975 A.D..* Jiddah, Saudi Arabia: Saudi Arabia. Directorate General of Mineral Resources, 1977. (Mineral resource bulletin 20) 98 pp.

ARCHITECTURE

1351 Doumato, Lamia. *The contemporary architecture of Saudi Arabia.* Monticello IL: Vance Bibliographies, 1980. (Architecture series bibliography A266) 8 pp.

Unannotated.

ECONOMIC DEVELOPMENT

1352 *Quarterly economic review of Saudi Arabia.* London: Economist Intelligence Unit, 1978-. Quarterly.

Includes summary of political and economic news as well as charts and statistics of selected economic indicators. Annual supplement.

HISTORICAL GEOGRAPHY

1353 Riley, Carroll L. *Historical and cultural dictionary of Saudi Arabia.* Metuchen NJ: Scarecrow, 1972. (Historical and cultural dictionaries of Asia 1) 133 pp.

Includes unannotated bibliography.

PUBLIC ADMINISTRATION

1354 Harmon, Robert B. *Politics and government in Saudi Arabia: a selected bibliography.* Monticello IL: Vance Bibliographies, 1981. (Public administration series bibliography P-739) 15 pp.

Brief annotations.

SYRIA
Al-Jamhouriya al Arabia as-Souriya
[The Syrian Arab Republic]
(a former mandated territory of France)

Syria occupies a diversity of favorable dryland environments extending from the Mediterranean coast to the steppelands beyond the Euphrates River. The nation's development problems stem from inconvenient arrangements of its resources.

The Latakia Coast. Syria's subhumid coast is rich in rainfall and soil but dense settlement has reduced the sizes of agricultural plots to the margins of economic viability. Nonetheless a variety of crops, including winter wheat, legumes, melons, citrus, olives and others, are cultivated. The city of Latakia is Syria's best port but is removed from the producing centers of Aleppo and Damascus, both of which send goods through Turkey and Lebanon to closer ports.

The Semiarid Steppe. The eastern inland slopes of the mountains that fringe the Mediterranean hold the bulk of Syria's agricultural land and urban population. Cities of ancient origin and continuing prosperity include Aleppo, Hamah, Homs, and Damascus (the capital and largest city). Rivers and springs issuing from the highlands supply water for extensive irrigated croplands, in some places relying on water wheels and canals dating from Roman times. Winter wheat, barley, olives, figs, sugar beets, cotton, and other crops are grown.

The Euphrates Steppe. Further east, the Euphrates River emerges from the Turkish highlands and slices through a semiarid to arid steppe. The region has high potential for irrigated agriculture but the deeply entrenched river requires costly engineering to raise the water from its gorge to the land above.

The Syrian Desert. This stony wasteland at the meeting of Syria, Jordan, Iraq, and Saudi Arabia is given over largely to Bedouin herders except in a few favored oases.

Dryland areas of SYRIA

DESERTIFICATION RISK	ARIDITY									
	Hyperarid		Arid		Semiarid		Subhumid		Aridity Totals	
	km²	%	km²	%	km²	%	km²	%	km²	%
Very High	By definition, desertification does not exist in hyperarid regions.									
High			93,583	50.4					93,583	50.4
Moderate					77,986	42.0			77,986	42.0
Desertification Totals			93,583	50.4	77,986	42.0			171,569	92.4
No Desertification							14,111	7.6	14,111	7.6
Total Drylands			93,583	50.4	77,986	42.0	14,111	7.6	185,680	100.0
Non-dryland										
Total Area of the Territory									185,680	100.0

HANDBOOKS

1355 Nyrop, Richard F. *Syria: a country study.* 3rd ed. Washington DC: U.S. Dept. of Defense, 1979. (American University foreign area studies) 268 pp.

Narrative summary of the nation's history, politics, geography, economy, social life; includes unannotated bibliography.

GAZETTEERS

1356 *Syria: official standard names approved by the United States Board on Geographic Names.* Washington DC: U.S. Board on Geographic Names, 1967. (Gazetteer 104) 460 pp.

List of 28,900 place-names with latitude/longitude coordinates.

CLIMATOLOGY

1357 *Climatic atlas of Syria.* Damascus: Syria. Meteorological Dept., 1977. Unpaged.

In English and Arabic. Scale of base map 1:2,000,000.

1358 **Weight, Marie L.** *An annotated bibliography of climatic maps of Syria.* Washington DC: U.S. Weather Bureau, 1963. (NTIS: AD 660-826) 26 pp.

Contains 66 citations.

SOILS

1359 **Butters, B.** *Soils of Syria and Lebanon (1973-1939).* Slough, U.K.: Commonwealth Bureau of Soils, 1976. (Annotated bibliography 1724) 5 pp.

ZOOLOGY

1360 **Kumerloeve, Hans.** "Die Saugetiere (Mammalia) Syriens und des Libanon" [the mammals (mammalia) of Syria and Lebanon: a preliminary review (as of 1974)]. Zoologischen Staatssammlung Munchen. *Veroffentlichungen* 18:3 (1975):159-225.

In German. Systematic survey with remarks on taxonomy and distribution of species. Chronologically arranged bibliography.

AGRICULTURE

1361 **Cigar, Norman.** *Agriculture and rural development in the Fertile Crescent: Lebanon, Syria, Jordan and Iraq.*

See entry 1323.

URBAN GEOGRAPHY

1362 **Tesdell, Lee S.** *Urbanization and economic growth in Jordan and Syria.*

See entry 1309.

ECONOMIC DEVELOPMENT

1363 *Quarterly economic review of Syria, Jordan.*

See entry 1312.

TOURISM AND RECREATION

1364 **al-Barbar, Aghil.** *Tourism in the Arab world: an introductory bibliography.*

See entry 1325.

TURKEY
Turkiye Cumhuriyeti
[The Republic of Turkey]

Turkey occupies the immensely strategic peninsula of Asia Minor and territory immediately adjacent in Europe. Although well-watered lands exist on the higher mountain ranges and along the Black Sea coast, most of the nation is dry.

The Mediterranean Coast. A belt of subhumid land runs along the Mediterranean and Aegean coasts, its width depending on the closeness of interior mountains to the sea and the strength of Mediterranean climatic influences on inland areas. Rainfed agriculture is generally possible, especially along the several lowland river valleys that flow westward into the Aegean; here rural population is most dense. The entire region—and Anatolia as a whole—suffers from deforestation, soil erosion, and overgrazing so adequate rainfall alone is often not enough to insure productivity.

The Anatolian Plateau. The geographic heartland of Turkey is a semiarid steppe with warm summers and cold winters; agriculture is practical only in irrigated valleys, and rural population is small. Despite its modest resources, the region was designated as the cultural homeland during the nation's modernization in the 1920's and, as a result, the capital was moved here to Ankara from Istanbul. Grazing is the principal land use and the mountain fringes hold important deposits of iron ore.

The East. Turkey occupies a complex highland region that forms the "root" of the peninsula; this semiarid to subhumid region includes mountains bordering the Soviet Union and Iran as well as the headwaters of the Tigris and Euphrates Rivers. Much of the area is a volcanic plateau punctuated by great cones, craters, and large lakes lacking outside drainage. Soils are good, but land is rarely level and crops are limited by aridity and cold; grazing is preferred. This is the homeland of the Armenians and Kurds.

Dryland areas of TURKEY

DESERTIFICATION RISK	Hyperarid km²	Hyperarid %	Arid km²	Arid %	Semiarid km²	Semiarid %	Subhumid km²	Subhumid %	Aridity Totals km²	Aridity Totals %
Very High	By definition, desertification does not exist in hyperarid regions.				120,815	15.5	27,281	3.5	148,096	19.0
High							42,870	5.5	42,870	5.5
Moderate					24,163	3.1			24,163	3.1
Desertification Totals					144,978	18.6	70,151	9.0	215,129	27.6
No Desertification					40,532	5.2	295,412	37.9	335,944	43.1
Total Drylands					185,510	23.8	365,563	46.9	551,073	70.7
Non-dryland									228,379	29.3
Total Area of the Territory									779,452	100.0

GENERAL BIBLIOGRAPHIES

1365 Guclu, Meral, comp. *Turkey.* Oxford, U.K.; Santa Barbara CA: Clio, 1981. (World bibliographical series 27) 331 pp.

Provides 993 annotated references to most major subjects.

HANDBOOKS

1366 Nyrop, Richard F. *Turkey: a country study.* 3rd ed. Washington DC: U.S. Dept. of Defense, 1980. (American University foreign area studies) 370 pp.

Narrative summary of the nation's history, politics, geography, economy, social life; includes unannotated bibliography.

STATISTICS

1367 *Tuerkiye istatistik yilligi [statistical yearbook of Turkey]*. Ankara: Turkey. State Institute of Statistics, 1928-. Annual.

In Turkish and English.

GEOGRAPHY

1368 Laking, Phyllis W. *The Black Sea, its geology, chemistry, biology: a bibliography*.

See entry 1613.

ATLASES

1369 Tanoglu, Ali; Erinc, Sirri; and Tumertekin, Erol. *Turkiye Atlas:/Atlas of Turkey*. Istanbul: Mulli Egitim Basimeui, 1961. Istanbul Universities. (Edebiyat Fakultesi yayinlari 903) 23 double leaves.

In English and Turkish. Scale 1:800,000.

GAZETTEERS

1370 *Turkey: official standard names approved by the United States Board on Geographic Names*. Washington DC: U.S. Board on Geographic Names, 1960. (Gazetteer 46) 665 pp.

List of 55,100 place-names with latitude/longitude coordinates.

GEOLOGY

1371 Ridge, John Drew. "Turkey." In *Annotated bibliographies of mineral deposits in Africa, Asia (exclusive of the USSR) and Australasia*, pp. 301-310. Oxford U.K.: Pergamon, 1976.

BOTANY

1372 Davis, Peter Hadland, and Edmondson, J. R. "Flora of Turkey: a floristic bibliography." Royal Botanic Garden, Edinburgh. *Notes* 37:2 (1979):273-283.

ZOOLOGY

1373 Kumerloeve, Hans. "Die Saugetiere (Mammalia) der Turkei" [the mammals (mammalia) of Turkey: attempt at a cursory review as of 1973-1974]. Zoologischen Staatssammlung (Munchen). *Veroffentlichungen* 18:3 (1975):69-158.

In German. Systematic survey with remarks on taxonomy and distribution of species. Chronologically arranged bibliography.

DEMOGRAPHY

1374 Balkan, Behire. *Turkiye ozetli nufus bibliografyasi [annotated bibliography of population studies in Turkey]*. Ankara: Hacettepe Universitesi, 1970-. Quarterly.

In Turkish and English. Covers a full range of information on demography, fertility, migration, reproduction and sex education.

WOMEN'S STUDIES

1375 Lytle, Elizabeth E. *Women in Turkey: a selected bibliography*. Monticello IL: Vance Bibliographies, 1979. (Public administration series bibliography P-172) 14 pp.

Unannotated. Covers the 20th century.

RURAL GEOGRAPHY

1376 Beeley, Brian W. *Koysel Turkiye bibliyografyasi/Bibliography of rural Turkey*. Ankara: Hacettepe Universitesi. Nufus Etutleri Enstitusu tarafindan Bastirilmistir, 1969. (Hacettepe University publications 10) 120 pp.

Covers many subjects from agriculture and rural sociology to economics and development.

1377 Franz, Erhard. *Die landliche Turkei im 20. Jahrhundert: eine bibliographishe Einfuhrung/Rural Turkey in the 20th century: a bibliographic introduction*. Hamburg, Federal Republic of Germany: Deutsches Orient-Institut. Dokumentations-Leitstelle Moderner Orient, 1974. 175 pp.

ECONOMIC DEVELOPMENT

1378 *Quarterly economic review of Turkey*. London: Economist Intelligence Unit, 1976-. Quarterly.

Includes summary of political and economic news as well as charts and statistics of selected economic indicators. Annual supplement.

1379 Vidergar, John J. *The modernization of Turkey: the 20th century*. Monticello IL: Council of Planning Librarians, 1977. (CPL exchange bibliography 1332) 12 pp.

Unannotated.

PUBLIC ADMINISTRATION

1380 Cigar, Norman. *Government and politics of the "Northern Tier": Turkey, Iran and Afghanistan*. Monticello IL: Vance Bibliographies, 1979. (Public administration series bibliography P-195) 9 pp.

Unannotated.

THE UNITED ARAB EMIRATES

(as the Trucial States or Trucial Coast, a former dependency of the United Kingdom)

The seven semiautonomous sheikhdoms that comprise the United Arab Emirates stretch from the Qatar peninsula to the Strait of Hormuz; all are located along the Gulf Coast and extend to undefined limits in the hyperarid interior. Before the discovery of oil the economy was limited to grazing, fishing, some coastal trade, and piracy. Now the U.A.E. boasts a very high per capita income and development of physical facilities, education, and social welfare has proceeded as fast as the money can be spent. One fertile oasis, at Buraimi (Al Ain) in Abu Dhabi, supplies vegetables and other crops, and local gardening has increased with the use of desalinated water.

Dryland areas of THE UNITED ARAB EMIRATES

DESERTIFICATION RISK	ARIDITY									
	Hyperarid		Arid		Semiarid		Subhumid		Aridity Totals	
	km²	%	km²	%	km²	%	km²	%	km²	%
Very High	By definition, desertification does not exist in hyperarid regions.									
High			7,276	7.9					7,276	7.9
Moderate					2,855	3.1			2,855	3.1
Desertification Totals			7,276	7.9	2,855	3.1			10,131	11.0
No Desertification	81,969	89.0							81,969	89.0
Total Drylands	81,969	89.0	7,276	7.9	2,855	3.1			92,100	100.0
Non-dryland										
Total Area of the Territory									92,100	100.0

HANDBOOKS

1381 Mostyn, Trevor, ed. *UAE: a MEED practical guide.* London: Middle East Economic Digest, 1982. 324 pp.

A detailed handbook covering various aspects of the nation of greatest use to businessmen, travellers, and those interested in modern cultural life. Includes maps, directories of addresses, statistics, bibliography.

1382 Nyrop, Richard F. *Area handbook for the Persian Gulf states.*

See entry 1314.

GAZETTEERS

1383 *Bahrain, Kuwait, Qatar, and United Arab Emirates: official standard names approved by the United States Board on Geographic Names.*

See entry 1225.

ECONOMIC DEVELOPMENT

1384 *Quarterly economic review of United Arab Emirates.* London: Economist Intelligence Unit, 1978-. Quarterly.

Includes summary of political and economic news as well as charts and statistics of selected economic indicators. Annual supplement.

HISTORICAL GEOGRAPHY

1385 Anthony, John Duke et al. *Historical and cultural dictionary of the Sultanate of Oman and the Emirates of Eastern Arabia.*

See entry 1338.

YEMEN ARAB REPUBLIC (North Yemen)
Al Jamhuriya al Arabiya al Yamaniya
[The Yemen Arab Republic]

North Yemen occupies the most favored position on the Arabian Peninsula in terms of water supply, climatic moderation, and agricultural potential. The peninsula's southwest corner is fringed by a high, dissected plateau plunging steeply to a narrow lowland along the Red Sea. The semiarid to subhumid climate permits crops of wheat, maize, barley, millet, fruits, citrus, coffee, and *qat* (a narcotic shrub), all grown on elaborately terraced plots. Along the hot, humid Red Sea coast (called the Tihama) grow irrigated date palms and cotton.

North Yemen has no oil and exports little of its agricultural produce. Instead, its rapidly expanding economy is fueled by its own people who migrate to the Arabian oil fields and send their earnings back home to their families. The nation is emerging from its isolation with a strong thirst for both consumer goods (imported at enormous cost) and *qat* (locally produced at the expense of exportable coffee that demands the same type of land).

Dryland areas of YEMEN (NORTH)

DESERTIFICATION RISK	ARIDITY									
	Hyperarid		Arid		Semiarid		Subhumid		Aridity Totals	
	km²	%	km²	%	km²	%	km²	%	km²	%
Very High	By definition, desertification does not exist in hyperarid regions.				53,235	27.3	19,110	9.8	72,345	37.1
High			100,620	51.6	7,995	4.1			108,615	55.7
Moderate										
Desertification Totals			100,620	51.6	61,230	31.4	19,110	9.8	180,960	92.8
No Desertification	14,040	7.2							14,040	7.2
Total Drylands	14,040	7.2	100,620	51.6	61,230	31.4	19,110	9.8	195,000	100.0
Non-dryland										
Total Area of the Territory									195,000	100.0

GENERAL BIBLIOGRAPHIES

1386 Macro, Eric. *Bibliography on Yemen and notes on Mocha.* Coral Gables FL: University of Miami Press, 1960. 63 pp.

Unannotated list of 894 citations; appended is a narrative description of the city of Mocha based on original sources. Covers present Yemen Arab Republic (North Yemen) only.

1387 Mondesir, Simone L. *A select bibliography of Yemen Arab Republic and People's Democratic Republic of Yemen.* Durham, U.K.: University of Durham. Centre for Middle Eastern and Islamic Studies, 1977. (Occasional Papers 5) 59 pp.

Unannotated list arranged by general subject. Approximately 750 citations.

HANDBOOKS

1388 *Area handbook for the Yemens.* Washington DC: U.S. Dept. of Defense, 1977. (American University foreign area studies) 265 pp.

Narrative summary of the nation's history, politics, geography, economy, social life; includes unannotated bibliography.

GEOGRAPHY

1389 Speece, Mark, comp. *Environmental report on Yemen (Yemen Arab Republic).* Rev. draft. Tucson AZ: University of Arizona. Office of Arid Lands Studies. Arid Lands Information Center, 1982. 208 pp.

A narrative summary of the nation's environment and development problems, with maps, charts, and unannotated bibliography.

1390 *Yemen Arab Republic.* Surrey, U.K.: United Kingdom. Ministry of Overseas Development. Land Resources Division, 1978. (Land resource bibliography 11) 87 pp.

GAZETTEERS

1391 *Yemen Arab Republic: official standard names approved by the United States Board on Geographic Names.* Washington DC: U.S. Board on Geographic Names, 1976. (Gazetteer 144) 124 pp.

List of 7,400 place-names with latitude/longitude coordinates.

ECONOMIC DEVELOPMENT

1392 *Quarterly economic review of Bahrain, Qatar, Oman, the Yemens.*
See entry 1226.

YEMEN, PEOPLE'S DEMOCRATIC REPUBLIC (South Yemen)
Jumhurijah al-Yemen al Dimuqratiyah al Sha'abijah
[The People's Democratic Republic of Yemen]

(as Aden and Aden Protectorate, former territories of the United Kingdom)

South Yemen occupies a large, ill-defined stretch of hyperarid to semiarid land along the south coast of Arabia. It owes its importance to the port of Aden, formed by a drowned volcanic cone and the finest harbor between Egypt and Pakistan.

Aden and the Coast. South Yemen's coastline is extremely dry, hot, and humid and its economic importance has always depended on the sea, whether on fishing or coastal trade. Aden itself was moribund until the British developed it as a coaling station on the route to India. On independence and the installation of a Marxist government the port has taken a strategic importance due to its proximity to oil shipping and to the Horn of Africa.

The Hadhramaut. This region is the site of a series of remarkable oasis cities located at the bottom of a steeply walled canyon that parallels the coast some 150 km (100 miles) inland. Upon crossing the arid windswept plateau, the visitor is confronted by thousand-foot cliffs bordering lush spring-fed cropland; skyscraper-like buildings of stone, brick, and mud rise from the canyon floor in architectural splendor. The people of the Hadhramaut are similar to their cousins in the North Yemen highlands in their physical isolation, their exportation of labor, and their level of luxury and culture. Although the canyon produces wheat, maize, millet, tobacco, sesame, dates, and other hot desert crops, food must be imported and paid for by remittances from migrant workers.

Socotra. Although politically tied to South Yemen, this island is discussed in the Indian Ocean section above.

Dryland areas of YEMEN (SOUTH)

DESERTIFICATION RISK	ARIDITY									
	Hyperarid		Arid		Semiarid		Subhumid		Aridity Totals	
	km²	%	km²	%	km²	%	km²	%	km²	%
Very High	By definition, desertification does not exist in hyperarid regions.				10,099	6.3			10,099	6.3
High			147,316	91.9					147,316	91.9
Moderate										
Desertification Totals			147,316	91.9	10,099	6.3			157,415	98.2
No Desertification	2,885	1.8							2,885	1.8
Total Drylands	2,885	1.8	147,316	91.9	10,099	6.3			160,300	100.0
Non-dryland										
Total Area of the Territory									160,300	100.0

GENERAL BIBLIOGRAPHIES

1393 Mondesir, Simone L. *A select bibliography of Yemen Arab Republic and People's Democratic Republic of Yemen.*
See entry 1387.

HANDBOOKS

1394 *Area handbook for the Yemens.*
See entry 1388.

GAZETTEERS

1395 *People's Democratic Republic of Yemen: official standard names approved by the United States Board on Geographic Names.* Washington DC: U.S. Board on Geographic Names, 1976. (Gazetteer 146) 204 pp.

List of 11,800 place-names with latitude/longitude coordinates.

SOILS

1396 *Bibliography on desert soils of Egypt, Eritrea, Somaliland, Aden (1964-1931).*

See entry 500.

ECONOMIC DEVELOPMENT

1397 *Quarterly economic review of Bahrain, Qatar, Oman, the Yemens.*

See entry 1226.

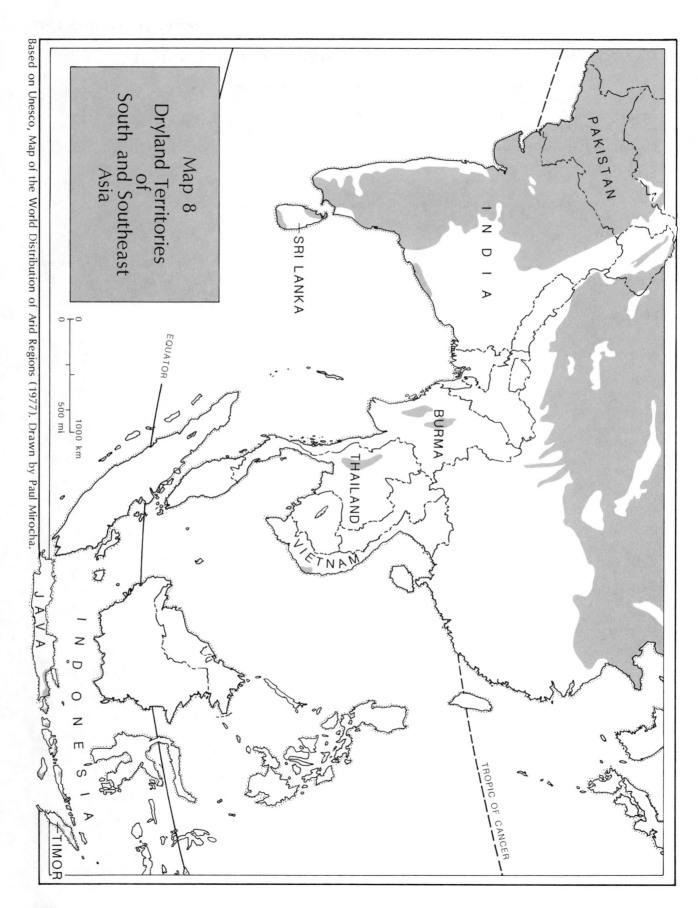

South and Southeast Asia

The area from Pakistan to Indonesia feels the full force of the monsoons that alternately bring cool dry air out of Siberia in the winter and warm moist air from the Indian Ocean in the summer. When the rains are delayed agriculture suffers as crops are exposed to the full force of the tropic summer sun; irrigation development can do much to spread out the water supply even in areas that, climatically, are well-watered.

South Asia's drylands are found almost exclusively in the plains and plateaus of Pakistan and western and southern India; these exist at the western edge of the monsoon belt and so are most affected by yearly variations. Although Southeast Asia lacks extensive drylands, local topography can produce rainshadows that induce significant aridity in the midst of rainforest environments.

GENERAL BIBLIOGRAPHIES

1398 Pearson, J. D. *South Asian bibliography: a handbook and guide.* Sussex, U.K.: Harvester, 1979. 381 pp.

Covers India, Pakistan, Bangladesh, Sri Lanka, Burma, Afghanistan, Tibet, Nepal, Bhutan, and the Maldive Islands. A comprehensive survey and bibliography prepared by the South Asia Library Group.

DIRECTORIES

1399 Fuller, S. C., ed. *South-east Asia.* 2nd ed. Guernsey, U.K.: Francis Hodgson, 1976. (Guide to world science 15) 289 pp.

Provides descriptions of scientific research activities and organizations. Coverage includes Burma, Kiribati, Indonesia, New Caledonia, Thailand, Vietnam.

1400 Winter, Alan; Kamm, Antony; and Khan, M. N. G. A., eds. *China, India, and Central Asia.*

See entry 1548.

GEOGRAPHY

1401 Sukhwal, B. L. *A bibliography of theses and dissertations in geography on South Asia.* Monticello IL: Council of Planning Librarians, 1973. (CPL exchange bibliography 438) 70 pp.

Unannotated. Includes unpublished theses and dissertations. Covers India, Pakistan, Bangladesh, Sri Lanka, Tibet, Nepal, Sikkim, Bhutan and the Indian Ocean and Islands.

1402 Sukhwal, B. L. *South Asia: a systematic geographic bibliography.* Metuchen NJ: Scarecrow, 1974. 827 pp.

A massive list of 10,346 unannotated citations covering most sub-disciplines of geography. Covers India, Pakistan, Bangladesh, Sri Lanka, Tibet, Nepal, Bhutan, Sikkim, and surrounding Indian Ocean. Arranged by subject but lacks subject index. Includes books, articles, and theses and dissertations.

GEOLOGY

1403 Wolfenden, E. B. et al. *Sources of geological information for Southeast Asia.* London: Geological Society of London, 1978. (Miscellaneous paper 9) 23 pp.

CLIMATOLOGY

1404 Gleeson, Thomas Alexander. *Bibliography of the meteorology of the Mediterranean, Middle East and South Asian area.*

See entry 122.

SOILS

1405 *Soil erosion in Africa, southern Asia, Australasia and Oceania (1957-1977).*

See entry 320.

LAND TENURE

1406 Anderson, Theresa J. *Land tenure and agrarian reform in East and Southeast Asia: an annotated bibliography.* Boston MA: G. K. Hall, 1980. 557 pp.

1407 *East and southeast Asia: a bibliography.* Madison WI: The University of Wisconsin. Land Tenure Center, 1971. (Training and methods 14) 88 pp.

Unannotated. Reflects holdings of the Land Tenure Center library; includes its call numbers. Covers China, Mongolia, Burma, Thailand, Vietnam, Indonesia, other east and southeast Asian nations. Continued by supplements 1 (1972), 2 (1979), 3 (1980).

1408 *Near East and South Asia: a bibliography.*

See entry 1174.

ANTHROPOLOGY

1409 Fuerer-Haimendorf, Elizabeth von, comp. *An anthropological bibliography of South Asia, together with a directory of recent anthropological field work.* Paris: Mouton, 1958, 1964, 1970. 3 vols.

Unannotated. Selective entries for pre-1940 works, as well as a bibliography covering 1940-1954. Most materials in western languages. Covers India, Pakistan, Sikkim, Bhutan, and Sri Lanka.

HISTORICAL GEOGRAPHY

1410 Schwartzberg, Joseph E., ed. *A historical atlas of South Asia.* Chicago IL: University of Chicago Press, 1978. (Association for Asian Studies. Reference series 2) 352 pp.

Covers India, Pakistan, Bangladesh, Afghanistan, Nepal, Bhutan, Sri Lanka, and the Maldives. Includes Burma from mid-19th century to 1948. Cursory view of Southeast Asia.

BURMA
Pyidaungsu Socialist Thammada Myanma Naingngandaw
[The Socialist Republic of the Union of Burma]
(a former colony of the United Kingdom)

Burma occupies the valley of the great Irrawaddy River and surrounding mountains with a long rocky coastline along the Indian Ocean. Although most of the nation lies in the path of tropical monsoons, moisture destined for the central Irrawaddy lowland is blocked by rugged mountains to the west (the Arabian Yomas), producing a subhumid region where the precipitation falls to as little as 500 mm (20 inches) annually. Vegetation consists of dry scrubland with little grass cover. With abundant irrigation water (derived, surprisingly, from wells and reservoirs rather than the river), crops include rice, beans, and cotton; goats are also raised specifically in the dry zone. Oil deposits in the region are of considerable importance.

Centered on the city of Mandalay, Burma's dry zone (called the Purple Plain for its heat and dust) is actually the cultural heartland of the Burmese people. Seven former capitals are located in the region and architectural jewels abound.

Dryland areas of BURMA

DESERTIFICATION RISK	ARIDITY								Aridity Totals	
	Hyperarid		Arid		Semiarid		Subhumid			
	km²	%	km²	%	km²	%	km²	%	km²	%
Very High	By definition, desertification does not exist in hyperarid regions.									
High										
Moderate										
Desertification Totals										
No Desertification							38,646	5.7	38,646	5.7
Total Drylands							38,646	5.7	38,646	5.7
Non-dryland									639,454	94.3
Total Area of the Territory									678,000	100.0

HANDBOOKS

1411 Roberts, T. D. et al. *Area handbook for Burma*. 2nd ed. Washington DC: U.S. Dept. of Defense, 1971. (American University foreign area studies) 341 pp.

Narrative summary of the nation's history, politics, geography, economy, social life; includes unannotated bibliography.

GEOGRAPHY

1412 Varady, Robert G., comp. *Draft environmental profile of Burma*. Tucson AZ: University of Arizona. Office of Arid Lands Studies, 1982. 233 pp.

A narrative summary of the nation's environment and development problems, with maps, charts, and an extensive unannotated bibliography.

GAZETTEERS

1413 *Burma: official standard names approved by the United States Board on Geographic Names*. Washington DC: U.S. Board on Geographic Names, 1966. (Gazetteer 96) 726 pp.

List of 52,000 place-names with latitude/longitude coordinates.

GEOLOGY

1414 Ridge, John Drew. "Burma." In *Annotated bibliographies of mineral deposits in Africa, Asia (exclusive of the USSR) and Australasia*, pp 211-219. Oxford, U.K.: Pergamon, 1976.

CLIMATOLOGY

1415 **Grimes, Annie E.** *An annotated bibliography of climatic maps of Burma.* Washington DC: U.S. Weather Bureau, 1963. (NTIS: AD 660-830; PB 185-879) 60 pp.

Contains 164 citations.

1416 **Grimes, Annie E.** *An annotated bibliography of climatic maps of Burma (supplement).* Silver Spring MD: Atmosphere Sciences Library, 1969. (NTIS: PB 185-879) 29 pp.

1417 **Nieuwolt, S.** "The climates of continental southeast Asia." In *Climates of southern and western Asia*, pp. 1-66. Edited by K. Takahashi and H. Arakawa. Amsterdam, The Netherlands: Elsevier Scientific Publishing Co., 1981. (World survey of climatology 9).

Covers Vietnam, Thailand, Burma, Laos, Kampuchea, Malaysia, Singapore. A narrative summary with maps, charts, and an unannotated bibliography.

1418 **Ohman, Howard L.** *Climatic atlas of Southeast Asia.* Natick MA: U.S. Army. Natick Laboratories, 1965. (Technical report ES-19) 7 pp. plus 87 maps.

Covers Burma, Thailand, Laos, Cambodia, Vietnam, the Malay Peninsula.

ECONOMIC DEVELOPMENT

1419 *Quarterly economic review of Thailand, Burma.* London: Economist Intelligence Unit, 1976-. Quarterly.

Includes summary of political and economic news as well as charts and statistics of selected economic indicators. Annual supplement.

INDIA
Bharat

(a former dominion of the United Kingdom)

As the world's second most populous nation and one occupying several extensive dryland regions, India probably has a higher number of people living in drylands than any other nation.

The Thar Desert. This is India's arid core and centers on the border with Pakistan; unlike Pakistan, India's share of the Thar lacks any exotic rivers save for the small Luni and its tributaries. The region is watered only erratically when the summer monsoons manage to penetrate far enough into the southwest. Most of the land can support only sheep and goats, although agricultural development has increased with irrigation and groundwater pumping; wheat is the primary crop. Despite the land's meager resources, settlement here is surprisingly dense with thousands of villages and at least three cities over 200,000 in population.

The North-Central Hills and Plateaus. The land between the arid Thar, the Ganges Valley, and the Narbada River to the south is a wide stretch of semiarid to subhumid territory consisting of relatively low relief and a complex network of agricultural settlements. Most of the region falls into the states of Madhya Pradesh and Rajasthan. Agriculture is practical wherever soils and water permit, although depletion of resources has been severe due to deforestation, erosion, overgrazing, and simply too much demand made of the land.

The Deccan Plateau. The Deccan is a great volcanic shield that forms the black-soiled core of southern India. From west to east the land is first very wet along the Arabian Sea, then semiarid across most of the plateau, then subhumid toward the Bay of Bengal, corresponding to the relative elevations. The natural vegetation is dry thorn scrub forest, though intensive cultivation and forest cutting have removed most of the tree cover. The extent of non-irrigated dry farming is surprising until one notices the unusual moisture-holding properties of the soil; chief crops are sorghum, wheat, cotton, and flaxseed. Increased water supplies would, however, increase yields dramatically and allay the periodic famines that are inevitable when resource exploitation is pushed to the limit.

The Carnatic. This coastal area along the Bay of Bengal is a subhumid extension of the Deccan dryland; since rainfall here comes only during the retreating monsoon (October to December), moisture deficiencies are high. Sugarcane and groundnuts, supported by irrigation along the river deltas, are the primary crops. The city of Madras—India's fourth largest in population, and third largest port—dominates the region.

The Indus Gorge. In the far north the Indus River rises in Tibet and slices deeply through northern Kashmir to create an arid gorge before entering Pakistan. Due to remoteness and inaccessibility, the region has little economic importance, though hydroelectric and water storage potential would appear to be high.

Dryland areas of INDIA

DESERTIFICATION RISK	ARIDITY									
	Hyperarid		Arid		Semiarid		Subhumid		Aridity Totals	
	km²	%	km²	%	km²	%	km²	%	km²	%
Very High	By definition, desertification does not exist in hyperarid regions.				78,591	2.5	13,711	0.4	92,302	2.9
High			95,106	3.0	313,860	9.9	122,142	3.9	531,108	16.8
Moderate					138,749	4.4	167	0.0	138,916	4.4
Desertification Totals			95,106	3.0	531,200	16.8	136,020	4.3	762,326	24.1
No Desertification			38,417	1.2	7,299	0.2	810,039	25.6	855,755	27.1
Total Drylands			133,523	4.2	538,499	17.0	946,059	29.9	1,618,081	51.2
Non-dryland									1,541,449	48.8
Total Area of the Territory									3,159,530	100.0

GENERAL BIBLIOGRAPHIES

1420 Gidwani, N. N., and Navalani, K. *A guide to reference materials on India.* Jaipur, India: Saraswati Publications, 1974. 2 volumes. 1,536 pp.

1421 Kalia, D. R., and Jain, M. K. *A bibliography of bibliographies on India.* Delhi: Concept, 1975. 204 pp.

INDEXES AND ABSTRACTS

1422 *Indian science abstracts.* Delhi: Indian National Science Documentation Centre, 1965-. Monthly.

THESES AND DISSERTATIONS

1423 *Bibliography of doctoral dissertations.* New Delhi: Association of Indian Universities, 1975-. Annual.

Appears annually in two volumes: social sciences and humanities, natural and applied sciences. Cumulative volumes cover period 1857-1970 and 1970-1975.

1424 *Indian dissertation abstracts.* Bombay: Indian Council of Social Science Research and the Inter-University Board of India, 1973-. Quarterly.

A source of recent theses in the social sciences.

1425 *Social sciences: a bibliography of doctoral dissertations accepted by Indian universities, 1857-1970.* New Delhi: Inter-University Board of India, 1974. 353 pp.

Subjects covered include library science, economics, commerce; some 2,820 theses are listed.

HANDBOOKS

1426 Nyrop, Richard F. *Area handbook for India.* 3rd ed. Washington DC: U.S. Dept. of Defense, 1975. (American University foreign area studies) 648 pp.

Narrative summary of the nation's history, politics, geography, economy, social life; includes unannotated bibliography.

GOVERNMENT DOCUMENTS

1427 Low, Donald A.; Iltis, J. C.; and Wainwright, M. D. *Government archives in South Asia: a guide to national and state archives in Ceylon, India, and Pakistan.* Cambridge, U.K.: Cambridge University Press, 1969. 355 pp.

GEOGRAPHY

1428 Aminullah. *Bibliography of C.A.Z.R.I. publications, 1959-1969.* Jodhpur: India. Central Arid Zone Research Institute, 1971. 38 pp.

Contains 714 citations.

1429 Singh, Ganda. *A bibliography of the Panjab.* Patiala, India: Punjabi University, 1966. 246 pp.

ATLASES

1430 Alam, S. Manzoor; Vithal, B. P. R.; and Sen, N. K. *Planning atlas of Andhra Pradesh.* Hyderabad, A.P.: India. Survey of India. Pilot Map Production Plant, 1974. 31 pp. plus 96 leaves of plates.

Includes natural resource, agriculture and irrigation, industry, and demographic maps. Each section preceded by a general outline. Presents statistical, rather than analytical, information. Scale 1:2,500,000.

1431 Dutt, Ashok K. *India in maps.* Dubuque IA: Kendall-Hunt Publishing Co., 1976. 136 pp.

Scale of most maps: 1:1,800,000.

1432 *National atlas of India.* Calcutta: National Atlas Organisation, 1977. 309 pp.

Base map scale of 1:1,000,000.

GAZETTEERS

1433 *India: official standard names approved by the United States Board on Geographic Names.* Washington DC: U.S. Board on Geographic Names, 1952 [1971?]. (Gazetteer 156) 2 vols.

List of 30,650 place-names with latitude/longitude coordinates (1952), and 370 place-name corrections (1971). Covers India, Bhutan, Nepal, French India, Jammu and Kashmir, Nepal and Portuguese India, as well as a gazetteer supplement for India, including Bhutan, former French India, Jammu and Kashmir, and Nepal.

GEOLOGY

1434 *Indian geological index.* Delhi: Indian Geological Index, 1971-1980. 4 volumes.

1435 *Indian minerals yearbook.* Nagpur, India: Indian Bureau of Mines. Economics Division, 1955-. Annual.

1436 Ridge, John Drew. "India." In *Annotated bibliographies of mineral deposits in Africa, Asia (exclusive of the USSR) and Australasia,* pp. 229-250. Oxford, U.K.: Pergamon, 1976.

1437 "Selected bibliography on Quaternary geology of India." *Indian Miner* (Calcutta) 31:4 (1977):75-80.

CLIMATOLOGY

1438 Boyer, Dennis L., and Smith, A. L., Jr. *An annotated climatological bibliography of India.* Washington DC: U.S. Air Force, 1969. (Environmental Technical Applications Center. Technical note 69-6; NTIS: AD 691-432) 55 pp.

1439 Grimes, Annie E. *An annotated bibliography of climatic maps of India.* Washington DC: U.S. Weather Bureau, 1964. (NTIS: AD 660-832) 90 pp.

Contains 266 citations.

1440 Grimes, Annie E. *An annotated bibliography of selected sources on the climate of India 1940-1971.* Washington DC: U.S. National Oceanic and Atmospheric Administration. Environmental Data Service, 1973. (NTIS: COM 74- 10271/6) 154 pp.

Contains 300 citations for the period 1940-1971.

1441 Grimes, Annie E. *An annotated bibliography of studies on surface and upper winds over India 1941-70.* Washington DC: Environmental Science Information Center, 1973. 98 pp.

1442 Rao, Y. P. "The climate of the Indian subcontinent." In *Climates of southern and western Asia*, pp. 67-182. Edited by K. Takahashi and H. Arakawa. Amsterdam, The Netherlands: Elsevier Scientific Publishing Co., 1981. (World survey of climatology 9).

Covers India, Pakistan, Bangladesh, Nepal, Bhutan, and Sri Lanka. A narrative summary with maps, charts, and an unannotated bibliography.

SOILS

1443 *Bibliography on desert soils of India, Pakistan and Afghanistan (1962-1945).* Harpenden, U.K.: Commonwealth Bureau of Soils, 196-. 3 pp.

Contains 15 citations.

1444 Gupta, I. C., and Pahwa, K. N. *A century of soil salinity research in India: an annotated bibliography 1863- 1976.* New Delhi: Oxford and IBH, 1978. 402 pp.

Contains 1,101 extensively annotated citations.

BIOLOGY

1445 Gupta, Raj Kumar. "Bibliography on the ecology (synecology and phytosociology) of the arid and semi-arid regions of India." *Excerpta botanica: sectio B: sociologica* 7:3 (1966):178-190.

Unannotated. Contains 186 items.

1446 Gupta, Raj Kumar. "Bibliography on the ecology (phytosociology and synecology) of the humid and tropical regions of India." *Excerpta botanica: sectio B: sociologica* 8:1 (1967):25-49.

Unannotated. Contains 346 items.

BOTANY

1447 Meher-Homji, V.M. et al. "Bibliography on vegetation science and physiological plant ecology in India." *Excerpta botanica: sectio B: sociologica* 5:1 (1963):54-79; 10:2 (1969):141-160; 10:3 (1969):161-203; 11:3 (1971):161-182; 12:2 (1972) :108-146; 13:3 (1974):194-240; 13:4 (1974):241-251; 15:3 (1976): 184-240; 17:4 (1978):309-320; 18:1 (1979):1-56; 20:1 (1980):17-37.

Some sections include annotations. Part nine forthcoming (1981).

AGRICULTURE

1448 Das Gupta, S. P., ed. *Atlas of agricultural resources of India.* Calcutta: India. National Atlas and Thematic Mapping Organisation, 1980. 36 plates.

Scale of base maps ranges from 1:2,000,000 to 1:6,000,000, with selected areas depicted at 1:1,000,000. Thematic maps cover crop regions, soils, land capability, and land use.

1449 *Indian agricultural atlas.* 3rd ed. New Delhi: India. Ministry of Agriculture, 1971. 76 pp.

Scale of base maps 1:15,000,000.

1450 Jain, Tara Chand. *Survey of Indian agro-bio-economic and allied literature, 1947-1975: a bibliography.* New Delhi: Agricole, 1978. 2 vols.

Cover 5,122 books, pamphlets, and other monographic literature on agriculture and related earth sciences.

1451 Kalia, D. R.; Jain, M. K.; and Guliani, T. D. *Statistical sources on Indian agriculture.* New Delhi: Marwah, 1978. 279 pp.

Extensively annotated compendium of 611 sources.

1452 Lal, Chhotey. *Reference sources in Indian agriculture and biology.* New Delhi: Ess Ess Publications, 1978. 213 pp.

Partially annotated. Includes theses, government documents, reference sources, serials. Primarily English-language items.

1453 Yadav, Shree Ram; Prasannalakshmi, S.; and Jain, T. C. *Bibliography of theses on ICRISAT specialities.* Hyderabad, India: International Crops Research Institute for the Semi-Arid Tropics, 1977. (NTIS: PB 81-215-139) 229 pp.

1454 Satyaprakash. *Agriculture: a bibliography.* New Delhi: Indian Documentation Service, 1977. (Subject bibliography series 3) 384 pp.

Unannotated. Lists about 10,000 articles, research papers, news, book reviews, and short monographs appearing primarily in 172 Indian English-language journals during the period 1962-1976.

1455 Sharma, Prakash C. *Green revolution in India: a selected research bibliography.* Monticello IL: Council of Planning Librarians, 1974. (CPL exchange bibliography 665) 11 pp.

Unannotated.

1456 Singh, Jasbir. *An agricultural atlas of India: a geographical analysis.* Kurukshetra, India: Vishal Publications, 1974. 365 pp.

1457 Singhvi, M. L., and Shrimali, D. S. *Reference sources in agriculture: an annotated bibliography.* Udaipure, India: Rajasthan College of Agriculture, 1962. 428 leaves.

IRRIGATION

1458 Choudhari, J. S. *Soils fertilizer responses and irrigation waters in arid and semi-arid regions: a bibliography (1833-1979).* Jodhpur, India: Latesh Prakashan, 1980. 140 pp.

Covers India only.

1459 *Irrigation and power abstracts.* New Delhi: India. Central Board of Irrigation and Power, 1943-. 6 per year.

1460 *Irrigation atlas of India.* Calcutta: India. Ministry of Education and Youth Services for Irrigation Commission, 1972. (Report of the Irrigation Commission 4) 35 pp.

Based on data compiled by the Ministry of Irrigation and Power. Scale 1:1,000,000. Shows irrigation projects in operation, under construction, and those completed. Additional plates on soils, climate, land use.

PLANT CULTURE

1461 Yadav, Shree Ram; Prasannalakshmi, S.; and Jadhav, P. S. *Indian theses on groundnut: a bibliography, 1948-1977.* Hyderabad, India: International Crops Research Institute for the Semi-Arid Tropics, 1978. (ICRISAT information bulletin 4; NTIS: PB 81-215-105) 54 pp.

Contains 303 entries.

FORESTRY

1462 Das Gupta, S. P., ed. *Atlas of forest resources of India.* Calcutta: India. National Atlas Organisation. Dept. of Science and Technology, 1976. 3 pp. and 36 leaves of maps.

1463 Taylor, George F., II. *Forestry in India: a selected bibliography of sources on forest history, forest policy and the forestry/agriculture interface.* Monticello IL: Vance Bibliographies, 1980. (Public administration series bibliography P-503) 15 pp.

Unannotated.

AGRICULTURAL ECONOMICS

1464 *Agricultural economics in India: a bibliography* 2nd ed. New Delhi: India. Directorate of Economics and Statistics, 1961. 342 pp.

Annotated. Updated version of 1952 bibliography. Covers important works published through 1959.

1465 Bose, P. R., and Vashist, V. N. *Rural development and technology: a status report-cum-bibliography.* New Delhi: Centre for the Study of Science Technology and Development, 1980. 373 pp.

An unannotated bibliography of 4,090 citations covering the full range of rural development topics: agriculture, demography, education, energy, financing, health, housing, migration, etc. Although coverage is worldwide the emphasis is strongly on India. Includes narrative descriptions of Indian development programs, tables of literature distributions, and a directory of Indian agencies.

1466 Sharma, Prakash C. *Agricultural planning and cooperatives in India (1944-1972): a selected research bibliography.* Monticello IL: Council of Planning Librarians, 1974. (CPL exchange bibliography 667) 23 pp.

Unannotated.

HUMAN GEOGRAPHY

1467 Chekki, Danesh A. *The social system and culture of modern India: a research bibliography.* New York: Garland, 1975. 843 pp.

1468 Kharbas, Datta Shankarrao. *Maharashtra and the Marathas: their history and culture, a bibliographic guide to western language material.* Rev. ed. Boston MA: G. K. Hall, 1975. 642 pp.

Maharashtra State lies inland of Bombay at the northern end of the Deccan Plateau.

1469 Patterson, Maureen L. P. *South Asian civilizations: a bibliographic synthesis.* Chicago IL: The University of Chicago Press, 1981.

Covers Pakistan, India, and Sri Lanka. A massive unannotated listing of sources on the region's history and culture, totalling some 28,000 citations.

1470 Sharma, Prakash C. *Tradition, social change and modernization in India: a selected research bibliography (1900-1972).* Monticello IL: Council of Planning Librarians, 1974. (CPL exchange bibliography 668) 42 pp.

Unannotated.

ANTHROPOLOGY

1471 Field, Henry, and Laird, Edith M. *Bibliography on the physical anthropology of the peoples of India.* Coconut Grove FL: Field Research Projects, 1968. 82 pp.

Unannotated.

1472 Mathur, U. B., comp. *Ethnographic atlas of Rajasthan; with reference to scheduled castes and scheduled tribes.* Delhi: India. Rajasthan. Superintendent of Census Operations, 1969. 138 pp.

India/1473-1481

Statistical report on various castes and tribes, including population, age, occupational and educational levels within the geographic region. Based on 1961 census. Rajasthan includes the Indian portion of the Thar Desert.

RURAL GEOGRAPHY

1473 Lambert, Claire M., ed. *Village studies data analysis and bibliography. Vol. 1: India, 1950-1975.* London: Mansell for IDS, 1978.

URBAN GEOGRAPHY

1474 Bose, Ashish. *Bibliography on urbanization in India, 1947-1976.* New Delhi: Tata McGraw-Hill, 1976. 129 pp.

Post-independence studies on urbanization. Includes unpublished material.

DEMOGRAPHY

1475 *Indian census centenary atlas.* Delhi: India. Controller of Publications, 1974. 198 pp.

Covers 1872-1971 with 11 censuses. Gives administrative boundaries from 1872 to the present, specimens of selected maps of earlier census reports, and some of the more important aspects of population variables. Emphasis on distributional aspects of demography.

WOMEN'S STUDIES

1476 Lytle, Elizabeth. *Women in India: a comprehensive bibliography.* Monticello IL: Vance Bibliographies, 1978. (Public administration series bibliography P-109) 29 pp.

Unannotated.

1477 Sakala, Carol *Women of South Asia: a guide to resources.* Millwood NY: Kraus International Publications, 1980. 517 pp.

Annotated. Covers historical and contemporary South Asia (India, Pakistan, Bangladesh, Sri Lanka, Nepal). Part I "Published resources" contains 4,600 entries to Western-language items. Part II "Libraries, archives and other local resources" contains reports identifying potential resources in South Asia.

ECONOMIC DEVELOPMENT

1478 *Asia/Australia data bank.*
See entry 1806.

1479 *Quarterly economic review of India, Nepal.* London: Economist Intelligence Unit, 1976-. Quarterly.

Includes summary of political and economic news as well as charts and statistics of selected economic indicators. Annual supplement.

HISTORICAL GEOGRAPHY

1480 Dey, Nundo Lal. *The geographical dictionary of ancient and mediaeval India.* 3rd ed. New Delhi: Oriental Books Reprint Corp., 1971. 262 pp.

First published 1927.

1481 Kurian, George Thomas. *Historical and cultural dictionary of India.* Metuchen NJ: Scarecrow, 1976. (Historical and cultural dictionaries of Asia 8) 307 pp.

Entries cover major topics, events, and personalities in the nation's history; includes extensive unannotated bibliography.

INDONESIA
Republik Indonesia

(as the Netherlands East Indies, a former colony of The Netherlands)

Indonesia is extremely well-watered overall but does hold some subhumid drylands on islands to the southeast.

Timor. The northeast and southeast coasts are dry enough to fall into the semiarid zone in places and the subhumid zone across much of their length. Steep mountains immediately inland force the relatively dry south and southwesterly winds to drop their precipitation on the upper slopes. Vegetation is savanna grassland and supports some horses and cattle; the island as a whole has a subsistence economy though discoveries of oil could alter the picture substantially. The eastern half of the island was a territory of Portugal until 1976 when it was invaded by Indonesia; it is now the province of Loro Sae.

Java. The extreme northeastern coast opposite the smaller island of Madura is subhumid and supports much more irrigated cropland than the rest of the island. Maize, sweet potatoes, groundnuts, and legumes contrast with the rice, rubber, sugarcane, coffee, and other tropical products grown in wetter regions. The island itself is perhaps the most densely populated large territory on earth, so the drylands must contribute to the agricultural base.

The Lesser Sunda Islands. Most of these islands stretching between Java and Timor have small coastal pockets of subhumid dryland, usually on their northeastern shores. Where the rugged topography permits, maize rather than rice is the principal crop.

Only selected references are provided.

Dryland areas of INDONESIA

DESERTIFICATION RISK	ARIDITY									
	Hyperarid		Arid		Semiarid		Subhumid		Aridity Totals	
	km²	%	km²	%	km²	%	km²	%	km²	%
Very High	By definition, desertification does not exist in hyperarid regions.									
High										
Moderate										
Desertification Totals										
No Desertification							17,216	0.9	17,216	0.9
Total Drylands							17,216	0.9	17,216	0.9
Non-dryland									1,886,434	99.1
Total Area of the Territory									1,903,650	100.0

HANDBOOKS

1482 Vreeland, Nena. *Area handbook for Indonesia.* 3rd ed. Washington DC: U.S. Dept. of Defense, 1975. (American University foreign area studies) 488 pp.

Narrative summary of the nation's history, politics, geography, economy, social life; includes unannotated bibliography.

GEOGRAPHY

1483 Kartawinata, Kuswata. "Indonesia." In *Handbook of contemporary developments in world ecology*, pp. 527-548. Edited by Edward John Kormondy and J. Frank McCormick. Westport CT: Greenwood, 1981.

A narrative summary of ecological research in Indonesia; with an annotated bibliography and unannotated list of references.

Indonesia/1484-1487

GAZETTEERS

1484 *Indonesia and Portuguese Timor: official standard names approved by the United States Board on Geographic Names.* 2nd ed. Washington DC: U.S. Board on Geographic Names, 1968. (Gazetteer 13) 901 pp.

List of 60,000 place-names with latitude/longitude coordinates. Covers Indonesia, Timor and Western New Guinea.

BOTANY

1485 Pinto da Silva, A. R., and Teles, A. N. "Bibliographia phytosociologica Portugaliae" [bibliography of the plant ecology of Portugal].

Coverage includes Timor.
See entry 191.

1486 Teles, A. N. "Bibliographie des cartes de vegetation du Portugal" [bibliography of vegetation maps of Portugal].

Coverage include Timor.
See entry 192.

ECONOMIC DEVELOPMENT

1487 *Quarterly economic review of Indonesia.* London: Economist Intelligence Unit, 1976-. Quarterly.

Includes summary of political and economic news as well as charts and statistics of selected economic indicators. Annual supplement.

PAKISTAN
The Islamic Republic of Pakistan
(formed from several territories of British India)

Pakistan comprises the old Indian provinces of Baluchistan, North-West Frontier, Sind, most of the Punjab, and western Kashmir. Save for Kashmir, all of the nation is dry.

The Indus Valley. The emergence of the Indus from the highlands of Kashmir has made possible an irrigated agricultural civilization as ancient as Egypt's and Mesopotamia's. The land is a similarly alluvial plain of high fertility and susceptibility to salt damage. Terraces above the river and its tributaries extend into the Thar Desert and limit agriculture wherever the land is out of reach of water. Wheat (winter), millet (summer), maize, barley, cotton, and many other crops are grown throughout the region.

Baluchistan. The uplands to the west of the Indus Valley form a sharp eco-climatic boundary between the cold winter steppes of south-central Asia and the warm winter monsoon lowlands of the Indian subcontinent. The Baluchis are a nomadic people existing independently of the national heartland, grazing flocks of sheep, goats, camels, cattle, and horses on sparse pasture and scrub. Agriculture is limited to a few favored villages with underground water. Deposits of natural gas, chromite, sulfur, and other minerals are important beyond the region.

Dryland areas of PAKISTAN

DESERTIFICATION RISK	ARIDITY									
	Hyperarid		Arid		Semiarid		Subhumid		Aridity Totals	
	km²	%	km²	%	km²	%	km²	%	km²	%
Very High	By definition, desertification does not exist in hyperarid regions.		2,145	0.3					2,145	0.3
High			267,750	30.3	83,307	9.4	11,998	1.4	363,055	41.1
Moderate					75,357	8.5			75,357	8.5
Desertification Totals			269,895	30.6	158,614	17.9	11,998	1.4	440,557	49.9
No Desertification										
Total Drylands			269,895	30.6	158,614	17.9	11,998	1.4	440,557	49.9
Non-dryland									442,318	50.1
Total Area of the Territory									882,875	100.1

GENERAL BIBLIOGRAPHIES

1488 Satyaprakash, ed. *Pakistan: a bibliography 1962-1974.* Gurgaon: Indian Documentation Service, 1975. 338 pp.

INDEXES AND ABSTRACTS

1489 *Pakistan science abstracts.* Karachi: Pakistan National Scientific and Technical Centre, 1961-. Quarterly.

Covers agriculture, zoology, geology, hydrology, engineering. Supersedes *Pakistan scientific literature: current bibliography.*

THESES AND DISSERTATIONS

1490 Anwar, Muhammad. *Doctoral dissertations on Pakistan.* Islamabad: Pakistan. National Commission on Historical and Cultural Research, 1976. (Bibliographical series 2) 124 pp.

HANDBOOKS

1491 Nyrop, Richard F. *Area handbook for Pakistan.* 4th ed. Washington DC: U.S. Dept. of Defense, 1975. (American University foreign area studies) 455 pp.

Narrative summary of the nation's history, politics, geography, economy, social life; includes unannotated bibliography.

GOVERNMENT DOCUMENTS

1492 Low, Donald A.; Iltis, J. C.; and Wainwright, M.D. *Government archives in South Asia: a guide to national and state archives in Ceylon, India, and Pakistan.*

See entry 1427.

STATISTICS

1493 *Pakistan statistical yearbook.* Karachi: Pakistan. Manager of Publications, 1952-. Annual.

GEOGRAPHY

1494 Beg, Abdur Rahman. "Pakistan." In *Handbook of contemporary developments in world ecology*, pp. 589-605. Edited by Edward John Kormondy and J. Frank McCormick. Westport CT: Greenwood, 1981.

A narrative summary of ecological research in Pakistan, with an unannotated list of references.

1495 Rahman, Mushtaqur. *Bibliography of Pakistan geography, 1947-1973.* Monticello IL: Council of Planning Librarians, 1974. (CPL exchange bibliography 655, 656) 117 pp.

Unannotated.

1496 Varady, Robert G., comp. *Environmental profile of The Islamic Republic of Pakistan.* Tucson AZ: The University of Arizona. Office of Arid Lands Studies, 1981.

A narrative summary of the nation's environment and development problems, with maps, charts, and unannotated bibliography.

GAZETTEERS

1497 *Pakistan: official standard names approved by the United States Board on Geographic Names.* 2nd ed. Washington DC: U.S. Board on Geographic Names, 1978. (Gazetteer 168) 522 pp.

List of 33,500 place-names with latitude/longitude coordinates.

GEOLOGY

1498 Ahmed, Zaki. *Bibliography and index of the Geological Society of Pakistan reports (unpublished).* Quetta: Pakistan. Geological Survey, 1977. (Records 42) 39 pp.

Unannotated. Includes only unpublished reports of the Geological Survey which are easily available. Several reports listed as "unpublished" have since been published. Emphasis is on reports from 1950-59, but contains items until 1967.

1499 Offield, Tarry W. *Preliminary bibliography and index of the geology of Pakistan.* Karachi: Pakistan. Geological Survey, 1964. (Records 12) 54 pp.

CLIMATOLOGY

1500 Grimes, Annie E. *An annotated bibliography of climatic maps of Pakistan.* Washington DC: U.S. Weather Bureau, 1965. (NTIS: AD 660-827; PB 194-754) 77 pp.

Contains 222 citations through April 1964.

1501 Grimes, Annie E. *An annotated bibliography of climatic maps of Pakistan (supplement).* Silver Spring MD: Environmental Data Service, 1970. (NTIS: PB 194-754) 42 pp.

Contains 117 citations.

1502 Rao, Y. P. "The climate of the Indian subcontinent."

See entry 1442.

HYDROLOGY

1503 Grey, D. R. C. *A bibliography and short review of the water resources development of Pakistan (with especial reference to the groundwater development of the Indus Plain).* London: Institute of Geological Sciences. Geophysical and Hydrological Division, 1979. (Report) 83 pp.

SOILS

1504 *Bibliography on desert soils of India, Pakistan and Afghanistan (1962-1945).*

See entry 1443.

BOTANY

1505 Gupta, Raj Kumar, and Meher-Homji, V. M. "Bibliography on 'plant-ecology' in Pakistan." *Excerpta botanica: sectio B: sociologica* 11:3 (1971):183-206.

Unannotated. Contains 295 items. Author index.

1506 Kazmi, S. M. A. *Bibliography on the botany of West Pakistan and Kashmir and adjacent regions.* Miami FL: Field Research Projects, 1970. 5 vols.

AGRICULTURE

1507 Siddiqui, Akhtar H. *Agriculture in Pakistan: a selected bibliography 1947-1969.* Karachi, Pakistan: 1969. 88 pp.

HUMAN GEOGRAPHY

1508 Gustafson, W. Eric, ed. *Pakistan and Bangladesh: bibliographic essays in social science.* Islamabad: University of Islamabad Press, 1976, 364 pp.

Specific subjects on Pakistan covered in essays include sociology and cultural anthropology, economics, and general bibliography.

1509 **Korson, J. Henry.** "Sociology in Pakistan, with a brief view of cultural anthropology." In *Pakistan and Bangladesh: bibliographic essays in social science*, pp. 156-208. Edited by Eric W. Gustafson. Islamabad: University of Islamabad Press, 1976.

A bibliographic essay with a 414-item unannotated bibliography covering demography, family life, women, sociology, anthropology and ethnography.

1510 **Patterson, Maureen L. P.** *South Asian civilizations: a bibliographic synthesis.*

See entry 1469.

1511 **Vidergar, John J.** *Bibliography on the social and economic development of Pakistan.* Monticello IL: Council of Planning Librarians, 1977. (CPL exchange bibliography 1404) 4 pp.

Unannotated.

WOMEN'S STUDIES

1512 **Sakala, Carol.** *Women of South Asia: a guide to resources.*

See entry 1477.

ECONOMIC DEVELOPMENT

1513 *Quarterly economic review of Pakistan, Bangladesh, Afghanistan.* London: Economist Intelligence Unit, 1976-. Quarterly.

Includes summary of political and economic news as well as charts and statistics of selected economic indicators. Annual supplement.

HISTORICAL GEOGRAPHY

1514 **Dey, Nundo Lal.** *The geographical dictionary of ancient and mediaeval India.*

See entry 1480.

SRI LANKA
The Democratic Republic of Sri Lanka
(as Ceylon, a former colony of the United Kingdom)

Sri Lanka has been described as a pearl dangling from Asia into the Indian Ocean. It is a mountainous island of considerable natural beauty with two disconnected dry zones. Both are formed by rainshadow effects: when the two seasonal monsoons blow from the northeast (winter) or southwest (summer) stretches of northwest and southeast coastline receive little rain. Vegetation is thorn scrub jungle and coarse grassland (called *patana*). With a rapidly increasing population and wetter areas almost totally cultivated, drylands agricultural development is crucial. Construction of irrigation schemes (often based on ancient remnants), water retention facilities, and hydroelectric generators have helped solve these problems. Maize rather than rice is the most promising crop.

Dryland areas of SRI LANKA

DESERTIFICATION RISK	ARIDITY										Aridity Totals	
	Hyperarid		Arid		Semiarid		Subhumid		Aridity Totals	
	km²	%	km²	%	km²	%	km²	%	km²	%
Very High	By definition, desertification does not exist in hyperarid regions.								
High									
Moderate									
Desertification Totals									
No Desertification							17,321	26.4	17,321	26.4
Total Drylands							17,321	26.4	17,321	26.4
Non-dryland									48,289	73.6
Total Area of the Territory									65,610	100.0

GENERAL BIBLIOGRAPHIES

1515 Goonetileke, H. A. I. *A bibliography of Ceylon; a systematic guide to the literature on the land, people, history and culture published in Western languages from the sixteenth century to the present day.* Zug, Switzerland: Inter Documentation, 1970. (Bibliotheca Asiatica 5) 2 vols.

HANDBOOKS

1516 Nyrop, Richard F. *Area handbook for Ceylon.* Washington DC: U.S. Dept. of Defense, 1971. (American University foreign area studies) 525 pp.

Narrative summary of the nation's history, politics, geography, economy, social life; includes unannotated bibliography.

GOVERNMENT DOCUMENTS

1517 Low, Donald A.; Iltis, J. C.; and Wainwright, M. D. *Government archives in South Asia: a guide to national and state archives in Ceylon, India, and Pakistan.*

See entry 1427.

STATISTICS

1518 Peebles, Patrick. *Sri Lanka: a handbook of historical statistics.* Boston MA: G. K. Hall, 1982. 357 pp.

Provides narrative and selected statistics on climate, geographical area, population, health, migration, education, labor force, agriculture, land use, mining, transport, commerce, finance, etc.

GAZETTEERS

1519 *Ceylon: official standard names approved by the United States Board on Geographic Names.* Washington DC: U.S. Board on Geographic Names, 1960. (Gazetteer 49) 359 pp.

List of 29,600 place-names with latitude/longitude coordinates.

GEOLOGY

1520 Ridge, John Drew. "Ceylon." In *Annotated bibliographies of mineral deposits in Africa, Asia (exclusive of the USSR) and Australasia*, pp. 219-222. Oxford, U.K.: Pergamon, 1976.

CLIMATOLOGY

1521 Grimes, Annie E. *An annotated bibliography of climatic maps of Ceylon.* Washington DC: U.S. Weather Bureau, 1965. (NTIS: AD 660-797) 53 pp.

Contains 159 citations through April 1964.

1522 Rao, Y. P. "The climate of the Indian subcontinent."

See entry 1442.

BOTANY

1523 Meher-Homji, V. M., and Punyasiri Perera, N. "Bibliography on plant-ecology in Ceylon." *Excerpta botanica: sectio B: sociologica* 12:2 (1972):147-157.

Unannotated. Contains 139 items. Author index.

ZOOLOGY

1524 De Silva, A. "An annotated bibliography of snakes of Sri Lanka." *Snake* 12:1-2 (1980):61-108.

HUMAN GEOGRAPHY

1525 Patterson, Maureen L. P. *South Asian civilizations: a bibliographic synthesis.*

See entry 1469.

WOMEN'S STUDIES

1526 Sakala, Carol. *Women of South Asia: a guide to resources.*

See entry 1477.

ECONOMIC DEVELOPMENT

1527 *Quarterly economic review of Sri Lanka (Ceylon).* London: Economist Intelligence Unit, 1976-. Quarterly.

Includes summary of political and economic news as well as charts and statistics of selected economic indicators. Annual supplement.

THAILAND
Prathes Thai/Muang-Thai
(formerly known as Siam)

Thailand's geography is roughly similar to Burma's: a central river valley holding most of the population is surrounded by high mountains on three sides and open to the sea on the south. Also like Burma, the mountains to the west of the river create a rainshadow dry zone along the valley floor.

The Ping Valley. Mae Nam Ping is the tributary river closest to the Dawna Range at the western border of the nation; its valley is thus relatively protected from the summer monsoons and exhibits subhumid characteristics. Irrigation makes possible rice, cotton, and tobacco cultivation. Tak (Rahaeng) is the region's principal town.

The Korat Plateau. This large rectangular upland comprises the eastern third of the nation and is surrounded on two sides by low mountains and on the other two sides by the Mekong River. Although soils are relatively thin and natural vegetation is dry monsoon forest, the region falls just outside the dryland zone as defined in this Guide by potential evapotranspiration data. This area is mentioned to illustrate the difficulties of defining in a precise way the extent of dryland areas, especially in climates with strong monsoonal precipitation patterns. Despite high water deficits before the onset of a monsoon, the rains that follow often erase the drought from annual precipitation figures.

Dryland areas of THAILAND

DESERTIFICATION RISK	ARIDITY									
	Hyperarid		Arid		Semiarid		Subhumid		Aridity Totals	
	km²	%	km²	%	km²	%	km²	%	km²	%
Very High	By definition, desertification does not exist in hyperarid regions.									
High										
Moderate										
Desertification Totals										
No Desertification							19,532	3.8	19,532	3.8
Total Drylands							19,532	3.8	19,532	3.8
Non-dryland									494,468	96.2
Total Area of the Territory									514,000	100.0

HANDBOOKS

1528 Bunge, Frederica M. *Thailand: a country study.* 4th ed. Washington DC: U.S. Dept. of Defense, 1981. (American University foreign area studies) 354 pp.

Narrative summary of the nation's history, politics, geography, economy, social life; includes unannotated bibliography.

GAZETTEERS

1529 *Thailand: official standard names approved by the United States Board on Geographic Names.* Washington DC: U.S. Board on Geographic Names, 1966. (Gazetteer 97) 675 pp.

List of 15,500 place-names with latitude/longitude coordinates.

CLIMATOLOGY

1530 Nieuwolt, S. "The climates of continental southeast Asia."

See entry 1417.

1531 Ohman, Howard L. *Climatic atlas of Southeast Asia.*

See entry 1418.

ECONOMIC DEVELOPMENT

1532 *Quarterly economic review of Thailand, Burma.*

See entry 1419.

HISTORICAL GEOGRAPHY

1533 Smith, Harold E. *Historical and cultural dictionary of Thailand.* Metuchen NJ: Scarecrow, 1976. (Historical and cultural dictionaries of Asia 6) 213 pp.

Entries cover major topics, events, and personalities in the nation's history; includes extensive unannotated bibliography.

VIETNAM
Cong Hoa Xa Hoi Chu Nghia Viet Nam
[The Socialist Republic of Viet Nam]

(as Tonkin, Annam, and Cochin-China, former colonies of France)

Vietnam is generally a very well-watered land; however, a small subhumid coastal belt exists south of Cape Padarin (Cape Dinh). Unlike neighboring areas where rice and sugarcane are dominant, here cotton is cultivated. The economic importance of the region to the nation is minimal.

Only selected references are provided.

Dryland areas of VIETNAM

DESERTIFICATION RISK	ARIDITY									
	Hyperarid		Arid		Semiarid		Subhumid		Aridity Totals	
	km²	%	km²	%	km²	%	km²	%	km²	%
Very High	By definition, desertification does not exist in hyperarid regions.									
High										
Moderate										
Desertification Totals										
No Desertification							8,239	2.5	8,239	2.5
Total Drylands							8,239	2.5	8,239	2.5
Non-dryland									321,327	97.5
Total Area of the Territory									329,566	100.0

GAZETTEERS

1534 *South Vietnam: official standard names approved by the United States Board on Geographic Names.* 2nd ed. Washington DC: U.S. Board on Geographic Names, 1971. (Gazetteer 58) 337 pp.

List of 24,000 place-names with latitude/longitude coordinates.

CLIMATOLOGY

1535 Nieuwolt, S. "The climates of continental southeast Asia."

See entry 1417.

1536 Ohman, Howard L. *Climatic atlas of southeast Asia.*

See entry 1418.

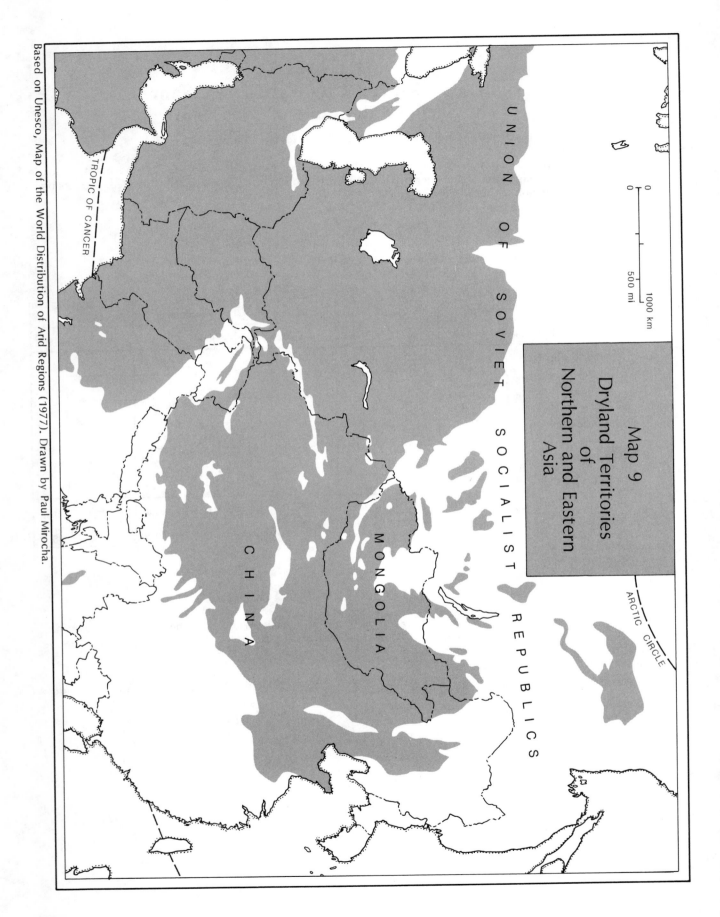

Northern and Eastern Asia

The dryland area extending from the Ukraine to Manchuria is characterized by extreme continentality in its climate; with little influence from maritime air masses, winters are extremely cold and dry while summers are often hot and scarcely less dry. Only three nations occupy this area—the Soviet Union, Mongolia, and China—and, while the two largest have extensive territories outside the dryland zones, the drylands themselves are relied on heavily as sources of food and mineral wealth. The common border also gives the area immense strategic importance.

BOTANY

1537   Merrill, E. D. and Walker, Egbert Hamilton. *A bibliography of eastern Asiatic botany.* Jamaica Plain MA: Harvard University. Arnold Arboretum, 1938. 719 pp.

Covers literature through 1936. Covers China, Japan, Formosa, Korea, Manchuria, Mongolia, Tibet, and eastern and southern Siberia. Partially annotated. Includes some sources in Asian languages. Continued by Walker (1960) (entry 1538).

1538   Walker, Egbert Hamilton. *Bibliography of eastern Asiatic botany, Supplement I.* Washington DC: American Institute of Biological Sciences, 1960. 552 pp.

Coverage from 1937-1958. Continues Merrill and Walker (1938) (entry 1537). Includes area of original volume with somewhat more emphasis on Soviet Far East. Gives titles in original language in original script and English translation.

LAND TENURE

1539   Anderson, Theresa J. *Land tenure and agrarian reform in East and Southeast Asia: an annotated bibliography.*

See entry 1406.

1540   *East and southeast Asia: a bibliography.*

See entry 1407.

CHINA
Zonghua Renmin Gonghe Guo
[The People's Republic of China]

China is the world's most populous nation and occupies a large proportion of the world's drylands. The nation's topography is extremely complex but six separate dryland regions can be identified.

The Northwestern Basins. China's northwestern provinces of Xinjiang (Sinkiang), Gansu, and Qinghai correspond to a large region of strikingly varied dryland topography. Two great interior basins alternate with very high mountain ranges: the smaller (Dzungarian) basin is arid and supports nomadic herds of cattle, sheep, horses, and camels, while the larger (Tarim) basin is a hyperarid sand sea. Nonetheless, nearby mountains send short torrential rivers into the basins around whose margins ancient cities have flourished as trading centers along the Silk Road across Central Asia. Agricultural oases still produce wheat, maize, cotton, and cold-winter fruits. Mineral wealth is believed substantial though development has barely begun. Despite its long history, for China the northwest is a largely untapped frontier.

Tibet. This great elevated plateau was formed when the drifting landmass of India collided with Asia and folded hunks of the earth's crust into the Himalayan Range and the broken plateau beyond. The most populated area of the region lies near the border with Nepal and is generally well-watered and outside the dryland zone; to the north stretches a virtually empty semiarid to arid land of mountain tundra and glaciers that can support only nomadic herds. Much is still unmapped and unexplored.

The Gobi Desert. China's share of this cold-winter steppe desert roughly corresponds to the Inner Mongolian Autonomous Region. Nomadic herding is the rule and crop production is confined to irrigated areas near the great bend of the Hwang Ho River; cold winter cereals, soybeans, and sugar beets are grown.

The Manchurian Plain. This semiarid to subhumid region has cold continental winters and summers warm enough to allow diversified agriculture. Nearby mountains supply quantities of water for hydropower and irrigation. In addition to being one of the world's most extensive soybean producers, wheat, millet, sorghum, sugar beets, tobacco, cotton, and flax are grown. Even more important is the region's heavy industry, developed first by the Japanese. Very large deposits of coal, iron, oil shale, and other metals led to concentrations of factories and mills at the southern margin of the plain nearest the rest of China.

The Loess Plateau. Along the middle Hwang Ho in the province of Ningxia, Shaanxi, and Shanxi great deposits of windblown soil have been deposited in formations hundreds of feet thick in some areas. The soil is easily eroded into deep gullies and canyons—made worse by extensive deforestation—but is also fertile whenever irrigation water is available. As in the rest of dryland China, cold-winter cereals are staple crops, supplemented by fruits, cotton, flax, tobacco, and others. Deposits of iron and coal have encouraged extensive industrial development.

The North China Plain. Formed where the Hwang Ho emerges from the loess plateau, this is the largest lowland area in the entire nation and holds perhaps a quarter of its total population. Ranging from semiarid to subhumid, the land can suffer from insufficient precipitation and, when rains bring floods, too much precipitation in the same growing season. Massive development schemes are being carried out to serve the triple needs for flood control, irrigation, and hydroelectric power. With the capital at Beijing (Peking), major manufacturing cities, and an extensive wheat producing agricultural base, the region is the nation's economic core.

Dryland areas of CHINA

DESERTIFICATION RISK	ARIDITY									
	Hyperarid		Arid		Semiarid		Subhumid		Aridity Totals	
	km²	%	km²	%	km²	%	km²	%	km²	%
Very High	By definition, desertification does not exist in hyperarid regions.		200,909	2.1					200,909	2.1
High			888,882	9.3	63,736	0.7	66,514	0.7	1,019,132	10.6
Moderate			136,684	1.4	760,117	7.9	348,192	3.6	1,244,993	13.0
Desertification Totals			1,226,475	12.8	823,853	8.6	414,706	4.3	2,465,034	25.7
No Desertification	378,766	4.0	771,126	8.0	636,666	6.6	1,092,242	11.4	2,878,800	30.0
Total Drylands	378,766	4.0	1,997,601	20.8	1,460,519	15.2	1,506,948	15.7	5,343,834	55.7
Non-dryland									4,253,166	44.3
Total Area of the Territory									9,597,000	100.0

GENERAL BIBLIOGRAPHIES

1541 McCone, Gary K. *China in U.S. Government documents: 1955-1975.* Tucson AZ: University of Arizona, 1976. 70 leaves.

Unannotated. Superintendent of Documents call numbers are provided.

1542 Moskowitz, Harry, and Roberts, Jack. *China: an analytical survey of literature.* 4th ed. Washington DC: U.S. Government Printing Office, 1978. 232 pp.

Covers period 1971-1976. Charts, tables and maps. Based on classified publications of "friendly and unfriendly" governments.

1543 Nathan, Andrew J. *Modern China, 1870-1972: an introduction to sources and research aids.* Ann Arbor MI: University of Michigan. Center for Chinese Studies, 1973. 95 pp.

1544 Pearson, J. D., ed. *South Asian bibliography: a handbook and guide.*

Coverage includes Xizang (Tibet).
See entry 1398.

1545 Tanis, Norman E.; Perkins, David L.; and Pinto, Justine, comp. *China in books: a basic bibliography in Western languages.* Greenwich CT: Jai Press, 1979. (Foundations in library and information science 4) 328 pp.

Unannotated.

HANDBOOKS

1546 Bunge, Frederica M., and Shinn, Rinn-Sup. *China: a country study.* 3rd ed. Washington DC: U.S. Dept. of Defense, 1981. (American University foreign area studies) 590 pp.

Narrative summary of the nation's history, politics, geography, economy, social life; includes unannotated bibliography.

DIRECTORIES

1547 *China handbook.* Hong Kong: Ta Kung Pao, 1980. 476 pp.

Includes names of political and administrative leaders in both Chinese characters and Pinyin, as well as addresses of major offices and government trading agencies.

1548 Winter, Alan; Kamm, Antony; and Khan, M. N. G. A., eds. *China, India, and Central Asia.* 2nd ed. Guernsey, U.K.: Francis Hodgson, 1976. (Guide to world science 14) 215 pp.

Provides descriptions of scientific research activities and organizations. Covers China, India, Afghanistan, Bangladesh, Iran, Pakistan, Sri Lanka.

STATISTICS

1549 *China: a statistical compendium: a reference aid.* Washington DC: U.S. Central Intelligence Agency, 1979. 13 pp.

Compares the current performance of mainland Chinese economy with that of the 1950's and 1960's. Data for 1965, 1970, and 1975 are U.S. estimates based on fragmentary Chinese data. Available on microfiche.

1550 *China facts and figures annual.* Gulf Breeze FL: Academic International Press, 1978-. Annual.

An excellent compendium of statistical data on many subjects.

GEOGRAPHY

1551 Paylore, P. "A selective list of references relating to Chinese deserts." *Arid lands newsletter* 11 (1980):16-21.

1552 Sukwal, B. L. *A bibliography of theses and dissertations in geography on South Asia.*

Coverage includes Xizang (Tibet).
See entry 1401.

1553 Sukhwal, B. L. *South Asia: a systematic geographic bibliography.*

See entry 1402.

ATLASES

1554 *Briefs on selected PRC cities.* Washington DC: U.S. Central Intelligence Agency, 1975. 63 leaves.

1555 Chang, Chi'-yun, ed. *National atlas of China.* Taiwan: National War College. Chinese Geographical Institute, 1960-63. 5 volumes.

In Chinese and English. Covers China and Mongolia.

1556 Geelan, P. J. M., and Twitchett, D. C. *The Times atlas of China.* New York: Quadrangle, 1974. 27 pp.

1557 Hsieh, Chiao-min. *Atlas of China.* New York: McGraw-Hill, 1973. 282 pp.

1558 *Peoples Republic of China: administrative atlas.* Washington DC: U.S. Central Intelligence Agency, 1975. 68 pp.

1559 *People's Republic of China atlas.* Washington DC: U.S. Central Intelligence Agency, 1971. 82 pp.

1560 *Zhonghua renmin gongheguo fen sheng diyuji [atlas of the People's Republic of China].* Beijing: Ditn Chubawshe, 1977. Approximately 300 pp.

In Chinese.

GAZETTEERS

1561 *Gazetter of the People's Republic of China: Pinyin to Wade-Giles, Wade-Giles to Pinyin.* Washington DC: U.S. Defense Mapping Agency, 1979. (Gazetteer 167) 919 pp.

List of 22,000 place-names with latitude/longitude coordinates.

SCIENCE AND TECHNOLOGY

1562 **Baark, Erik; Jonsen, Roar; and Wagner, Donald B.** *Survey of PRC literature on science and technology: a bibliography, partly annotated.* Lund, Sweden: University of Lund, 1977. 36 pp.

Though small, covers a variety of subjects including geology, biology, agriculture, engineering. All titles transliterated using Pinyin.

GEOLOGY

1563 **Li, Hsiao-Fang.** *Bibliography of Chinese geology: bibliography of geology and geography of Sinkiang.* Nanking: China. National Geological Survey, 1947. 213 pp.

1564 **Tseng, T. C.** *Bibliography of Chinese geology: bibliography of geology and allied sciences of Tibet and regions to the west of the Chinshachiang.* Nanking: China. National Geological Survey, 1946. 114 pp.

CLIMATOLOGY

1565 *An atlas of Chinese climatology - Communist China.* Washington DC: U.S. Joint Publications Research Service, 1962. (JPRS 16,321) 582 pp.

Translation from the Chinese "Chung-Kuo Ch'i T'u" (Peiping, 1960). Most data used ended in 1950.

1566 **Vitale, C. S.** *Bibliography on the climate of Sinkiang, China.* Washington DC: 1963. (NTIS: AD 660-813) 46 pp.

1567 **Wallace, J. A., Jr.** *An annotated bibliography of climatic maps of People's Republic of China.* Washington DC: U.S. Weather Bureau, 1961. (NTIS: PB 176-047) 21 pp.

Contains 50 references for the period 1901-1961.

1568 **Wallace, J. A., Jr.** *An annotated bibliography on cloud information and data for mainland China.* Washington DC: U.S. Weather Bureau, 1966. (NTIS: PB 176-043; AD 664-720) 45 pp.

Contains 108 citations.

1569 **Watts, I. E. M.** "Climates of China and Korea." In *Climates of northern and eastern Asia*, pp. 1-117. Edited by H. Arakawa. Amsterdam, The Netherlands: Elsevier, 1969. (World survey of climatology 8).

A narrative summary of these nations' climates, with maps, charts, and an unannotated bibliography.

BOTANY

1570 **Gupta, Raj Kumar.** "Bibliography on 'plant ecology' in Bhutan, Sikkim and Tibet." *Excerpta botanica: sectio B: sociologica* 12:3 (1972):226-237.

Unannotated. Contains 189 items.

1571 **Hanelt, P.** "Bibliographia phytosociologicae: China" [bibliography of plant ecology: China]. *Excerpta botanica: sectio B: sociologica* 6:2 (1964):106-134.

Unannotated.

AGRICULTURE

1572 *Current bibliography of agriculture in China.* Wageningen, The Netherlands: Centre for Agricultural Publishing and Documentation/Centrum voor Landbouwpublikaties en Landbouwdocumentatie, 1979-. Quarterly.

In English, French and German.

1573 **Logan, William J. C., and Schroeder, Peter B.** *Publications on Chinese agriculture prior to 1949.* Washington DC: U.S. Dept. of Agriculture, 1966. (Library list 85) 142 pp.

Includes serials and monographs.

AGRICULTURAL ECONOMICS

1574 **Broadbent, Kieran P.** *A Chinese/English dictionary of China's rural economy.* Farnham Royal, Bucks, U.K.: Commonwealth Agricultural Bureaux, 1978. 406 pp.

An unusual dictionary of 9,000 Chinese terms relating to agricultural development and Chinese development theory. Based on recent sources published since the 1949 revolution. Arranged phonetically similar to Matthew's Chinese-English dictionary; Wade-Giles and Pinyin are also used. The researcher ignorant of Chinese should use the English text provided. Especially valuable are informed explanations of terms having particular Chinese significance and no simple English equivalents. Includes full bibliography (unannotated) and key character index. An important technical supplement to standard Chinese language reference sources.

PETROLEUM

1575 **Chang, Raymond J.** *Chinese petroleum: an annotated bibliography.* Boston MA: G. K. Hall, 1982. 204 pp.

Extensively annotated. Includes English, Chinese, and Japanese language publications; the latter two are provided with original character titles, transliterations, and translations.

1576 **Chang, Raymond J.** *The petroleum industry in China: its planning in the development of energy resources: a bibliography with selective annotations.* Mon-

HUMAN GEOGRAPHY

1577 **Emerson, John Philip et al.** *The provinces of the People's Republic of China: a political and economic bibliography.* Washington DC: U.S. Dept. of Commerce. Bureau of Economic Analysis, 1976. (International population statistics reports series P-90 25) 734 pp.

Covers source material in both English and Chinese.

PUBLIC HEALTH

1578 **Akhtar, Shahid.** *Health care in the People's Republic of China: a bibliography.with abstracts.* Ottawa: International Development Research Centre, 1975. 182 pp.

URBAN GEOGRAPHY

1579 **Ma, Laurence J. C.** *Cities and city planning in the People's Republic of China.* Washington DC: U.S. Dept. of Housing and Urban Development. Office of Policy Development and Research, 1980. (HUD user bibliography series) 62 pp.

ECONOMIC DEVELOPMENT

1580 *Asia/Australia data bank.*
See entry 1806.

ticello IL: Council of Planning Librarians, 1977. (CPL exchange bibliography 1285) 28 pp.

Brief annotations.

1581 **Blair, Patricia W., and Barnett, A. Doak.** *Development in the People's Republic of China.* Washington DC: Overseas Development Council, 1976. 94 pp.

1582 *China: economic indicators.* Washington DC: U.S. Central Intelligence Agency, 1978. 50 pp.

The fourth in an annual series on the People's Republic of China. Contains statistics to 1977. (Data not published systematically since 1957.)

1583 *Quarterly economic review of China, North Korea.* London: Economist Intelligence Unit, 1982-. Quarterly.

Includes summary of political and economic news as well as charts and statistics of selected economic indicators. Annual supplement.

HISTORICAL GEOGRAPHY

1584 **Herrmann, Albert.** *An historical atlas of China.* Chicago IL: Aldine Publishing Co, 1966. 88 pp.

Most plates at a scale of 1:15,000,000. Based on Herrmann's "Historical and commercial atlas of China" (1935).

PUBLIC ADMINISTRATION

1585 **Harmon, Robert B.** *A selected and annotated guide to the government and politics of China.* Monticello IL: Council of Planning Librarians, 1976. (CPL exchange bibliography 132) 19 pp.

Annotated.

MONGOLIA
Bugd Nayramdakh Mongol Ard Uls
[The Mongolian People's Republic]
(a former colony of China and the Soviet Union)

Mongolia is a remote, landlocked nation sandwiched between the great powers of the Soviet Union and China. It has a long history of economic and political dependence on both neighbors. Although much of the land is dry, well-watered forests exist toward the Soviet border and on the higher peaks of the Altai Mountains.

The Steppe Grasslands. This semiarid steppe is characterized by low rolling hills and sparse grassland. Nomadic flocks of sheep (predominant), cattle, horses, and camels are pastured here and provide virtually the only means of livelihood.

The Gobi Desert. Mongolia's portion of this cold-winter desert is somewhat wetter than China's to the south, though both are arid with summer precipitation and wide extremes of temperature. Winter snowfall is very light so patches of dried grass can be grazed.

Dryland areas of MONGOLIA

DESERTIFICATION RISK	Hyperarid km²	Hyperarid %	Arid km²	Arid %	Semiarid km²	Semiarid %	Subhumid km²	Subhumid %	Aridity Totals km²	Aridity Totals %
Very High	By definition, desertification does not exist in hyperarid regions.									
High			228,490	14.6	3,130	0.2			231,620	14.8
Moderate					838,840	53.6	150,240	9.6	989,080	63.2
Desertification Totals			228,490	14.6	841,970	53.8	150,240	9.6	1,220,700	78.0
No Desertification			6,260	0.4	15,650	1.0	46,950	3.0	68,860	4.4
Total Drylands			234,750	15.0	857,620	54.8	197,190	12.6	1,289,560	82.4
Non-dryland									285,440	17.6
Total Area of the Territory									1,565,000	100.0

HANDBOOKS

1586 Dupuy, Trevor N., and Blanchard, Wendell. *Area handbook for Mongolia*. Washington DC: U.S. Dept. of Defense, 1970. (American University Foreign area studies) 500 pp.

Narrative summary of the nation's history, politics, geography, economy, social life; includes unannotated bibliography.

ATLASES

1587 Chang, Chi'-yun, ed. *National atlas of China*.
See entry 1555.

GAZETTEERS

1588 *Mongolia: official standard names approved by the United States Board on Geographic Names*. Rev. ed. Washington DC: U.S. Board on Geographic Names, 1970. (Gazetteer 116) 256 pp.

List of 13,000 place-names with latitude/longitude coordinates.

CLIMATOLOGY

1589 Wallace, J. Allen, Jr. *An annotated bibliography on the climate of Mongolia*. Washington DC: U.S. Environmental Data Service, 1965. (WB/BC-89; NTIS: PB 176-001; AD 664-702) 14 pp.

Covers the period 1895-1965.

SOILS

1590 Bespalov, N. D. *Soils of Outer Mongolia*. Jerusalem: Israel Program for Scientific Translations, 1964. (Akademiia Nauk SSR. Mongol'skaia Komissiia. *Trudy* 41) 320 pp.

Translation of title originally published by the Akademiia Nauk SSR in 1951. Bibliography: pp. 305-314.

THE UNION OF SOVIET SOCIALIST REPUBLICS
Soyuz Sovyetskikh Sotsialisticheskikh Respublik

The Soviet Union is the world's largest nation, the third most populous, and together with the United States shares the responsibilities of a superpower in its ability to project its influence and military strength around the world and into outer space. Although less than one-third of the nation is dry, the Soviet Union occupies more dryland than any other except Australia. These lands are especially important to a nation that has always suffered from a lack, not of resources, but of convenient arrangement of resources for efficient exploitation. Unable to feed itself, with less than 20 percent of its land located south of 50 degrees north latitude, the Soviet Union is forced to increase drastically its development of dryland agriculture.

The Ukrainian Steppe. This subhumid to semiarid region extends vaguely from the central Ukraine eastward toward the Volga River. Already a primary producer of wheat, barley, millet, and livestock, the region is a focus of efforts to increase yields of existing lands. The area is generally well-watered by major southward flowing rivers (the Dnieper, the Don, and many others), but erratic precipitation and dessicating winds (called *sukhovei*) make irrigation increasingly essential.

The Black Sea Coast. With warmer winters and patches of Mediterranean-type climate, the coast of the Ukraine and the Crimean Peninsula is best suited to fruit crops, grapes, and tobacco. Perhaps of equal importance to the Soviet people is the region's status as a summer resort, a "Russian Riviera" that is the nation's nearest thing to a subtropical seacoast.

The Caspian Basin. The Caspian Sea occupies most of a large arid structural depression; the water surface fluctuates between 26 and 28 meters (85 and 92 feet) below sea level, depending on inflow from the Volga and other large rivers. Sharp relief confines the sea to a fairly stable shoreline along the southeast, south, and western shores but in the north and northeast relief is low and vast tracts are alternately revealed and flooded. Where rivers such as the Volga and the Ural have built deltas the land is very productive and supports intensive irrigated fields of fruits and vegetables. The waters themselves harbor caviar-producing sturgeon. The peninsula at Baku is a major oil-producing area.

The Kura Valley. The Kura River flows eastward between the Caucasus Mountains and the Armenian Highlands to the Caspian Sea. Its valley becomes increasingly drier and warmer downstream. Irrigation is essential and supports fruit crops and cotton.

The Central Asian Deserts. These vast expanses of semiarid to arid desert comprise the Soviet republics of Kazakh, Uzbek, and Turkmen. Most of the land supports little more than nomadic flocks of sheep, though important irrigation districts appear along the rivers draining into the Aral Sea. With continued development of water supplies the region has great potential.

The Central Asian Mountains. The long complex ranges bordering China are well-watered on their upper slopes and send their captured moisture westward to the deserts below. Along the mountain front, in deep canyons and narrow valleys, the caravan cities of Samarkand, Tashkent, and others are centers of irrigated agriculture, trade, and Moslem culture. The Soviet republics of Kirgiz and Tadzhik occupy the fertile sheep pastures and include important deposits of oil as well. Although less dry than the lowlands, winter temperatures here are extreme and limit crops to the hardiest varieties.

The Siberian Steppes. Scattered and disconnected pockets of subhumid dryland exist in a number of locations along the southern margins of Siberia and along the Lena River. Generally, these pockets exist where steppe grassland has opened up within the prevailing taiga forests; where possible, as along the Lena, spring wheat, barley, and other hardy cereals are grown by the native Yakuts, and beef cattle are reared.

U.S.S.R./1591-1601

Dryland areas of THE UNION OF THE SOVIET SOCIALIST REPUBLICS

DESERTIFICATION RISK	ARIDITY									
	Hyperarid		Arid		Semiarid		Subhumid		Aridity Totals	
	km²	%	km²	%	km²	%	km²	%	km²	%
Very High	By definition, desertification does not exist in hyperarid regions.		52,586	0.2	25,081	0.1			77,667	0.3
High	^		1,957,940	8.8	76,469	0.3			2,034,409	9.1
Moderate	^		8,502	0.03	1,721,296	7.7	647,102	2.9	2,376,900	10.6
Desertification Totals	^		2,019,028	9.0	1,822,846	8.1	647,102	2.9	4,488,976	20.0
No Desertification			17,315	0.1			1,394,391	6.2	1,411,706	6.3
Total Drylands			2,036,343	9.1	1,822,846	8.1	2,041,493	9.1	5,900,682	26.3
Non-dryland									16,499,318	73.7
Total Area of the Territory									22,400,000	100.0

GENERAL BIBLIOGRAPHIES

1591 *Bibliography of recent Soviet source material on Soviet Central Asia and the borderlands.* London: Central Asian Research Centre, 1958-62. 6 vols.

Annotated. All Cyrillic has been transliterated into Roman characters. Covers most general subjects; good for geography, economics, earth sciences, history of the Kazakh SSR and surrounding Soviet Central Asia. Issued as supplements to *Central Asian Review*.

1592 Pierce, Richard A. *Soviet Central Asia: a bibliography (1558-1966).* Berkeley CA: University of California. Center for Slavic and East European Studies, 1966. 3 vols.

Only transliterated (romanized) titles are given; no Cyrillic is used.

1593 Thompson, Anthony. *Russia/U.S.S.R.: a selective annotated bibliography of books in English.* Oxford, U.K.; Santa Barbara CA: Clio Press, 1979. (World bibliographical series 6) 287 pp.

An annotated list of reference publications in English, totalling 1,247 citations. Covers most major subjects; best for history, politics, and literature.

THESES AND DISSERTATIONS

1594 Dossick, Jesse John. *Doctoral research on Russia and the Soviet Union.* New York: New York University Press, 1960. 248 pp.

Covers U.S., Canadian, and U.K. dissertations from the period 1876-1960.

1595 Dossick, Jesse John. *Doctoral research on Russia and the Soviet Union, 1960-1975: a classified list of 3,150 American, Canadian and British dissertations, with some critical and statistical analysis.* New York: Garland, 1976. 345 pp.

Updated in *Slavic Review* (Columbus OH: American Association for the Advancement of Slavic Studies).

HANDBOOKS

1596 Keefe, Eugene K. *Area handbook for the Soviet Union.* Washington DC: U.S. Dept. of Defense, 1971. (American University foreign area studies) 827 pp.

Narrative summary of the nation's history, politics, geography, economy, social life; includes unannotated bibliography.

ENCYCLOPEDIAS

1597 *The Cambridge encyclopedia of Russia and the Soviet Union.* Cambridge, U.K.: Cambridge University Press, 1982. 492 pp.

1598 Prokhorov, Aleksandr Mikhailovich, ed. *Great Soviet encyclopedia.* New York: Macmillan, 1973-. 31 vols.

The standard national encyclopedia. A translation of the 3rd edition of the *Bolshaia sovetskaia entsiklopediia* (Moskva: Izd-vo "Sovetskaia entsiklopediia," 1970-81), 31 vols. 28 volumes of the translation have appeared to date (July 1982).

DIRECTORIES

1599 *Directory of Soviet research organizations.* Seattle WA: University Press of the Pacific, 1979. 290 pp.

1600 *Directory of USSR Academy of Sciences.* Wooster OH: Transdex, 1976. (JPRS 66,592) 167 pp.

Directory of members and organizations of the U.S.S.R. Academy of Sciences (Akademiia Nauk SSR).

1601 Pernet, Ann, ed. *U.S.S.R.* 2nd ed. Guernsey, U.K.: Francis Hodgson, 1976. (Guide to world science 11) 214 pp.

Provides descriptions of scientific research activities and organizations.

1602 *Reference aid: membership, U.S.S.R. Academy of Sciences*. Washington DC: U.S. Central Intelligence Agency, 1977. (CR 77-11360) 42 pp.

Biographical information and status in the Academy included.

1603 **Shabad, Theodore.** *Directory of Soviet geographers*. Washington DC: Scripta, 1977. Variously paged.

1604 **White, Sarah, ed.** *Guide to science and technology in the USSR*. Guernsey, U.K.: Frances Hodgson, 1971. 300 pp.

STATISTICS

1605 *USSR facts and figures annual*. Gulf Breeze FL: Academic International Press, 1977-. Annual.

1606 *The USSR in figures for 1979: statistical handbook*. Moscow: Statistika, 1980. 224 pp.

INFORMATION SCIENCE

1607 *Biblioteki SSSR spravochnik [reference guide to Soviet libraries]*. Moskva: Kniga, 1973. 2 vols.

In Russian. Brief descriptions of Soviet university and technical libraries.

GEOGRAPHY

1608 **Bogomolov, G. V., ed.** *Problemy kompleksnogo izucheniia zasuschlykh zon SSSR [problems of composite study of arid zones of the U.S.S.R.]*. Moscow: Akademiia Nauk SSR, 1963. 242 pp.

In Russian. Contains bibliographies covering natural resources, including underground water, of Soviet Central Asia.

1609 **Giul, K. K.; Lappalainen, T. N.; and Polushkin, V. A.** *Kaspiiskoe More [Caspian Sea]*. Moskva: Vsesoiuznyi Institut Nauchnoi Tekholcheskoi Informatisii, 1970. 236 pp.

In Russian. Annotated. Contains 1,850 items.

1610 **Harris, Chauncy Dennison.** *Guide to geographical bibliographies and reference works in Russian or on the Soviet Union*. Chicago IL: University of Chicago. Dept. of Geography, 1975. (Research paper 164) 478 pp.

1611 **Johnson, W. Carter, and French, Norman R.** "Soviet Union." In *Handbook of contemporary developments in world ecology*, pp. 343-383. Edited by Edward John Kormondy and J. Frank McCormick. Westport CT: Greenwood Press, 1981.

A good narrative summary of recent research on ecology in the Soviet Union with an extensive, though unannotated, list of references.

1612 **Kirkpatrick, Meredith.** *Environmental problems and policies in Eastern Europe and the U.S.S.R.*

See entry 210.

1613 **Laking, Phyllis N.** *The Black Sea, its geology, chemistry, biology: a bibliography*. Woods Hole MA: Woods Hole Oceanographic Institution, 1974. (Contributions 3330) 368 pp.

1614 **Lamprecht, Sandra J.** *The Soviet Union and Scandinavia: a bibliography of theses and dissertations in geography*. Monticello IL: Council of Planning Librarians, 1974. (CPL exchange bibliography 609) 22 pp.

Unannotated. Covers degrees granted by North American and British institutions only.

1615 **Nazarevskiy, A., ed.** *Kazakhskaya SSR: ekonomiko-geograficheskaya kharakteristika [Kazakh SSR: economic and geographical characteristics]*. Moskva: Gosudarstvennoye Izdatel'stvo Geograficheskoy Literatury, 1957. 733 pp.

1616 **Petrov, Mikhail Platonovich.** "Kritika i bibliografiia: obzor literatury po izucheniio pustyn i polupustyn SSSR i fitomelioratsii ikh za 1975 g." [review of literature published in 1975 on the study of deserts and semi-deserts of the U.S.S.R. and their reclamation via planting]. *Problemy osveniia pustyn* no. 5 (1976):73-84.

In Russian. Lists over 180 titles, one-third of which were published in *Problemy osvoeniia pustyn*.

1617 **Petrov, Mikhail Platonovich.** "Kritika i bibliografiia: obzor literatury po izucheniio i osvoeniio pustyn i polupustyn SSSR za 1977 g." [review of literature on the study and reclamation of deserts and semi-deserts of the U.S.S.R. for 1977]. *Problemy osvoeniia pustyn* no. 5 (1978):75-88.

In Russian. Annotated. Contains approximately 200 titles.

1618 **Petrov, Mikhail Platonovich.** *Pustyni Turkmenii i ikh khoziaistvennoe osvoenie: literatury (1950-1965) [the Turkmen deserts and their agricultural development]*. Ashkhabad, U.S.S.R.: Akademiia Nauk Turkmenskoi SSR. 435 pp.

1619 **Petrov, Mikhail Platonovich, and Tatarsky, I. V.** "Bibliografiya po izueheniyu, osvoeniyu i okhrane prirody i prirodykh resursov rustyn i bor be s orustynivaniem za 1965-1977 gg" [bibliography of works on research, development, and conservation of nature and natural resources of deserts and desertification control in 1965-1975]. *Okhrana Prirody i Vosproizbodstvo Prirodnyka Resursov* 3 (1977):61-146.

In Russian. Most items are on the Soviet Union. Also contains a smaller multilingual bibliography worldwide in scope.

1620 *Soviet geography: a bibliography.* New York: Greenwood Press, 1969. 668 pp.

Reprinted from U.S. government documents published in 1951.

1621 *Soviet geography: review and translation.* New York: American Geographical Society, 1960-. 10 per year.

ATLASES

1622 Abdullaev, I. K., ed. *Atlas Azerbaidzhanskoi Sovietskoi Sotsialisticheskoi Respubliki [atlas of Azerbaijan Soviet Socialist Republic].* Baku, U.S.S.R.: Glavnoe Upravlenie Geodezii i Kartografii, 1963. 213 pp.

In Russian. Covers agriculture, industry, and communications in Soviet Azerbaijan. Azerbaijan SSR lies between the Caucasus Mountains, the Caspian Sea, and Iran.

1623 Aliev, S. D., ed. *Atlas Azerbaidzhanskoi SSR [atlas of Azerbaijan SSR].* Moskva: Glavnoe Upravlenie Geodezii i Kartografii, 1979. 40 pp.

In Russian. General atlas with most thematic maps at 1:2,000,000. Does *not* supersede 1963 edition (entry 1622).

1624 *Atlas Armianskoi Sovetskoi Sotsialisticheskoi Respubliki [atlas of Armenian Soviet Socialist Republic].* Yerevan U.S.S.R.: Akademia Nauk Armianskoi SSR, 1961. 111 pp.

In Russian. Armenian SSR lies between the Caucasus Mountains and the Turkey-Iran border.

1625 *Atlas Kustanaiskoi oblasti [atlas of Kustanay province].* Moskva: Glavnoe Upravlenie Geodezii i Kartografii, 1963. 79 pp.

In Russian. Kustanay province lies in northern Kazakh SSR east of the Ural Mountains.

1626 *Atlas Novosibirskoi oblasti [atlas of Novosibirsk province].* Moskva: Glavnoe Upravlenie Geodezii i Kartografii, 1979. (G-2557) 32 pp.

In Russian. Novosibirsk province lies in western Siberia along the Upper Ob River.

1627 *Atlas Voronezhskoi oblasti [atlas of Voronezh province].* Moskva: Glavnoe Upravlenie Geodezii i Kartografii, 1968. (3-655) 32 pp.

In Russian. Voronezh province lies along the Don River bordering the eastern Ukraine.

1628 Baranov, A. N., and Teplova, S. N., eds. *Atlas SSSR.* 2nd ed. Moskva: Glavnoe Upravlenie Geodezii i Kartograffii, 1969. (I01360) 199 pp.

In Russian.

1629 Dzotsenidze, G. S., ed. *Atlas Gruzinskoi Sovetskoi Sotsialisticheskoi Respubliki [atlas of Georgia Soviet Socialist Republic].* Tbilisi, U.S.S.R.: Glavnoe Upravlenie Geodezii i Kartografii, 1964. 269 pp.

In Russian. Physical, climatic, economic, historical, and demographic maps of Soviet Georgia. Georgian SSR lies between the Caucasus Mountains and Turkey.

1630 Galazii, G. I. et al. *Atlas Baikala [atlas of Lake Baikal].* Irkutsk-Moskva: Glavnoe Upravlenie Geodezii i Kartografii, 1969. (Akademiia Nauk SSR. Sibirskoe Otdelenie. Limnologicheskii Institut. V-837) 30 pp.

In Russian. Covers Lake Baikal and surrounding territory. Lake Baikal is in south-central Siberia near the border with Mongolia.

1631 Khukin, N. V. et al., eds. *Atlas Astrakhanskoi oblasti [atlas of Astrakhan province].* Moskva: Glavnoe Upravlenie Geodezii i Kartografii, 1968. 38 pp.

In Russian. Astrakhan province borders Kazakh SSR along the lower Volga River and includes the delta as it enters the Caspian Sea.

1632 Liakhova, A. G., ed. *Atlas Volgogradskoi oblasti [atlas of Volgograd province].* Moskva: Glavnoe Upravlenie Geodezii i Kartografii, 1967. 32 pp.

In Russian. Volgograd province borders Kazakh SSR along the lower Volga River.

1633 Narzikulov, I. K.; Stanukovich, K. V., eds. *Atlas Tadzhikskoi SSR [atlas of Tadzhik SSR].* Dushanbe-Moskva: Glavnoe Upravlenie Geodezii i Kartografii; 1968. 200 pp.

In Russian. Physical, demographic, historical, and cultural maps of Tadzhik SSR, which embraces the Alay and Pamir Mountains along the Afghanistan-China border.

1634 Plummer, Thomas F. et al. *Landscape atlas of the U.S.S.R..* West Point NY: U.S. Military Academy, 1971. 197 pp.

Selected areas are portrayed at 1:250,000 to illustrate various geographical regions of the nation.

1635 Semenova, M. I. et al. *Atlas Karagandinskoi oblasti [atlas of Karaganda province].* Moskva: Glavnoe Upravlenie Geodezii i Kartografii, 1969. 48 pp.

In Russian. Karaganda province lies within Kazakh SSR between Lake Balkhash and the Irtysh River.

1636 Sochava, V. B. *Atlas Zabaiklia: Buriatskaia ASSR i Chitinskaya oblast [atlas of Zabaykalye: Buriat ASSR and Chitinskaya province].* Moskva: Glavnoe Upravlenie Geodezii i Kartografii, 1967. 176 pp.

In Russian. Covers Transbaikalia; that is, the territory to the east of Lake Baikal in southern Siberia bordering Mongolia.

1637 Vetrov, A. S. et al. *Atlas Orenburgskoi oblasti [atlas of Orenburg province].* Moskva: Glavnoe Upravlenie Geodezii i Kartografii, 1969. 36 pp.

In Russian. Orenburg province borders Kazakh SSR at the southern end of the Ural Mountains.

GAZETTEERS

1638 **Birkenmayer, Sigmund S.** *An accented dictionary of place names in the Soviet Union.* University Park PA: Pennsylvania State University. Dept. of Slavic Languages, 1967. 97 pp.

Contains nearly 3,000 place-names arranged in the order of the Russian alphabet, Russian to English. Includes lists of common geographic terms, adjectives, and abbreviations as well as lists of Soviet administrative divisions and recent place-name changes. An English to Russian list of the best-known places is also provided.

1639 *Europe and U.S.S.R.: official standard names approved by the United States Board on Geographic Names.*

See entry 112.

1640 *U.S.S.R.: official standard names approved by the United States Board on Geographic Names.* 2nd ed. Washington DC: U.S. Board on Geographic Names, 1970. (Gazetteer 42) 7 vols.

List of 400,000 place-names with latitude/longitude coordinates.

GEOLOGY

1641 **Alexandrov, Eugene A., comp.** *Mineral and energy resources of the U.S.S.R.: a selected bibliography of sources in English.* Falls Church VA: American Geological Institute, 1980. 91 pp.

Unannotated.

1642 *Geologicheskaia literatura SSSR [geological literature of the U.S.S.R.].* Leningrad: Tsentralnaia Geologicheskaia Biblioteka, 1880-. 2 vols./year.

Unannotated. In Russian.

SAND CONTROL

1643 **Petrov, Mikhail Platonovich.** "Kritika i bibliografiia: obzor literatury po fitomelioratsii peskov pustyn i polupustyn" [review of literature on the reclamation of desert- and semi-desert sands in the U.S.S.R. for 1965]. Akademiia Nauk Turkmenskoi S.S.R. *Izvestiya. Seriia Biologicheskikh Nauk* 6 (1966):79-87.

In Russian. "This is a list of 106 annotated references to the natural conditions and vegetation of desert- and semi-desert sands and to their reclamation by the planting of windbreaks or sowing herbage plants." (Paylore, 1969.)

CLIMATOLOGY

1644 **Borisov, Anatolii Aleksandrovich.** *Climates of the U.S.S.R..* Edinburgh: Oliver and Boyd, 1965. 255 pp.

"The most authoritative and detailed work on climate." (Thompson, *Russia/U.S.S.R.*)

1645 *Klimaticheskii atlas Ukrainskoi SSR [climatic atlas of the Ukrainian SSR].* Leningrad: Gidrometeorologicheskoe Izdatelstvo, 1968. 232 pp.

In Russian.

1646 **Lydolph, Paul E.** *Climates of the Soviet Union.* Amsterdam, The Netherlands: Elsevier Scientific, 1977. (World survey of climatology 7) 443 pp.

A narrative summary with maps, charts, and unannotated bibliographies.

1647 **Lydolph, Paul E.** "Soviet work and writing in climatology." *Soviet geography: review and translation* 12:10 (1971):637-665.

Contains a selected bibliography of 207 citations.

1648 **Mingkov, N. I.** *Bibliograficheskii ukazatel literatury po limatu Turkmenii [bibliography on the climate of the Turkmen Republic].* Ashkabad, U.S.S.R.: Upravlenie Gidrometsluzhby Turkmenskoi SSR. Ashkhabadskaia Gidrometeorologicheskaia Observatoriia, 1957.

In Russian. Contains 1,439 citations.

1649 **Thran, P., and Broekhuizen, Simon.** *Agroclimatic atlas of Europe.*

See entry 114.

WEATHER MODIFICATION

1650 **Zikeev, Nikolay T., and Doumani, George A.** *Weather modification in the Soviet Union, 1946-1966, a selected annotated bibliography.* Washington DC: Library of Congress, 1967. 78 pp.

Brief annotations.

BOTANY

1651 **Kuechler, August Wilhelm.** *International bibliography of vegetation maps: Volume 3: U.S.S.R., Asia, and Australia.* Lawrence KS: University of Kansas Libraries, 1968. (Publication 29) 389 pp.

AGRICULTURE

1652 *Atlas silskogo gospodarstva Ukrainskoi RSR [agricultural atlas of the Ukrainian SSR].* Kiev, U.S.S.R.: Vidavntstvo Kiirskogo Universitetu, 1958.

1653 *Selskokhoziaistvennaia literatura SSSR [agricultural literature of the U.S.S.R.].* Moskva: Izdatelstvo Ministerstva Selskogo Khoziaistva SSSR, 1960-. Monthly.

In Russian. Annotated. Arranged by subject area.

U.S.S.R.

1654 Thran, P., and Broekhuizen, Simon. *Atlas of the cereal-growing areas in Europe.*

See entry 118.

1655 Tulupnikov, A. I. et al., eds. *Atlas selskogo Khoziaistva SSSR [atlas of agriculture in the U.S.S.R.].* Moskva: Glavnoe Upravlenie Geodezii i Kartografii. Ministerstva Geologii Okhrany nedr SSSR, 1960. 308 pp.

In Russian. Base maps at 1:20,000,000 and 1:30,000,000. Supplemented by *English key to the Agricultural atlas of USSR* (New York: Telberg, 1962).

1656 *USSR agriculture atlas.* Washington DC: U.S. Central Intelligence Agency, 1974. 59 pp.

IRRIGATION

1657 Ataeva, E. Sh., and Kogan, Sh. I. "Kritika i bibliografiia: annotirovannyi spisok literatury po gidrobiologii i gidrokhimii Karakumskogo Kanala za 1955-1974 gg." [annotated bibliography on hydrology and hydrochemistry of the Kara-Kum Canal, 1955-1974]. Akademiia Nauk Turkmenskoi S.S.R. *Izvestiia. Seriia biologicheskikh nauk* 2 (1976):89-94.

In Russian.

1658 Bystriakov, O. V., and Alpatiev, S. M. *Zroshennia ta obvodnennia na pivdni Ukrainskoi RSR: Bibliografichnyi pokazhchyk, 1952-1965 rr. [irrigation and watering in the south of the Ukrainian SSR].* Kiev, U.S.S.R.: Naukova dumka, 1968. 217 pp.

AGRICULTURAL ECONOMICS

1659 *Soviet agricultural commodity trade, 1960-76: a statistical survey: a reference aid.* McLean VA: U.S. Central Intelligence Agency, 1978. (ER 78-10516) 241 pp.

Supplements FAO *Trade yearbook* by providing breakdowns by country of origin and destination.

ENERGY

1660 Kirkpatrick, Meredith. *Energy resources and energy policies in Eastern Europe and the U.S.S.R.: a bibliography.*

See entry 211.

HUMAN GEOGRAPHY

1661 Gabrovska, Svobodozarya. *European guide to social science information and documentation services.*

See entry 119.

ANTHROPOLOGY

1662 Allworth, Edward. *Nationalities of the Soviet East: publications and writing systems.* New York: Columbia University Press, 1971. (Modern Middle East series 3) 440 pp.

Subtitled "Bibliographic directory and transliteration titles for Iranian- and Turkic-language publications, 1818-1945, located in U.S. libraries."

1663 Katz, Zev, ed. *Handbook of major Soviet nationalities.* New York: Free Press, 1975. 481 pp.

Chapter by chapter discussion, each with selected statistics and unannotated references.

1664 Lytle, Elizabeth E. *A bibliography of the Kurds, Kurdistan, and the Kurdish question.*

See entry 1178.

MEDICINE

1665 Perkins, Lee. *A bibliography of Soviet sources on medicine and public health in the U.S.S.R..* Bethesda MD: U.S. Public Health Service, 1975. 235 pp.

ECONOMIC DEVELOPMENT

1666 Inch, Peter. *Bibliography of regional economic planning in the USSR.* Monticello IL: Council of Planning Librarians, 1972. (CPL exchange bibliography 295) 27 pp.

Briefly annotated with narrative summaries of various topics.

1667 Kazmer, Daniel R., and Kazmer, Vera, eds. *Russian economic history: a guide to information sources.* Detroit MI: Gale Research Co., 1976. (Economic information guide series 4) 520 pp.

Annotated bibliography of materials in English. Covers works on pre-Revolutionary Russian economy as well as Soviet economy. Theses and most U.S. government documents omitted.

1668 Kish, George. *Economic atlas of the Soviet Union.* 2nd ed. Ann Arbor MI: University of Michigan Press, 1971. 90 pp.

Consists of four general maps and 60 regional maps, showing agriculture, resources, and industry. Includes brief summaries of regions. Information included based primarily on Soviet government publications and Soviet journals.

1669 *Quarterly economic review of USSR.* London: Economist Intelligence Unit, 1976-. Quarterly.

Includes summary of political and economic news as well as charts and statistics of selected economic indicators. Annual supplement.

HISTORICAL GEOGRAPHY

1670 *Obrazovanie i razvitie Soiuza SSR; atlas [formation and development of the USSR; atlas]*. Moscow: Glavnoe Upravlenie Geodezii i Kartografii, 1972. 112 pp.

In Russian with foreward and table of contents in English. Historical atlas.

1671 *Ruskii Istoriko-etnograficheskii atlas: zemledelie, Krestiaskaia odezhda [Russian historico-ethnographic atlas: farming, farmers' clothes]*. Moskva: Izdatelstvo Nauke, 1967. 40 sheets in portfolio.

In Russian. Covers the 19th-20th centuries of Russia west of the Ural Mountains.

1672 Wieczynski, Joseph L., ed. *Modern encyclopedia of Russian and Soviet history*. Gulf Breeze FL: Academic International, 1976-.

A monumental work to be completed in about 50 volumes. Includes many articles translated from the *Sovetskaia istoricheskaia entsiklopediia* (1961-).

PUBLIC ADMINISTRATION

1673 Goehlert, Robert. *Reference sources for the study of Soviet politics*. Monticello IL: Vance Bibliographies, 1980. (Public administration series P-450) 9 pp.

Unannotated.

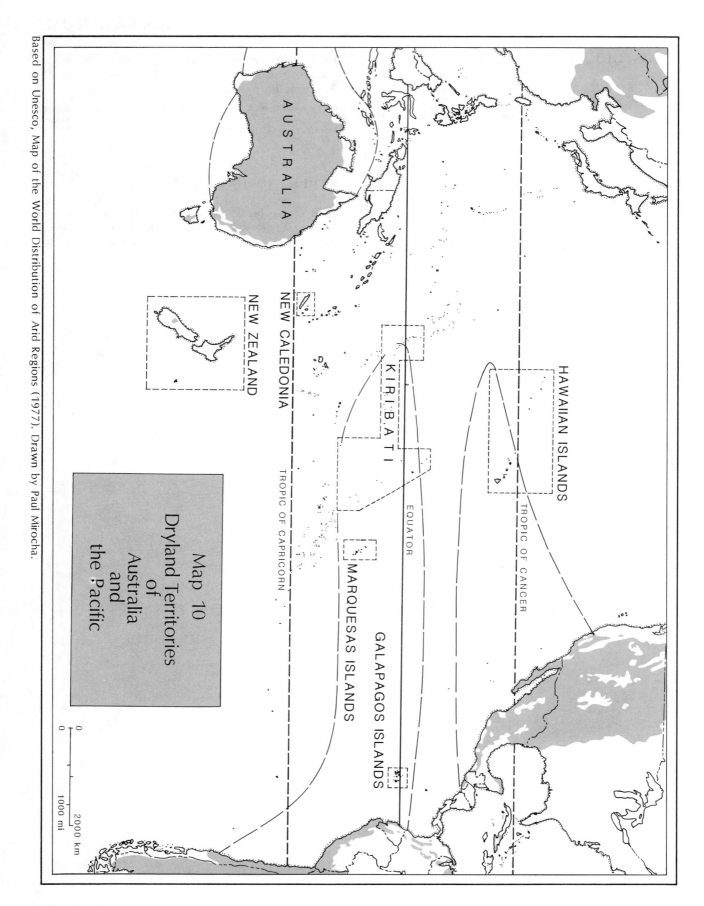

AUSTRALIA AND THE PACIFIC

The continent of Australia and the islands of the Pacific Ocean are physiographically unrelated yet they share several important bonds between them. Both are remote from the world's populated centers and hold only small populations themselves; these settlements are scattered over immense distances of sea and land. Transport and communication, both internally and with the outside, are expectedly difficult and unreliable. Both continent and islands straddle the tropics and are little influenced by polar air masses or mid-latitude cyclonic storms. Temperature variations between summer and winter are remarkably low, due to the overpowering influence of the sea; precipitation, however, can be highly variable between seasons and years so that even areas outside the drylands can experience extreme drought exacerbated by high rates of evapotranspiration.

Dryland areas of THE PACIFIC ISLANDS (EXCLUDES AUSTRALIA)

DESERTIFICATION RISK	ARIDITY									
	Hyperarid		Arid		Semiarid		Subhumid		Aridity Totals	
	km²	%	km²	%	km²	%	km²	%	km²	%
Very High	By definition, desertification does not exist in hyperarid regions.									
High			7,844	0.9					7,844	0.9
Moderate										
Desertification Totals			7,844	0.9					7,844	0.9
No Desertification			50	<0.1	1,834	0.2	11,413	1.4	13,297	1.6
Total Drylands			7,894	0.9	1,834	0.2	11,413	1.4	21,141	2.5
Non-dryland									809,660	97.5
Total Area of the Territory									830,801	100.0

HANDBOOKS

1674 Carter, John, ed. *Pacific Islands year book.* Sydney; New York: Pacific Publications, 1981. 14th ed. 560 pp.

Provides narratives and statistical data on all the islands, updated every three years.

STATISTICS

1675 Harvey, Joan M. *Statistics Asia and Australasia: sources for market research.*

See entry 1130.

1676 *Statistical yearbook for Asia and the Pacific.*

See entry 1131.

GEOGRAPHY

1677 Marsden, B., and Tugby, E. *Bibliography of Australasian geography theses 1933-71.* St. Lucia, Australia: University of Queensland. Dept. of Geography, 19??. 77 pp.

Continued by a second volume covering 1972-80 by D. Wadley and E. Tugby (300 pp.).

1678 *Pacific Islands.* London: United Kingdom. Naval Intelligence Division, 1943-45. 4 vols.

A detailed narrative survey with frequent chapter bibliographies covering all the Pacific Islands. Though dated, an essential source for physical geography, climate, biota, and social conditions.

ATLASES

1679 Stevenson, Merritt R. *Marine atlas of the Pacific coastal waters of South America.* Berkeley CA: University of California Press, 1970. 23 pp.

Text in English and Spanish. Covers the coast from Panama to Chile (at 20° south latitude), with an extension from the Ecuador coast to the Galapagos Islands.

GAZETTEERS

1680 *Australia, New Zealand and Oceania: official standard names approved by the United States Board on Geographic Names.* Washington DC: U.S. Board on Geographic Names, 1972. (Gazetteer supplement 126) 48 pp.

Pacific Islands/1681-1688

List of 700 place-name corrections with latitude/longitude coordinates. Covers Australia, New Zealand, the South Pacific, West Pacific, South Atlantic, and Indian Ocean Islands. Does not include the Southwest Pacific Islands.

GEOLOGY

1681 **Kroenke, Loren W., and Bardsley, Elaine, eds.** *Bibliography of geology and geophysics of the South Pacific.* Canberra: Australian Development Assistance Agency by the Australian Government Publishing Service, 1975. (Technical bulletin 1) 91 pp.

1682 **Ridge, John Drew.** *Annotated bibliographies of mineral deposits in Africa, Asia (exclusive of the USSR) and Australasia.*

See entry 312.

CLIMATOLOGY

1683 **Hastenrath, Stefan, and Lamb, Peter J.** *Climatic atlas of the tropical Atlantic and eastern Pacific Oceans.*

See entry 250.

1684 **Hastenrath, Stefan, and Lamb, Peter J.** *Heat budget atlas of the tropical Atlantic and eastern Pacific Oceans.*

See entry 251.

SOILS

1685 *Soil erosion in Africa, southern Asia, Australasia and Oceania (1957-1977).*

See entry 320.

BOTANY

1686 **Merrill, Elmer Drew.** *Polynesian botanical bibliography 1773-1936.* Honolulu HI: Bernice P. Bishop Museum, 1937. (Bulletin 144) 194 pp.

Annotated. Covers the entire Pacific basin from 30° N to 30° S latitude. A revised edition of Merrill's *Bibliography of Polynesian botany* (1924).

ENERGY

1687 *Energy atlas of Asia and the Far East.*

See entry 1139.

DEMOGRAPHY

1688 **Ware, Helen Ruth E.** *Fertility and family formation: Australasian bibliography and essays.* Canberra A.C.T.: Australian National University. Institute of Advanced Studies. Dept. of Demography, 1973. 358 pp.

AUSTRALIA
The Commonwealth of Australia
(a former colony of the United Kingdom)

Australia is, at the same time, the driest continent (by percentage) and the driest nation (by area). The nation straddles the Tropic of Capricorn so that nearly all the land lies within the belt of subtropical high pressure systems that do little to draw moisture-laden air masses into the area. Only where the land mass extends into the great wind belts does the precipitation increase beyond the dry zones; the Cape York and Arnhem Peninsulas intercept the northeast trades, and the southwest corner and Tasmania capture frontal systems from the west.

Australia's topography increases its aridity. The surface is remarkably stable and free of earthquakes, vulcanism, and associated mountain-building. Where mountains do exist—especially the Great Dividing Range along the east coast—prevailing tropical winds dump their moisture on the windward slopes and leave little for the interior. Similar conditions where coastal mountains block rain-bearing winds occur along the southwest corner and the southeast mainland and Tasmania. To make matters worse from the water storage point of view, none of the mountains is high enough to maintain permanent snowfields or glaciers. Although nature appears to have conspired to make Australia as dry as possible, in fact no area on the continent is dry enough to fall into the hyperarid zone; also fortunate are the huge artesian basins underlying the uniform topography.

The Northern Savannas. A great belt of semiarid to subhumid grassland savanna stretches across the northern tier from the Kimberley Range in the west to the lee side of the Dividing Range. Summers are hot and fairly wet while winters are dry and almost as hot. The best use of the land is range for beef cattle though the long winters produce dry unpalatable grasses and, hence, low yield per areal unit. Agricultural soils are generally poor, though where river alluvium is present—as along the lower Ord—irrigation allows cotton and fodder crops to flourish. Despite great potential, development schemes must contend with distant markets and long transportation routes; no rails, for example, link the Ord River area with the rest of the nation. Mining still provides the strongest economic base.

The Northeast Coast. Although much of the Queensland coastline is wet enough to support tropical rainforest (and is occasionally struck by hurricanes), a stretch of subhumid dry eucalypt forest exists from roughly Townsville to Rockhampton, with the driest portion located around Cape Upstart. Tracts have been cleared for the cultivation of sugarcane, tropical fruits, and fodder crops for beef and dairy cattle. A coastal railway links the region with centers of population to the south; with the fabulously beautiful Great Barrier Reef offshore, the region offers tremendous potential for tourism and recreation.

Riverina Savanna. The semiarid grassland savanna to the lee of the Dividing Range from southern Queensland south into Victoria is the nation's primary agricultural region; wheat and sheep are mixed on the same land. Adequate all-season rainfall, sophisticated cropping practices, and mechanization insure yields among the highest in the world. Rail links to ports on the coast are essential to the strong export trade.

The Gulflands Coast. The peninsula and interior lands around Adelaide in South Australia exhibit a classic Mediterranean-type climate regime: semiarid with winter precipitation and summer drought. Most of the land supports mixed wheat-wool farming together with some dairying. The Barossa Valley near Adelaide is the home of Australia's expanding wine industry.

The Southwest. The hinterland of Perth in the nation's southwest corner is the second Mediterranean-type area. The extreme southwest supports dense forests and lies outside the dry zone but a wide semiarid belt stretches across the corner reaching both coasts. This is mixed wheat-wool land and export production is channelled through Perth.

The Central and Western Deserts. The arid heart of the nation occupies some two-thirds of the total land area. Generally, the region consists of low tablelands and sandy basins only rarely punctuated by mountains and almost totally devoid of perennial surface water. Only the exotic Darling-Murray system in the southeast offers extensive water storage and irrigation potential, though pipelines and deep wells are used elsewhere to enhance the scanty and erratic precipitation. Where sufficient grassland exists around the desert margins the principal use is extensive grazing of sheep for wool. The environment poses major problems: dingoes (wild dogs) feed on the flocks, rabbits (introduced) are a menace, and overstocking has led to deterioration of pasture and soil erosion. Mineral resources in several areas are an important economic mainstay.

Tasmania. This rugged island extends southward into the belt of westerlies and so intercepts sufficient moisture for intensive mixed farming wherever the soils permit. A small marginally subhumid dryland area exists along the east coast where mountains cause a rainshadow effect; however, due to relatively low average temperatures the water deficit is minimal and scarcely noticed.

Australia/1689-1700

Dryland areas of AUSTRALIA

DESERTIFICATION RISK	ARIDITY									
	Hyperarid		Arid		Semiarid		Subhumid		Aridity Totals	
	km²	%	km²	%	km²	%	km²	%	km²	%
Very High	By definition, desertification does not exist in hyperarid regions.				312,465	4.1	38,038	0.5	350,503	4.6
High			2,160,104	28.1	563,335	7.3	80,623	1.1	2,804,062	36.5
Moderate			1,581,577	20.6	1,559,404	20.3	149,635	1.9	3,290,616	42.8
Desertification Totals			3,741,681	48.7	2,435,204	31.7	268,296	3.5	6,445,181	83.9
No Desertification					37,915	0.5	541,206	7.0	579,121	7.5
Total Drylands			3,741,681	48.7	2,473,119	32.2	809,502	10.5	7,024,302	91.4
Non-dryland									658,235	8.6
Total Area of the Territory									7,682,537	100.0

GENERAL BIBLIOGRAPHIES

1689 Borchardt, Dietrich Hans. *Australian bibliography: a guide to printed sources of information.* 3rd ed. Rushcutters Bay, N.S.W.: Pergamon, 1976. 270 pp.

An excellent narrative guide to Australian reference works with useful and critical descriptions of the major sources.

1690 *CSIRO published papers: subject index 1916-1968.* East Melbourne, Victoria: CSIRO Central Library, 1970-73. 16 vols.

An index to material held in the CSIRO archival collection.

1691 Moyal, Ann Mozley. *Guide to the manuscript records of Australian science.* Canberra A.C.T.: Australian Academy of Science and Australian National University Press, 1966. 127 pp.

Updated irregularly in the *Records* of the Australian Academy of Science.

ONLINE DATABASES

1692 ABSDATA. St. Leonards, N.S.W.: Computer Sciences of Australia; I. P. Sharp Associates, 19—. Updated quarterly.

Available only online. A nonbibliographic database covering statistics on agricultural production, mining, and other economic and financial time series relating to Australia since 1953. Vendors: I. P. Sharp, Computer Sciences of Australia.

INDEXES AND ABSTRACTS

1693 *Australian science index: index to articles published in Australian scientific and technical serials.* Melbourne: CSIRO, 1957-. Monthly.

"...the most comprehensive source of reference to Australian scientific literature..." (Borchardt, 1976). Published on microfiche beginning with vol. 22 (1978).

THESES AND DISSERTATIONS

1694 Wylie, Enid, ed. *Union list of higher degree theses in Australian university libraries: cumulative edition to 1965.* Hobart: University of Tasmania Library, 1967. 568 pp.

First edition compiled by M. J. Marshall (Hobart, 1959). Supplement covers period 1966-68 (Hobart, 1971).

HANDBOOKS

1695 *The Far East and Australia: a survey and directory of Asia and the Pacific.*

See entry 1127.

1696 *Official year book of New South Wales.* Sydney: Australian Bureau of Statistics, 1904/05-. Annual.

1697 *Queensland year book.* Brisbane: Australian Bureau of Statistics. Queensland Office, 1964-. Annual.

1698 *South Australian year book.* Adelaide: Australian Bureau of Statistics. South Australian Office, 1966-. Annual.

1699 *Tasmanian year book.* Hobart: Australian Bureau of Statistics. Tasmanian Office, 1967-. Annual.

1700 *Victorian year book.* Melbourne: Australian Bureau of Statistics, Victorian Office, 1873-. Annual.

1701 *Western Australian year book*. Perth: Australian Bureau of Statistics. Western Australian Office, 1957-. Annual.

1702 Whitaker, Donald P. *Area handbook for Australia*. Washington DC: U.S. Dept. of Defense, 1974. (American University foreign area studies) 458 pp.

Narrative summary of the nation's history, politics, geography, economy, social life; includes unannotated bibliography.

ENCYCLOPEDIAS

1703 *The Australian encyclopaedia*. East Lansing MI: Michigan State University Press, 1958. 10 vols.

Perhaps the most reliable general encyclopedia of the nation but now too old for many subjects. *The Encyclopaedia of Australia*, last revised in 1973 may help, but the best sources of current general information and statistics are the national and state official yearbooks listed above.

GOVERNMENT DOCUMENTS

1704 Borchardt, Dietrich Hans, ed. *Australian official publications*. Melbourne: Longman Cheshire, 1979. 365 pp.

A narrative summary of national and state government publishing in Australia.

1705 Coxon, Howard. *Australian official publications*. Oxford, U.K.; New York: Pergamon Press, 1981. (Guide to official publications 5) 211 pp.

Includes a bibliography of Australian official publications: pp. 185-199.

DIRECTORIES

1706 Crump, Ian A. *Australian scientific societies and professional associations*. 2nd ed. East Melbourne, Victoria: CSIRO, 1978. 226 pp.

1707 Crump, Ian A. *Scientific and technical research centres in Australia*. 2nd ed. East Melbourne, Victoria: CSIRO Information Service, 1975. 224 pp.

1708 *The Far East and Australia: a survey and directory of Asia and the Pacific*.

See entry 1127.

1709 Ronayne, J., and Nede, G. J., eds. *Australia and New Zealand*. 2nd ed. Guernsey, U.K.: Francis Hodgson, 1975. (Guide to world science 18) 346 pp.

Provides descriptions of scientific research activities and organizations.

STATISTICS

1710 Cameron, R. J. *Year book Australia*. Canberra A.C.T.: Australian Bureau of Statistics, 1977-. Annual.

Continues *Official yearbook of Australia* (1973-76).

1711 Finlayson, Jennifer A. S. *Historical statistics of Australia: a select list of official sources*. Canberra A.C.T.: Australian Historical University. Research School of Social Sciences. Dept. of Economic History, 1970. 55 pp.

1712 Harvey, Joan M. *Statistics Asia and Australasia: sources for market research*.

See entry 1130.

INFORMATION SCIENCE

1713 Wicks, Vera M. *Directory of special libraries in Australia*. 4th ed. Sydney: Library Association of Australia, 1976. 310 pp.

GEOGRAPHY

1714 Ives, Alan. *The Mallee of south-eastern Australia: a short bibliography*. Clayton, Queensland: Australia. Monash University. Dept. of Geography, 1973. 66 pp.

1715 Marsden, B. S., and Tugby, E. E., comps. *Bibliography of Australasian geography theses, 1933-1971*. Brisbane: University of Queensland. Dept. of Geography, 1972. 77 pp.

Covers the results of research on both Australia and New Zealand for the period 1933-1971; the bulk of citations, however, are for bachelor's degree theses.

1716 Nuttonson, Michael Y. *The physical environment and agriculture of Australia with special reference to its winter rainfall regions and to climatic and latitudinal areas analogous to Israel*. Washington DC: American Institute of Crop Ecology, 1958. 1,124 pp.

1717 Reiner, Ernst. "Literaturbericht uber Australien und Neuseeland, 1938-1963" [literature report on Australia and New Zealand, 1938-1963].*Geographisches Jahrbuch* 62 (1967).

"The only large scale survey of the geographical literature on Australia..." (Borchardt, *Australian bibliography*, p. 64). Covers the geographical and geological literature in English and German in nearly 3,000 entries. A continuation is in preparation for years 1962-72, as of 1975.

1718 Specht, Raymond. "Australia." In *Handbook of contemporary developments in world ecology*, pp. 387-415. Edited by Edward J. Kormondy and J. Frank McCormick. Westport CT: Greenwood Press, 1981.

A good narrative summary of ecological research in Australia; includes unannotated list of references.

1719 Twindale, C. R.; Tyler, M. J.; and Webb, B. P. *Natural history of the Adelaide region*. Adelaide: Royal Society of South Australia, 1976. 189 pp.

Australia/1720-1741

A series of separately authored articles on the geology, soils, climate, hydrology, vegetation, fauna, oceanography, ecology, and Aborigines of Adelaide and nearby peninsulas. Includes bibliographies.

ATLASES

1720 *Atlas of Australian resources*. Canberra A.C.T.: Australia. Dept. of National Development. Geographic Section, 1962-79. (2nd series) 35 leaves.

Physical geography. First series published 1959-60. Scale 1:6,000,000.

1721 *BMR earth science atlas of Australia*. Canberra A.C.T.: Australia. Bureau of Mineral Resources Geology and Geophysics, 1979-. Looseleaf.

Each map accompanied by commentary. Scale 1:10,000,000.

1722 **Davies, J. L.** *Atlas of Tasmania*. Hobart: Tasmania. Dept. of Lands and Surveys, 1965. 128 pp.

Scale 1:1,800,000.

1723 *Queensland resources atlas*. Brisbane: Queensland. State Relations Bureau, 1976. 120 pp.

General commentary on Queensland with maps. Scale 1:8,000,000.

1724 *Reader's Digest atlas of Australia*. Sydney: Reader's Digest Services, 1977. 287 pp.

Scale 1:1,000,000.

GAZETTEERS

1725 *Australia 1:250,000 maps series and gazetteer*. Canberra A.C.T.: Australia. Dept. of Minerals and Energy. Division of National Mapping, 1975. 101 pp.

This gazetteer contains all names of places and features which appear on the original series of 1:250,000 maps of Australia, completed in 1968.

1726 *Australia: official standard names approved by the United States Board on Geographic Names*. Washington DC: U.S. Board on Geographic Names, 1957. (Gazetteer 40) 75 pp.

List of 63,000 place-names with latitude/longitude coordinates.

1727 **Praite, R., and Tolley, J. C.** *Place names of South Australia*. Adelaide: Rigby, 1970. 203 pp.

1728 **Reed, Alexander Wyclif.** *Place names of Australia*. Sydney: A. H. and A. W. Reed, 1973. 271 pp.

Popular account of the continent's major place names. Should be supplemented by place name dictionaries of the individual states.

1729 **Reed, Alexander Wycliff.** *Place names of New South Wales*. Sydney: Reed, 1969. 144 pp.

MAP COLLECTIONS

1730 **Alonso, Patrica Ann Greechie.** "Australian cartographic bibliography (ACB)." *Cartography* 11:2 (1979):108-113.

1731 *Index to Australian resources maps of 1940-59*. Canberra A.C.T.: Australia. Dept. of National Development, 1961. 241 pp.

Covers maps illustrating economic and general resources. Continued by Supplement for period 1960-64 (Canberra, 1966).

1732 **Rauchle, Nancy M., and Alonso, P. A. G.** *Map collections in Australia: a directory*. 3rd ed. Canberra A.C.T.: Australia. National Library. 1980. 141 pp.

GEOLOGY

1733 *Australian mining yearbook*. Chippendale, New South Wales: Australia. Thomson Publications, n.d.-. Annual.

An annual review of mineral production with lists of mining companies, personnel, suppliers, products.

1734 **Bambrick, S.** *Minerals processing in Australia, a select bibliography*. Canberra A.C.T.: Australia. National Library, 1972. 189 pp.

1735 "A bibliography of the Quaternary of Australia." *Australian Quaternary newsletter* 12 (1978):39-54.

1736 **Bridge, Peter J.** *Combined index to the publications of the Geological Survey of Western Australia, 1910-1970*. 2nd ed. Mt. Lawley, W. A.: Hesperian, 1972. 341 pp.

1737 **Colhoun, Eric A.** *Bibliography of Tasmanian geology*. Hobart: University of Tasmania, 1973.

Covers the period 1893-1973.

1738 **David, Tannatt William Edgeworth, Sir.** *The geology of the Commonwealth of Australia*. Rev. ed. London: E. Arnold, 1950. 3 vols.

Extensive bibliographies at the end of each chapter.

1739 **Hill, Dorothy.** *Bibliography of Australian geological serials and of other Australian periodicals that include geological papers*. St. Lucia, Queensland: University of Queensland. Dept. of Geology and Mineralogy, 1980. (Papers 9:3) 76 pp.

1740 *List of publications and reports, 1879-1972*. Brisbane: Australia. Dept. of Mines, 1973. 37 pp.

Bibliography of the mines and mineral resources of Queensland.

1741 **Ridge, John Drew.** "Australia." In *Annotated bibliographies of mineral deposits in Africa, Asia (exclusive of the USSR) and Australasia*, pp 311-483. Oxford U.K.: Pergamon, 1976.

1742 Teesdale-Smith, E. N. *Bibliography of South Australian geology.* Adelaide: South Australia. Geological Survey. Dept. of Mines, 1959. 240 pp.

Includes all literature published up to and including June 1958.

CLIMATOLOGY

1743 *Bibliography on urban meteorological studies in Australia.* Mordialloc, Victoria: Australia. Royal Meteorological Society. Australian Branch. Mordialloc, 1978. 201 pp.

1744 *Climatic atlas of Australia.* Canberra A.C.T.: Australian Government Publishing Service, 1975-.

Prepared by the staff of the Bureau of Meteorology, Dept. of Science and the Environment, Commonwealth of Australia. Issued in map sets; to be completed in 143 maps. Standard base scale 1:12,500,000.

1745 Creasi, Vincent J. *Bibliography of climatic maps for Australia.* Washington DC: U.S. Weather Bureau, 1960. (NTIS: AD 665-178; previously PB 176-007) 44 pp.

Annotated list of 86 reports.

1746 Creasi, Vincent J. *A selected bibliography on the climate of Australia.* Washington DC: U.S. Weather Bureau, 1960. (NTIS: AD 664-746) 112 pp.

1747 Gentilli, J., ed. *Climates of Australia and New Zealand.* Amsterdam, The Netherlands: Elsevier, 1971. (World survey of climatology 13) 405 pp.

A narrative summary with maps, charts, and an unannotated bibliography.

HYDROLOGY

1748 *A bibliography of groundwater recharge in Australia.* Canberra A.C.T.: Australia. Bureau of Mineral Resources, Geology and Geophysics, 1977. (Open-file record.)

1749 Davey, Lois. *CSIRO water research bibliography, 1923-1963.* Melbourne: CSIRO, 1964. 98 pp.

1750 *Inventory of water resources research in Australia.* 7th ed. Kingston, South Australia: Australia. Government Publishing Service, 1977. 157 pp.

Includes government bodies, universities, and private organizations.

1751 Sharpe, Jane. *Bibliography of Murray Valley hydrology, 1946-1967.* Melbourne: CSIRO, 1969. 117 pp.

The Murray River flows through the states of Victoria, New South Wales, and South Australia.

SOILS

1752 *Bibliography on arid soils of Australia (1965-1950).* Harpenden, U.K.: Commonwealth Bureau of Soils, 196-. 27 pp.

Contains 137 citations.

1753 Lang, A. R. G., and Hicks, Margaret Rowland. *Bibliography on soils in the N. S. W. Riverina.* Canberra A.C.T.: CSIRO. Division of Irrigation Research, 1975. 126 pp.

BIOLOGY

1754 Keast, Allen; Crocker, R. L.; and Christian, C. J., eds. *Biogeography and ecology in Australia.* The Hague: W. Junk, 1959. (Monographiae biologicae 8) 640 pp.

BOTANY

1755 *Bibliography on eucalypts, 1956-1961.* Canberra A.C.T.: CSIRO. Division of Forest Products, 1961.

1756 Cooper, C. F. *An annotated bibliography of the effects of fire on Australian vegetation.* Melbourne: Soil Conservation Authority, 1963. 21 leaves.

1757 Kraehenbuehl, D. N. "A botanical bibliography of the Flinders Range, South Australia, 1800-1970." *Victorian naturalist* 88:8 (1971):231-237.

1758 Kuechler, August Wilhelm. *International bibliography of vegetation maps: volume 3: U.S.S.R., Asia, and Australia.*

See entry 1651.

1759 Pickard, J. "Annotated bibliography of floristic lists of New South Wales." New South Wales National Herbarium. *Contributions* 4:5 (1972):291-317.

1760 Specht, M. M., and Specht, R. L. "Bibliographia phytosociologica: Australia" [bibliography of plant ecology: Australia]. *Excerpta botanica: sectio B: sociologica* 4:1 (1962):1-58.

Unannotated. A major retrospective bibliography on Australian plant ecology; over 700 articles, documents, and books are cited.

ZOOLOGY

1761 Cowling, S. S. *Bibliography on waterbird research and management in Australia.* Rev. ed. Canberra A.C.T.: Australian Committee on Waterbirds for the Australian, 1972. (Special publication 1).

1762 Kloot, T. "Ornithological bibliographies: biographies and archives." *South Australian ornithologist* 26:3 (1972):55.

1763 Moulds, Maxwell Sydney, comp. "An accumulative bibliography of Australian entomology." *Australian entomological magazine* 5:1 (1978):16- 20; 5:3 (1978):60; 5:4 (1978):80; 5:5 (1979):94-99; 6:1 (1979):18-20; 6:2 (1979):39-40; 6:4 (1979):78-79; 6:5 (1980):96-100; 6:6 (1980):104; 7:5 (1981):79-80; 8:1 (1981):18-20.

1764 Moulds, Maxwell Sydney. *Bibliography of the Australian butterflies.* Greenwich, N.S.W.: Australian Entomological Press, 1977. 239 pp.

1765 Musgrave, A. *Bibliography of Australian entomology, 1775-1930.* Sydney: New South Wales. Royal Zoological Society, 1932. 380 pp.

1766 Whittell, Hubert Massey. *The literature of Australian birds: a history and a bibliography of Australian ornithology.* Perth: Paterson Brokensha, 1954. 788 pp.

Part I is a history of Australian ornithology from 1618-1950. Part II is an unannotated bibliography of Australian ornithology for the same period.

AGRICULTURE

1767 *Agricultural data bases.* St. Leonards, N.S.W.: Computer Sciences of Australia, 1976-.

An online database providing access to annual agricultural censuses. Vendor: Computer Sciences of Australia Pty. Ltd.

1768 *Bibliography of research and extension in the North Eastern region of Victoria.* Melbourne: Victoria. Dept. of Agriculture, 1973-. Freq. unknown.

1769 *Bibliography of research and extension libraries, North West region of Victoria.* Melbourne: Victoria. Dept. of Agriculture, 1973-.

1770 Brien, J. P. *A select bibliography of Australian agricultural extension research..* Sydney: University of Sydney. Dept. of Agricultural Economics, 1977. (Agricultural extension bulletin 2) 62 pp.

PLANT CULTURE

1771 Evenson, J. P., and Basinski, J. J. "Bibliography of cotton pests and diseases in Australia." *Cotton growing review* 50:1 (1973):79-86.

1772 *South Australia and Western Australia: grassland research, 1973-78.* Hurley, U.K.: Commonwealth Bureau of Pastures and Field Crops, 1978. (Annotated bibliography) 20 pp.

FORESTRY

1773 *Australian forest resources.* Canberra A.C.T.: Australia. Forestry and Timber Bureau, 1930-. Annual.

Formerly *Annual report of the Forestry and Timber Bureau* (until 1977).

1774 Boughton, Valerie H. *A survey of the literature concerning the effects of fire on the forests of Australia.* Gordon, N.S.W.: Ku-ring-gai Council, 1970. 40 pp.

A narrative bibliography covering hazard reduction and the effects of fire on soil, seeds, animals, runoff, plant communities, and human settlements.

1775 Hulme, M. I. *Bibliography on the utilization of eucalypts.* South Melbourne, Victoria: CSIRO. Division of Forest Products, 1956. 129 pp.

Prepared for the FAO World Eucalypt Conference in Rome, 1956.

1776 Marris, Bernice. *A bibliography of Australian references on eucalypts, 1956-June 1966.* Canberra A.C.T.: Australia. Forestry and Timber Bureau, 1966. 90 pp.

Covers the timbering and cultivation of eucalypts.

ANIMAL CULTURE

1777 Culey, Alma G., comp. *An Australian bibliography on the biology of the sheep and the sheep industry, 1925-1967.* Melbourne: CSIRO, 1970. 514 pp.

Unannotated. Includes published papers, books, and non-serial materials and theses. Limited to sources which directly discuss sheep.

1778 Culey, Alma G. *Bibliography of beef production in Australia: c. 1930-1958.* Sydney: CSIRO. Division of Animal Health. McMaster Animal Health Laboratory, 1961. 239 pp.

Supplement covers period 1959-1963 (1965), 169 pp.

1779 Powell, S. C. *Some references to research and extension publications on pastures in Victoria, 1903-June 1970.* Melbourne: Victoria. Dept. of Agriculture. Division of Animal Husbandry, (1970). 42 leaves.

1780 Richardson, R. A.; Powys, J. J.; and Nankivell, P. S. *A bibliography relating to the marketing of Australian wool.* Armidale, N.S.W.: University of New England. Dept. of Agricultural Economics and Business Management, 1976. (Miscellaneous publication 4) 50 pp.

AGRICULTURAL ECONOMICS

1781 Dillon, John L. *Australian bibliography of agricultural economics, 1788-1960.* Sydney: New South Wales. Dept. of Agriculture, 1967. 433 pp.

LAND TENURE

1782 Archer, R. W. *A bibliography on rural land subdivision for small holding land uses in Australia.* Monticello IL: Vance Bibliographies, 1979. Rev. ed. (Public administration series bibliography P-174) 20 pp.

Unannotated.

ENERGY

1783 *Energy atlas of Asia and the Far East.*
See entry 1139.

SOLAR ENERGY

1784 **Case, Glenna L.** *Solar energy in Australia: a profile of renewable energy activity in its national context.* Golden CO: U.S. Solar Energy Research Institute, 1980. 75 pp.

A summary and directory of solar energy research activities and organizations in Australia.

HUMAN GEOGRAPHY

1785 *Australian public affairs information service: a subject index to current literature.* Canberra A.C.T.: Australia. National Library, 1945-. 11 per year.

A periodical index of Australian journals and certain "composite" monographs covering the social sciences.

1786 *Select bibliography on economic and social conditions in Australia, 1918-1953.* Canberra A.C.T.: Australia. National Library, 1953. 11 leaves.

"...still one of the fundamental documents because it covers a very vital period in Australian social history." (Borchardt, *Australian bibliography*, p. 75)

DEMOGRAPHY

1787 **Lancaster, Henry Oliver.** "Bibliography of vital statistics in Australia and New Zealand." *Australian journal of statistics* 6:2 (1964):33-99.

Primarily English-language works, Includes medical and statistical sources, reports, and journal articles. Reprinted as a separate by the Australasian Medical Publishing Co. (Sydney, 1964), 67 pp. Updated by entry 1788.

1788 **Lancaster, Henry Oliver.** "Bibliography of vital statistics in Australia: a second list." *Australian journal of statistics* 15:1 (1973):1-26.

Appendix to Lancaster's 1964 "Bibliography." (entry 1787).

ANTHROPOLOGY

1789 *Australian Aborigines: annual bibliography.* Canberra A.C.T.: Australian Institute of Aboriginal Studies, 1975-. Annual.

Covers ethnography, linguistics, art, archeology, economic development, and education.

1790 **Coppell, W. G.** *World catalogue of theses and dissertations about the Australian Aborigines and Torres Strait Islanders.* Sydney: Sydney University Press, 1977. 113 pp.

Unannotated. Covers masters' theses and doctoral dissertations produced worldwide through June 1976.

1791 **Craig, Beryl F.** *Arnhem Land peninsular region (including Bathurst and Melville Islands).* Canberra A.C.T.: Australian Institute of Aboriginal Studies, 1966. (Occasional papers 8; Bibliography series 1) 205 pp.

Annotated bibliography of 1,044 items through 1965. Includes unpublished material. Omits fictional materials.

1792 **Craig, Beryl F.** *Cape York.* Canberra A.C.T.: Australian Institute of Aboriginal Studies, 1967. (Occasional paper 9; Bibliography series 2) 233 pp.

Cape York is in northern Queensland.

1793 **Craig, Beryl F.** *Central Australian and western desert regions: an annotated bibliography.* Canberra A.C.T.: Australian Institute of Aboriginal Studies, 1969. (Australian Aboriginal studies 31; Bibliography series 5) 351 pp.

Covers sources on anthropology, archeology and linguistics for Aboriginal peoples of central and western Australia through mid-1969. Contains 2,205 annotated entries.

1794 **Craig, Beryl F.** *Kimberly region: an annotated bibliography.* Canberra A.C.T.: Australian Institute of Aboriginal Studies, 1968. (Bibliography series 3) 209 pp.

1795 **Craig, Beryl F.** *North-west-central Queensland: an annotated bibliography.* Canberra A.C.T.: Australian Institute of Aboriginal Studies, 1970. (Bibliography series 6) 137 pp.

Covers Queensland west of the Great Dividing Range; "as complete as possible up to June, 1970."

1796 **Denham, Woodrow W.** "Introduction to the Alyawara ethnographic data base." *Behavior science research* 14:2 (1979):133-153.

The Alyawara live in central Australia some 300 km northeast of Alice Springs. This database consists of some 440,000 items of raw ethnographic data coded for online retrieval. Includes bibliography on the Alyawara.

1797 **Greenway, John.** *Bibliography of the Australian Aborigines and the native peoples of Torres Strait to 1959.* Sydney: Angus and Robertson, 1963. 420 pp.

The major retrospective bibliography on the Australian Aborigine. Unannotated.

1798 **Hill, Marji, and Barlow, Alex.** *Black Australia: an annotated bibliography and teacher's guide to resources on Aborigines and Torres Strait Islanders.* Canberra A.C.T.: Australian Institute of Aboriginal Studies; Atlantic Highland NJ: Humanities Press, 1978. (Bibliography series 7) 200 pp.

A model bibliography covering teaching materials from primary to university level. The introductory essays by Warwick Dix and Alex Barlow—on the restrictions placed on information by the Aborigines themselves and on racism and ethnocentric stereotypes—show an uncommon concern for the ethical implications of bibliographical information. Annotations are often sharply critical. Covers films and recordings as well as printed matter; selection of material was based at least in part on current availability.

Australia/1799-1810

1799 Houston, Carol A., comp. *A selected regional bibliography of the Aboriginals of South Australia.* Adelaide: South Australian Museum, 1973. 78 leaves.

Intended as a preliminary work.

1800 Massola, Aldo. *Bibliography of the Victorian Aborigines from earliest manuscripts to 31 December 1970.* Melbourne: Hawthorn Press, 1971. 95 pp.

1801 Moodie, Peter M., and Pedersen, E. B. *The health of Australian Aborigines: an annotated bibliography.* Canberra A.C.T.: Australia. Dept. of Health. School of Public Health and Tropical Medicine, 1971. (Service publication 8) 248 pp.

1802 Peterson, Nicolas, ed. *Aboriginal land rights: a handbook.* Canberra A.C.T.: Australian Institute of Aboriginal Studies, 1981. 297 pp.

1803 Plomley, Norman James Brian. *Annotated bibliography of the Tasmanian Aborigines.* London: Royal Anthropological Institute, 1969. (Occasional paper 28) 143 pp.

Covers the period through 1965.

1804 Tindale, Norman Barnett. *Aboriginal tribes of Australia.* Berkeley CA: University of California Press, 1974. 2 vols.

Vol. 1: text, 104 pp.; vol. 2: 4 sheet maps. Scale 1:2,534,400.

URBAN GEOGRAPHY

1805 Goodhew, Barbara, ed. *Bibliography of urban studies in Australia: cumulated volume, 1966-1973.* Canberra A.C.T.: Australian Institute of Urban Studies, 1978. (AIUS publication 71) 231 pp.

ECONOMIC DEVELOPMENT

1806 *Asia/Australia data bank.* Washington DC: Data Resources, 19—.

An online database providing economic data on Australia, India, China, and other East Asian nations. Supersedes *East Asian Data Bank.* Vendor: Data Resources.

1807 *KOMPASS Australia: register of Australian industry and commerce.* Prahran, Victoria: Peter Isaacson Publications, 1970-. Annual.

1808 *Quarterly economic review of Australia, Papua New Guinea.* London: Economist Intelligence Unit, 1976-. Quarterly.

Includes summary of political and economic news as well as charts and statistics of selected economic indicators. Annual supplement.

TOURISM AND RECREATION

1809 *Bibliography of tourism and recreation research.* Melbourne: Australian Tourist Commission, 1973. 141 leaves.

HISTORICAL GEOGRAPHY

1810 Clark, Charles Manning Hope. *A history of Australia.* Parkville, Australia: Melbourne University Press; New York: Cambridge University Press, 1962-.

"The most comprehensive survey of our past..." (Borchardt, *Australian bibliography*, p. 54). Includes bibliographies. 5 volumes to date (August 1982).

GALAPAGOS ISLANDS
Archipelago de Colon
(constitutes a province of Ecuador)

The Galapagos Islands are a group of volcanic islands emerging from the Pacific some 1,050 km (650 miles) west of Ecuador. Although they straddle the Equator they are part of the dryland belt that stretches along the western coast of South America. Climatic conditions here are similar to those of the arid mainland: relatively cool temperatures, frequent fog, clouds and rain only at the higher elevations of the volcanic cones. The islands' importance derives from its strategic position near the Panama Canal, and, more significantly, as an unusual natural laboratory illustrating the process of evolution and natural selection. Darwin's visit in the 1830's prompted him to make important observations concerning variation between species of the islands' ubiquitous lizards, tortoises, and birds.

Natural vegetation varies considerably from the coast to mountain peaks: sparse desert shrubs and volcanic scablands give way to dense thorn forest and then to low bracken cover on the highest slopes. The older, eastern islands of Santa Cruz and San Cristobal have fertile well-developed soils and support crops of coffee, as well as dairy and beef cattle. The principal industry, however, is tourism, and, secondarily, support for international scientific research teams.

Note: Additional citations can be found later under Ecuador.

Dryland areas of THE GALAPAGOS ISLANDS

DESERTIFICATION RISK	ARIDITY									
	Hyperarid		Arid		Semiarid		Subhumid		Aridity Totals	
	km²	%	km²	%	km²	%	km²	%	km²	%
Very High	By definition, desertification does not exist in hyperarid regions.									
High			7,844	100.0					7,844	100.0
Moderate										
Desertification Totals			7,844	100.0					7,844	100.0
No Desertification										
Total Drylands			7,844	100.0					7,844	100.0
Non-dryland										
Total Area of the Territory									7,844	100.0

GAZETTEERS

1811 *Ecuador: official standard names approved by the U.S. Board on Geographic Names.*
See entry 2658.

BOTANY

1812 Schofield, E. K. "Annotated bibliography of Galapagos, Ecuador, botany: supplement 1." *Brittonia* 32:4 (1980):537-547.

THE HAWAIIAN ISLANDS
The State of Hawaii
(a state of the United States of America)

The State of Hawaii includes eight large and many smaller islands and reefs stretching across 2,800 km (1,700 miles) of ocean on either side of the Tropic of Cancer. Honolulu lies due west of Guadalajara, Mexico. All eight islands in the main southeastern group hold dryland areas of varying sizes and degrees. All these drylands are formed by rainshadow effects to the lee of the prevailing Northeast Trades, though local topography and interference from neighboring islands can alter the pattern.

Hawaii. Locally called The Big Island, Hawaii is the largest and has the largest and driest drylands. The core is an arid pocket along the coast of Kawaihae Bay around Hapuna Beach Park; a semiarid zone stretches up several steep gulches and a much larger subhumid zone spreads out to encompass the entire northwest quadrant of the island. The Mamalahoa Highway (Route 19) traverses the lava flows and cattle ranges through the heart of the region.

Maui. The second largest island is actually two volcanic islands joined by a narrow lowland isthmus. This isthmus and a stretch of southwest coastline comprise a semiarid zone which supports irrigated sugarcane and cattle ranching. Water is made available from the wet windward slopes through a series of ditches and tunnels.

Kahoolawe. The smallest of the eight, this island is almost entirely subhumid and serves as a gunnery target for the United States Navy. Though cattle used to be run on the grasslands, the island is now littered with live ammunition; local efforts have been made to turn it into a natural preserve.

Lanai. In the lee of Maui and Molokai this island is subhumid except for a fringe of wetland on its highest slopes. Once its grasslands supported cattle but now pineapple cultivation is the principal economic activity.

Molokai. The western half is subhumid and supports cattle ranges. An exotic animal breeding ranch delights tourists from nearby beach resorts.

Oahu. The third largest and by far the most populated, Oahu holds the city of Honolulu and important military installations. Much of the west and south coasts, including Pearl Harbor and downtown Honolulu, are semiarid with a wide stretch of subhumid land to the north through Wahiawa Valley. Irrigated pineapple and sugarcane are grown.

Kauai. Although the central peak of Mt. Waialeale is regarded as the wettest spot on earth, the coast to the southwest around Waimea extends into the semiarid zone. Additional subhumid drylands stretch along the entire west coast.

Niihau. This small island is in the lee of Kauai and thus entirely subhumid; grassland supports cattle and sheep.

Note: Many additional citations can be found later under United States.

Dryland areas of THE HAWAIIAN ISLANDS

DESERTIFICATION RISK	Hyperarid km²	Hyperarid %	Arid km²	Arid %	Semiarid km²	Semiarid %	Subhumid km²	Subhumid %	Aridity Totals km²	Aridity Totals %
Very High	By definition, desertification does not exist in hyperarid regions.									
High										
Moderate										
Desertification Totals										
No Desertification			50	0.3	1,419	8.5	5,060	30.3	6,529	39.1
Total Drylands			50	0.3	1,419	8.5	5,060	30.3	6,529	39.1
Non-dryland									10,173	60.9
Total Area of the Territory									16,702	100.0

ATLASES

1813 **Armstrong, R. Warwick,** ed. *Atlases of Hawaii.* Honolulu HI: The University Press of Hawaii, 1973. 222 pp.

GEOLOGY

1814 **MacDonald, Gordon A.** *Bibliography of the geology and water resources of the Island of Hawaii: annotated and indexed.* Honolulu HI: Hawaii. Division of Hydrography, 1947. (Bulletin 10) 191 pp.

HYDROLOGY

1815 **Mink, John F.** *Handbook-index of Hawai'i groundwater and water resources data.* Honolulu HI: University of Hawaii. Water Resources Research Center, 1977. (Technical report 113) 119 pp.

A compendium of charts and statistical data covering the island of Hawai'i, with an index to various data sets available.

1816 **Pfund, Rose T., and Steller, Dorothy L.** *Bibliography of water resources of the Hawaiian Islands.* Honolulu HI: University of Hawaii. Water Resources Research Center, 1971. (Technical report 50) 142 pp.

Annotated. Covers the period through 1966. For period 1967-1971, supplemented by: Pfund, Rose T. and Wickes, James W. (Technical report 88; 1975).

KIRIBATI
The Republic of Kiribati

(as The Gilbert Islands, a former colony of the United Kingdom)

Kiribati consists of a number of islands and island chains stretching nearly 4,900 km (3,000 miles) across the central Pacific on either side of the Equator. Temperatures are remarkably uniform but precipitation varies widely from island to island and, on the dry islands, from year to year. The relatively wet Ocean Island, with its huge phosphate deposits, contains the largest single resource, though presently copra is the only export.

The Southern Gilbert Islands (Kingsmill Group). The Gilbert Islands north of the Equator are quite wet overall, despite occasional years of drought, while those to the south of the line fall clearly into the subhumid zone. Nonouti, Tabiteuea, Beru, Nikunau, Onotoa, Tamana, and Arorae are all reef-fringed coral islands sufficiently large to support a few thousand persons each. The local economy is dependent upon fishing and copra production from the coconut palm.

The Phoenix Islands. This group lies roughly in the center of the nation; together with nearby Baker and Howland Islands (which are administered by the United States) they range from subhumid (Gardner, Hull/Orora, Sydney/Mauru) to semiarid (McKean, Birnie, Phoenix/Rawaki, Enderbury, Canton/Abariringa, Baker, Howland). Vegetation is scanty, limited to low grasses and shrubs. In the middle nineteenth century most of the islands were stripped of their phosphate deposits by American and British firms; attempts to plant coconut palms have succeeded only where drought has not intervened. Attempts to colonize the Phoenix group from the Gilberts have not succeeded, and the islands are presently uninhabited.

The Line Islands. Kiribati's easternmost chain contains several wet and seven dry islands: Christmas/Kiribati, Malden, Starbuck, Vostok, Caroline, Flint, and Jarvis (the last administered by the United States). Their exploitation follows the Phoenix pattern of phosphate, coconut palm, and ultimate abandonment (save for Christmas/Kiribati).

Dryland areas of KIRIBATI

DESERTIFICATION RISK	ARIDITY									
	Hyperarid		Arid		Semiarid		Subhumid		Aridity Totals	
	km²	%	km²	%	km²	%	km²	%	km²	%
Very High	By definition, desertification does not exist in hyperarid regions.									
High										
Moderate										
Desertification Totals										
No Desertification					305	44.6	70	10.2	375	54.8
Total Drylands					305	44.6	70	10.2	375	54.8
Non-dryland									309	45.2
Total Area of the Territory									684	100.0

DIRECTORIES

1817 Fuller, S. C., ed. *South-east Asia.*
See entry 1399.

GAZETTEERS

1818 *British Solomon Islands Protectorate and Gilbert and Ellice Islands Colony: official standard names approved by the United States Board on Geographic Names.* Washington DC: U.S. Board on Geographic Names, 1974. (Gazetteer 136) 202 pp.

List of 12,450 place-names with latitude/longitude coordinates. Includes 850 entries for Kiribati. Does not include Jarvis Island, Palmyra Island, or Kingman Reef in the Line Islands group.

MARQUESAS ISLANDS
Iles Marquises

(this island group forms one of five *circonscriptions* of French Polynesia (Polynesie Francaise), an overseas territory of France)

The Marquesas chain is the northernmost unit of French Polynesia, lying roughly 10 degrees south of the Equator and east of Kiribati's Line Islands. The islands are all volcanic and heavily eroded into spectacular peaks and gorges. The prevailing southeast trades provide the windward eastern slopes with generally sufficient rainfall but the rainshadows to leeward give each of the islands a sizeable extent of dryland. Desertification has occurred where sheep and goats have stripped the vegetation cover and periodic drought has had disastrous effects on the coconut plantations; copra is a primary export.

Dryland areas of THE MARQUESAS ISLANDS

DESERTIFICATION RISK	ARIDITY									
	Hyperarid		Arid		Semiarid		Subhumid		Aridity Totals	
	km²	%	km²	%	km²	%	km²	%	km²	%
Very High	By definition, desertification does not exist in hyperarid regions.									
High										
Moderate										
Desertification Totals										
No Desertification					110	8.6	1,164	91.4	1,274	100.0
Total Drylands					110	8.6	1,164	91.4	1,274	100.0
Non-dryland										
Total Area of the Territory									1,274	100.0

GAZETTEERS

1819 *South Pacific: official standard names approved by the United States Board on Geographic Names.* Washington DC: U.S. Government Printing Office, 1957. (Gazetteer 39) 68 pp.

List of 5,400 place-names with latitude/longitude coordinates for places and features in the Pacific Ocean. Generally south of 15° north latitude. Includes the Marquesas Islands of French Polynesia.

NEW CALEDONIA
Nouvelle Caledonie

(an overseas territory of France)

New Caledonia is a long narrow mountainous island located just north of the Tropic of Capricorn some 1,600 km (1,000 miles) opposite the Queensland coast of Australia. The Northeast Trades bring steady rains along the island's mountain front; to leeward on the west coast precipitation drops dramatically and open savanna prevails. Pockets of subhumid land can be found along the coast. The island's economy is based almost totally on mining of strategic minerals such as nickel and chromium, but agriculture is important and has considerable development potential. The drylands contribute by providing pasturage for beef cattle.

Only selected reference sources are provided.

Dryland areas of NEW CALEDONIA

DESERTIFICATION RISK	ARIDITY									
	Hyperarid		Arid		Semiarid		Subhumid		Aridity Totals	
	km²	%	km²	%	km²	%	km²	%	km²	%
Very High	By definition, desertification does not exist in hyperarid regions.									
High										
Moderate										
Desertification Totals										
No Desertification							1,088	5.7	1,088	5.7
Total Drylands							1,088	5.7	1,088	5.7
Non-dryland									18,015	94.3
Total Area of the Territory									19,103	100.0

DIRECTORIES

1820 Fuller, S. C., ed. *South-east Asia.*
See entry 1399.

GEOGRAPHY

1821 Greenway, M. E., comp. *New Hebrides and New Caledonia.* Surbiton, U.K.: United Kingdom. Overseas Development Administration. Land Resources Division, 1974. (Land resource bibliography 5) 84 pp.

ATLASES

1822 *Atlas de la nouvelle Caledonie et dependances [atlas of New Caledonia and Dependencies].* Paris: O.R.-S.T.O.M., 1981. 53 plates.

In French with English summaries. Standard base map scale 1:1,000,000. Atlas plates and text cover a wide range of subjects in physical and cultural geography.

GAZETTEERS

1823 *New Caledonia and dependencies and Wallis and Futuna: official standard names approved by the United States Board on Geographic Names.* Washington DC: U.S. Defense Mapping Agency Topographic Center, 1974. (Gazetteer supplement 137) 100 pp.

List of 5,950 place-names with latitude/longitude coordinates.

NEW ZEALAND
The Dominion of New Zealand
(a former colony of the United Kingdom)

Although generally well-watered, New Zealand possesses a small subhumid dryland zone in the Otago region of the South Island. Due primarily to the rainshadow effect of mountains to the west, reduced precipitation has formed a dry grassland; nonetheless, the region has proved ideal for irrigated orchard crops such as cherries, apricots, and peaches. These are flown out by air to markets elsewhere in the nation.

Only selected reference sources are provided.

Dryland areas of NEW ZEALAND

DESERTIFICATION RISK	ARIDITY									
	Hyperarid		Arid		Semiarid		Subhumid		Aridity Totals	
	km²	%	km²	%	km²	%	km²	%	km²	%
Very High	By definition, desertification does not exist in hyperarid regions.									
High										
Moderate										
Desertification Totals										
No Desertification							4,031	1.5	4,031	1.5
Total Drylands							4,031	1.5	4,031	1.5
Non-dryland									264,673	98.5
Total Area of the Territory									268,704	100.0

GENERAL BIBLIOGRAPHIES

1824 Grover, Ray. *New Zealand.* Oxford, U.K.; Santa Barbara CA: Clio Press, 1980. (World bibliographical series 18) 257 pp.

Annotated. Contains 878 citations to most major subjects.

DIRECTORIES

1825 Ronayne, J., and Nede, C. J., eds. *Australia and New Zealand.*

See entry 1709.

GEOGRAPHY

1826 Mark, Alan F. "New Zealand." In *Handbook of contemporary developments in world ecology*, pp. 417-444. Edited by Edward John Kormondy and J. Frank McCormick. Westport CT: Greenwood, 1981.

A narrative summary of ecological research in New Zealand; with an annotated list of references.

1827 *Otago region.* Wellington: New Zealand. Ministry of Works. Town and Country Planning Division, 1967. (National resources survey V) 272 pp.

ATLASES

1828 Anderson, A. Grant. *New Zealand in maps.* New York: Holmes and Meier, 1978. 141 pp.

1829 Wards, Ian McLean. *New Zealand atlas.* Wellington: A. R. Shearer, 1976. 291 pp.

Scale 1:1,000,000.

GAZETTEERS

1830 *New Zealand: official standard names approved by the United States Board on Geographic Names.* Washington DC: U.S. Board on Geographic Names, 1954. (Gazetteer 150) 454 pp.

List of 18,500 place-names with latitude/longitude coordinates.

CLIMATOLOGY

1831 Gentilli, J., ed. *Climates of Australia and New Zealand.*

See entry 1747.

THE AMERICAS

North America, the Caribbean Islands, and South America mirror the diversity of dryland types found in the Old World of Eurasia, Africa, and the Pacific. In the Americas, one encounters climate and terrain analogous to the Central Asian steppes, the Middle Eastern uplands, the Mediterranean coast, the Saharan sand seas, the African scrub forests, the trade wind rainshadow islands, and the fog desert of the Namib. Most of these types exist in a much smaller area and in much closer juxtaposition; the great size and uniformity of Old World dryland landscapes are missing in the Americas. In some ways, this enhances development and exploitation: no area is far removed from the sea, overland transport is generally easier, water supplies are closer to dryland centers, and a complex network of climates increases the variety of the resource base. Actual development, however, varies widely between nations and regions. Several Latin American nations are as poor as their African counterparts, while in the Sonoran Desert, for example, the same dryland is both an expansive economic mecca north of the international border and a hardscrabble pocket of poverty to the south. The resource base alone is often a poor indicator of economic viability.

Reference sources in this section cover either the Americas as a whole or Latin America in general.

Dryland areas of THE AMERICAS (INCLUDES GREENLAND)

DESERTIFICATION RISK	ARIDITY								Aridity Totals	
	Hyperarid		Arid		Semiarid		Subhumid			
	km²	%	km²	%	km²	%	km²	%	km²	%
Very High	By definition, desertification does not exist in hyperarid regions.		57,791	0.1	503,681	1.2			561,472	1.3
High			1,599,350	3.8	963,268	2.3	146,346	0.4	2,708,964	6.5
Moderate			265,614	0.7	3,012,303	7.1	1,466,030	3.5	4,743,947	11.3
Desertification Totals			1,922,755	4.6	4,479,252	10.6	1,612,376	3.9	8,014,383	19.1
No Desertification	229,033	0.6	15,227	<0.1	71,365	0.2	2,148,182	5.1	2,463,807	5.9
Total Drylands	229,033	0.6	1,937,982	4.6	4,550,617	10.8	3,760,558	9.0	10,478,190	25.0
Non-dryland									31,510,196	75.0
Total Area of the Territory									41,988,386	100.0

GENERAL BIBLIOGRAPHIES

1832 Ulibarri, George S., and Harrison, John P. *Guide to materials on Latin American in the National Archives of the United States.* Washington DC: U.S. National Archives and Records Services, 1974. 489 pp.

INDEXES AND ABSTRACTS

1833 *Index to Latin American periodical literature, 1929-1960.* Boston MA: G. K. Hall, 1962-. 8 vols.

Supplement 1 (1968) covers years 1961-65, 2 volumes. Supplement 2 (1980) covers years 1966-70, 2 volumes. Contains material processed at the Organization of American States Library. Primarily of Latin American origin. Includes government decrees.

DIRECTORIES

1834 *National directory of Latin Americanists.* 2nd ed. Washington DC: U.S. Library of Congress, 1972. (Hispanic Foundation bibliographical series 12) 684 pp.

Covers researchers primarily in the social sciences and humanities. A 3rd edition is in progress (August 1982).

1835 Richards, Robert A. C., ed. *Latin America.* 2nd ed. Guernsey, U.K.: Francis Hodgson, 1975. (Guide to world science 21) 331 pp.

Provides descriptions of scientific research activities and organizations. Covers Mexico, Central America, the West Indies (Cuba, Dominican Republic, Haiti, Puerto Rico), and South America.

The Americas/1836-1850

STATISTICS

1836 *Anuario estadistico de America Latina/ Statistical yearbook for Latin America.* Santiago, Chile: U.N. Economic Commission for Latin America, 1973-. Annual.

In Spanish and English.

1837 **Harvey, Joan M.** *Statistics America: sources for market research (North, Central and South America).* 2nd ed. Beckenham, U.K.: CBD Research, 1980. 300 pp.

1838 **Wilkie, James Wallace, and Haber, Stephen,** eds. *Statistical abstract of Latin America*, vol. 21. Los Angeles CA: University of California at Los Angeles, 1981. 671 pp.

Contains economic, socioeconomic, political, international and geographic data. Includes maps and several essays.

GAZETTEERS

1839 *The Americas: official standard names approved by the United States Board on Geographic Names.* Washington DC: U.S. Army Topographic Command, 1971. (Gazetteer supplement 121) 86 pp.

List of place-name corrections with latitude/longitude coordinates. Does not include the islands of St. Pierre and Miquelon, the United States and its possessions, or the Dominican Republic.

MAP COLLECTIONS

1840 **Monteiro, Palmyra V. M.** *A catalogue of Latin American flat maps, 1926-1964.* Austin TX: University of Texas at Austin. Institute of Latin American Studies, 1967, 1969. (Guides and bibliographies 2) 2 vols.

Covers Mexico, West Indies, all of Central and South America.

SCIENCE AND TECHNOLOGY

1841 **James, Dilmus D.** "Bibliography on science and technology policy in Latin America." *Latin American research review* 12:3 (1977):71-101.

Unannotated.

BOTANY

1842 **Bogusch, E. R.** "A bibliography on mesquite." *Texas journal of science* 2:4 (1950):528-538.

A narrative bibliography and list of some 100 references to the genus *Prosopis*, which occurs throughout the Americas from Kansas to Patagonia.

1843 **Graham, Alan,** ed. *Vegetation and vegetational history of Northern Latin America. Papers presented as part of a symposium at the American Institute of Biological Sciences meetings, Bloomington, Ind. (USA), 1970.* Amsterdam, The Netherlands: Elsevier Scientific, 1973. 393 pp.

Contains approximately 1,000 items. Covers Mexico, Central America, the Antilles, and South America. Updated in 1979 (entry 1844).

1844 **Graham, Alan,** ed. "Literature on vegetational history in Latin America. Supplement 1." Kent State University. Dept. of Biological Sciences. *Review of palaeobotany and palynology* 27:1 (1979):29-52.

Unannotated. Contains 445 items. Covers Mexico, Central America, the Antilles, and South America. Supplement to Graham's 1973 publication (entry 1843).

1845 **Huber, Otto.** *Le Savane neotropicali: bibliografia sulla loro ecologia vegetale e fitogeografia/The Neotropical savanna: select bibliography on their plant ecology and phyto-geography.* Rome: Istituto Italo-Latino Americano, 1974. 855 pp.

In Italian, Spanish, English, and Portuguese.

1846 *Index to American botanical literature, 1886-1966.* Boston MA: G. K. Hall, 1969. 4 vols.

Covers the Western Hemisphere. A reproduction of an index published serially since 1866 in the *Bulletin* of the Torrey Botanical Club. Taxonomic and non-author items listed separately. *Supplement* One for period 1967-76 (1977), 740 pp.

1847 **Salas de Leon, Sonia.** "Bibliografia del nopal" [bibliography of prickly pear]. *Acta cientifica Potosina* 6:2 (1977):205-225.

In Spanish. Unannotated. Covers the years 1704-1975 with more than 260 references to the genus *Opuntia*.

AGRICULTURE

1848 *Agriculture of the American Indian: a select bibliography.* Beltsville MD: U.S. Dept. of Agriculture, 1979. (Bibliographies and literature of agriculture 3) 64 pp.

Updates 1941 work by Edwards and Rasmussen. Covers North and South America.

1849 **Alvear, Alfredo.** *Bibliografia de bibliografias agricolas de American Latina [bibliography of agricultural bibliographies of Latin America].* 2nd ed. Turrialba, Costa Rica: Instituto Interamericano de Ciencias Agricolas. Biblioteca y Servicio de Documentacion, 1969. (Bibliotecologia y documentacion 10) 121 pp.

In Spanish. Annotated. Includes books, pamphlets, journal articles and government documents. Omits bibliographies which form parts of monographs.

1850 *Indice agricola de America Latina y el Caribe [agricultural index to Latin America and the Caribbean].* Turrialba, Costa Rica: Centro Interamericano de Documentacion e Informacion Agricola, 1974-. Quarterly.

In Spanish, Portuguese, English, French. Formerly *Bibliografia agricola latinoamericana* (1966-73).

1851 *Indices of agricultural production for the Western hemisphere*. Washington DC: U.S. Dept. of Agriculture, 1979. (Statistical bulletin 622) 33 pp.

Excludes the United States and Cuba.

1852 **Woehlcke, Manfred, comp.** *Agrarstruktur und laendliche Entwicklung in Lateinamerika: Auswahlbibliographie/Estructura agraria y desarrollo rural: bibliografia selecta [agrarian structure and development in Latin America: selected bibliography]*. Hamburg, Federal Republic of Germany: Institut fur Iberoamerika-Kunde. Dokumentations-Leitstelle Lateinamerika, 1978. (Dokumentationsdienst Lateinamerika Reihe A: 4) 109 pp.

Annotated. Covers the Caribbean, Central and South America. Text in German and Spanish.

LAND TENURE

1853 **Anderson, Theresa J.** *Sources for legal and social science research on Latin America: land tenure and agrarian reform*. Madison WI: University of Wisconsin. Land Tenure Center, 1970. (Training and methods 11) 34 pp.

An unannotated bibliography with lists of organizations.

MARKETS

1854 **Smith, Robert H. T.** *Periodic markets in Africa, Asia, and Latin America*.

See entry 341.

AGRICULTURAL ECONOMICS

1855 *Agrarian reform in Latin America: an annotated bibliography*. Madison WI: University of Wisconsin-Madison. Land Tenure Center. Library, 1974. (Land economics monographs 5) 667 pp.

Includes published and unpublished material on Latin America and the Caribbean.

1856 **LeBaron, Allen.** *Bibliography of Latin American agricultural production and development*. Logan UT: Utah State University. Economic Research Center, 1973. 2 vols.

In Spanish and English. Most entries in Spanish. Unannotated. Volume I "Inputs and outputs in crop and livestock production"; volume II "General agricultural planning, background and statistical studies." Covers most of the Caribbean and Latin America. No index.

1857 **Uribe Contreras, Maruja, and Isaza Velez, Guillermo.** *Bibliografia selectiva sobre reforma agraria en America Latina [selective bibliography on agrarian reform in Latin America]*. Bogota: Instituto Interamericano de Ciencias Agricolas de la OEA, 1972. 381 pp.

Unannotated. In Spanish. Covers publications from 1964-71. Topics include traditional and new forms of land tenure systems, campesino organization, production, and law and legislation concerning agrarian reform in Latin America.

PETROLEUM

1858 *Latin America petroleum directory*. Tulsa OK: Petroleum Publishing, 1971-. Annual.

HUMAN GEOGRAPHY

1859 *CLASE: citas Latinoamericanas en sociologia, economia y humanidades [Latin American citations in sociology, economy, and humanities]*. Mexico, DF: Centro de Informacion Cientifica y Humanistica, 1976-. Quarterly.

In Spanish. Indexes Spanish- and Portuguese-language journals covering economic development, area studies, education, and sociology originating in Latin America.

1860 *Colonization and settlement: a bibliography*. Madison WI: University of Wisconsin. Land Tenure Center, 1969. (Training and methods 8) 41 pp.

Unannotated. Coverage is worldwide with strongest emphasis on Latin America. Continued by *Supplement* 1 (1971), *Supplement* 2 (1972), *Supplement* 3 (1977).

1861 **Delorme, Robert L.** *Latin America: social science information sources, 1967-1979*. Santa Barbara CA: ABC-Clio, 1981. 262 pp.

Unannotated bibliography of 5,602 citations. Covers Latin America and the Caribbean.

1862 **Grossman, Jorge, ed.** *Index to Latin American periodicals: humanities and social sciences*. Metuchen NJ: Scarecrow, 1961-70. 9 vols.

Annual cumulations.

1863 *Hispanic American periodical index (Hapi)*. Los Angeles CA: University of California, Los Angeles, 1974-. Annual.

Indexes English-, Spanish-, and Portuguese-language journals covering the social sciences in Latin America and the Caribbean. All citations are derived from the library holdings at Arizona State University, Tempe AZ.

DEMOGRAPHY

1864 **Edmonston, Barry, ed.** *Population research in Latin America and the Caribbean, a reference bibliography*. Ann Arbor MI: University Microfilms International, 1979. 161 pp.

Unannotated.

1865 **Thomas, Robert N., ed.** *Population dynamics of Latin America: a review and bibliography*. East Lansing MI: CLAG Publications, 1973. 200 pp.

Consists of papers presented at the Conference of Latin American Geographers, 2nd General Session (Boston MA, 17 April 1971). Includes reviews of available census data and an extensive unannotated bibliography.

ANTHROPOLOGY

1866 *Boletin bibliografico de antropologia Americana [bibliographic bulletin of American anthropology].* Mexico, DF: Pan American Institute of Geography and History, 1937-. Irreg.

Includes recent studies, official documents, and bibliographies, primarily in Spanish.

WOMEN'S STUDIES

1867 Knaster, Meri. *Women in Spanish America: an annotated bibliography from pre-Conquest to contemporary times.* Boston MA: G. K. Hall, 1977. 696 pp.

Extensive annotations. Includes only works in Spanish or English. Does not cover foreign-born women, or Latin American women in the United States, unless as part of a comparative study. Includes women of various indigenous groups, whether Spanish-speaking or not. Publications from 17th century to 1974. Mainly secondary sources. Includes some government documents.

1868 Saulniers, Suzanne Smith, and Rakowski, Cathy A. *Women in the development process: a select bibliography on women in Sub-Saharan Africa and Latin America.*

See entry 350.

URBAN GEOGRAPHY

1869 Sable, Martin Howard. *Latin American urbanization: a guide to the literature, organizations, and personnel.* Metuchen NJ: Scarecrow, 1971. 1,077 pp.

In English and Spanish. Part one is an unannotated bibliography; Part two a directory of organizations and individuals concerned with urbanization.

HOUSING

1870 Porzecanski, Leopoldo. *A selected bibliography on urban housing in Latin America.* Monticello IL: Council of Planning Librarians, 1973. (CPL exchange bibliography 412) 31 pp.

Unannotated. Covers Mexico, Central America, the Caribbean, and South America.

ECONOMIC DEVELOPMENT

1871 *Directorio Latino Americano de instituciones financieras de desarrollo, 1979-1980 [Latin American directory of financial and development institutions, 1979-1980].* Lima: ALIDE, 1980.

Covers Mexico, Central America, the Caribbean, and South America. In Spanish. ALIDE = Associacion Latinoamericana de Instituciones Financieras de Desarrollo.

1872 *Economic survey of Latin America.* Santiago, Chile: United Nations. Economic Commission for Latin America, 1948-. Annual.

Covers all of the Caribbean, Central and South America, omitting the colonies of the Netherlands Antilles, Puerto Rico, Virgin Islands, etc.

1873 *Latin American data bank.* Washington DC: Data Resources, 197?.

An online database providing economic data on selected Latin American and Caribbean nations. Vendor: Data Resources.

1874 Pesoa, Lillian, ed. *Bibliografia de la CEPAL 1948-1972 [bibliography of CEPAL, 1948-1972].* Santiago, Chile: United Nations. Economic Council for Latin America, 1973. 165 pp.

Unannotated. In Spanish. Does not include preliminary works or speeches, even if they were later printed. CEPAL = Comision Economica para America Latina.

1875 Shea, Donald R., ed. *Reference manual on doing business in Latin America.* Milwaukee WI: University of Wisconsin Milwaukee. Center for Latin America, 1979. 210 pp.

Basic guide for lawyers and executives involved in international business transactions in Latin America. Identifies information sources, reference data, reports on assistance from domestic and foreign sources. Covers Central and South America.

TOURISM AND RECREATION

1876 Mings, Robert C., and Quello, Steve. *The tourist industry in Latin America, 1974-1979: a bibliography for planning and research.* Monticello IL: Vance Bibliographies, 1979. (Public administration series bibliography P-333) 32 pp.

Unannotated. Covers Mexico, Central America, the Caribbean, and South America.

HISTORICAL GEOGRAPHY

1877 Condarco Morales, Ramiro. *Atlas historico de America [historical atlas of America].* La Paz, Bolivia: Ediciones Condarco, 1968. 185 pp. plus 25 maps.

In Spanish. Bibliography. Includes speculative maps depicting prehistoric demographic movements. Scale varies.

1878 Denevan, William M. *A bibliography of Latin American historical geography.* Washington DC: Pan American Institute of Geography and History. U.S. National Section, 1971. (Special publication 6) 32 pp.

An unannotated listing of 560 items covering the Americas from Mexico south; primarily journal articles and monographs in English only.

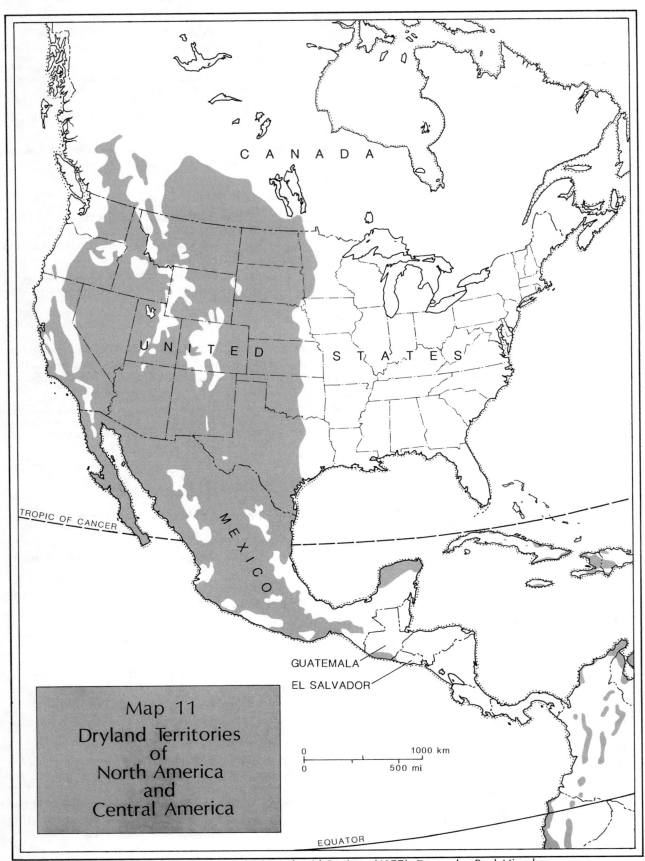

Based on Unesco, Map of the World Distribution of Arid Regions (1977). Drawn by Paul Mirocha.

North and Central America

The drylands of the United States and Mexico comprise significantly large fractions of their national territories; those of Canada, El Salvador, and Guatemala play smaller roles but are still significant to their overall economies. The range of dryland environments is wide: deep rainshadow basins inland of the northwest coast, the cold winter steppes of the Great Plains, the Mediterranean-type coastal climate of California, the fog desert of Baja California, and the hot summer subtropical deserts of the Southwestern United States and northern Mexico all offer a tremendous variety of climates and agricultural potential. Where sufficient capital and technical expertise exists, the scarcity of local water supplies can be overcome and even hyperarid regions can be made to produce abundant crops. Desertification, however, can be as extreme in highly developed North America as anywhere, and require the same attention to carrying capacity, soil depletion, erosion, and water conservation.

Dryland areas of NORTH AND CENTRAL AMERICA
(Includes Greenland, excludes Caribbean Islands)

DESERTIFICATION RISK	Hyperarid km²	Hyperarid %	Arid km²	Arid %	Semiarid km²	Semiarid %	Subhumid km²	Subhumid %	Aridity Totals km²	Aridity Totals %
Very High	By definition, desertification does not exist in hyperarid regions.		15,999	0.1	155,916	0.6			171,915	0.7
High			724,459	3.0	593,659	2.5	83,005	0.3	1,401,123	5.8
Moderate			193,008	0.8	2,120,017	8.9	765,979	3.2	3,079,004	12.9
Desertification Totals			933,466	3.9	2,869,592	12.0	848,984	3.5	4,652,042	19.4
No Desertification	30,667	0.1	15,227	0.1	19,619	0.1	1,301,692	5.5	1,367,205	5.7
Total Drylands	30,667	0.1	948,693	4.0	2,889,211	12.1	2,150,676	9.0	6,019,247	25.2
Non-dryland									17,755,919	74.8
Total Area of the Territory									23,775,166	100.0

DIRECTORIES

1879 Bender, Gordon Lawrence; Paylore, Patricia; and Medellon-Leal, Fernando. *A directory of North American arid lands research scientists*. Washington DC: American Association for the Advancement of Science, 1977. 250 pp.

Covers Canada, the United States, and Mexico. Contains 954 entries. Prepared for the UNEP Conference on Desertification.

GEOGRAPHY

1880 Bender, Gordon Lawrence, ed. *Reference handbook on the deserts of North America*. Westport CT: Greenwood Press, 1982. 594 pp.

A set of separately-authored articles covers: the physical geography of the Great Basin, Mojave, Sonoran, Chihuahuan, and Arctic deserts; includes checklists of species and supplementary material on geomorphology and research areas. Includes unannotated bibliographies.

1881 Jaeger, Edmund Carroll. *The North American deserts*. Stanford CA: Stanford University Press, 1957. 308 pp.

A narrative discussion with bibliography.

GEOLOGY

1882 *Abstracts of North American geology*. Washington DC: U.S. Geological Survey, 1966-71. 11 vols.

Covers North America including the West Indies and Central America; also includes Hawaii, Guam, and other U.S. possessions.

1883 *Bibliography of North American geology*. Washington DC: U.S. Geological Survey, 1923-. Annual.

Covers North America; also Hawaii, Guam, and other U.S. island possessions. Unannotated.

1884 *Geologic field trip guidebooks of North America: a union list incorporating monographic titles, 1968*

North and Central America/1885-1900

edition. Houston TX: Geoscience Information Society. Guidebook and Ephemeral Materials Committee, 1968. 152 pp.

Lists field guides prepared for excursions in the continental United States and Canada; some in Mexico, Cuba, Guatemala, and Puerto Rico.

1885 *Stratigraphic atlas of North and Central America.* Princeton NJ: Princeton University Press, 1975. 272 pp.

Scale 1:25,000,000 for the base map used for each plate. Covers North and Central America and the western Caribbean (to Puerto Rico). Prepared by Exploration Dept., Shell Oil Company.

CLIMATOLOGY

1886 *Climatic atlas of North and Central America. I: Maps of mean temperature and precipitation.* Geneva: WMO, 1979-. 30 maps.

Covers North and Central America and the Caribbean. Text in English, French, Spanish. Scale 1:10,000,000 and 1:5,000,000.

1887 Grayson, Donald K. *A bibliography of the literature on North American climates of the past 13,000 years.* New York: Garland, 1975. 206 pp.

HYDROLOGY

1888 *Catalog of information on water data: index to water-data acquisition.* Reston VA: U.S. Geological Survey, 1974-. Irreg.

Covers the United States with some information on Canada and Mexico.

BOTANY

1889 Beetle, Alan Ackerman. *Buffalograss, native of the shortgrass plains.* Laramie WY: University of Wyoming. Wyoming Agricultural Experiment Station, 1950. (Bulletin 293) 31 pp.

Buchoe dactyloides is a native of the Great Plains from Canada to southern Mexico. Contains a bibliography of 177 citations.

1890 Inglis, Michael; Glore, Denise; and Price, Lynn. *Bibliographic literature search concerning the relationships between soils and plants in arid and semi-arid regions in North America. Final report.* Albuquerque NM: University of New Mexico. Technology Application Center, 1978. (NTIS: AD A071-751) 378 pp.

Contains around 400 citations covering the period 1950-1978.

1891 Kuechler, August Wilhelm. *International bibliography of vegetation maps: vol. 1: North America.* Lawrence KS: University of Kansas, 1965. (Library series publication 21) 453 pp.

Part I of a four volume set. Covers North and Central America and the Caribbean.

1892 Phillips, W. Louis, and Stuckey, Ronald L. *Index to plant distribution maps in North American periodicals through 1972.* Boston MA: G. K. Hall, 1976. 686 pp.

Provides over 28,500 entries arranged by taxon. Emphasis on North America (Canada, the United States, Mexico).

1893 Schuster, Joseph L., ed. *Literature of the mesquite (Prosopis L.) of North America: an annotated bibliography* Lubbock TX: Texas Tech University. International Center for Arid and Semi-Arid Land Studies, 1969. (Special report 26) 84 pp.

ZOOLOGY

1894 Arnett, Ross H., ed. *Bibliography of Coleoptera of North America, north of Mexico, 1758-1948.* Baltimore MD: World Natural History Publications for Biological Research Institute of America, 1978. (North American beetle fauna project) looseleaf.

Reference source on beetles of North America, Mexico, Central America and the West Indies. Reprinted from the Leng *Catalog* and its supplements (1920-1948). Author index and updated necrology added.

1895 Douglass, John F. *Bibliography of the North American land tortoises (Genus Gopherus).* Washington DC: U.S. Dept. of the Interior. Fish and Wildlife Service, 1975. (Special scientific report - Wildlife 190) 60 pp.

1896 Hull, William B., ed. *Directory of North American entomologists and acarologists.* Hyattsville MD: Entomological Society of America, 1979. 189 pp.

Covers Canada, the United States, the Canal Zone, the Antilles, and American researchers overseas.

1897 Jaeger, Edmund Carroll. *Desert wildlife.* Stanford CA: Stanford University Press, 1961. 308 pp.

Covers the major North American deserts: Great Basin, Mohave, Sonora, and Chihuahua. Revision of *Our desert neighbors* (1950).

1898 Post, George, ed. *An annotated bibliography of the wild sheep of North America.* Ft. Collins CO: Colorado State University. Dept. of Fish and Wildlife Biology, 1971. (Rachelwood Wildlife Research Preserve publication 1) 86 pp.

Selective bibliography of literature pertaining to wild sheep and diseases affecting them. Includes unpublished theses and reports.

1899 Rickert, Jon E. *A guide to North American bird clubs.* Elizabethtown KY: Avian Publications, 1978. 564 pp.

FOODS

1900 Robson, John R. K., and Elias, Joel N. *The nutritional value of indigenous wild plants: an annotated bibliography.* Troy NY: Whitson, 1978. 232 pp.

Covers North America, both wild and introduced species.

HISTORICAL GEOGRAPHY

1901 **Lamar, Howard R., ed.** *The reader's encyclopedia of the American West.* New York: Thomas Y. Crowell, 1977. 1,306 pp.

The American West is defined here as the area encompassed by the moving frontier along which Europeans clashed with native peoples for land and power; topics covered in this volume touch all of North America though, naturally, the traditional "West" is emphasized. Unusually full coverage of broad themes and movements make this more than a quick reference work. See "reclamation and irrigation," "vegetation," "physiography of the United States," "mining metal," "Great American Desert," etc. for extended but succinct discussions in the context of historical geography. Articles are signed and include reliable bibliographies.

CANADA
The Dominion of Canada
(a former colony of the United Kingdom)

Canada ranks second in size but only 30th in population among the nations of the world. Canada resembles Australia in its great undeveloped resource base, its remote and underpopulated interior, and its industrial economy and relatively good access to investment capital. Although Canada is favored with extensive humid temperate agricultural lands, it must, together with Australia, rely on its drylands for much of its agricultural production.

The Prairies. The Great Plains of North America stretch northward into portions of Manitoba, Saskatchewan, and Alberta. As in the United States' portion, the climate ranges from semiarid to subhumid and supports extensive grasslands as its natural vegetation. With rainfall heaviest in the early summer growing season, the region is ideal for the production of beef cattle and cereals; huge tracts and level land favor highly mechanized production methods. Yields are generally very high, though the environment continually poses challenges such as drought, cold winters, and soil erosion. Mineral wealth is also great with large deposits of coal, oil, and natural gas located in numerous places throughout the region.

The Western Valleys. Several of the river valleys in southern British Columbia are deep enough to produce subhumid conditions, and one—the Okanagan Valley—extends into the semiarid range. Surrounded by the high Fraser Plateau, these valleys support concentrations of agriculture and population. The southern Okanagan is noted for its irrigated fruit orchards (similar to those in Washington State). A long lake at the bottom of the valley moderates winter temperatures and provides recreation potential.

Dryland areas of CANADA

DESERTIFICATION RISK	Hyperarid km²	Hyperarid %	Arid km²	Arid %	Semiarid km²	Semiarid %	Subhumid km²	Subhumid %	Aridity Totals km²	Aridity Totals %
Very High	By definition, desertification does not exist in hyperarid regions.									
High										
Moderate					92,093	0.9	215,767	2.2	307,860	3.1
Desertification Totals					92,093	0.9	215,767	2.2	307,860	3.1
No Desertification			15,177	0.2			271,111	2.7	286,288	2.9
Total Drylands			15,177	0.2	92,093	0.9	486,878	4.9	594,148	6.0
Non-dryland									9,381,992	94.0
Total Area of the Territory									9,976,140	100.0

GENERAL BIBLIOGRAPHIES

1902 Ryder, Dorothy E. *Canadian reference sources: a selective guide.* 2nd ed. Ottawa: Canadian Library Association, 1981. 311 pp.

Covers material published through December 1980. The standard annotated bibliography.

INDEXES AND ABSTRACTS

1903 *Microlog index: an index and document delivery service for Canadian report literature.* Toronto: Micro Media, 1979-. Monthly.

Covers federal, provincial, and selected local government reference publications.

THESES AND DISSERTATIONS

1904 Bruchet, Susan, and Evans, Gwynneth. *Theses in Canada: a guide to sources of information about theses compiled or in preparation.* Ottawa: Canada. National Library, 1978. 25 pp.

Provides lists of general and subject bibliographies.

1905 *Canadian theses/Theses canadiennes: 1947-1960.* Ottawa: Canada. National Library, 1973. 2 vols.

Continued by *Canadian Theses/Theses canadiennes* (Ottawa: National Library, 1960/61-). In French and English.

ENCYCLOPEDIAS

1906 *Encyclopedia Canadiana: the encyclopedia of Canada.* Toronto: Grolier of Canada, 1977. 10 vols.

The standard general reference. Watch for a new Canadian encyclopedia scheduled for 1985 publication by Hurtig (Edmonton).

GOVERNMENT DOCUMENTS

1907 Bishop, Olga B. *Canadian official publications.* Oxford, U.K.; New York: Pergamon, 1981. 297 pp.

Narrative explanation of Canadian government publishing with instructions for obtaining publications.

DIRECTORIES

1908 *Canadian conservation directory.* Ottawa: Canadian Nature Federation, 1973-. Irreg.

Lists around 4,000 organizations concerned with conservation and natural resources.

1909 *Directory of Canadian environmental experts/ Repertoire des specialistes canadiens de l'environment.* Rev. ed. Ottawa: Canada Institute for Scientific and Technical Information, 1981. 482 pp.

In English and French.

1910 Doucet, Cyril D., ed. *Canada.* 2nd ed. Guernsey, U.K.: Francis Hodgson, 1975. (Guide to world science 24) 277 pp.

Provides descriptions of scientific research activities and organizations.

1911 *Scientific and technical societies of Canada/ Societes scientifiques et techniques du Canada.* 4th ed. Ottawa: Canada Institute for Scientific and Technical Information, 1976. 77 pp.

Lists 272 societies with information in English or in French. Does not include trade associations, undergraduate groups, or fund-raising organizations.

STATISTICS

1912 *Canadian statistical review.* Ottawa: Statistics Canada, 1926-. Monthly.

Weekly supplement, 1948-.

GEOGRAPHY

1913 *Canadian environment (CENV).* Ottawa: Environment Canada, 1970-. Updated monthly.

An online database indexing environmental literature of Canada. Vendor: QL Systems.

1914 *Environment source book: a guide to environmental information in Canada.* Ottawa: Canada. Minister of Supply and Services, 1978. 115 pp.

A French edition is also available.

1915 Fraser, John Keith, and Hynes, Mary C., comps. *List of theses and dissertations on Canadian geography/Liste des theses et dissertations sur la geographie du Canada.* 3rd ed. Ottawa: Canada. Dept. of the Environment. Lands Directorate, 1972. (Geographical paper 51) 114 pp.

Over 2,400 entries. Annual supplement.

1916 Sawchuk, John P. *A natural resources bibliography for Manitoba.* Winnipeg: Manitoba. Dept. of Mines, Resources and Environmental Management, 1972. Various pagings.

1917 *Selected bibliography on Canadian geography/ Bibliographie choisie d'ouvrage sur la geographie au Canada.* Ottawa: Canada. Dept. of Mines and Technical Surveys. Geographical Branch, 1950-66. (Bibliographical series 1-34) 34 vols.

MAP COLLECTIONS

1918 Carrington, David K. *Map collections in the United States and Canada: a directory.*

See entry 2073.

1919 Dubreuil, Lorraine. *Directory of Canadian map collections.* 4th ed. Ottawa: Association of Canadian Map Libraries, 1980. 144 pp.

In English and French.

1920 Nicholson, Norman L. *The maps of Canada: a guide to official Canadian maps, charts, atlases and gazetteers.* Folkstone, U.K.: Dawson, 1981. 251 pp.

ATLASES

1921 *Atlas of Alberta.* Edmonton: University of Alberta. Dept. of Geography, 1969. 158 pp.

1922 Farley, Albert L. *Atlas of British Columbia: people, environment, and resource use.* Vancouver: University of British Columbia, 1979. 36 pp.

1923 *The national atlas of Canada.* 4th ed. Toronto: Macmillan of Canada, 1974. 254 pp.

Scale 1:15,000,000.

1924 Richards, J. Howard. *Atlas of Saskatchewan.* Saskatoon: University of Saskatchewan Press, 1969. 236 pp.

Includes bibliography.

Canada/1925-1944

1925 Weir, Thomas R. *Atlas of the prairie provinces.* Toronto: Oxford University Press, 1971. 31 leaves.

Covers Manitoba, Saskatchewan, Alberta.

1926 Weir, Thomas R. *Economic atlas of Manitoba.* Winnipeg: Manitoba. Dept. of Industry and Commerce, 1960. 81 pp.

Data from 1870-1959.

GAZETTEERS

1927 *Alberta.* 2nd ed. Ottawa: Canadian Permanent Committee on Geographical Names, 1974. (Gazetteer of Canada 6) 153 pp.

Annual supplements since 1958.

1928 *British Columbia.* 2nd ed. Ottawa: Canadian Permanent Committee on Geographical Names, 1966. (Gazetteer of Canada 1) 739 pp.

In English and French. Series supplemented semi-annually. List of more than 35,000 place-names with latitude/longitude coordinates.

1929 *Canada: official standard names approved by the United States Board on Geographic Names.* Washington DC: U.S. Board on Geographic Names, 1953. (Gazetteer 151) 415 pp.

List of 16,700 place-names with latitude/longitude coordinates.

1930 *Manitoba.* 2nd ed. Ottawa: Canadian Permanent Committee on Geographical Names, 1968. (Gazetteer of Canada 3) 93 pp.

In English and French. Series supplemented semi-annually. List of more than 7,000 place-names with latitude/longitude coordinates.

1931 *Saskatchewan.* 2nd ed. Ottawa: Canadian Permanent Committee on Geographical Names, 1969. 173 pp.

In English and French. Series supplemented semi-annually. List of 12,500 place-names with latitude/longitude coordinates.

1932 Sealock, Richard B.; Sealock, Margaret M.; and Powell, Margaret S. *Bibliography of place-name literature: United States and Canada.*

See entry 2077.

GEOLOGY

1933 Crockfield, W. E.; Hall, E.; and Wright, J. F., comps. *Index to reports of Geological Survey of Canada, from 1927-50.* Ottawa: Canada. Dept. of Mines and Technical Surveys, 1962. 723 pp.

Continued by: *Index to reports of Geological Survey of Canada from 1951-59* (entry 1943). Index to annual reports, economic geology series, geological bulletins, maps without reports, museum bulletins; geological papers, 1935-50; and summary reports, 1927-33.

1934 *Directory of geoscience departments: United States and Canada.*

See entry 2091.

1935 Ferrier, Walter Frederick. *Annotated catalogue of and guide to the publications of the Geological Survey Canada, 1845-1917.* Ottawa: J. de Labroquerie Tache, 1920. 544 pp.

Appendix: list of separates of some papers in publications other than those of the Geological Survey.

1936 *Geological survey of Canada: index to publications 1959-1974.* Ottawa: Canada. Geological Survey, 1975. 138 pp.

In English and French. Continues Johnston's *Index* (entry 1940).

1937 *Geological survey of Canada: index to publications 1975-1979.* Ottawa: Canada. Geological Survey, 1980. 247 pp.

1938 Gleason, V. E. *Bibliography on mined-land reclamation.*

See entry 2100.

1939 *Index: Geological Survey of Canada.* Ottawa: Canada. Geological Survey, 1977-. Annual.

In English and French.

1940 Johnston, A. G. *Index of publications of the Geological Survey of Canada, 1845-1958.* Ottawa: Canada. Dept. of Mines and Technical Surveys, 1961. 378 pp.

Updates but does not replace the 1920 volume (entry 1935).

1941 Kupsch, Walter Oscar. *Annotated bibliography of Saskatchewan geology, 1823-1970.* Rev. ed. Regina: Saskatchewan. Dept. of Mineral Resources, 1973. (Report 9) 421 pp.

Supplement covering the years 1970-1976 appeared in 1979.

1942 McCrossen, Robert G., comp. *Annotated bibliography of geology of the sedimentary basin of Alberta and adjacent parts of British Columbia and Northwest Territories, 1845-1955.* Calgary: Alberta Society of Petroleum Geologists, 1958. 499 pp.

1943 Wright, J.F., comp. *Index to reports of Geological Survey of Canada from 1951-59.* Ottawa: Canada. Dept. of Mines and Technical Surveys, 1965. 379 pp.

Continues: *Index to reports of Geological Survey of Canada, from 1927-50* (entry 1933).

CLIMATOLOGY

1944 *Atlas of climatic maps, series 1-10/Atlas de cartes climatologiques: serie 1 a 10.* Toronto: Canada. Meteorological Branch, 1966-70. 62 pp.

1945 Hare, Frederick Kenneth, and Hay, John E. "The climate of Canada and Alaska." In *Climates of North America*, pp. 49-192. Edited by Reid A. Bryson and F. Kenneth Hare. Amsterdam, The Netherlands: Elsevier Scientific, 1974. (World survey of climatology 11).

A narrative summary with maps, charts, and an unannotated bibliography.

1946 Hare, Frederick Kenneth, and Thomas, Morley K. *Climate Canada*. 2nd ed. Toronto: Wiley Canada, 1979. 230 pp.

Narrative text on the nation's climate, with maps, chapter bibliographies, and statistical appendices.

1947 Longley, Richard Wilberforce, and Powell, John M., eds. *Bibliography of climatology for the Prairie provinces 1957-1969*. Edmonton: University of Alberta Press, 1971. (Studies in geography. Bibliographies 1) 64 pp.

Unannotated. Updates Thomas' *Bibliography of Canadian climate, 1763-1957* (1962). Contains 665 items.

1948 Phillips, D. W., comp. *Handbook on climatological data sources of the Atmospheric Environment Service*. Downsview, Ontario: Canada. Atmospheric Environment Service, 1979. Looseleaf.

1949 Thomas, Morley K., and Phillips, David W. *Bibliography of Canadian climate/Bibliographie du climat canadien*. Ottawa: Canada. Dept. of Transport, 1961-79. 3 vols.

Covers period 1763-1976. Volume one in English only; volumes two and three in English and French. Volume two entitled *Environment Canada*.

HYDROLOGY

1950 *Hydrological atlas of Canada/Atlas hydrologique du Canada*. Ottawa: Canada. Fisheries and Environment, 1978. Looseleaf.

In English and French. Scale 1:10,000,000.

1951 Nielsen, G. L.; Hackbarth, D. A.; and Bainey, S. *Bibliography of groundwater studies in Alberta, 1912-1971*. Edmonton: Alberta. Research Council, 1972. (Contribution).

1952 Randolph, J. R.; Baker, N. M.; and Deike, R. G. *Bibliography of hydrology of the U.S. and Canada*.

See entry 2119.

1953 Toth, J. *The hydrological reconnaissance maps of Alberta*. Edmonton: Alberta. Research Council. Geological Division, 1977. (Bulletin 35) 11 pp.

SOILS

1954 Miska, John P., comp. *Soil science: Research Station CDA, Lethbridge, Alberta, 1906-1972*. Lethbridge: Research Station, 1973. 35 leaves.

BOTANY

1955 Adams, John; Norwell, M. H.; and Senn, Harold Archie, comps. "Bibliography of Canadian plant geography."

The first 8 parts have appeared in *Transactions of the Royal Canadian Institute*, 1928-50. The final part appeared in Canada. Dept. of Agriculture. Economics Branch. *Publications, a list of material published*, 1951. Unannotated. Covers the period 1635-1945.

1957 Campbell, R. Wayne et al. *A bibliography of Pacific Northwest herpetology*.

See entry 2462.

1957 Field, William Dewitt; dos Passos, Cyril F.; and Masters, John H. *A bibliography of the catalogs, lists, faunal and other papers on the butterflies of North America north of Mexico arranged by state and province*.

See entry 2156.

AGRICULTURE

1958 Minter, Ella S. G., comp. *Publications of the Canada Dept. of Agriculture, 1867-1959*. Ottawa: Queen's Printer, 1963. 387 pp.

Does not include papers published in scientific journals or items issued from units of the Department outside Ottawa.

1959 Miska, John P., comp. *Agriculture 1906-1972: a bibliography of research, Research Station, Agriculture Canada, Lethbridge, Alberta*. Lethbridge: Agriculture Canada. Research Station, 1973. (Bibliographic series 1-5) 193 leaves.

1960 *Selected agricultural statistics for Canada/Donnees statistiques agricoles canadiennes*. Ottawa: Canada. Dept. of Agriculture, 1972-. Annual.

In English and French.

1961 Troughton, Michael J. *An atlas of Canadian agriculture: portfolio of maps based on the 1971 census of agriculture*. London: University of Western Ontario. Dept. of Geography, 1979. 104 pp.

RANGE MANAGEMENT

1962 Vallentine, John F. *U.S.-Canadian range management, 1935-1977: a selected bibliography on ranges, pastures, wildlife, livestock, and ranching*.

See entry 2173.

Canada/1963-1974

ANTHROPOLOGY

1963 Murdock, George Peter, and O'Leary, Timothy J. *Ethnographic bibliography of North America.*

See entry 2194.

1964 Sturtevant, William C., ed. *Handbook of North American Indians.*

See entry 2196.

RURAL GEOGRAPHY

1965 Gill, Dhara S. *A bibliography of socio-economic studies on rural Alberta, Canada.* Monticello IL: Council of Planning Librarians, 1977. (CPL exchange bibliography 1260-1261-1262) 206 pp.

Unannotated.

URBAN GEOGRAPHY

1966 Armstrong, Frederick H.; Artibise, Alan F. J.; and Baker, Melvin. *Bibliography of Canadian urban history—Part V: Western Canada.* Monticello IL: Vance Bibliographies, 1980. (Public administration series bibliography P-542) 72 pp.

Unannotated. Covers Manitoba, Saskatchewan, Alberta, British Columbia.

1967 *Index to current urban documents.*

See entry 2197.

ECONOMIC DEVELOPMENT

1968 Brown, Barbara E., ed. *Canadian business and economics: a guide to sources of information/Sources d'information economique et commerciales canadiennes.* Ottawa: Canadian Library Association, 1976. 636 pp.

Annotated. A second edition is in preparation. In English and French.

1969 *Canadian business index.* Toronto: Micromedia, 1980-. Monthly.

Indexes around 150 Canadian business periodicals. Also available online. Previous title: *Canadian business periodicals index* (1975-80).

1970 Dick, Trevor J. O. *Economic history of Canada: a guide to information sources.* Detroit MI: Gale, 1978. (Economics information guide series 9) 174 pp.

1971 Land, Brian. *Sources of information for Canadian business.* 3rd ed. Montreal: Canadian Chamber of Commerce, 1978. 76 pp.

1972 *Quarterly economic review of Canada.* London: Economist Intelligence Unit.

Includes summary of political and economic news as well as charts and statistics of selected economic indicators. Annual supplement.

HISTORICAL GEOGRAPHY

1973 *America: history and life.*

See entry 2207.

1974 Smith, Dwight L. *The American and Canadian West: a bibliography.*

See entry 2211.

EL SALVADOR
Republica de El Salvador
(a former colony of Spain)

El Salvador—a tiny nation on the western coast of Central America between Guatemala and Honduras—contains a sliver of subhumid dryland in its northwest corner bordering Guatemala. Despite its small size, this region is the nation's largest single area of level land and supports cotton and beef cattle.

Only selected references are provided.

Dryland areas of EL SALVADOR

DESERTIFICATION RISK	ARIDITY									
	Hyperarid		Arid		Semiarid		Subhumid		Aridity Totals	
	km²	%	km²	%	km²	%	km²	%	km²	%
Very High	By definition, desertification does not exist in hyperarid regions.									
High										
Moderate										
Desertification Totals										
No Desertification							1,347	6.3	1,347	6.3
Total Drylands							1,347	6.3	1,347	6.3
Non-dryland									20,046	93.7
Total Area of the Territory									21,393	100.0

HANDBOOKS

1975 **Blutstein, Howard I. et al.** *El Salvador: a country study.* Washington DC: U.S. Dept. of Defense, 1971. (American University foreign area studies) 260 pp.

Narrative summary of the nation's history, politics, geography, economy, social life; includes unannotated bibliography.

ATLASES

1976 **Arbingast, Stanley Alan et al.** *Atlas of Central America.*

See entry 1985.

GAZETTEERS

1977 *El Salvador: official standard names approved by the United States Board on Geographic Names.* Washington DC: U.S. Board on Geographic Names, 1956. (Gazetteer 26) 65 pp.

List of 4,860 place-names with latitude/longitude coordinates.

CLIMATOLOGY

1978 *Atlas climatologico e hidrologico del Istmo Centroamericano [climatological and hydrological atlas of the Central American isthmus].*

See entry 1988.

1979 **Portig, W. H.** "The climate of Central America."

See entry 2483.

HUMAN GEOGRAPHY

1980 **Nuhn, H.; Krieg, P.,; and Schlick, W., eds.** *Zentralamerika: Karten zur Bevolkerungs- und Wirtschaftsstruktur [Central America: maps of population and economic structure].*

See entry 1992.

ECONOMIC DEVELOPMENT

1981 *Quarterly economic review of Guatemala, El Salvador, Honduras.*

See entry 1994.

HISTORICAL GEOGRAPHY

1982 **Flemion, Philip F.** *Historical dictionary of El Salvador.* Metuchen NJ: Scarecrow, 1972. (Latin American historical dictionaries 5) 157 pp.

Entries cover major topics, events, and personalities in the nation's history; includes extensive unannotated bibliography.

GUATEMALA
Republica de Guatemala
(a former colony of Spain)

The drylands of Guatemala are limited to a narrow strip of lowland along the nation's Pacific coast. The area is considered subhumid with warm wet summers and warm dry winters.

The soils here are of volcanic origin, fertile and porous. The dominant vegetation formation at the time of the European conquest was woodland savanna, which the Spanish continued to maintain through annual dry-season burning to facilitate open-range cattle ranching. Small areas were given over to slave-produced cacao and sugarcane during the Colonial period, but due to the presence of mosquito-induced disease, the area was largely neglected until the 1960's.

The region was opened to modern agriculture by the United Fruit Company, which attempted to move its banana plantations from the Caribbean coast to the Pacific slope in an effort to avoid infestations of banana diseases. Between 1939 and 1964 the company's Tequisate plantation was a prime producer, but new methods of disease-control and new disease-resistant banana varieties sent the center of production back across the mountains. United Fruit's abandonment of Tequisate and surrounding lands in 1964 made available sizeable acreages for colonization.

Since the 1950's the Pacific coast region has enjoyed a great boom in settlement and agricultural productivity. Presently cotton and beef are favored. The yield here for cotton is perhaps the highest in the world on unirrigated land; most is exported to Japan. Beef cattle are now run on cultivated pasture rather than open range. A packing plant at Esquintla prepares the meat for shipment to other Central American nations as well as to the United States.

Dryland areas of GUATEMALA

DESERTIFICATION RISK	ARIDITY									
	Hyperarid		Arid		Semiarid		Subhumid		Aridity Totals	
	km²	%	km²	%	km²	%	km²	%	km²	%
Very High	By definition, desertification does not exist in hyperarid regions.									
High										
Moderate										
Desertification Totals										
No Desertification								8,820	8.1	8,820	8.1
Total Drylands								8,820	8.1	8,820	8.1
Non-dryland										100,069	91.9
Total Area of the Territory										108,889	100.0

HANDBOOKS

1983   Dombrowski, John. *Area handbook for Guatemala.* Washington DC: U.S. Dept. of Defense, 1970. (American University foreign area studies) 361 pp.

Narrative summary of the nation's history, politics, geography, economy, social life; includes unannotated bibliography.

GEOGRAPHY

1984   *Guia geografica de Guatemala [geographic guide to Guatemala].* Guatemala City: Guatemala. Instituto Geografico Nacional, 1978. (Publicacion IPGH 319) 157 pp.

In Spanish and English. Publication of the Insituto Panamericano de Geografia e Historia (=IPGH).

ATLASES

1985   Arbingast, Stanley Alan et al. *Atlas of Central America.* Austin TX: Bureau of Business Research, 1979. 62 pp.

Covers Guatemala, Belize, Honduras, El Salvador, Nicaragua, Costa Rica, and Panama; thematic maps depict administration, population, economics, transportation, and geology of each nation.

1986   *Atlas nacional de Guatemala [national atlas of Guatemala].* Guatemala City: Guatemala. Instituto Geografico Nacional, 1972. Various pagings.

Includes Belize. Historical maps. In Spanish.

GAZETTEERS

1987 *Guatemala: official standard names approved by the United States Board on Geographic Names.* Washington DC: U.S. Board on Geographic Names, 1965. (Gazetteer 94) 213 pp.

List of 14,900 place-names with latitude/longitude coordinates.

CLIMATOLOGY

1988 *Atlas climatologico e hidrologico del Istmo Centroamericano [climatological and hydrological atlas of the Central American isthmus].* Guatemala City: Instituto Panamericano de Geografia e Historica, 1976. (Publicacion 367) 9 pp.

Covers Guatemala to Panama. Scale 1:2,000,000. In Spanish.

1989 **Portig, W. H.** "The climate of Central America."

See entry 2483.

AGRICULTURE

1990 **Ramirez de Amaya, Ruth, comp.** *Bibliografías agricolas de America Central: Guatemala [agricultural bibliographies of Central America: Guatemala].* Turrialba, Costa Rica: IICA, Centro Interamericano de Documentacion e Informacion Agricola, 1975. 244 pp.

AGRICULTURAL ECONOMICS

1991 **Graber, Eric S.** *An annotated bibliography of rural development and levels of living in Guatemala.* Washington DC: U.S. Agency for International Development. Bureau for Latin America and the Caribbean. Rural Development Division, 1979. (Guatemala, general working document 1) 81 pp.

HUMAN GEOGRAPHY

1992 **Nuhn, H.; Krieg, P.; and Schlick, W., eds.** *Zentralamerika: Karten zur Bevolkerungs- und Wirtschaftsstruktur [Central America: maps of population and economic structure].* Hamburg, Federal Republic of Germany: Wirtschaftsgeographische Abt., Institut fur Geographie und Wirtschafts-geographie der Universitat Hamburg, 1975. 180 pp.

In German. Covers Guatemala to Panama. Bibliography. Scale 1:2,000,000.

ANTHROPOLOGY

1993 **Bernal, Ignacio.** *Bibliografia de arqueologia y etnografia: Mesoamerica y Norte de Mexico 1514-1960 [bibliography of archeology and ethnography: Mesoamerica and northern Mexico, 1514-1960].*

See entry 2041.

ECONOMIC DEVELOPMENT

1994 *Quarterly economic review of Guatemala, El Salvador, Honduras.* London: Economist Intelligence Unit, 1976-. Quarterly.

Includes summary of political and economic news as well as charts and statistics of selected economic indicators. Annual supplement.

MEXICO
Estados Unidos Mexicanos
[The United States of Mexico]
(a former colony of Spain)

Mexico, with a large and fast growing population and a high percentage of dry and desertified lands, is in a position where dryland development is crucial to the nation's economic survival.

The Sonoran Desert. The Baja Peninsula, the delta of the Colorado River, and the states of Sonora and Sinaloa form Mexico's dryland core. When compared to and contrasted with the Sonoran Desert of the United States, the realities and possibilities of development in this region are startling and diverse. Since the international border follows no natural divide, virtually all differences between each nation's response to the Sonoran Desert are due to human factors (and thus are, presumably, under human control). The resource base itself is not inconsiderable. Nearly all the lands are within 320 km (200 miles) of the Pacific coast; drought is not extreme except at the hyperarid Colorado River delta; two dependable rainy seasons bring enough precipitation to support a remarkable arborescent desert flora, especially along the mountain slopes found throughout the region. Topographically the prevailing basin-range pattern divides the desert into numerous catchment areas which concentrate rainfall into vast underground pools that are relied on heavily as sources of water. In addition, the great North American divide to the east sends numerous exotic rivers through the desert to the sea; the greatest of these is, of course, the Colorado River, but others also provide for water storage, irrigation, and hydropower. Only the Baja Peninsula lacks extensive surface water resources.

Exploitation of the region has depended on making essential capital improvements in irrigation works and the transportation network. With water and roads, the agricultural base has expanded so that the region produces the bulk of Mexico's wheat and cotton and also exports quantities of fruits and vegetables northward to the United States. Mining, fishing, and light manufacturing—as well as a significant tourist trade and an uncountable amount of smuggling—have given the Sonoran Desert a diversified economy but one that is, nevertheless, tied strongly to the sometimes capricious policies of its northern neighbor. Despite border difficulties ranging from the quality of irrigation water to the mass migration of undocumented workers the unity of the Sonoran Desert tends to erase the international line.

The Chihuahuan Desert. This arid region is also sliced by the international border and exhibits the same differences in resource development as the Sonoran Desert. The Chihuahuan Desert is a landlocked intermontane plateau accessible most easily from its northern and southern ends. To the west and east the great ranges of the Sierra Madre Occidental and Oriental inhibit transportation to the oceans to either side. Within the desert is a typical basin-range pattern with interior drainage except for the Rio Grande (called Rio Bravo del Norte in Mexico) at the international border. The summer rainfall maximum supports adequate grasslands for cattle and this form of land use has predominated since Spanish times. Irrigated agriculture appears in scattered localities but the shorter growing season precludes the high-value fruit and vegetable crops grown so successfully in the Sonoran and Tamaulipan deserts; cotton is dominant, especially along the Rio Bravo del Norte below Ciudad Juarez. The region also supplies the greatest share of Mexico's non-petroleum mineral wealth.

The Tamaulipan Desert. This is a semiarid to subhumid region that stretches along the Gulf of Mexico from the valley of the Rio Bravo del Norte south to Tampico and inland up the slopes of the Sierra Madre Oriental. Like the Sonoran Desert, its agricultural potential is aided by warm winters and easy access to markets in the United States. Most of the irrigated lands, in fact, extend along the international border. Cotton is also a significant crop grown along both the river and the coastal plain. Extensive grazing lands for cattle and goats are also typical. At Monterey and Monclova are to be found most of Mexico's steel industry; the proximity of iron, coal, limestone, natural gas, and manganese deposits makes possible a strong local manufacturing base seldom seen in Latin America.

Central Mexico. Although this region lies outside the three arid cores of the north, it too has extensive semiarid and subhumid drylands. Central Mexico is topographically diverse and best characterized as a series of distinct basins separated by hills, punctuated by volcanic peaks, all sitting on a high plateau between the oceans. Climate, rainfall, vegetation, and soil cover all vary markedly across short distances. Human settlement is concentrated in clusters in the basins with the Basin of Mexico holding the capital and largest city. Maize is the predominant crop while wheat and cattle are also important. However, historically the mining of incredibly rich deposits of silver and other minerals has imprinted the land and social system and still remains of great importance today.

The Yucatan Peninsula. This marginally dry region is climatically and physiographically part of the Caribbean area. The Yucatan Peninsula is a limestone plateau riven with complex subsurface drainage channels, solution caves, and sinkholes (called locally *cenotes*) that are, like the Florida peninsula, characteristic of karst topography. A semiarid strip along the northwest coast gives way to extensive subhumid scrub forest backlands and eventually to wet tropical forest toward Guatemala and Belize. The peninsula was the center of Mayan civilization and so once supported a highly developed agricultural base. Today the jungle areas are virtually uninhabited and the drylands manage to support only cattle and agave. At one time the cultivation of *Agave farcroydes* for its henequen fiber was an important industry, but competition from large African plantations and synthetic fibers has crippled the Yucatan's primary agricultural crop.

Dryland areas of MEXICO

DESERTIFICATION RISK	ARIDITY									Aridity Totals	
	Hyperarid		Arid		Semiarid		Subhumid			
	km²	%	km²	%	km²	%	km²	%	km²	%
Very High	By definition, desertification does not exist in hyperarid regions.		3,040	0.2	67,227	3.4			70,267	3.6
High	^		358,912	18.2	142,946	7.3	6,705	0.3	508,564	25.8
Moderate	^		20,459	1.0	261,258	13.3	162,822	8.3	444,539	22.6
Desertification Totals	^		382,411	19.4	471,431	24.0	169,527	8.6	1,023,370	52.0
No Desertification	17,383	0.9			18,200	0.9	487,696	24.8	523,279	26.6
Total Drylands	17,383	0.9	382,411	19.4	489,631	24.9	657,223	33.4	1,546,649	78.6
Non-dryland									420,534	21.4
Total Area of the Territory									1,967,183	100.0

HANDBOOKS

1995 Weil, Thomas E. *Area handbook for Mexico.* 2nd ed. Washington DC: U.S. Dept. of Defense, 1975. (American University foreign area studies) 450 pp.

Narrative summary of the nation's history, politics, geography, economy, social life; includes unannotated bibliography.

ENCYCLOPEDIAS

1996 *Enciclopedia de Mexico [encyclopedia of Mexico].* Mexico DF: Instituto de la Enciclopedia de Mexico, 1966-77. 12 vols.

DIRECTORIES

1997 Jamail, Milton Henry. *The United States-Mexico border: a guide to institutions, organizations and scholars.*

See entry 2059.

GEOGRAPHY

1998 *An annotated bibliography for the Mexican desert.* Los Angeles CA: University of Southern California, 1958.

1999 Barrett, Ellen C. *Baja California: a bibliography of historical, geographical, and scientific literature relating to the peninsula of Baja, California and to the adjacent islands....* Vol. 1: Los Angeles CA: Bennett and Marshall, 1957; Vol. 2: Los Angeles CA: Westernlore, 1967.

Some annotations. *Supplement* (La Jolla CA: Friends of the U.C. San Diego Library, 1978).

2000 Linden, George L. *Baja California, Mexico: the physical, socio-economic, and political environment for planning.* Monticello IL: Vance Bibliographies, 1978. (Public administration series bibliography P-20) 9 pp.

Unannotated.

2001 *Los recursos naturales de Mexico [the natural resources of Mexico].* Mexico DF: Instituto Mexicano de Recursos Naturales Renovables, 1955-61. 3 vols.

In Spanish. Contains extensive bibliographies in each volume.

2002 Sarukhan, Jose. "Mexico." In *Handbook of contemporary developments in world ecology*, pp. 35-51. Edited by Edward J. Kormondy and J. Frank McCormick. Westport CT: Greenwood, 1981.

A recent survey of research on Mexican ecology, most valuable for its citations of recent literature and annotated bibliography.

2003 "The Sonoran Desert: a retrospective bibliography."

See entry 2223.

2004 Yates, Richard, and Marshall, Mary. *The Lower Colorado River: a bibliography.*

See entry 2225.

MAP COLLECTIONS

2005 *Guia de informacion cartografica para investigadores (de Mexico) [guide to cartographic information for researchers].* Mexico DF: Instituto Panamericano de Geografia e Historia. Comision de Cartografia, 1980.

In Spanish. Consists of descriptions and index maps of various carto- and photo-graphic series covering natural resources, population, economics, anthropology, and history.

Mexico/2006-2021

ATLASES

2006 Arbingast, Stanley A. et al. *Atlas of Mexico.* 2nd ed. Austin TX: Bureau of Business Research, 1975. 165 pp.

Historical, demographic, and industrial maps.

2007 Garcia de Miranda, Enriqueta. *Atlas: nuevo atlas Porrua de la Republica Mexicana [new Porrua atlas of the Mexican Republic].* 4th ed. Mexico DF: Editorial Porrua, 1979. 197 pp.

In Spanish.

2008 Miller, Tom. *The Baja book: a complete map-guide to today's Baja.* Santa Ana CA: Baja Travel Publications, 1974. 180 pp.

2009 Tamayo, Jorge L. *Atlas geografico general de Mexico; con cartas fisicas, biologicas, demograficas, sociales, economicas y cartogramas [general geographic atlas of Mexico, with physical, biological, demographic, and economic maps].* 2nd ed. Mexico DF: Instituto Mexicano de Investigaciones Economicas, 1962. Unpaged.

In Spanish. Includes data to 1960.

GAZETTEERS

2010 Henrickson, James, and Straw, Richard M. *A gazetteer of the Chihuahuan Desert region.*

See entry 2371.

2011 *Mexico: official standard names approved by the United States Board on Geographic Names.* Washington DC: U.S. Board on Geographic Names, 1956. (Gazetteer 15) 2 vols.

List of 53,000 place-names with latitude/longitude coordinates.

GEOLOGY

2012 Aguilar y Santillan, Rafael. *Bibliografia geologica y minera de la Republica Mexicana (hasta el ano 1896) [geologic and mining bibliography of the Mexican Republic until 1896].* Mexico DF: Mexico. Instituto Geologico, 1898. (Boletin 10) 158 pp.

In Spanish. English translation available as "Bibliography of Mexican geology and mining," American Institute of Mining, Metalurgical and Petroleum Engineers. *Transactions* 32 (1902):605-680.

2013 Aguilar y Santillan, Rafael. *Bibliografia geologica y minera de la Republica Mexicana (hasta el ano 1904) [geologic and mining bibliography of the Mexican Republic until 1904].* Mexico DF: Mexico. Instituto Geologico, 1908. (Boletin 17) 330 pp.

In Spanish.

2014 Aguilar y Santillan, Rafael. *Bibliografia geologica y minera de la Republica Mexicana, 1905-1918 [geologic and mining bibliography of the Mexican Republic, 1905-1918].* Mexico DF: Mexico. Instituto Geologico, 1918. 97 pp.

Contains 1,635 references. In Spanish.

2015 Aguilar y Santillan, Rafael. *Bibliografia geologica y minera de la Republica Mexicana, 1919-1930 [geologic and mining bibliography of the Mexican Republic, 1919-1930].* Mexico DF: Mexico. Secretaria de la Economia Nacional, 1936. 83 pp.

In Spanish.

2016 Enciso de la Vega, Salvador. *Bibliografia mexicana de tesis en geologia [bibliography of geologic theses on Mexico].* Mexico DF: Universidad Nacional Autonoma de Mexico. Instituto de Geologia, 1976. (Serie divulgacion 4) 93 pp.

In Spanish.

CLIMATOLOGY

2017 Durrenberger, Robert W. *Clima del estado de Sonora, Mexico/Climatological data for the state of Sonora, Mexico.* Tempe AZ: Arizona. Office of the State Climatologist, 1978. (Climatological publications, Mexican climatology series 3) 42 pp.

In English and Spanish. Consists of tables of temperature and precipitation data for various reporting stations in Sonora from 1961 to 1977.

2018 Mosino Aleman, Pedro A., and Garcia, Enriqueta. "The climate of Mexico." In *Climates of North America*, pp. 345-404. Edited by Reid A. Bryson and F. Kenneth Hare. Amsterdam, The Netherlands: Elsevier Scientific, 1974. (World survey of climatology 11).

A narrative summary with maps, charts, and unannotated bibliography.

2019 Weight, M. L., and Saltzmann, E. J. *An annotated bibliography of climatic maps for Mexico.* Washington DC: U.S. Weather Bureau, 1963. (NTIS: AD 660-829) 30 pp.

Contains 82 citations.

HYDROLOGY

2020 Blasquez Lopez, Luis. *Hidrogeologia de las regiones deserticas de Mexico [hydrogeology of the desert regions of Mexico].* Mexico DF: Universidad Nacional Autonoma de Mexico. Instituto Geologia, 1959. (Anales 15) 172 pp.

In Spanish. Bibliography: pp. 169-172.

2021 Blasquez Lopez, Luis, and Loehnberg, Alfredo. *Bibliografia hidrogeologica de la Republica Mexicana [hydrogeological bibliography of the Mexican Republic].* Mexico DF: Universidad Nacional Autonoma de Mexico. Instituto Geologia, 1961. (Anales 17) 99 pp.

In Spanish.

WATER LAW

2022 Jamail, Milton H., and Ullery, Scott J. *International water use relations along the Sonoran Desert borderlands.*

See entry 2242.

BOTANY

2023 Elias-Cesnik, Anna. *Jojoba: guide to the literature.*

See entry 2247.

2024 Fontana, Bernard L. "Ethnobotany of the saguaro: an annotated bibliography."

See entry 2248.

2025 Jones, George Neville. *An annotated bibliography of Mexican ferns.* Urbana IL: University of Illinois Press, 1966. 297 pp.

2026 Langman, I. K. *A selected guide to the literature on the flowering plants of Mexico.* Philadelphia PA: University of Pennsylvania Press, 1964. 1,015 pp.

Contains bibliography of 359 citations.

2027 McGinnies, William G., and Haase, Edward F. *Guayule: a rubber-producing shrub for arid and semi-arid regions.* Tucson AZ: University of Arizona. Office of Arid Lands Studies, 1975. (Resource information paper 7) 267 pp.

Contains 929 references; Guayule (*Parthenium argentatum*) is native to the Chihuahuan Desert of southwestern Texas, Chihuahua, Coahuila, Durango, Zacatecas, Nuevo Leon, and San Luis Potosi.

2028 Sherbrooke, Wade C. *Jojoba: an annotated bibliography.*

See entry 2250.

2029 Sherbrooke, Wade C., and Haase, E. F. *Jojoba: a wax-producing shrub of the Sonoran Desert: literature review and annotated bibliography.*

See entry 2251.

2030 Shreve, Forrest. *Vegetation and flora of the Sonoran Desert.*

See entry 2252.

2031 Steenberg, Warren F., and Hendrickson, Lupe P. "The Saguaro giant cactus: a bibliography."

See entry 2253.

ZOOLOGY

2032 Skoglund, C. "An annotated bibliography of references to marine mollusca from the northern state of Sonora, Mexico." *Veliger* 12:4 (1970):427-432.

Contains 73 citations.

AGRICULTURE

2033 Velasquez Gallardo, Pablo, and Zerlucke, Raul. *Bibliografia agricola nacional, 1946-1970 [national agricultural bibliography, 1946-1970].* Mexico DF: Mexico. Instituto Nacional de Investigaciones Agricolas, 1973. 2 vols.

In Spanish. Unannotated. Includes theses, books, and serials. Includes relevant works published by Mexicans and others in international sources.

PLANT CULTURE

2034 Florescano, Enrique. *Bibliografia del maiz en Mexico [bibliography of maize in Mexico].* Xalapa, Mexico: Universidad Veracruzana, 1966. (Biblioteca de la Facultad de Filosofia, Letras y Ciencias 20) 359 pp.

In Spanish. Annotated. Part one covers corn in Mexico, the origin of the plant. Parts two through four cover corn in prehistoric, colonial and contemporary Mexico.

FORESTRY

2035 Caballero Deloya, Miguel, and Castellanos Jarquin, S. "Bibliografia selecta sobre publicaciones relacionadas con inventarios forestales en Mexico" [select bibliography on publications related to forest inventories in Mexico]. Mexico. Inventario Nacional Forestal. Direccion General. *Notas I.N.F.* 2:28 (1974). 38 pp.

In Spanish.

AGRICULTURAL ECONOMICS

2036 Emery, Sarah Snell. *Mexico's rural development and education: a select bibliography.* Monticello IL: Vance Bibliographies, 1978. (Public administration series bibliography P-97) 33 pp.

Extensive annotations.

LAND TENURE

2037 *Land tenure and agrarian reform in Mexico: a bibliography.* Madison WI: University of Wisconsin. Land Tenure Center, 1969. (Training and methods 10) 51 pp.

Unannotated. Reflects holdings of the Land Tenure Center library; includes its call numbers. Continued by *Supplement* one (1971) and *Supplement* two (1976).

2038 Martinez Rios, Jorge. *Tenencia de la tierra y desarrollo agrario en Mexico: bibliografia selectiva y comentada, 1522-1968 [land tenure and agrarian development in Mexico: selective annotated bibliography, 1522-1968].* Mexico DF: Instituto de Investigaciones Sociales, 1970. 305 pp.

In Spanish.

SOLAR ENERGY

2039 **Hawkins, Donna.** *Energy in Mexico: a profile of solar energy activity in its national context.* Golden CO: U.S. Solar Energy Research Institute, 1980. 35 pp.

A summary and directory of solar energy research activities and organizations in Mexico.

MIGRATION

2040 **Corwin, Arthur F.** "Mexican emigration history, 1900-1970: literature and research." *Latin American research review* 8:2 (1973):3-24.

A narrative bibliography covering Mexican emigation into the Southwestern United States.

ANTHROPOLOGY

2041 **Bernal, Ignacio.** *Bibliografia de arqueologia y etnografia: Mesoamerica y Norte de Mexico, 1514-1960 [bibliography of archeology and ethnography: Mesoamerica and northern Mexico, 1514-1960].* Mexico DF: Instituto Nacional de Antropologia e Historia, 1962. (Memorias 7) 634 pp.

Unannotated. Contains 13,990 citations. Detailed map.

2042 **Murdock, George Peter, and O'Leary, Timothy J.** *Ethnographic bibliography of North America.*

See entry 2194.

2043 **Ortiz, Alfonso, ed.** *Southwest.*

See entry 2265.

2044 **Sturtevant, William C., ed.** *Handbook of North American Indians.*

See entry 2196.

2045 **Wauchope, Robert, ed.** *Handbook of Middle American Indians.* Austin TX: University of Texas Press, 1964-76. 16 vols.

Separate volumes focus on ethnology, physical anthropology, etc. Subdivided geographically. *Supplement* (1981-) edited by Victoria Reifler Bricker.

ECONOMIC DEVELOPMENT

2046 *La bibliografia economica de Mexico [economic bibliography of Mexico].* Mexico DF: Mexico. Departamento de Estudios Economicos, 1954/55-. Annual.

2047 **Foster, Stephanie.** *Economic development in the U.S.-Mexico border region: a review of the literature.*

See entry 2202.

2048 **Longrigg, S. J., ed.** *Major companies of Argentina, Brazil, Mexico and Venezuela 1982.*

See entry 2570.

2049 *Mapas de los mercados mexicanos/Mexican marketing maps.* Mexico DF: Marynka Olizar, 1974? 11 leaves.

In Spanish and English. Scale 1:12,000,000.

2050 *Quarterly economic review of Mexico.* London: Economist Intelligence Unit, 1976-. Quarterly.

Includes summary of political and economic news as well as charts and statistics of selected economic indicators. Annual supplement.

PUBLIC ADMINISTRATION

2051 **Harmon, Robert B.** *A selected and annotated guide to the government and politics of Mexico.* Monticello IL: Council of Planning Librarians, 1977. (CPL exchange bibliography 1242) 13 pp.

Brief annotations.

HISTORICAL GEOGRAPHY

2052 **Barnard, Joseph D., and Rasmussen, Randall.** "A bibliography of bibliographies for the history of Mexico." *Latin American research review* 13:2 (1978):229-235.

Unannotated.

2053 **Briggs, Donald C., and Alisky, Marvin.** *Historical dictionary of Mexico.* Metuchen NJ: Scarecrow, 1981. (Latin American historical dictionaries 21) 259 pp.

Entries cover major topics, events, and personalities in the nation's history; includes extensive unannotated bibliography.

2054 **Gerhard, Peter.** *A guide to the historical geography of New Spain.* Cambridge, U.K.: Cambridge University Press, 1972. 476 pp.

Coverage is confined to the area of the *gobierno* of Nueva Espana, that is, central Mexico from San Luis Potosi to the narrowing of the isthmus of Tehuantepec.

THE UNITED STATES OF AMERICA

(the dryland areas were once colonies of Spain, France, and the United Kingdom; much territory was taken or purchased from Mexico)

The United States is the world's greatest economic and military power. With access to enormous amounts of investment capital, technical expertise, and trained administrators, one would expect the nation to meet the challenges of dryland development more successfully than any other. In some ways this has indeed occurred, but in others it becomes clear how formidable the challenges are and how difficult it is to put research results into practice.

The drylands of the United States are sizeable in extent (fourth among nations of the world) but also more favored in their physical arrangement and resource base. Overall, the topography is very complex and divides the dryland half of the nation into a number of much smaller regions, all of which have access to water supplies stored in the mountain highlands separating each. Climates, soils, and vegetation are diverse; many sizeable rivers course through the drylands to the sea, while interior drainage is rare. With a relatively dense network of railways and roads there are few large areas of total isolation. Although vast tracts are uninhabited, urban populations are large and cities are characteristically extensive and suburban. The historic pattern of migration into the drylands has increased in recent years to become a flood of new residents fleeing the harsh winters and declining economic base of the nation's northeastern heartland. The West evokes powerful and compelling images for Americans: self-reliant pioneers, frontier justice, and a contradictory attitude toward the land as both a source of immediate wealth and a vision of sublime beauty.

The Sonoran Desert. This region's portion occupied by the United States is geographically similar to Mexico's (which see) but differs markedly in economic development. The hot summer/warm winter areas of extreme southern Nevada, southeastern California, and southern Arizona have become especially appealing meccas for immigrants from both the northeastern United States and northern Mexico in their search for jobs and a mild climate. Although mining (primarily along the mountains bordering the Colorado Plateau) and irrigated agriculture (especially in the Salt River, Gila, and Imperial Valleys) have long been important, not until the introduction of domestic air conditioning after 1945 were the summer months made bearable to potential immigrants. Electronics and other light manufacturing firms then moved to take advantage of the pre-existing labor supply; their expansion in the 1960's fueled the growth of the cities of Phoenix, Tucson, and Las Vegas. The region has now become an urbanized desert, with a diversified economy based on manufacturing, tourism, agriculture, military bases, education, and related service industries.

Many problems, however, remain to be solved. Water supplies are substantial (from both exotic river storage and groundwater) but are being depleted with abandon. The city of Phoenix and its suburbs, for example, sit upon unusually fertile alluvial land complete with a reservoir-based irrigation system; yet, as agricultural plots are subdivided for residential and commercial use the right to water remains with the land, and now provides cheap water for lawns, ornamental shrubbery, and swimming pools. Agricultural producers have had increasingly to rely on groundwater to bring in new croplands while a giant costly water diversion system (the Central Arizona Project) will eventually tap the already overburdened Colorado River. Although this summary cannot do justice to the complexities of the situation, it must be mentioned that the region still lacks a coherent and comprehensive water policy based on renewing rather than mining existing sources. This failure to make use of current knowledge of and research on the environment extends to other problems as well: urban congestion, air pollution, endangered biota, toxic wastes, unrestricted in-migration, overburdened social services have all contributed to a declining quality of life in what is an unusually beautiful and fragile environment.

The Colorado Plateau. This great uplifted sedimentary shield is fringed by higher mountains and scored by canyons (one of which, the Grand Canyon of the Colorado River, is a breath-takingly awesome natural wonder). The area is a cold winter/warm summer semiarid steppe with pockets of pasture, rain-fed cropland, and timber. Winter snow cover is important for water storage for the surrounding lowland deserts. Except for coal and uranium, mineral resources are found not on the plateau itself but in the surrounding mountains. Settlement is sparse. The native Hopi and Navaho peoples maintain their largely traditional economies of cultivation and sheep-raising despite land degradation and—among the Navaho—an expanding population. Lands beyond the native reservations support cattle, lumbering, and small urban service centers; tourism is a major industry in the summer months.

The Chihuahuan Desert. Again, like the Sonoran Desert, this arid region differs little in geography on either side of the international border. To the north of the border, the desert extends up the Rio Grande Valley through west Texas and New Mexico into southern Colorado, with small extensions westward into Arizona. Although large urban populations are found in El Paso, Albuquerque, and other centers—based on similar light manufacturing industries—the region as a whole has not experienced the same rate of development as the Sonoran Desert. This is due to natural and social factors: winters are colder and the exotic cactus flora is absent, the distance to the California coast is greater, the cities lack the palm-studded oasis appearance of Phoenix, and the diversity of cultures (nowhere more apparent than northern New Mexico) creates a confused market for commercial interests and a psychological barrier for potential migrants from the eastern half of the nation. Only recently have people wealthy enough to fuel a real estate boom discovered the considerable attractions of the northern end of the Desert, where pine-covered mesas meet the grasslands and where Pueblo, Hispanic, and Anglo cultures have produced a fruitful intermingling of the arts.

The United States

The Great Plains. This vast semiarid to subhumid grassland has been turned into the nation's grainery despite insufficient and erratic precipitation and occasionally unsound cropping practices. The "plains" are, in fact, multiple series of hills and river terraces crossed by eastward-flowing streams emerging from the Rocky Mountain front. The deep soils, gently rolling terrain, hot summers, and easy access to transcontinental transportation routes favor highly mechanized farming and very high yields per areal unit and per agricultural worker. Wheat is the principal crop, but sorghum, barley, rye, sunflowers, flax, cotton, beef cattle, sheep, and many other products are produced. Mineral wealth is primarily in petroleum products: coal, natural gas, oil. Settlement patterns are different from other American dryland regions: farming towns predominate and even the largest cities maintain a rural Anglo Midwestern flavor that turns distinctly Western the closer one travels toward the Rockies or into Texas. The region—excepting Texas—is perhaps the least ethnically diverse in the nation.

The Tamaulipan Desert. The extent of this region outside Mexico is confined to the lower Rio Grande Valley of Texas. Intensive irrigated agriculture supports citrus and winter truck crops. With warm winters and hot summers, the "Winter Garden" area resembles parts of the Sonoran Desert as an attractive subtropical oasis. McAllen and Brownsville are the chief centers.

The California Coast. Here is found North America's example of the Mediterranean-type climate regime with its mild wet winters and almost equally mild summers. The coastline from a point south of San Francisco to just beyond the international border with Mexico, including the offshore Channel Islands, is a semiarid region of beaches, bold headlands, rolling grass-covered hills, and enormous sprawling cities. Once a sleepy backland of cattle ranches first of Mexico, then the United States, the area was transformed by a series of real estate booms beginning in the late 1800's to become one of the most highly desired places to live in the world.

The economy was based first on cattle ranching and farming (still practised on what little open land is left) but the Second World War created massive defense-related manufacturing plants and this, with associated aircraft, aerospace, and related industries, fueled the region's growth into a major industrial center. The Los Angeles basin—the widest central plain in the region—received the greatest influx of people and is now the second largest urban area in the country. San Diego, with a better port and an important naval base, also flourished. Of equally great, but more intangible, economic and social influence is the notoriety of the Los Angeles area as the center of motion picture and television production for the world. Greatly distorted but still compelling images of "Hollywood, California" have become talismans of twentieth century civilization as well as performing the more mundane function of publicizing a publicity-dependent industry.

The environment has suffered terribly for this unchecked growth, especially along the highly urban coast and interior valleys from Santa Barbara east and south to Ensenada, Mexico. Air pollution is most noticeable; the source is primarily auto exhaust that winds from the sea concentrate in the interior valleys. Water supplies for the vast population have been an issue for decades. Giant aqueducts and other diversion works draw water for the cities from desert basins and the Colorado River; one basin—the Owens Valley—was the scene of open warfare in the 1920's when Los Angeles authorities purchased water rights and ruined the local farming economy. Electrical power to the megalopolis is also largely imported; restrictions on air pollution and nuclear power in California have forced utilities to construct and operate plants in the Sonoran Desert and Colorado Plateau, adding to air pollution and radiation hazards in those areas. Despite these and many other problems associated with populations too dense for the local resource base, much natural beauty remains, especially along the coast west and north of Santa Barbara.

California's Central Valley. A deep structural trough exists between the low coast ranges and the high fault-block Sierra Nevada; this Central Valley is the watershed of the Sacramento River (in the northern third) and San Joaquin River (southern two-thirds). The land is semiarid to arid (in the extreme southern end) with cooler winters and warmer summers than the coast. Natural vegetation was grassland and patches of riparian forest and tule swamp, all of which have largely disappeared under the plow. Today, a large irrigation system carries water from storage reservoirs in the northern mountain fringe and sends it out to the valley to supply one of the nation's most intensely cultivated regions. Barley, wheat, rice, maize, sorghum, cotton, alfalfa, sugar beets, wine and raisin grapes, and many fruits, vegetables, and nuts are produced; citrus is scarce except in the warmer-winter southeast corner. Beef cattle, chickens, and sheep are also important.

Settlement in the Central Valley itself is mostly urban despite the agrarian character of the landscape; the farms themselves are mostly highly mechanized corporate enterprises requiring few workers except during the harvests when migrant laborers are employed. Most valley inhabitants look to the coastal megalopoli around San Francisco or Los Angeles for commercial and cultural contacts.

The Great Basin. This large interior desert is occupied by the states of Nevada, extreme eastern California, western Utah, southeastern Oregon, and southern Idaho. The land is colder in the winter and not as dry as the Sonoran Desert to the south; winter snowfall provides a significant fraction of the precipitation. Topography is classically mountain/bolson (basin and range), with alternating mountain ranges and long valleys trending north-south. Drainage is confined to the basins so salt flats are common; the largest basin is occupied by the Great Salt Lake, a shallow remnant of a much larger freshwater body that fluctuates widely due to mountain and irrigation runoff. Most of the region is given over to cattle and sheep range, though irrigated agriculture is intensive along the Wasatch Front of Utah, Idaho's Snake River

The United States/2055-2059

Plain, and in small pockets elsewhere. Rich mineral resources have been exploited for nearly a century and a half. A more recent development has been the creation of several large military reservations in southern Nevada and western Utah for the testing and storage of weapons.

Settlement is very sparse and confined to small ranching centers, though urban corridors exist on the periphery along the Wasatch Front (Logan to Provo) and, to a lesser extent, in western Nevada (Reno to Carson City) and southern Idaho (in a broad curve from Boise to Idaho Falls). The Utah cities have the largest industrial and manufacturing base. This is also the center of the Mormon culture region which historically embraced the Great Basin and still plays an important part in the social pattern of the American West. Mormon pioneers entering the valley of the Great Salt Lake from 1847 onwards developed irrigated agriculture early on and a relatively dense rural population of small settlements unique to the North American drylands; rural Utah and southeastern Idaho are still largely composed of small Mormon towns.

The Columbia Plateau. Central Washington State holds a semiarid volcanic plateau encircled by great bends in the Columbia, Spokane, and Snake Rivers. These exotic streams are fed by high mountain snowfields surrounding the plateau and send their accumulated flow westward to cut through the Cascade Range to the sea; steep gradients make the region ideal for hydroelectric power development and irrigation. On the plateau itself, below the great Grand Coulee Dam, irrigated croplands produce alfalfa, vegetables, sugar beets, and other crops. To the west, in the rainshadow of the Cascades, deep valley oases produce much of the nation's apples and other fruits. To the east a series of rounded upland hills supports the nation's richest wheatlands, dry-farmed rather than irrigated. Settlement is primarily rural with few large towns and no large cities.

Dryland areas of THE UNITED STATES

DESERTIFICATION RISK	ARIDITY									
	Hyperarid		Arid		Semiarid		Subhumid		Aridity Totals	
	km²	%	km²	%	km²	%	km²	%	km²	%
Very High	By definition, desertification does not exist in hyperarid regions.		12,959	0.1	88,689	1.0			101,648	1.1
High			365,546	4.0	450,713	4.9	76,300	0.9	892,559	9.8
Moderate			172,549	1.9	1,766,666	19.3	387,390	4.2	2,326,605	25.4
Desertification Totals			551,054	6.0	2,306,068	25.2	463,690	5.1	3,320,812	36.3
No Desertification	13,284	0.1	50	<0.1	1,419	<0.1	532,718	5.8	547,471	5.9
Total Drylands	13,284	0.1	551,104	6.0	2,307,487	25.2	996,408	10.9	3,868,283	42.2
Non-dryland									5,292,180	57.8
Total Area of the Territory									9,160,463	100.0

GOVERNMENT DOCUMENTS

2055 *Monthly checklist of state publications.* Washington DC: U.S. Government Printing Office, 1910-. Monthly.

Includes monographs, university bulletin series and periodicals.

2056 Morehead, Joe. *Introduction to United States public documents.* 2nd ed. Littleton CO: Libraries Unlimited, 1978. (Library science text series) 377 pp.

The best and most recent introductory guide to U.S. government publications.

2057 *State publications.* Englewood CO: Information Handling Services, 1976-. Updated quarterly.

An online database that corresponds to printed *State publications index.* Covers official publications of state agencies and state universities. Vendor: BRS.

DIRECTORIES

2058 Atkins, A. G., ed. *United States of America.* 2nd ed. Guernsey, U.K.: Francis Hodgson, 1975. (Guide to world science 22-23) 2 parts.

Provides descriptions of scientific research activities and organizations.

2059 Jamail, Milton Henry. *The United States-Mexico border: a guide to institutions, organizations and scholars.* Tucson AZ: University of Arizona. Latin America Area Center, 1980. 153 pp.

Covers governmental and private agencies dealing with administration, public health, tourism, communications, social sciences, and education.

The United States/2060-2077

STATISTICS

2060 **Gordon, Mary S. J.** *Directory of federal statistics for local areas: a guide to sources 1976.* Washington DC: U.S. Dept. of Commerce. Bureau of the Census, 1978. 359 pp.

2061 *Statistical abstract of the United States.* Washington DC: U.S. Dept. of Commerce. Bureau of the Census, 1878-. Annual.

GEOGRAPHY

2062 *An annotated bibliography for the desert areas of the United States.* Los Angeles CA: University of Southern California, 1959. 387 pp.

2063 **Burgess, Robert L.** "United States." In *Handbook of contemporary developments in world ecology*, pp. 67-101. Edited by Edward J. Kormondy and J. Frank McCormick. Westport CT: Greenwood, 1981.

A narrative summary of the structure of recent ecological research in the United States, focusing on the dynamics of the journal literature, United States participation in the International Biological Program, the growth of mathematical ecology, and other broad trends. Includes an unannotated bibliography.

2064 *A directory of research natural areas on federal lands of the United States of America.* Washington DC: U.S. Forest Service, 1977. 280 pp.

Entries provide locations, administering agencies, and primary functions of several hundred sites in the United States (including Hawaii), and Puerto Rico; many are located in the arid west.

2065 **Dunn, James E., Jr.** *Land use planning directory of the 17 Western states.* Denver CO: U.S. Dept. of the Interior. Bureau of Reclamation, 1976. 266 pp.

Includes local, state, federal, and tribal agencies; includes universities. Covers Arizona, California, Colorado, Idaho, Kansas, Montana, Nebraska, Nevada, New Mexico, North Dakota, Oklahoma, Oregon, South Dakota, Texas, Utah, Washington, Wyoming.

2066 *Federal environmental data: a directory of selected sources.* Washington DC: U.S. National Science Foundation, 1977. (NTIS: PB 275-902/5) 136 pp.

2067 **Moyer, D. David.** *Land information systems: an annotated bibliography.* Washington DC: U.S. Dept. of Agriculture. National Resource Economics Div., Economics, Statistics, and Cooperative Service, 1978. 195 pp.

Emphasizes legal aspects of land information (titles, transfers, assessment for taxation, etc.) and includes remote sensing and surveying. Annotated.

2068 *Profile of environmental quality: region 8, Colorado, Montana, North Dakota, South Dakota, Utah, Wyoming.* Denver CO: U.S. Environmental Protection Agency. Region 8, 1978. 28 pp.

2069 **Rogoff, Marc Jay.** *Statewide computer-based land information systems: an annotated bibliography of an emergent field.* Monticello IL: Council of Planning Librarians, 1978. (CPL exchange bibliography 1490) 20 pp.

Annotated with narrative explanation of the topic. Covers the efforts of five states: Minnesota, Nebraska, New York, North Dakota, Ohio, and South Dakota.

2070 *Soil, water, air sciences directory.* Washington DC: U.S. Dept. of Agriculture. Agricultural Research, 1979. 54 pp.

2071 *Southwest environment index: a Kwoc (keyword-out-of-context) index to information on the Southwest environment.* Tempe AZ: Arizona State University, 1973-. Annual.

Covers Utah, Colorado, Arizona, New Mexico, Southeastern Nevada, and Southern California.

2072 *U.S. directory of environmental sources.* 3rd ed. Washington DC: U.S. Environmental Protection Agency, 1979. (EPA: 840-79-010) 861 pp.

Authored by United Nations Environment Program (UNEP).

MAP COLLECTIONS

2073 **Carrington, David K.** *Map collections in the United States and Canada: a directory.* 3rd ed. New York: Special Libraries Association, 1978. 230 pp.

2074 *Map data catalog.* Reston VA: National Cartographic Information Center, 1981. 48 pp.

A catalog of cartographic products available from the U.S. Geological Survey as part of its National Mapping Program. Includes information on topographic maps (both current and out-of-print), land-use and slope maps, orthophotoquads, digital terrain types, and various forms of aerial and space imagery.

ATLASES

2075 *The national atlas of the United States of America.* Washington DC: U.S. Geological Survey, 1970. 417 pp.

GAZETTEERS

2076 *Decisions on geographic names in the United States.* Washington DC: U.S. Board on Geographic Names, 1980-. Quarterly.

2077 **Sealock, Richard B.; Sealock, Margaret M.; and Powell, Margaret S.** *Bibliography of place-name literature: United States and Canada.* 3rd ed. Chicago IL: American Library Association, 1982. 435 pp.

Lightly annotated listing of 4,830 citations on place-names, gazetteers, origins, and other references to geographical onomastics.

REMOTE SENSING

2078 *Aerial Photography Summary Record System (APSRS).* Washington DC: U.S. Geological Survey. National Cartographic Information Center, 1977. 2 per year.

A series of 34 catalogs describing the extent of aerial photography of the United States done by various United States government agencies. Index maps key dates and scales to 7.5 minute quadrangles as well as to county boundaries. Each catalog is published twice a year. Covers the entire United States.

2079 **Baker, Simon, and Dill, Henry W., Jr.** *The look of our land: an airphoto atlas of the rural United States.* Washington DC: U.S. Dept. of Agriculture, 1970-71. (Agriculture handbook) 5 vols.

Volume one is titled *The East and the South*, 99 pp; volume two *The far West*, 48 pp.; volume three *The mountains and deserts*, 68 pp.; volume four *The plains and prairies*, 84 pp; and volume five *North Central*, 64 pp. Contain lists of selected references.

ENVIRONMENTAL IMPACT STATEMENTS

2080 **Askow, Catherine.** *Primary source materials on environmental impact studies.* Chicago IL: CPL Bibliographies, 1979. (CPL bibliography 10) 21 pp.

Unannotated.

2081 **Clark, Brian Drummond; Bisset, Ronald; and Wathern, Peter.** *Environmental impact assessment: a bibliography with abstracts.* London: Mansell, 1980. 516 pp.

2082 *E.I.S. cumulative: environmental impact statements cumulative.* Arlington VA: Information Resources Press, 1977. Annual.

Available on computer-searchable tape.

2083 *EIS: digests of environmental impact statements.* Washington DC: Information Resources Press, 1977-. Monthly.

2084 **Landy, Marc, ed.** *Environmental impact statement directory: the national network of EIS-related agencies and organizations.* New York: IFI/Plenum, 1981. 367 pp.

2085 **Worsham, John P., Jr.** *Environmental impact statements: a selected bibliography of the report and periodical literature.* Monticello IL: Vance Bibliographies, 1978. (Public administration series bibliography P-17) 10 pp.

Unannotated.

SCIENCE AND TECHNOLOGY

2086 *Directory of information resources in the United States: physical sciences and engineering.* Washington DC: U.S. Library of Congress, 1971. 803 pp.

GEOLOGY

2087 *Analogs of Yuma terrain in the southwest United States desert.* Vicksburg MS: U.S. Army Engineer Waterways Experiment Station. Army Corps of Engineers, 1963. 2 vols.

Covers portions of Oregon, Idaho, Nevada, Utah, California, Arizona, New Mexico, and Texas.

2088 **Andriot, Dona.** *Guide to U.S. Government maps: geologic and hydrologic maps.* Preliminary ed. McLean VA: Documents Index, 1975. 2 vols.

Covers geologic and hydrologic maps published by the U.S. Geological Survey through December 1974. Limited to the United States except for some entries on Arabian Peninsula, Saudi Arabia, Libya, Puerto Rico, the Moon. Supplemented by updates through June 1979.

2089 *Comprehensive index of the publications of the American Association of Petroleum Geologists.*

See entry 2789.

2090 **Corbin, John Boyd, comp.** *An index of state geological survey publications issued in series.* New York: Scarecrow, 1965. 667 pp.

Covers monographs but omits maps. Companion volume to Jane Clapp's *Museum publications* (1962), 2 volumes.

2091 *Directory of geoscience department: United States and Canada.* Washington DC: American Geological Institute, 1952-. Annual.

Covers faculties and degrees issued, field courses, camps, etc. of U.S. and Canadian institutions.

2092 **Eister, Margaret F.** *Selected bibliography and index of earth science reports and maps relating to land-resource planning and management published by the U.S. Geological Survey through October 1976.* Reston VA: U.S. Geological Survey, 1978. (Geological Survey bulletin 1442) 76 pp.

2093 *Geologic atlas of the Rocky Mountain region, United States of America.* Denver CO: Rocky Mountain Association of Geologists, 1972. 331 pp.

Covers Montana, Wyoming, Idaho, Utah, Colorado, Arizona, New Mexico, and portions of North Dakota, South Dakota, Nebraska, Kansas, Oklahoma, Texas, and Nevada. (Omits Oregon, Washington, California, and much of Nevada of the western United States.)

2094 *A guide to obtaining information from the USGS.* Rev. ed. Washington DC: U.S. Geological Survey, 1979. 42 pp.

2095 **Pampe, William R., comp.** *Maps and geological publications of the United States: a layman's guide.* Falls Church VA: American Geological Institute, 1978. 57 pp.

2096 *Publications of the Geological Survey.*

See entry 2791.

The United States/2097-2112

2097 *State geologic maps.* Reston VA: U.S. Geological Survey, 1975. 20 pp.

LAND RECLAMATION

2098 **Colling, Gene, comp.** *Bibliography of SEAM (surface environment and mining) publications.* Ogden UT: U.S. Forest Service. Intermountain Forest and Range Experiment Station, 1980. (USDA Forest Service general technical report INT-88; NTIS: PB 81-184-625) 25 pp.

Covers period 1974-1979.

2099 **Evans, A. Kent; Uhleman, E. W.; and Eby, P. A.** *Atlas of Western surface-mined lands: coal, uranium and phosphate.* Ft. Collins CO: U.S. Fish and Wildlife Service, 1978. (FWS/OBS-78/20) 373 pp.

Contains information on all coal, uranium and phosphate surface operations in excess of 10 acres which were in operation prior to 1976. Covers Arizona, New Mexico, Colorado, Montana, North Dakota, South Dakota, Wyoming, Idaho, Utah, California, Nevada, Oregon, Washington.

2100 **Gleason, V. E.** *Bibliography on mined-land reclamation.* Cincinnati OH: U.S. Environmental Protection Agency, 1979. (Coal and the environment abstract series) 373 pp.

Annotated. Covers mostly United States and Canada.

2101 **Ralston, Sally et al.** *The ecological effects of coal strip-mining: a bibliography with abstracts.* Fort Collins CO[?]: U.S. Fish and Wildlife Service. Office of Biological Studies. Western Energy and Land Use Team, 1977. 617 pp.

Covers the Western United States with emphasis on the Northern Great Plains.

CLIMATOLOGY

2102 **Baldwin, John L.** *Weather atlas of the United States.* Washington DC: U.S. Dept of Commerce. Environmental Science Services Administration. Environmental Data Service, 1968; reprint ed., Detroit MI: Gale Research, 1975. 262 pp.

Originally titled *Climatic atlas of the United States.*

2103 **Bradley, Raymond S.** *Precipitation history of the Rocky Mountain states.* Boulder CO: Westview, 1976. 334 pp.

Bibliography, pp. 323-334. Covers Montana, Idaho, Wyoming, Utah, Nevada, Colorado, northern New Mexico and northern Arizona.

2104 *Climates of the states.* 2nd ed. Detroit MI: Gale Research, 1980. 2 vols.

A variety of calculated values of United States reporting stations, organized by state. Data for each state is prefaced by a brief summary of that state's climate. Prepared by National Oceanic and Atmospheric Administration.

2105 **Court, Arnold.** "The climate of the conterminous United States." In *Climates of North America*, pp. 183-343. Edited by Reid A. Bryson and F. Kenneth Hare. Amsterdam, The Netherlands: Elsevier Scientific, 1974. (World survey of climatology 11).

A narrative summary with maps, charts and an unannotated bibliography.

2106 *Directory of federal drought assistance 1977.* Washington DC: U.S. Dept. of Agriculture. Western Region Drought Action Task Force, 1977. 67 pp.

2107 **Hershfield, David M.** *Rainfall frequency atlas of the United States, for durations from 30 minutes to 24 hours and return periods from 1 to 100 years.* Washington DC: U.S. Soil Conservation Service. Engineering Division, 1961. (U.S. Weather Bureau. Technical paper 40) 61 pp.

2108 **London, Elizabeth B. H.** *Bibliography of precipitation and runoff in the arid region of North America.* Tempe AZ: Office of the State Climatologist. Arizona State University, 1978. (Climatological publications. Bibliography series 5) 77 pp.

Covers the western United States.

2109 **Miller, John F.** *Precipitation-frequency atlas of the Western United States.* Silver Spring MD: U.S. Weather Service, 1973. 11 vols.

Covers Montana, Wyoming, Colorado, New Mexico, Idaho, Utah, Nevada, Arizona, Washington, Oregon, California.

2110 *Weather data handbook: for HVAC and cooling equipment design.* New York: McGraw-Hill, 1980. 285 pp.

Prepared by Ecodyne Corporation.

HYDROLOGY

2111 **Berry, George W.** *Thermal springs list for the United States.* Boulder CO: U.S. Dept. of Commerce. National Geophysical and Solar-Terrestrial Data Center, 1980. 59 pp.

Nearly all the hot springs in the United States occur in the Western—and largely arid—third of the nation. The springs themselves form important micro-environmental niches; they are of interest both for their biology and their indication of underlying geothermal resources. This title presents a list of thermal springs, their temperatures, and their exact locations keyed to latitude/longitude and topographic quadrangles. Index maps at 1:5,000,000 summarize locations for the 48 states and Alaska. This work also serves as an index to a set of large-scale overlay maps.

2112 **Burford, James B., and Clark, John M.** *Hydrologic data for experimental agricultural watersheds in the United States 1968.* Washington DC: U.S. Dept. of Agriculture. Agricultural Research Service, 1976. 542 pp.

2113 Carrigan, Philip Hadley. *Index of flood maps prepared by the U.S. Geological Survey.* Reston VA: U.S. Geological Survey, 1974. (Water resources investigations 57-73) 332 pp.

Available on microfiche.

2114 *Federal reclamation projects, water and land resource accomplishments.* Washington DC: U.S. Dept. of Interior. Bureau of Reclamation, 1949-. Annual.

Tables of frequently-sought financial and physical data.

2115 Geraghty, James J. et al. *Water atlas of the United States.* Port Washington NY: Water Information Center, 1973. 122 plates.

Includes maps on precipitation distribution, potential evapotranspiration, ground water use, withdrawal of water for various purposes, etc.

2116 Giefer, Gerald J. *Water publications of state agencies: a bibliography of publications on water resources and their management published by the states of the United States.* Port Washington NY: Water Information Center, 1972. 319 pp.

2117 Giefer, Gerald J. *Water publications of state agencies: a bibliography of publications on water resources and their management: first supplement, 1971-1974.* Huntington NY: Water Information Center, 1976. 189 pp.

2118 *Guide to water quality standards of the United States.* Washington DC: American Petroleum Institute, 1980. (API publication 4321) looseleaf.

2119 Randolph, J. R.; Baker, N. M; and Deike, R. G. *Bibliography of hydrology of the U.S. and Canada.* Washington DC: U.S. Geological Survey, 1969. (Water-supply paper 1863; 1864) 232 pp.

2120 *River basins of the United States; drainage areas of more than 700 square miles, land resources regions, and major land resources areas.* 2nd ed. Washington DC: U.S. Soil Conservation Service, 1970. 82 leaves.

2121 Vorhis, Robert Corson. *Bibliography of publications relating to ground water prepared by the Geological Survey and cooperating agencies, 1946-1955.* Washington DC: U.S. Geological Survey, 1957. (Water supply paper 1492) 203 pp.

2122 Waring, Gerald Ashley. *Bibliography and index of publications relating to ground water prepared by the Geological Survey and cooperating agencies.* Washington DC: U.S. Geological Survey, 1947. (Water supply paper 992) 412 pp.

Covers the years 1879-1946.

2123 *Water quality management directory.* 4th ed. Washington DC: Environmental Protection Agency. Water Planning Division, 1980. 176 pp.

Lists state and local agencies. Includes Guam, Puerto Rico, Trust Territories of the Pacific and Virgin Islands.

2124 *Water resources coordination directory.* Washington DC: United States Water Resources Council, 1980. Unpaginated; about 250 pp.

Includes federal, state and interagency commissions. Covers the 50 states and Guam, Puerto Rico and the Virgin Islands.

WATER LAW

2125 Chalmers, John R. *Southwestern groundwater law: a textual and bibliographic interpretation.* Tucson AZ: University of Arizona. Office of Arid Lands Studies, 1974. (NTIS: PB 228-130) 229 pp.

Contains 180 annotated citations for the states of Arizona, California, Colorado, Nevada, New Mexico, Texas, and Utah.

2126 Holmes, Beatrice H.; Simons, George G.; and Ellis, Harold H., comp. *State water-rights law and related subjects: a supplemental bibliography 1959 to mid-1967.* Washington DC: U.S. Dept of Agriculture, 1973. 268 pp.

2127 Jacobstein, J. Myron, and Mersky, Roy M. *Water law bibliography, 1847-1965: source book on U.S. water and irrigation studies: legal, economic and political.* Silver Spring MD: Jefferson Law Book Co., 1966. 249 pp.

Supplement one (1969) covers period 1966-67, 138 pp.; *Supplement* two (1974) covers period 1968-73, 167 pp., and *Supplement* three (1978) covers period 1974-77, 174 pp. Unannotated.

2128 Jamail, Milton H.; McCain, John R.; and Ullery, Scott J. *Federal-state water use relations in the American West: an evolutionary guide to future equilibrium.* Tucson AZ: University of Arizona. Office of Arid Lands Studies, 1978. (Resource information paper 11; NTIS: PB 286-309/OST) 155 pp.

Covers the states of the Colorado River Basin: Arizona, Colorado, Wyoming, California, Nevada, Utah, New Mexico. Includes a 166-item bibliography.

2129 Maloney, Frank Edward, ed. *Interstate water concepts: a bibliography.* Washington DC: U.S. Water Resources Scientific Information Center. Office of Water Research and Technology, 1975. (OWRT/WRSIC 75-205) 488 pp.

2130 Nelson, Michael C., and Booke, Bradley L. *The Winters doctrine: seventy years of application of reserved water rights to Indian reservations.* Tucson AZ: University of Arizona. Office of Arid Lands Studies, 1977. (Resource information paper 9; NTIS: PB 272-299/9ST) 168 pp.

A survey of this important concept in U.S. water law with lists of cases and an 87-item annotated bibliography.

2131 Turney, Jack R., and Ellis, Harold H., comps. *State water-rights laws and related subjects, a bibliography*. Washington DC: U.S. Dept. of Agriculture, 1962. 199 pp.

Covers period to mid-1959.

2132 White, Anthony G. *Arid lands: government impact, water and the public: a selected bibliography*. Monticello IL: Vance Bibliographies, 1980. (Public administration series bibliography P-585) 7 pp.

Unannotated. Covers arid United States.

SOILS

2133 *List of published soil surveys*. Washington DC: U.S. Soil Conservation Service, 1981. 17 pp.

Covers published soil surveys of the United States and Puerto Rico. Entries are arranged by state and give name of survey and year. A list of state conservation offices is provided as sources of the most recent information. The surveys themselves usually consist of texts plus maps at a scale of 1:24,000 or better. As depository items they are usually available in United States government documents collections.

BIOLOGY

2134 *Abstracts: U.S.-International Biological Program ecosystem analysis studies*. Oak Ridge TN: Oak Ridge National Laboratory, 1970-1980. 5 vols.

2135 Burke, Hubert D. *Bibliography of manuals and handbooks from natural resource agencies*. 2nd ed. Ft. Collins CO: U.S. Dept. of the Interior. Fish and Wildlife Service. Office of Biological Services. Western Energy and Land Use Team, 1978. 79 pp.

An annotated bibliography of unpublished manuals and handbooks relating to wildlife covering the western United States (North Dakota to Texas and westward from these). Contains 454 references; subject index. Many of the references are not otherwise retrievable through standard information systems.

2136 Cochran, Anita. *A selected, annotated bibliography on fish and wildlife implications of Missouri basin water allocation*. Boulder CO: University of Colorado. Institute of Behavioral Science, 1975. 71 pp.

Covers the Upper Missouri River system upstream of a point just below the confluence of the Missouri and the Platte; that is, Nebraska, Iowa, South Dakota, Minnesota, North Dakota, Colorado, Wyoming, Montana.

2137 *Directory of information resources in the United States: biological sciences*. Rev. ed. Washington DC: U.S. Library of Congress. National Referral Center, 1972. 577 pp.

Includes government agencies, professional societies, university institutes, individuals, etc.

2138 Eisenman, Eric, et al. *Economic data for wildland planning and management in the western United States: a source guide*. Berkeley CA: U.S. Forest Service. Pacific Southwest Forest and Range Experiment Station, 1980. (General technical report PSW-42) 125 pp.

2139 *Habitat preservation abstracts*. Washington DC: U.S. Fish and Wildlife Service. Office of Biological Sciences, 1979-. Irreg.

Contains abstracts of current literature covering the effects of energy development, water allocation, and contaminants of wildlife. Covers the United States. First two issues cover period 1976-79 in 288 citations.

2140 Hinckley, Dexter A., and Haug, Peter T. *User's guide to IBP biome information from the United States International Biological Program (IBP)*. Indianapolis IN: Institute of Ecology, 1977. 48 pp.

Two of the five generalized biomes are arid: the Grassland (covering the High Plains) and the Desert (covering the drylands from central Washington south to the border with Mexico).

BOTANY

2141 Axelton, Elvera A. *Ponderosa pine bibliography through 1965*. Ogden UT: U.S. Forest Service, 1967. (Research paper INT-40) 150 pp.

Contains 2,275 unannotated citations. Continues entry 2153.

2142 Axelton, Elvera A. *Ponderosa pine bibliography II, 1966-1970*. Ogden UT: U.S. Forest Service, 1974. (General technical report INT-12) 63 pp.

Contains over 732 unannotated citations. Updates entry 2141.

2143 Axelton, Elvera A. *Ponderosa pine bibliography III: 1971 through 1975*. Ogden UT: U.S. Forest Service. Intermountain Forest and Range Experiment Station, 1977. (General technical report INT-33) 54 pp.

Includes 590 references to published material on *Pinus ponderosa*, (Laws.) issued from 1971-1975. Updates entry 2142.

2144 Ayensu, Edward S. *Endangered and threatened plants of the United States*. Washington DC: Smithsonian Institution, 1978. 403 pp.

Covers all 50 states, Puerto Rico, and the Virgin Islands. Includes extensive unannotated bibliographies and maps. Based on *Report on endangered and threatened plant species of the United States* (1975), 200 pp.

2145 Benson, Lyman David. *Trees and shrubs of the southwestern deserts*. 3rd ed. Tucson AZ: University of Arizona Press, 1981. 416 pp.

Covers the deserts of California from Death Valley south, of the southern tip of Nevada, of the Virgin River Valley of Utah, of the land west and south of the Mogollon Rim of Arizona, of the Rio Grande Valley of New Mexico south of Albuquerque, and of El Paso County in Texas. Distribution maps end sharply at the border with Mexico.

2146 Burgess, R. L. "Utilization of desert plants by native peoples: an overview of southwestern North America." In *Native plants and animals as resources, in arid lands of the southwestern U.S.*, pp. 6-21. Edited by J. L. Gardner. Flagstaff AZ: Arizona State College, 1965. (American Association for the Advancement of Science. Committee on Desert and Arid Zones Research. Contribution 8).

Includes bibliography.

2147 "Creosote bush." *Arid lands abstracts* 3 (1972): items 137-211.

The creosote bush (*Larrea tridentata*) is a native of the Mojave, Chihuahuan, and Sonoran deserts of North America. This bibliography gives 74 annotated references from 1905-1972.

2148 **Ffolliott, Peter Frederick, and Clary, W. P.** *A selected and annotated bibliography of understory-overstory vegetation relationships.* Tucson AZ: Arizona Agricultural Experiment Station, 1972. (Technical bulletin 198) 33 pp.

Covers continental United States with slight emphasis on Arizona. Includes 262 citations through 1971.

2149 **Harniss, Roy O.; Harvey, Stephen J.; and Murray, Robert B., comps.** *A computerized bibliography of selected sagebrush species (genus Artemisia) in western North America.* Ogden UT: U.S. Forest Service. Intermountain Forest and Range Experiment Station, 1981. (General technical report INT-102; NTIS: PB 81-177-818) 111 pp.

2150 **Little, Elbert L., Jr.** *Atlas of United States trees: Volume 3. Minor western hardwoods.* Washington DC: U.S. Forest Service, 1976. (Miscellaneous publication 1314) 13 pp., 290 maps.

Contains 290 colored maps of 210 tree species.

2151 **Little, Elbert L., Jr., and Honkala, Barbara H.** *Trees and shrubs of the United States: a bibliography for identification.* Washington DC: U.S. Forest Service, 1977 (Miscellaneous publication 1336) 56 pp.

Bibliography of more than 470 titles lists general references as well as those of specific geographic regions, all 50 states, Puerto Rico, Virgin Islands and Guam. Covers 1950-1975, as well as older publications.

2152 *Pinyon-juniper project, 5/24/72, bibliography listing.* Merced CA: Merced College. Library, 1972. 36 pp.

Contains some 430 citations.

2153 **Roe, Arthur L., and Boe, Kenneth N.** *Ponderosa pine bibliography.* Ogden UT(?): U.S. Forest Service. Northern Rocky Mountain Forest and Range Experiment Station, 1950. (Station paper 22) 74 pp.

Original bibliography continued by entries 2141, 2142, and 2143.

2154 **Whiting, A. F.** "The present status of ethnobotany in the Southwest." *Economic botany.* 20:3 (1966):316-325.

Contains 123 citations.

ZOOLOGY

2155 **Berger, Thomas J.** *Directory of federally controlled species.* Lawrence KS: Association of Systematics Collections, 1979. Looseleaf.

Covers animal species of the United States.

2156 **Field, William Dewitt; dos Passos, Cyril F.; and Masters, John H.** *A bibliography of the catalogs, lists, faunal and other papers on the butterflies of North America north of Mexico arranged by state and province (Lepidoptera: Rhopalocera).* Washington DC: Smithsonian Institution Press, 1974. 104 pp.

2157 **Morgan, John, ed.** *Mammalian status manual: a state by state survey of the endangered and threatened mammals of the United States.* N. Eastham MA: Linton, 1980. Looseleaf.

Gives species distribution by state. Includes list according to order and family. Brief bibliography.

AGRICULTURE

2158 *Agricultural statistics.* Washington DC: U.S. Dept. of Agriculture, 1936-. Annual.

Most data from the U.S. Dept. of Agriculture. Includes foreign agricultural trade statistics. Historical series limited to data beginning with 1965.

2159 **Olsen, Wallace C.** *Directory of information resources in agriculture and biology.* Beltsville MD: U.S. Dept. of Agriculture, 1971. 523 pp.

Directory of institutions and agencies in the United States giving addresses, personnel, areas of expertise, library holdings, etc.

2160 **Rogers, Earl M.** *A list of references for the history of agriculture in the Great Plains.* Davis CA: University of California. Agricultural History Center, 1976. 90 pp.

Covers North Dakota, South Dakota, Nebraska, Kansas, and Oklahoma.

2161 **Rogers, Earl M.** *A list of references for the history of agriculture in the Mountain States.* Davis CA: University of California. Agricultural History Center, 1972. 91 pp.

Covers Arizona, Colorado, Idaho, Montana, Nevada, New Mexico, Utah, and Wyoming.

2162 *Services available through the U.S. Department of Agriculture.* Rev. ed. Washington DC: U.S. Dept. of Agriculture, 1980. 39 pp.

A directory of services available to other government agencies, businesses, and the general public. Descriptions of many of the programs are extremely brief; this should be used as a first source only.

2163 **Wimberly, Ronald C., comp.** *Structure of U.S. agriculture bibliography.* Beltsville MD: U.S. Dept. of Agriculture. Science and Education Administration. Technical information System, 1981. (Bibliographies and literature of agriculture 16) 514 pp.

Unannotated. Covers period 1970-1979.

PLANT CULTURE

2164 **McKiernan, Gerard.** *Desert gardening: desert plants and their cultivation, an annotated bibliography.* Lisle IL: Morton Arboretum. Council on Botanical and Horticultural Libraries, 1978. (Plant bibliography 1) 28 pp.

Intended for the novice desert gardener. All works in English; most deal with the Southwest U.S. No index.

FORESTRY

2165 **Evans, Peter A.** *Directory of selected forestry-related bibliographic data bases.* Berkeley CA: U.S. Forest Service. Pacific Southwest Forest and Range Experiment Station, 1979. (General technical report PSW-34) 42 pp.

"Lists 117 bibliographic data bases maintained by scientists of the Forest Service."

2166 **Ogden, Gerald R., comp.** *The United States Forest Service: a historical bibliography, 1876-1972.* Davis CA: University of California. Agricultural History Center; U.S. Dept. of Agriculture. Agricultural History Group; U.S. Forest Service, 1976. 439 pp.

IRRIGATION

2167 **Sloggett, Gordon et al.** *An annotated bibliography of publications related to Great Plains irrigation.* Stillwater OK: U.S. Dept. of Agriculture. Economic Research Service, 1976. (Oklahoma State University. Dept. of Agricultural Economics 7601, supplement 3) 90 leaves.

2168 *1969 supplement to an annotated bibliography of publications related to Great Plains irrigation.* Lincoln NE: Nebraska Agricultural Experiment Station, 1970. (University of Nebraska. College of Agriculture and Home Economics. Report 55) 16 pp.

2169 **White, Anthony G.** *Major irrigation projects; with an emphasis on public administration impacts: a selected bibliography.* Monticello IL: Vance Bibliographies, 1980. (Public administration series bibliography P-604) 18 pp.

Unannotated. Covers primarily the southwestern United States. Many general works listed.

ANIMAL CULTURE

2170 **Wydoski, Richard S.** *Annotated bibliography for aquatic resource management of the upper Colorado River ecosystem.* Washington DC: U.S. Fish and Wildlife Service, 1980. (Resource publication 135) 186 pp.

RANGE MANAGEMENT

2171 *Economic research in the use and development of range resources: range and range economics bibliography.* Las Cruces NM: Western Agricultural Economics Research Council. Committee on Economics of Range Use and Development, 1969. (Report 11) 199 pp.

Unannotated. Covers western United States.

2172 *National range handbook: rangeland, grazable woodland, native pasture.* Washington DC: U.S. Dept. of Agriculture. Soil Conservation Service, 1976. 143 pp.

2173 **Vallentine, John F.** *U.S.-Canadian range management, 1935-1977: a selected bibliography on ranges, pastures, wildlife, livestock, and ranching.* Phoenix AZ: The Oryx Press, 1978. 337 pp.

Unannotated. English language material. Covers the United States, Canada, and Mexico. Updates Frederic G. Renner's *A selected bibliography on management of Western ranges, livestock, and wildlife* (Washington DC: U.S. Government Printing Office, 1938), U.S. Dept. of Agriculture miscellaneous publication 281, 468 pp. *Supplement* (1981) covers period 1978-80, 166 pp. Also available online (vendor: SDC).

ENERGY

2174 **Cohen, Sanford F.** *The impact of western energy development on water resources: a selected bibliography.* Monticello IL: Vance Bibliographies, 1981. (Public administration series bibliography P-644) 5 pp.

Unannotated.

2175 **Cohen, Sanford F.** *National energy projections and implications for western energy development: a selected bibliography.* Monticello IL: Vance Bibliographies, 1981. (Public administration series bibliography P-645) 5 pp.

Unannotated.

2176 **Cohen, Sanford F.** *Socioeconomic impacts of western energy development: a selected bibliography of an emerging national concern.* Monticello IL: Vance Bibliographies, 1981. (Public administration series bibliography P-643) 5 pp.

Unannotated.

2177 **Cohen, Sanford F.** *Western energy and land use: a selected bibliography.* Monticello IL: Vance Bibliographies, 1981. (Public administration series bibliography P-642) 6 pp.

Unannotated.

2178 **Cortese, Charles F., and Cortese, Jane Archer.** *The social effects of energy boomtowns in the West: a partially annotated bibliography.* Monticello IL: Council of Planning Librarians, 1978. (CPL exchange bibliography 1557) 30 pp.

Annotated.

2179 Cuff, David J., and Young, William J. *The United States energy atlas*. New York: The Free Press, 1980. 416 pp.

Covers non-renewable and renewable energy sources in maps, narrative text, and statistical charts. Brief unannotated bibliography.

2180 *Energy: a guide to organizations and information resources in the United States.* 2nd ed. Claremont CA: Public Affairs Clearinghouse, 1978. (Who's doing what series 1) 221 pp.

Prepared by the Center for California Public Affairs.

2181 *The energy directory.* New York: Environment Information Center, 1974-. Quarterly.

A directory of energy-related organizations, government agencies, and private companies for the United States. Updated quarterly in print and immediately online (as ENERGY NET; available on DIALOG). The final issue is cumulative for the year.

2182 Hamilton, Michael S., and Townsend, Ruth J. *Power plant siting in the American Southwest: an annotated bibliography.* Chicago IL: CPL Bibliographies, 1981. (CPL bibliography 52) 58 pp.

Annotated. Covers Arizona, California, Colorado, Nevada, New Mexico, and Utah.

2183 Hatcher, Suzanne, ed. *Energy executive directory.* Washington DC: Carroll Publishing, 1980. 255 pp.

A directory of state and federal government agencies and individuals involved in energy matters in the United States, giving addresses, titles, and responsibilities.

2184 *Western energy and land use team publications: an annotated bibliography.* Ft. Collins CO: U.S. Dept. of Interior, 1979. (WELUT-79/10) 35 pp.

SOLAR ENERGY

2185 *Solar census: the directory for the 80s.* Ann Arbor MI: Aafec Publications, 1980. 484 pp.

2186 Vidich, Charles. *Solar access and land use planning: a bibliography.* Chicago IL: CPL Bibliographies, 1981. (CPL bibliography 47) 8 pp.

Unannotated.

HYDROPOWER

2187 *Hydropower sites of the United States, developed and undeveloped.* Washington DC: U.S. Federal Energy Regulatory Commission, 1981. 178 pp.

DEMOGRAPHY

2188 Andriot, John L., ed. *Population abstract of the United States.* McLean VA: Andriot Associates, 1980. 925 pp.

A companion to the *Township atlas of the United States* (1979). Provides a summary of all United States decennial population census totals for the nation's counties and cities through the preliminary census of 1976.

2189 *Census geography.* Washington DC: U.S. Bureau of the Census, 1979. (Data access descriptions, DAD series) 33 pp.

Contains information about how the census is taken.

2190 Klosterman, Richard E. *Demographic and economic data sources: an annotated bibliography and source guide.* Chicago IL: CPL Bibliographies, 1980. (CPL bibliography 32) 46 pp.

Annotated. Covers the United States with emphasis on the sub-state level.

2191 *Profile of census programs: source document for water resource planners.* Washington DC: U.S. Bureau of the Census, 1978. 88 pp.

MIGRATION

2192 Weiss, Joseph E. *A bibliography on migration, with special emphasis on Sunbelt migration.* Monticello IL: Vance Bibliographies, 1979. (Public administration series bibliography P-371) 23 pp.

Annotated.

ANTHROPOLOGY

2193 Fowler, Catherine S., comp. *Great Basin anthropology...a bibliography.* Reno NV: University of Nevada System. Desert Research Institute, 1970. (Social sciences and humanities publication 5) 418 pp.

Unannotated. Covers Native Americans of the Great Basin.

2194 Murdock, George Peter, and O'Leary, Timothy J. *Ethnographic bibliography of North America.* 4th ed. New Haven CT: Human Relations Area Files, 1975. (Behavior science bibliographies) 5 vols.

Massive unannotated bibliography covering the literature through 1972 for Canada, the United States, and northern Mexico.

2195 Stewart, Omer C. *Indians of the Great Basin: a critical bibliography.* Bloomington IN: Indiana University Press, 1982. 138 pp.

A narrative bibliography of 364 citations. Covers Nevada, eastern California, northwest Utah, southern Idaho, eastern Oregon and Washington.

2196 Sturtevant, William C., ed. *Handbook of North American Indians.* Washington DC: Smithsonian Institution, 1978-. To be completed in 20 vols.

As of Fall 1982, three volumes of the projected 20 have appeared; these cover the Northeast (volume 15), California (volume 8), and the Colorado Plateau of the Southwest (volume 9). No short summary can do justice to the breadth and depth of coverage contained in these volumes. Suffice it to say that careful, systematic scholarship is wedded to graceful presentation and design. Regional volumes are listed separately in their appropriate places in this Guide; when complete, the set will cover Canada, the United States, and northern Mexico.

URBAN GEOGRAPHY

2197 *Index to current urban documents.* Westport CT: Greenwood, 1972-. Quarterly.

Contains bibliographic descriptions of the majority of known local government documents, issued annually by the largest cities and counties in the United States. Provides access to the Urban Documents Microfiche Collection.

2198 **Robinson, G. D., and Spieker, Andrew M.,** eds. *"Nature to be commanded...": earth-science maps applied to land and water management.* Washington DC: U.S. Geological Survey, 1978. (Professional paper 950) 95 pp.

Designed for persons involved in urban planning, design, management and development. Deals with unique problems of differing environments—controlling hillside development, gravel deposits, etc.

2199 *Urban atlas: tract data for standard metropolitan statistical areas.* Washington DC: U.S. Bureau of the Census, 1974-. ([United States Maps] GE-80).

Includes SMSA's in the western United States. Based on data from 65 largest metropolitan areas, atlases show selected data characteristics from 1970 Census of Population and Housing.

ARCHITECTURE

2200 **Miller, James D.** *Design and the desert environment: landscape architecture and the American Southwest.* Tucson AZ: University of Arizona. Office of Arid Lands Studies, 1978. (Resource information paper 13; NTIS: PB 289-955/7GA) 216 pp.

Reviews methods of solar radiation control, wind control, and water conservation.

HOUSING

2201 **Eribes, Richard A.** *Housing problems of our Spanish heritage populations.* Monticello IL: Council of Planning Librarians, 1977. (CPL exchange bibliography 1274) 9 pp.

Unannotated. Coverage focuses on Mexican-American populations in the southwestern United States.

ECONOMIC DEVELOPMENT

2202 **Foster, Stephanie.** *Economic development in the U.S.-Mexico border region: a review of the literature.* Monticello IL: Vance Bibliographies, 1981. (Public administration series bibliography P-692) 17 pp.

Extensive annotations.

2203 *Quarterly economic review of U.S.A.* London: Economist Intelligence Unit, 1976-. Quarterly.

Includes summary of political and economic news as well as charts and statistics of selected economic indicators. Annual supplement.

2204 **Snodgrass, Marjorie P.** *Economic development of American Indians and Eskimos, 1930 through 1967: a bibliography.* Washington DC: U.S. Bureau of Indian Affairs, 1968. (Bibliography series 10) 263 pp.

TOURISM AND RECREATION

2205 *A bibliography of USTS research publications.* Washington DC: U.S. Dept. of Commerce, 1979. 12 pp.

Includes United States Travel Service basic data and market research data.

2206 **Slankey, George, and Lime, David W.,** comp. *Recreational carrying capacity: an annotated bibliography.* Ogden UT: U.S. Forest Service. Intermountain Forest and Range Experiment Station, 1973. 45 pp.

References cover the impact of recreation on government lands in the United States.

HISTORICAL GEOGRAPHY

2207 *America: history and life.* Santa Barbara CA: Clio Press, 1964-. Annual.

Covers the United States, Canada, Puerto Rico, Virgin Islands. A retrospective volume 0 covers the period 1953-63.

2208 **Billington, Ray Allen, and Ridge, Martin.** "Bibliography." In *Westward expansion: a history of the American frontier*, pp. 699-858. 5th ed. New York: Macmillan, 1982.

A fine narrative bibliography covering the major monographs and journal articles on the historical geography of the West.

2209 **Lee, Lawrence B.** *Reclaiming the American West: an historiography and guide.* Santa Barbara CA: ABC-Clio, 1980. 131 pp.

2210 **Schmeckebier, Laurence Frederick.** *Catalogue and index of the publications of the Hayden, King, Powell, and Wheeler Surveys.* Washington DC: U.S. Government Printing Office, 1904; reprint ed., Portland OR: Northwest Books, 1970. (Bulletin 222; Series G, Miscellaneous 26) 208 pp.

2211 **Smith, Dwight L.** *The American and Canadian West: a bibliography.* Santa Barbara CA: ABC-Clio Press, 1979. (Clio bibliography series 6) 558 pp.

Annotated. 4,157 citations. Arranged by topics with subject and author index.

PUBLIC ADMINISTRATION

2212 **Coons, William.** *The Sagebrush Rebellion: legitimate assertion of states' rights or retrograde land grab? a selected subject bibliography and resource guide.* Monticello IL: Vance Bibliographies, 1981. (Public administration series bibliography P-667) 18 pp.

Unannotated. The Sagebrush Rebellion consists of the actions of various Western states to gain control of the federal land within their jurisdictions.

Arizona

GENERAL BIBLIOGRAPHIES

2213 **Powell, Donald M.** *The Arizona index: a subject index to periodical articles about the state.* Boston MA: G. K. Hall, 1978. 2 vols.

STATISTICS

2214 **de Gennaro, Nat.** *Arizona statistical abstract: a data handbook.* 2nd ed. [Flagstaff AZ]: Northland Press, 1979. 640 pp.

GEOGRAPHY

2215 *BLM Facts: Arizona national resource lands digest.* Phoenix AZ: U.S. Bureau of Land Management. Arizona State Office, 1976. 113 pp.

A statistical compendium of data concerning the 12.5 million acres of Arizona land administered by the Bureau of Land Management (nearly one-sixth of the State's total area). Latest figures are for 1975.

2216 **Brooks, Wahner F.** *Desert testing environmental bibliography.*

See entry 62.

2217 **de Kok, David A.; Worden, Marshall A.; and Gibson, Lay James.** *Bibliography of references and data sources on the Arizona lands bordering the lower Colorado River.* Tucson AZ: University of Arizona. Dept. of Geography, Regional Development, and Urban Planning, 1978. 120 pp.

Some annotations.

2218 **LaVoie, Joseph R., and McGarvin, Thomas G.** *Index to road logs and river logs in Arizona 1950-1980.* Tucson AZ: Arizona. Bureau of Geology and Mineral Technology. Geological Survey Branch, 1981. (Circular 22) 14 pp.

Unannotated list with index maps.

2219 **McGinnies, W. G., comp.** *Publications related to the work of the Desert Botanical Laboratory of the Carnegie Institution of Washington, 1903-1940.* Tucson AZ: University of Arizona. Office of Arid Lands Studies, 1968. 50 pp.

Contains 530 citations.

2220 **Paylore, Patricia, comp.** *Seventy-five years of arid-lands research at The University of Arizona: a selective bibliography, 1891-1965.* Tucson AZ: University of Arizona. Office of Arid Lands Studies, 1966. 95 pp.

Contains 1,609 citations.

2221 **Ruder, Ruth L.** *Selected list of references on the Grand Canyon area.* Flagstaff AZ: Museum of Northern Arizona Library, 1970. 42 leaves.

2222 **Smith, Ernest Linwood.** *Established natural areas in Arizona: a guidebook for scientists and educators.* Phoenix AZ: Arizona Academy of Science, 1974. 300 pp.

2223 "The Sonoran Desert: a retrospective bibliography." *Arid Lands abstracts* 8 (1976). 800 pp.

Contains 777 annotated citations on the desert's physical environment.

2224 **Spamer, Earle E. et al.** *Bibliography of the Grand Canyon and the Lower Colorado River 1540-1980.* Grand Canyon AZ: Grand Canyon Natural History Association, 1981. 119 pp.

Unannotated. Covers earth sciences, biology, archeology, anthropology, history, literature.

2225 **Yates, Richard, and Marshall, Mary.** *The Lower Colorado River: a bibliography.* Yuma AZ: Arizona Western College Press, 1974. 153 pp.

Unannotated. Covers Indians, early exploration, social history, navigation, the Delta, mining, and the politics of water rights and reclamation.

MAP COLLECTIONS

2226 **Coumans, Cheryl Louise.** *A directory of map and aerial photo resources in Arizona.* Tucson AZ: University of Arizona Library. Map Collection, 1978. 47 pp.

Includes agencies, university, museum, and public libraries.

ATLASES

2227 **Hecht, Melvin E.** *The Arizona atlas.* Tucson AZ: University of Arizona. Office of Arid Lands Studies, 1981. 164 pp.

GAZETTEERS

2228 **Henrickson, James, and Straw, Richard M.** *A gazetteer of the Chihuahuan Desert region.*

See entry 2371.

GEOLOGY

2229 **Anthony, John W.** *Mineralogy of Arizona.* Tucson AZ: University of Arizona Press, 1977. 254 pp.

Contains unannotated bibliography.

2230 "Index of geologic maps of Arizona." Arizona. Bureau of Mines. *Fieldnotes* 7:1 (1977):6-18.

Brief annotations. Contains 402 items. Covers period through 1974. Continued by entry 2233.

2231 Jett, John H., comp. *Directory of earth sciences clubs in Arizona*. 2nd ed. Phoenix AZ: Arizona. Dept. of Mineral Resources, 1980. (Annual directory 10) 30 pp.

2232 Moore, Richard Thomas, and Wilson, E. D. *Bibliography of the geology and mineral resources of Arizona, 1848-1964*. Tucson AZ: Arizona. Bureau of Mines, 1965. (Bulletin 173) 321 pp.

Coverage includes published and unpublished materials. Continued by entry 2234.

2233 "New geologic maps of Arizona, 1975-1978." Arizona. Bureau of Geology and Mineral Technology. *Fieldnotes* 8:3 (1978):4-12.

2234 Vuich, J. S., and Wilt, J. C. *Bibliography of the geology and mineral resources of Arizona 1965-1970*. Tucson AZ: Arizona. Bureau of Mines, 1974. (Bulletin 190) 155 pp.

Continues entry 2232.

2235 Wright, Ann Finley. *Bibliography of geology and hydrology, San Juan Basin, New Mexico, Colorado, Arizona, and Utah*.

See entry 2378.

CLIMATOLOGY

2236 Hasemeier, Amanda N. *Drought: a selected bibliography*. Tempe AZ: Arizona. Office of the State Climatologist, 1977. (Bibliography series 3) 40 pp.

Covers the southwestern United States with an emphasis on Arizona.

2237 Sellers, William D., and Hill, Richard H. *Arizona climate: 1931-1972*. Tucson AZ: University of Arizona Press, 1974. 616 pp.

A compilation of average monthly and total precipitation for each station. A useful station history precedes the data for that station. Tables prepared from data provided by the National Climatic Center.

HYDROLOGY

2238 Babcock, H. M. *Bibliography of U.S. Geological Survey water resources reports for Arizona, May 1965 through June 1971*. Phoenix AZ: Arizona. Water Commission, 1972. (Bulletin 2) 60 pp.

Continues entry 2239.

2239 *Bibliography of U.S. Geological Survey water resources reports, Arizona, 1891 to 1965*. Phoenix AZ: Arizona. State Land Dept., 1965. (Water resources report 22) 59 pp.

Continued by entry 2238.

2240 *Bibliography of water resources of Arizona*. Tucson AZ: U.S. Geological Survey. Arizona District. Ground Water Branch, 1962. 42 leaves.

2241 Cooper, E. Nathan et al. *Water resources research in the Lower Colorado River Basin, 1972-1976*. Las Vegas NV: University of Nevada. Desert Research Institute, 1976. (NTIS: PB 263-487) 186 pp.

Prepared for the Board of Reclamation and the Office of Water Research and Technology, U.S. Dept. of the Interior, by the University of Nevada, the University of California at Los Angeles, and the University of Arizona. Contains 283 citations.

WATER LAW

2242 Jamail, Milton H., and Ullery, Scott J. *International water use relations along the Sonoran Desert borderlands*. Tucson AZ: University of Arizona. Office of Arid Lands Studies, 1979. (Resource information paper 14; NTIS: PB 80-134-976) 139 pp.

Contains 355 references.

BIOLOGY

2243 Patton, David R. *RUN WILD II: data files for wildlife species and habitat for Arizona and New Mexico*. Washington DC: U.S. Forest Service, 1979. (NTIS: PB 296-984) 1,582 pp.

Describes a storage and retrieval system available to biologists in the Southwest.

2244 Patton, David R., and Ffolliott, Peter F. *Selected bibliography of wildlife and habitats for the Southwest*. Ft. Collins CO: U.S. Forest Service, 1975. (General technical report RM-16) 39 pp.

Unannotated. Covers Arizona and New Mexico for the years 1913 to early 1975.

BOTANY

2245 Benson, Lyman David. *The cacti of Arizona*. 3rd ed. Tucson AZ: University of Arizona Press, 1969. 218 pp.

2246 Bowers, J. E. "Local floras of Arizona: an annotated bibliography." *Madrono* 28:4 (1981):193-209.

2247 Elias-Cesnik, Anna. *Jojoba: guide to the literature*. Tucson AZ: University of Arizona. Office of Arid Lands Studies, 1982. 221 pp.

Partly annotated. Includes some 685 citations integrating material from two previous bibliographies (entries 2251 and 2250) with some 300 new items.

2248 Fontana, Bernard L. "Ethnobotany of the saguaro: an annotated bibliography." *Desert plants* 2:1 (1980):62-78.

Includes popular magazine articles, scholarly material, and government documents. Primarily English language material.

2249 Schmutz, Ervin M. *Classified bibliography on native plants of Arizona.* Tucson AZ: University of Arizona Press, 1978. 160 pp.

Unannotated. Citations cover period from 1838-1976. Includes publications on individual species, ecological factors, poisonous plants. Does not include taxonomic, herbicidal, or physiological publications which do not contain data on ecology, distribution, or composition. Omits introduced species.

2250 Sherbrooke, Wade C. *Jojoba: an annotated bibliography.* Tucson AZ: University of Arizona. Office of Arid Lands Studies, 1978. (Resource information paper 5, supplement) 80 pp.

Annotated bibliography. Reviews various topics in biology and natural history, chemistry and economic utilization of jojoba wax; literature on the production of jojoba seed.

2251 Sherbrooke, Wade C., and Haase, E. F. *Jojoba: a wax-producing shrub of the Sonoran Desert: literature review and annotated bibliography.* Tucson AZ: University of Arizona. Office of Arid Lands Studies, 1974. (Resource information paper 5) 141 pp.

Jojoba is native to southern California, southern Arizona, the Baja California peninsula, and Sonora, Mexico.

2252 Shreve, Forrest. *Vegetation and flora of the Sonoran Desert.* Stanford CA: Stanford University Press, 1964. 2 vols.

Covers the Sonoran biotic province of California, Arizona, Baja California, Sonora. Ignores the Mohave desert of California.

2253 Steenberg, Warren F., and Hendrickson, Lupe P. "The saguaro giant cactus: a bibliography." In *Ecology of the saguaro: II*, pp. 223-38. Edited by Warren F. Steenberg and Charles H. Lowe. Washington DC: U.S. National Park Service, 1977.

Unannotated. Includes selected historical and non-technical reports as well as the botanical literature.

ZOOLOGY

2254 Anderson, Anders H. *A bibliography of Arizona ornithology: annotated.* Tucson AZ: University of Arizona Press, 1972. 241 pp.

2255 McDonald, J. L., and Olton, G. S. "A list and bibliography of the mosquitoes in Arizona." *Mosquito systematics newsletter* 6:2 (1974):89-92.

2256 *Threatened and unique wildlife of Arizona.* Phoenix AZ: Arizona. Game and Fish Commission, 1978. 7 pp.

Covers animal life of Arizona which is extinct, endangered, threatened, or of special interest because of limited distribution in Arizona.

2257 Zarn, Mark; Heller, Thomas; and Collins, Kay. *Wild, free-roaming burros - an annotated bibliography.* Denver CO: U.S. Bureau of Land Management, 1977. (Technical note 297) 29 pp.

An annotated bibliography of articles, books, and manuscripts on wild burros, domestic burros, and other *Equidae*. Information on equine science, populations, habitats, competition, legislation, and other general data are included. Libraries or agencies from which the information may be obtained are listed. Covers Arizona and California.

AGRICULTURE

2258 *Cropland atlas of Arizona.* Phoenix AZ: Arizona Crop and Livestock Reporting Service, 1974. 58 pp.

Gives location, size, and shape of agricultural fields in Arizona. Mapped from cloud-free aerial photography. Includes citrus groves, deciduous fruit and nut trees, vineyards, and pastureland as well as field crops. Irrigated and non-irrigated fields delineated.

RANGE MANAGEMENT

2259 Reynolds, Hudson Gillis. "Selected bibliography on range research in Arizona." American Society of Range Management, Arizona Section. *Proceedings* (1962):50-68.

Reprinted as a separate for the U.S. Forest Service. Contains 237 citations.

ENERGY

2260 Cohen, Sanford F. *A selected bibliography: energy related issues in the southwest United States.* Monticello IL: Vance Bibliographies, 1981. (Public administration series bibliography P-641) 6 pp.

Unannotated. Covers Arizona, California, and New Mexico.

GEOTHERMAL ENERGY

2261 Calvo, Susanna. *Geothermal resources in Arizona: a bibliography.* Tucson AZ: Arizona. Bureau of Geology and Mineral Technology, 1982. (Circular 23) 23 pp.

HYDROPOWER

2262 Stahler, Gerald. *The Grand Canyon dams controversy, 1963-1968: a bibliographic research guide.* Monticello IL: Vance Bibliographies, 1979. (Public administration series bibliography P-274) 41 pp.

Unannotated.

ANTHROPOLOGY

2263 Kluckhohn, Clyde, and Spencer, Katherine. *A bibliography of the Navaho Indians.* New York: J. J. Augustin, 1940. 93 pp.

Reprint edition New York: AMS Press, 1972. Though old, still an excellent annotated bibliography.

2264 Laird, W. David. *Hopi bibliography.* Tucson AZ: University of Arizona Press, 1977. 735 pp.

Annotated. 2,935 citations.

2265 **Ortiz, Alfonso, ed.** "Southwest." In *Handbook of North American Indians*, vol. 9. Edited by William C. Sturtevant. Washington DC: Smithsonian Institution, 1979. 701 pp.

An encyclopedic treatise on the native peoples of Arizona, New Mexico, and northern Mexico, covering the prehistory of the entire region and the historical period of the Pueblo peoples of northern Arizona and New Mexico. Includes maps, illustrations, and extensive unannotated bibliography. A companion volume (10) is in press; it will cover the historical period for the remainder of the tribes in Arizona, New Mexico, and northern Mexico.

ARCHITECTURE

2266 **Vance, Mary.** *Paolo Soleri: a bibliography.* Monticello IL: Vance Bibliographies, 1980. (Architecture series bibliography A200) 9 pp.

Unannotated. This internationally-known architect is constructing an ecologically-based city in central Arizona.

HISTORICAL GEOGRAPHY

2267 **Walker, Henry P.** *Historical atlas of Arizona.* Norman OK: University of Oklahoma Press, 1979. 150 pp.

California

GENERAL BIBLIOGRAPHIES

2268 *The California handbook: a comprehensive guide to sources of current information and action, with selected background material.* 4th ed. Claremont CA: Center for California Public Affairs, 1981. 178 pp.

Includes a directory of organizations and selected bibliography.

GEOGRAPHY

2269 **Heaser, Eileen, comp.** *The American River: a bibliography.* Sacramento CA: California State University. The Library, 1973. 40 pp.

2270 **Lamprecht, Sandra J.** *California: a bibliography of theses and dissertations in geography.* Monticello IL: Council of Planning Librarians, 1975. (CPL exchange bibliography 753) 53 pp.

Unannotated.

2271 *Maps and publications for areas of potential environmental critical concern.* Sacramento CA: California Governor's Office. Office of Planning and Research, 1974. 58 pp.

Covers California.

2272 "The Sonoran Desert: a retrospective bibliography."

See entry 2223.

2273 **Yates, Richard, and Marshall, Mary.** *The Lower Colorado River: a bibliography.*

See entry 2225.

ATLASES

2274 **Donley, Michael W. et al.** *Atlas of California.* Culver City CA: Pacific Book Center, 1979. 191 pp.

2275 **Durrenberger, Robert W.** *California: patterns on the land.* 5th ed. Palo Alto CA: Mayfield, 1976. 134 pp.

GAZETTEERS

2276 **Gudde, Erwin Gustav.** *California place names: the origin and etymology of current geographical names.* 3rd ed. Berkeley CA: University of California Press, 1969. 416 pp.

GEOLOGY

2277 **Oakeshott, Gordon B.** "A guide to information on the geology of California." California. Division of Mines and Geology. *Mineral Information Service* 23:10 (1970):195-199.

2278 **Rapp, John S. et al.** *Mines and mineral producers active in California 1981.* Sacramento CA: California. Division of Mines and Geology, 1981. (Special publication 58) 59 pp.

Lists active mining companies and mines with addresses and names of owner-operators, current to early 1981.

2279 **Roberts, Albert E.** *Selected geologic literature on the California continental borderland and adjacent areas, to January 1, 1976.* Reston VA: U.S. Geological Survey, 1975. (Circular 714) 116 pp.

2280 **Strand, Rudolph G. et al.** *Index to geologic maps of California.* San Francisco CA: California. Division of Mines, 1958. (Special report).

Special report 52 covers items through 1956, 128 pp.; *Special report* 52A (1962) covers items from 1957-60, 60 pp.; *Special report* 52B (1968) covers items from 1961-64, 72 pp.; *Special report* 102 (1972) covers items from 1965-68, 78 pp.; and *Special report* 130 (1977) covers items from 1969-75, 121 pp.

HYDROLOGY

2281 "Abstracts of DWR publications, 1922-." California. Dept. of Water Resources. *Bulletin* 170-69-. 2 per year.

Arranged by bulletin number. Index. Irregular cumulative editions.

2282 **Cooper, E. Nathan et al.** *Water resources research in the Lower Colorado River Basin, 1972-1976.*

See entry 2241.

2283 Crippen, J. R. *Index of flood maps for California prepared by the Geological Survey through 1974.* Menlo Park CA: U.S. Geological Survey, 1975. (Open-file report 1002-37) 29 pp.

2284 Kahrl, William L. *The California water atlas.* Sacramento CA: California. The Governor's Office of Planning and Research, 1979. 119 pp.

An impressive atlas covering the full range of the state's hydrology, water resources, and water project development. Excellent layout and reproduction; a model for what other states should produce.

WATER LAW

2285 Jamail, Milton H., and Ullery, Scott J. *International water use relations along the Sonoran Desert borderlands.*

See entry 2242.

BOTANY

2286 Benson, Lyman David. *The native cacti of California.* Stanford CA: Stanford University Press, 1969. 243 pp.

2287 Elias-Cesnik, Anna. *Jojoba: guide to the literature.*

See entry 2247.

2288 Sherbrooke, Wade C. *Jojoba: an annotated bibliography.*

See entry 2250.

2289 Sherbrooke, Wade C., and Haase, E. F. *Jojoba: a wax-producing shrub of the Sonoran Desert: literature review and annotated bibliography.*

See entry 2251.

2290 Shreve, Forrest. *Vegetation and flora of the Sonoran Desert.*

See entry 2252.

ZOOLOGY

2291 Zarn, Mark; Heller, Thomas; and Collins, Kay. *Wild, free roaming burros - an annotated bibliography.*

See entry 2257.

AGRICULTURE

2292 Orsi, Richard J., comp. *A list of references for the history of agriculture in California.* Davis CA: University of California. Agricultural History Center, 1974. 141 pp.

Evolved from a 1967 list compiled by James H. Shideler and Lawrence B. Lee, published by the Agricultural History Center. Adds material from 1964-67.

FORESTRY

2293 Aitro, Vincent P., comp. *Fifty years of forestry research: annotated bibliography of the Pacific Southwest Forest and Range Experiment Station, 1926-1975.* Berkeley CA: U.S. Forest Service. Pacific Southwest Forest and Range Experiment Station, 1977. 250 pp.

Annotated. Lists 2,905 publications issued during the first 50 years of the Station's work. Covers California and Hawaii (1959-75).

PLANT CULTURE

2294 Guttadauro, Guy J., comp. *A list of references for the history of grapes, wines, and raisins in America.* Davis CA: University of California, Davis. Agricultural History Center, 1976. 70 pp.

Emphasis on California. Includes labor, technology, and economics. Unannotated

IRRIGATION

2295 Rada, Edward L., and Berquist, Richard J. *Irrigation efficiencies in producing calories and proteins: an annotated bibliography.* Davis CA: University of California at Davis, 1975. (NTIS: PB 275-057/8ST) 65 pp.

Contains 704 citations.

ENERGY

2296 Cohen, Sanford F. *A selected bibliography: energy related issues in the southwest United States.*

See entry 2260.

ANTHROPOLOGY

2297 Branstedt, Wayne G. *A bibliography of North American Indians in the Los Angeles metropolitan area: the urban Indian capital.* Monticello IL: Vance Bibliographies, 1979. (Public administration series bibliography P-233) 14 pp.

Unannotated.

2298 Heizer, Robert Fleming, ed. "California." In *Handbook of North American Indians*, vol. 8. Edited by William C. Sturtevant. Washington DC: Smithsonian Institution, 1978. 800 pp.

An encyclopedic treatise on the native peoples of California, covering all of the state except the deserts of the eastern region. Includes maps, illustrations, and an unannotated bibliography.

2299 Heizer, Robert Fleming, and Elsasser, Albert B. *A bibliography of California Indians: archaeology, ethnography, Indian history.* New York: Garland, 1977. (Reference library of social science 48) 267 pp.

URBAN GEOGRAPHY

2300 **Cohen, Phyllis.** *San Diego: an introductory bibliography to the region.* Monticello IL: Vance Bibliographies, 1981. (Public administration series bibliography P-774) 8 pp.

Unannotated.

2301 **Heaser, Eileen, and Kong, Les.** *The Sacramento region: planning, growth, development: a bibliographic guide.* Monticello IL: Vance Bibliographies, 1981. (Public administration series bibliography P-673) 61 pp.

Unannotated.

2302 *Los Angeles and its environs in the twentieth century: a bibliography of a metropolis.* Los Angeles CA: The Ward Ritchie Press, 1973. 501 pp.

Unannotated. 9,895 citations arranged by topics with author and subject indexes.

2303 **Mowery, M. Kay, and Mintier, J. Laurence.** *Local planning in California: a bibliography of reference materials.* Chicago IL: CPL Bibliographies, 1981. (CPL bibliography 44) 43 pp.

Unannotated.

2304 **Trotta, Victoria K.** *Documentary information sources of Los Angeles County: a classified and annotated list.* Monticello IL: Council of Planning Librarians, 1977. (CPL exchange bibliography 1345) 74 pp.

Extensively annotated. Covers county-government publications on a variety of subjects including pollution, demography, government, earthquakes, flood control, housing, water resources, etc.

ARCHITECTURE

2305 *Architecture and preservation in California: a guide to historic sites, historic homes and churches.* Monticello IL: Vance Bibliographies, 1981. (Architecture series bibliography A462) 11 pp.

Unannotated. Prepared by Coppa and Avery Consultants.

2306 **Harmon, Robert B.** *Development of architecture in Southern California: a selected bibliography.* Monticello IL: Vance Bibliographies, 1980. (Architecture series bibliography A325) 9 pp.

Briefly annotated.

HISTORICAL GEOGRAPHY

2307 **Beck, Warren A.** *Historical atlas of California.* Norman OK: University of Oklahoma Press, 1974. Unpaged.

Contains 101 maps with commentary.

Colorado

GENERAL BIBLIOGRAPHIES

2308 **Wilcox, Virginia Lee.** *Colorado: a selected bibliography of its literature, 1858-1952.* Denver CO: Sage Books, 1954. 151 pp.

Annotated.

GEOGRAPHY

2309 **Joseph, Timothy W.** *Annotated bibliography of natural resource information: northwestern Colorado.* Ft. Collins CO: U.S. Dept. of Interior, 1977. (FWS/OBS-77/35) 185 pp.

GAZETTEERS

2310 **Eichler, George R.** *Colorado place-names: communities, counties, peaks, passes: with historical lore and facts, plus a pronunciation guide.* Boulder CO: Johnson Publishing, 1977. 109 pp.

2311 **Shaffer, Ray.** *A guide to places on the Colorado prairie, 1540-1975.* Boulder CO: Pruet, 1978. 386 pp.

ENVIRONMENTAL IMPACT STATEMENTS

2312 **Worsham, John P., Jr.** *Checklist of major federal actions significantly affecting the state environment: an information resource survey of environmental impact statements for Colorado, 1970-79.* Monticello IL: Vance Bibliographies, 1980. (Public administration series bibliography P-485) 25 pp.

Unannotated.

GEOLOGY

2313 *Bibliography and index of Colorado geology, 1875-1975.* Denver CO: Colorado. Geological Survey. Dept. of Natural Resources, 1976. (Bulletin 37) 488 pp.

Prepared by the staff of the American Geological Institute, Falls Church VA.

2314 *Bibliography of the geology of the Green River formation, Colorado, Utah, and Wyoming, to March 1, 1977.* Arlington VA: U.S. Geological Survey, 1977. (Circular 754) 52 pp.

Supersedes Geological Survey Circular 675.

2315 **Chronic, Felicie.** *Bibliography and index of geology and hydrology, Front Range Urban Corridor, Colorado.* Washington DC: U.S. Geological Survey, 1975. (Bulletin 1306) 102 pp.

2316 Wright, Ann Finley. *Bibliography of geology and hydrology, San Juan Basin, New Mexico, Colorado, Arizona, and Utah.*

See entry 2378.

HYDROLOGY

2317 Pearl, Richard Howard. *Bibliography of hydrogeologic reports in Colorado.* Denver CO: Colorado. Geological Survey, 1971. (Bulletin 33) 39 pp.

PUBLIC ADMINISTRATION

2318 Harmon, Robert B. *Government and politics in Colorado: an information source survey.* Monticello IL: Vance Bibliographies, 1978. (Public administration series bibliography P-33) 15 pp.

Brief annotations.

Idaho

ATLASES

2319 Highsmith, Richard M., and Kimmerling, Jon. *Atlas of the Pacific Northwest.*

See entry 2450.

GAZETTEERS

2320 *Gazetteer of cities, villages, unincorporated communities, and landmark sites in the state of Idaho.* 3rd ed. Boise ID: Idaho Highway Planning Survey, 1966. 54 leaves.

ENVIRONMENTAL IMPACT STATEMENTS

2321 Worsham, John P., Jr. *Checklist of major federal actions significantly affecting the state environment: an information resource survey of environmental impact statements for Idaho, 1970-79.* Monticello IL: Vance Bibliographies, 1980. (Public administration series bibliography P-616) 23 pp.

Unannotated.

GEOLOGY

2322 Bettis, M. G. *Bibliography of the U.S. Geological Survey's open-file reports on Idaho, 1941-1978.* Moscow ID: Idaho. Bureau of Mines and Geology, 1980. (Open-file report 80-1) 37 pp.

2323 Bryant, Mark. *Mineral atlas of the Pacific Northwest.*

See entry 2455.

2324 Gaston, M. P. *Graduate theses on the geology of Idaho, 1900-1977.* Moscow ID: Idaho. Bureau of Mines and Geology, 1979. (Information circular 32) 25 pp.

2325 Hansen, M. W. et al. "Bibliography of theses and dissertations on Idaho and North Dakota, 1968-1980." *Mountain geologist* 17:4 (1980):108-125.

PUBLIC ADMINISTRATION

2326 Harmon, Robert B. *Government and politics in Idaho: an information source survey.* Monticello IL: Vance Bibliographies, 1978. (Public administration series bibliography P-38) 16 pp.

Extensive annotations.

Kansas

MAP COLLECTIONS

2327 McClain, Thomas. *Contemporary Kansas maps: selected products for map users.* Lawrence KS: Kansas. Geological Survey, 1977. (Environmental geology series 1) 60 pp.

GAZETTEERS

2328 Rydjord, John. *Kansas place-names.* Norman OK: University of Oklahoma Press, 1972. 613 pp.

GEOLOGY

2329 *Bibliography and index of Kansas geology through 1974.* Lawrence KS: Kansas. Geological Survey, 1977. (Bulletin 213) 183 pp.

Produced by the staff of American Geological Institute, Falls Church VA.

HYDROLOGY

2330 Roberts, Robert S., and Hodson, Warren G. *Ground water in Kansas: bibliography and subject index.* Lawrence KS: Kansas. Geological Survey, 1966. (Bulletin 182) 41 pp.

BOTANY

2331 Kuechler, August Wilhelm. "Bibliography on vegetation ecology of Kansas, U.S.A." *Excerpta botanica: sectio B: sociologica* 20:1 (1980):53-60.

Unannotated.

HISTORICAL GEOGRAPHY

2332 Socolofsky, Homer Edward, and Self, Huber. *Historical atlas of Kansas.* Norman OK: University of Oklahoma Press, 1972. 70 pp.

PUBLIC ADMINISTRATION

2333 Harmon, Robert B. *Government and politics in Kansas: an information source survey.* Monticello IL: Vance Bibliographies, 1978. (Public administration series bibliography P-81) 22 pp.

Annotated.

Minnesota

ATLASES

2334 Borchert, John R., and Gustafson, Neil C. *Atlas of Minnesota resources and settlement.* 3rd ed. Minneapolis MN: University of Minnesota. Center for Urban and Regional Affairs, 1980. 309 pp.

GEOLOGY

2335 Melone, Theodore G., and Weis, Leonard W. *Bibliography of Minnesota geology.* Minneapolis MN: University of Minnesota Press, 1951. (Bulletin - Minnesota Geological Survey 34) 124 pp.

Updated by *Supplement*, 1951-80, compiled by G. B. Morey, Nancy Balaban, and Lynn Swanson (St. Paul MN: University of Minnesota. Minnesota Geological Survey, 1981), 143 pp. (Bulletin - Minnesota Geological Survey 46.)

FORESTRY

2336 Merz, Robert W. *Forest atlas of the Midwest.* St. Paul MN: U.S. Forest Service. North Central Forest Experiment Station, [1978]. 48 pp.

Covers Minnesota, Iowa, Missouri east to Ohio and Kentucky.

PUBLIC ADMINISTRATION

2337 Harmon, Robert B. *Government and politics in Minnesota: an information source survey.* Monticello IL: Vance Bibliographies, 1978. (Public administration series bibliography P-126) 17 pp.

Annotated.

Montana

GEOGRAPHY

2338 Joseph, Timothy W. *Annotated bibliography of natural resource information: Powder River Basin, northeastern Wyoming/southeastern Montana.*

See entry 2464.

ATLASES

2339 Taylor, Robert L.; Edie, Milton J.; and Gritzner, Charles F. *Montana in maps 1974.* Bozeman MT: Big Sky Books/Montana State University, 1974. 76 pp.

GAZETTEERS

2340 Cheney, Roberta Carkeek. *Names on the face of Montana: the story of Montana's place names.* Missoula MT: University of Montana. Printing Dept., 1971. (Publications in history) 275 pp.

ENVIRONMENTAL IMPACT STATEMENTS

2341 Worsham, John P., Jr. *Checklist of major federal actions significantly affecting the state environment: an information resource survey of environmental impact statements for Montana, 1970-79.* Monticello IL: Vance Bibliographies, 1980. (Public administration series bibliography P-614) 21 pp.

Unannotated.

HYDROLOGY

2342 *Bibliography of Montana water resources and related publications.* Helena MT: Montana. Water Resources Board, 1969. (Inventory series report 10) 91 pp.

2343 Botz, Maxwell K., and Bord, Ernest W., comps. *Index map and bibliography of ground-water studies in Montana.* Butte MT: Montana. Bureau of Mines and Geology, 1966. (Special publication 37) 1 sheet map.

Map lists 45 unannotated bibliographic items.

2344 Rautio, S. A., and Sonderegger, J. L., comps. *Annotated bibliography of the geothermal resources of Montana.* Butte MT: Montana. Bureau of Mines and Geology, 1980. (Bulletin 110) 25 pp.

PUBLIC ADMINISTRATION

2345 Harmon, Robert B. *Government and politics in Montana: an information source survey.* Monticello IL: Vance Bibliographies, 1979. (Public administration series bibliography P-147) 18 pp.

Annotated.

Nebraska

GAZETTEERS

2346 Fitzpatrick, Lilian Linder. *Nebraska place-names, including selections from the origin of the place-names of Nebraska.* Rev. ed. Lincoln NE: University of Nebraska Press, 1960. (Bison book BB107) 227 pp.

CLIMATOLOGY

2347 Lawson, Merlin P. *Climatic atlas of Nebraska.* Lincoln NE: University of Nebraska Press, 1977. 88 pp.

Appendix: State and federal climatological publications.

HYDROLOGY

2348 Stork, Karen E., and Thomsen, Nyla R. *Water resources publications related to the state of Nebraska.* 2nd ed. Lincoln NE: University of Nebraska. Water Resources Research Institute, 1972. (Publication 7) 89 pp.

2349 Lawson, Merlin P. *Agricultural atlas of Nebraska.* Lincoln NE: University of Nebraska Press, 1977. 110 pp.

Shows the growth of Nebraska during the past century, and serves as a baseline against which future trends can be measured. Maps illustrate the present state of agriculture by counties.

ECONOMIC DEVELOPMENT

2350 *Economic atlas of Nebraska.* Lincoln NE: University of Nebraska Press, 1977. 165 pp.

Provides a broad introduction to the state's economy and focuses on historical development as well as detailing the nature and geographic distribution of economic activities, such as agriculture, mining, transportation. Utilizes United States government documents for data.

PUBLIC ADMINISTRATION

2351 Harmon, Robert B. *Government and politics in Nebraska: an information source survey.* Monticello IL: Vance Bibliographies, 1979. (Public administration series bibliography P-148) 16 pp.

Annotated.

Nevada

GEOGRAPHY

2352 O'Farrell, T. P., and Emery, L. A. *Ecology of the Nevada test site: a narrative summary and annotated bibliography.* Boulder City NV: University of Nevada. Desert Research Institute, 1976. (NTIS: NVO-167) 264 pp.

Includes 333 annotated citations.

2353 Romney, E. M. et al. *Some characteristics of soil and perennial vegetation in northern Mojave desert areas of the Nevada test site.* Los Angeles CA: University of California, Los Angeles. Laboratory of Nuclear Medicine and Radiation Biology, 1973. (NTIS: UCLA-12-916) 340 pp.

MAP COLLECTIONS

2354 Ansari, Mary B. *Nevada collections of maps and aerial photographs.* Reno NV: Camp Nevada, 1976. (Monograph 2) 39 pp.

GAZETTEERS

2355 Carlson, Helen S. *Nevada place names: a geographical dictionary.* Reno NV: University of Nevada Press, 1974. 282 pp.

2356 *Directory of geographic names.* Carson City NV: Nevada. Dept. of Highways. Cartographic Section, 1971. 192 pp.

ENVIRONMENTAL IMPACT STATEMENTS

2357 Worsham, John P., Jr. *Checklist of major Federal actions significantly affecting the State environment: an information resource survey of environmental impact statements for Nevada, 1970-79.* Monticello IL: Vance Bibliographies, 1980. (Public administration series bibliography P-615) 11 pp.

Unannotated.

GEOLOGY

2358 Ansari, Mary B. *Bibliography of Nevada mining and geology, 1966-1970.* Reno NV: University of Nevada. Mackay School of Mines, 1975. (Nevada Bureau of Mines and Geology-Report 24) 61 pp.

2359 Lutsey, Ira A. *Bibliography of graduate theses on Nevada geology to 1976.* Reno NV: University of Nevada. Mackay School of Mines, 1978. (Nevada Bureau of Mines and Geology-Report 31) 20 pp.

CLIMATOLOGY

2360 Houghton, John G. *Nevada's weather and climate.* Reno NV: University of Nevada. Mackay School of Mines. Nevada Bureau of Mines and Geology, 1975. (Special publication 2) 78 pp.

HYDROLOGY

2361 *Nevada state water planning references.* Carson City NV: Nevada. Dept. of Conservation and Natural Resources. Division of Water Resources, 1976. 101 pp.

Prepared by the Office of the Nevada State Engineer.

2362 *Water for Nevada: hydrologic atlas.* Carson City NV: Nevada. Division of Water Resources, 1972. 4 pp.

BOTANY

2363 Tueller, P. T., and Robertson, J. H. *The vegetation of Nevada: a bibliography.* Reno NV: Nevada. Agricultural Experiment Station, 1978. 28 pp.

2364 Wallace, A., and Romney, E. M. *Radioecology and ecophysiology of desert plants at the Nevada test site.* Los Angeles CA: University of California at Los Angeles. Laboratory of Nuclear Medicine and Radiation Biology, 1972. (NTIS: TIC-25954) 446 pp.

Includes 740 citations.

ZOOLOGY

2365 Banks, R. C. "Annotated bibliography of Nevada, U.S.A., ornithology since 1951." *Great Basin naturalist* 28:2 (1968):49-60.

2366 Banta, B. H. "An annotated chronological bibliography of the herpetology of the state of Nevada." *Wasmann journal of biology* 23:1-2 (1965):1-224.

Contains 300 citations from 1852 to 1965.

PUBLIC ADMINISTRATION

2367 Harmon, Robert B. *Government and politics in Nevada: an information source survey.* Monticello IL: Vance Bibliographies, 1979. (Public administration series bibliography P-149) 16 pp.

Brief annotations.

New Mexico

GEOGRAPHY

2368 Joseph, Timothy W., and Wood, John. *Annotated bibliography of natural resource information: northwestern New Mexico.* Ft. Collins CO: U.S. Dept. of the Interior, 1977. (FWS/OBS-77/33) 141 pp.

2369 Smith-Sanclare, Shelby. *A selected annotated bibliography on physical planning in arid lands.* Monticello IL: Council of Planning Librarians, 1973. (CPL exchange bibliography 423) 46 pp.

Annotated. Covers general geography, geology, climate, hydrology, biology, with an emphasis on central New Mexico.

ATLASES

2370 Williams, Jerry L., and McAllister, Paul E. *New Mexico in maps.* Albuquerque NM: Technology Application Center, 1979. 177 pp.

GAZETTEERS

2371 Henrickson, James, and Straw, Richard M. *A gazetteer of the Chihuahuan Desert region.* Los Angeles CA: California State University at Los Angeles, 1976. 271 pp.

2372 Pearce, Thomas Matthews, ed. *New Mexico place names: a geographical dictionary.* Albuquerque NM: University of New Mexico Press, 1965. 187 pp.

ENVIRONMENTAL IMPACT STATEMENTS

2373 Worsham, John P., Jr. *Checklist of major Federal actions significantly affecting the State environment: an information resource survey of environmental impact statements for New Mexico, 1970-79.* Monticello IL: Vance Bibliographies, 1980. (Public administration series bibliography P-525) 18 pp.

Unannotated.

GEOLOGY

2374 *Bibliography of New Mexico geology and mineral technology.* Socorro NM: New Mexico. Bureau of Mines and Mineral Resources, 1961-81.

Unannotated. *Bulletin* 74 (1961) covers the period 1956-60, 124 pp.; *Bulletin* 90 (1966) covers the period 1961-65, 124 pp.; *Bulletin* 99 (1973) covers the period 1966-70, 288 pp.; *Bulletin* 106 (1977) covers the period 1971-75, 137 pp.; and *Bulletin* 108 supplements the period through 1975, 136 pp.

2375 *Bibliography of Permian Basin geology: West Texas and Southeastern New Mexico.* Midland TX: West Texas Geological Society. Bibliographical Committee, 1967. 163 pp.

2376 Wright, Ann Finley. *Bibliography of geology and hydrology, eastern New Mexico.* Albuquerque NM: U.S. Geological Survey, 1979. (Water resources investigations 79-76) 170 pp.

Contains over 1,900 citations on physical sciences of the High Plains of eastern New Mexico.

2377 Wright, Ann Finley. *Bibliography of the geology and hydrology of the Albuquerque greater urban area, Bernalillo and parts of Sandoval, Santa Fe, Socorro, Torrance, and Valencia counties, New Mexico.* Washington DC: U.S. Geological Survey, 1978. (Bulletin 1458) 31 pp.

2378 Wright, Ann Finley. *Bibliography of geology and hydrology, San Juan Basin, New Mexico, Colorado, Arizona, and Utah.* Washington DC: U.S. Geological Survey, 1979. (Bulletin 1481) 123 pp.

2379 Wright, Ann Finley. *Bibliography of geology and hydrology, southwestern New Mexico.* Albuquerque: U.S. Geological Survey. Water Resources Division, 1980. (Water resources investigations 80-20) 255 pp.

HYDROLOGY

2380 Borton, R. L. *Bibliography of ground-water studies in New Mexico, 1873-1977.* Santa Fe NM: New Mexico. State Engineer, 1978. (Special publication) 121 pp.

Supplemented by Borton's *Bibliography of ground-water studies in New Mexico, 1848-1979* (1980), 46 pp.

2381 Hernandez, John W., and Eaton, T. J. *A bibliography pertaining to the Pecos River Basin in New Mexico.* University Park NM: New Mexico State University. Water Resources Research Institute, 1967. (Publication 2; SWRA W69-01133) 50 pp.

Contains 398 citations.

2382 Moe, Christine E. *Rio Grande flood control and drainage.* Monticello IL: Vance Bibliographies, 1981. (Public administration series bibliography P-769) 27 pp.

Unannotated. Covers New Mexico's portion of the river.

BIOLOGY

2383 Patton, David R. *RUN WILD II: data files for wildlife species and habitat for Arizona and New Mexico.*

See entry 2243.

2384 Patton, David R., and Ffolliott, Peter F. *Selected bibliography of wildlife and habitats for the Southwest.*

See entry 2244.

BOTANY

2385 Weniger, Del. *Cacti of the Southwest: Texas, New Mexico, Oklahoma, Arkansas, and Louisiana.*

See entry 2430.

ZOOLOGY

2386 Wolff, T. A.; Nielsen, L. T.; and Hayes, R. O. "A current list and bibliography of the mosquitoes of New Mexico, U.S.A." *Mosquito systematics* 7:1 (1975):13-18.

ENERGY

2387 Cohen, Sanford F. *A selected bibliography: energy related issued in the southwest United States.*

See entry 2260.

ANTHROPOLOGY

2388 Ortiz, Alfonso, ed. *Southwest.*

See entry 2265.

ARCHITECTURE

2389 *Architecture and preservation in New Mexico: a guide to historic sites, churches and homes.* Monticello IL: Vance Bibliographies, 1981. (Architecture series bibliography A407) 6 pp.

Unannotated. Prepared by Coppa and Avery Consultants.

2390 Moe, Christine E. *Preservation of the regional architecture and historic buildings of New Mexico.* Monticello IL: Vance Bibliographies, 1981. (Architecture series bibliography A415) 28 pp.

Unannotated.

ECONOMIC DEVELOPMENT

2391 Worsham, John P., Jr. *Select economic conditions in the Southwest as compiled from the literature of the Federal Reserve Bank of Dallas.* Monticello IL: Vance Bibliographies, 1980. (Public administration series bibliography P-399) 24 pp.

Unannotated. Covers the four states of the Eleventh Federal Reserve District: Oklahoma, Texas, Louisiana, and New Mexico.

PUBLIC ADMINISTRATION

2392 Harmon, Robert B. *Government and politics in New Mexico: an information source survey.* Monticello IL: Vance Bibliographies, 1979. (Public administration series bibliography P-198) 21 pp.

Annotated.

North Dakota

GEOGRAPHY

2393 Lynott, Bill; Ryckman, L. Frederick; and Joseph, Timothy W. *Annotated bibliography of natural resource information: southwestern North Dakota.* Ft. Collins CO: U.S. Dept. of Interior, 1977. (FWS/OBS-77/32) 215 pp.

ATLASES

2394 Goodman, Lowell Robert. *The atlas of North Dakota.* Fargo ND: North Dakota Studies, 1976. 112 pp.

GEOLOGY

2395 Hansen, M. W. et al. "Bibliography of theses and dissertations on Idaho and North Dakota, 1968-1980."

See entry 2325.

2396 Scott, Mary Woods. *Annotated bibliography of the geology of North Dakota, 1806-1959.* Grand Forks ND: North Dakota. Geological Survey, 1972. (Miscellaneous series 49) 132 pp.

PUBLIC ADMINISTRATION

2397 Harmon, Robert B. *Government and politics in North Dakota: an information source survey.* Monticello IL: Vance Bibliographies, 1979. (Public administration series bibliography P-230) 18 pp.

Annotated.

Oklahoma

GAZETTEERS

2398 Shirk, George H. *Oklahoma place names*. 2nd ed. Norman OK: University of Oklahoma Press, 1974. 268 pp.

GEOLOGY

2399 Curtis, Neville M., Jr. "Bibliography and index of Oklahoma geology." *Oklahoma geology notes* 21:3- (1961-).

Appears annually from 1961 to date.

HYDROLOGY

2400 Stoner, J. D. *Index of published surface water quality data for Oklahoma, 1946-1975*. Washington DC: U.S. Geological Survey, 1977. (Open file report 77/204) 218 pp.

BOTANY

2401 Weniger, Del. *Cacti of the Southwest: Texas, New Mexico, Oklahoma, Arkansas, and Louisiana*.

See entry 2430.

ECONOMIC DEVELOPMENT

2402 Worsham, John P., Jr. *Select economic conditions in the Southwest as compiled from the literature of the Federal Reserve Bank of Dallas*.

See entry 2391.

HISTORICAL GEOGRAPHY

2403 Morris, John W.; Goins, Charles R.; and McReynolds, Edwin C. *Historical atlas of Oklahoma*. 2nd ed. Norman OK: University of Oklahoma Press, 1976. Unpaged.

PUBLIC ADMINISTRATION

2404 Harmon, Robert B. *Government and politics in Oklahoma: an information source survey*. Monticello IL: Vance Bibliographies, 1979. (Public administration series bibliography P-291) 24 pp.

Brief annotations.

Oregon

ATLASES

2405 Highsmith, Richard M., and Kimmerling, Jon. *Atlas of the Pacific Northwest*.

See entry 2450.

2406 Loy, William G. *Atlas of Oregon*. Eugene OR: University of Oregon, 1976. 215 pp.

GAZETTEERS

2407 McArthur, Lewis Ankeny. *Oregon geographic names*. 4th ed. Portland OR: Oregon Historical Society, 1974. 835 pp.

GEOLOGY

2408 Bela, James. *Annotated bibliography of the geology of the Columbia Plateau (Columbia River Basalt) and adjacent areas of Oregon*.

See entry 2453.

2409 Bryant, Mark. *Mineral atlas of the Pacific Northwest*.

See entry 2455.

2410 Corcoran, R. E. *Index to published geologic mapping in Oregon 1898-1967*. Portland OR: Oregon. Dept. of Geology and Mineral Resources, 1968. 20 pp.

2411 Hodge, Edwin T. *Bibliography of the geology and mineral resources of Oregon with digests and index to July 1, 1936*. Portland OR: Conger Printing, 1936. 224 pp.

Supplements appear in the *Bulletin* of the Oregon Dept. of Geology and Mineral Industries, 1937-. *Bulletin* 33/1st supplement (1947) covers period July, 1936 through 1945, 108 pp.; *Bulletin* 44/2nd supplement (1953) covers period 1946-50, 61 pp.; *Bulletin* 53/3rd supplement (1962) covers period 1951-55, 97 pp.; *Bulletin* 67/4th supplement (1970) covers period 1956-60, 88 pp.' *Bulletin* 78/5th supplement (1973) covers period 1961-70, 199 pp.; *Bulletin* 97/6th supplement (1978) covers period 1971-75, 74 pp.

HYDROLOGY

2412 Sweet, H. R. *Bibliography of available ground-water information in Oregon*. Salem: Oregon State Engineer, 1974. 26 pp.

SOILS

2413 Klock, Glen O., comp. *Forest and range soils research in Oregon and Washington, a bibliography with abstracts from 1964 through 1968*. Portland OR: U.S. Forest Service. Pacific Northwest Forest and Range Experiment Station, 1969. (Research paper PNW-90) 28 pp.

Updated by Klock in 1976 (entry 2414).

2414 Klock, Glen O., comp. *Forest and range soils research in Oregon and Washington: a bibliography with abstracts from 1969 through 1974*. Portland OR: U.S. Forest Service. Pacific Northwest Forest and Range Experiment Station, 1976. (Technical report PNW-47) 36 pp.

Contains 230 items. Annotated bibliography supplementing Tarrant (1964) covering period 1956-1963, and Klock (1969) (entry 2413), covering period 1964-68.

ZOOLOGY

2415 **Scott, J. M.; Haislip, T. W., Jr.; and Thompson, M.** "A bibliography of Oregon ornithology, 1935-1970, with a cross reference list of the birds of Oregon." *Northwest science* 46:2 (1972):122-139.

PUBLIC ADMINISTRATION

2416 **Harmon, Robert B.** *Government and politics in Oregon: an information source survey.* Monticello IL: Vance Publications, 1979. (Public administration series bibliography P-292) 21 pp.

Brief annotations.

South Dakota

ATLASES

2417 **Hogan, Edward Patrick; Opheim, Lee A.; and Zieske, Scott H., eds.** *Atlas of South Dakota.* Dubuque IA: Kendall/Hunt, 1970. 137 pp.

Covers physical, cultural, and economic aspects of the state.

PUBLIC ADMINISTRATION

2418 **Harmon, Robert B.** *Government and politics in South Dakota: an information source survey.* Monticello IL: Vance Bibliographies, 1979. (Public administration series bibliography P-296) 21 pp.

Brief annotations.

Texas

GENERAL BIBLIOGRAPHIES

2419 *Texas reference sources: a selective guide.* Austin TX: Texas Library Association. Reference Round Table, 1975. 134 pp.

GEOGRAPHY

2420 **Bagur, Jacques D.** *Barrier islands of the Atlantic and Gulf coasts of the United States: an annotated bibliography.* Baton Rouge LA: Gulf South Research Institute for the United States, 1978. 215 pp.

2421 **Gunn, Clare A.** *Texas gulf coast: annotated bibliography of resource use.* College Station TX. 1969. 387 pp.

ATLASES

2422 **Arbingast, Stanley A. et al.** *Atlas of Texas.* 5th ed. Austin TX: The University of Texas at Austin, 1976. 179 pp.

GAZETTEERS

2423 **Henrickson, James, and Straw, Richard M.** *A Gazetteer of the Chihuahuan Desert region.*

See entry 2371.

GEOLOGY

2424 *Bibliography of Permian Basin geology: west Texas and southeastern New Mexico.*

See entry 2375.

2425 **Girard, Roselle M.** *Bibliography and index of Texas geology, 1933-1950.* Austin TX: University of Texas. Bureau of Economic Geology, 1959. 238 pp.

Supplemented by entries 2426 and 2427.

2426 **Moore, Elizabeth T.** *Bibliography and index of Texas geology, 1951-1960.* Austin TX: University of Texas. Bureau of Economic Geology, 1972. 575 pp.

Continues entry 2425; continued by entry 2427.

2427 **Moore, Elizabeth T.** *Bibliography and index of Texas geology, 1961-1974.* Austin TX: University of Texas. Bureau of Economic Geology, 1976. 446 pp.

Continues entry 2426.

BOTANY

2428 **McGinnies, William G., and Haase, Edward F.** *Guayule: a rubber-producing shrub for arid and semi-arid regions.*

See entry 2027.

2429 **Smeins, Fred E., and Shaw, Robert Blaine.** *Natural vegetation of Texas and adjacent areas, 1675-1975.* College Station TX: Texas Agricultural Experiment Station. Texas A&M University System, 1978. (Miscellaneous publication 1399) 36 pp.

2430 **Weniger, Del.** *Cacti of the southwest: Texas, New Mexico, Oklahoma, Arkansas, and Louisiana.* Austin TX: University of Texas Press, 1969. (The Elma Dill Russell Spencer Foundation series 4) 249 pp.

ARCHITECTURE

2431 *Architecture and preservation in Texas: a guide to historic sites, churches and homes.* Monticello IL: Vance Bibliographies, 1981. (Architecture series bibliography A404) 105 pp.

Unannotated. Prepared by Coppa and Avery Consultants.

The United States-Texas/2432-2448

ECONOMIC DEVELOPMENT

2432 Worsham, John P., Jr. *Select economic conditions in the Southwest as compiled from the literature of the Federal Reserve Bank of Dallas.*

See entry 2391.

HISTORICAL GEOGRAPHY

2433 Pool, William C. *A historical atlas of Texas.* Austin TX: Encino Press, 1975. 190 pp.

PUBLIC ADMINISTRATION

2434 Harmon, Robert B. *Government and politics in Texas: an information source survey.* Monticello IL: Vance Bibliographies, 1980 (Public administration series bibliography P-401) 37 pp.

Annotated.

Utah

GEOGRAPHY

2435 Low, Jessop, and Joseph, Timothy W. *Annotated bibliography of natural resource information: southern Utah.* Fort Collins CO: U.S. Dept. of Interior, 1977. (FWS/OBS-77/34) 246 pp.

ATLASES

2436 *Atlas of Utah.* Provo UT: Brigham Young University Press, 1981. 300 pp.

An excellent full-color, large-format atlas with extensive text and bibliography.

ENVIRONMENTAL IMPACT STATEMENTS

2437 Worsham, John P., Jr. *Checklist of major Federal actions significantly affecting the State environment: an information resource survey of environmental impact statements for Utah, 1970-79.* Monticello IL: Vance Bibliographies, 1980. (Public administration series bibliography P-509) 14 pp.

Unannotated.

GEOLOGY

2438 *Abstracts of theses concerning the geology of Utah to 1966.* Salt Lake City UT: Utah. Geological and Mineral Survey, 1970. (Bulletin 86)

2439 *Bibliography of the geology of the Green River formation, Colorado, Utah, and Wyoming, to March 1, 1977.*

See entry 2314.

2440 "Bibliography of Utah geology." Salt Lake City UT: Utah Dept. of Natural Resources, 1974-79. *Utah geology* (appear annually from 1974 to date).

2441 Buss, W. R., and Goeltz, N. S. *Bibliography of Utah geology 1950 to 1970.* Salt Lake City UT: Utah. Geological and Mineral Survey, 1974. (Bulletin 103) 285 pp.

2442 Hansen, M. W. et al. "Bibliography of theses and dissertations on Utah, 1968-1979." *Montana geologist* 17:3 (1980):71-87.

Contains 416 references on geosciences. Unannotated.

2443 Wright, Ann Finley. *Bibliography of geology and hydrology, San Juan Basin, New Mexico, Colorado, Arizona, Utah.*

See entry 2378.

HYDROLOGY

2444 LaPray, Barbara A., and Hamblin, L. S. *Bibliography of U.S. Geological Survey water resources reports for Utah.* Salt Lake City UT: Utah. Dept. of Natural Resources, 1980. (Information bulletin 27) 75 pp.

BOTANY

2445 Christensen, Earl M. "Bibliography of Utah botany and wildland conservation." Brigham Young University. *Science bulletin, Biological series* 9:1 (1967):1-136.

2446 Welsh, S. L. et al. "Preliminary index of Utah, U.S.A., vascular plant names." *Great Basin naturalist* 41:1 (1981):1-108.

ZOOLOGY

2447 Hayward, C. L. et al. "Birds of Utah, U.S.A." Brigham Young University. *Great Basin naturalist Memoirs* 1 (1976).

ENERGY

2448 *Utah renewable energy directory.* Portland OR: Western Solar Utilization Network, 1981. 108 pp.

Contains a listing of model solar energy projects, commercial firms interested in solar applications, sources of financial aid, and organizations involved in solar and other alternative energy technologies, all in the state of Utah. Emphasis is on solar but also has entries for biomass, geothermal, hydrogen, wind, and others. Useful for names, addresses, and descriptions of relevant activities. Produced by the Utah Energy Office for *Western SUN*.

PUBLIC ADMINISTRATION

2449 Harmon, Robert B. *Government and politics in Utah: an information source survey.* Monticello IL: Vance Bibliographies, 1980. (Public administration series bibliography P-467) 20 pp.

Annotated.

Washington

ATLASES

2450 Highsmith, Richard M., and Kimmerling, Jon. *Atlas of the Pacific Northwest.* 6th ed. Corvallis OR: Oregon State University Press, 1979. 135 pp.

Covers Washington, Oregon, and Idaho.

2451 *Washington environmental atlas.* 2nd ed. Seattle WA: U.S. Army Corps of Engineers. Seattle, Washington, District. Experimental Resources Section, 1975. 114 pp.

Originally published as *Provisional U.S. Army Corps of Engineers environmental reconnaissance inventory of the State of Washington* (1972).

GAZETTEERS

2452 Phillips, James Wendell. *Washington state place names.* Seattle WA: University of Washington Press, 1971. 167 pp.

GEOLOGY

2453 Bela, James. *Annotated bibliography of the geology of the Columbia Plateau (Columbia River Basalt) and adjacent areas of Oregon.* Portland OR: Oregon. Dept. of Geology and Natural Industries, 1979. (Open-file report 0-79-1) 748 pp.

2454 Bennett, William Alfred Glen. *Bibliography and index of the geology and mineral resources of Washington, 1814-1936.* Olympia WA: Washington. Division of Mines and Geology, 1939. (Bulletin 35) 140 pp.

Supplemented by *Bulletin* 46, covering 1937-1956 (1960), 721 pp., and *Bulletin* 59, covering 1957-1962 (1969), 375 pp., by William Henry Reichert.

2455 Bryant, Mark. *Mineral atlas of the Pacific Northwest.* Moscow ID: University Press of Idaho, 1980. 27 pp.

Covers Washington, Oregon, and Idaho.

2456 Manson, C., comp. *Theses on Washington geology: a comprehensive bibliography, 1901-1979.* Washington. Division of Geology and Earth Resources, 1980. (Information circular 70) 212 pp.

HYDROLOGY

2457 *Bibliography of bibliographies on water resources.* Olympia WA: State of Washington. Dept. of Ecology. Water Resources Information System, 1973. 16 leaves.

2458 *A selected annotated bibliography on water resources of the State of Washington.* Olympia WA: State of Washington. Dept. of Ecology, 1973. (Water resources information system. Information bulletin 7; Dept. of Ecology technical report 73-018-7) 548 pp.

SOILS

2459 Klock, Glen O., comp. *Forest and range soils research in Oregon and Washington, a bibliography with abstracts from 1964 through 1968.*

See entry 2413.

2460 Klock, Glen O., comp. *Forest and range soils research in Oregon and Washington: a bibliography with abstracts from 1969 through 1974.*

See entry 2414.

BOTANY

2461 Daubenmire, Rexford. "Vegetation of the State of Washington: a bibliography." *Northwest science* 36:2 (1962):50-54.

Unannotated list of 106 titles. Includes published accounts only. "Supplement" 1, 51:2 (1977):111-113, contains 86 titles.

ZOOLOGY

2462 Campbell, R. Wayne et al. *A bibliography of Pacific Northwest herpetology.* Victoria, British Columbia: The British Columbia Provincial Museum, 1982. (Heritage record 14) 152 pp.

Covers Washington, British Columbia, Yukon Territory, and Alaska. Unannotated.

PUBLIC ADMINISTRATION

2463 Harmon, Robert B. *Government and politics in Washington [state]: an information source survey.* Monticello IL: Vance Bibliographies, 1980. (Public administration series bibliography P-484) 22 pp.

Annotated.

Wyoming

GEOGRAPHY

2464 Joseph, Timothy W. *Annotated bibliography of natural resource information: Powder River Basin, northeastern Wyoming/southeastern Montana.* Ft. Collins CO: U.S. Dept. of Interior, 1977. (FWS/OBS-77/31) 240 pp.

GAZETTEERS

2465 Urbanek, Mae Bobb. *Wyoming place names.* 2nd ed. Boulder CO: Johnson Publishing, 1969. 224 pp.

ENVIRONMENTAL IMPACT STATEMENTS

2466 Worsham, John P., Jr. *Checklist of major Federal actions significantly affecting the State environment: an information resource survey of environmental impact statements for Wyoming, 1970-79.* Monticello IL: Vance Bibliographies, 1980. (Public administration series bibliography P-524) 15 pp.

Unannotated.

GEOLOGY

2467 *Bibliography of the geology of the Green River formation, Colorado, Utah, and Wyoming, to March 1, 1977.*

See entry 2314.

2468 Bovee, Gladys G. *Bibliography and index of Wyoming geology, 1823-1916.* Cheyenne WY: Wyoming. Geological Survey, 1918. (Bulletin 17) 446 pp.

Updated by Bulletin 53, covering the years 1917-1945 (Max Lorain Troyer: 1969), 73 pp.; Bulletin 57, covering the years 1945-1949 (Jane M. Love: 1973): and Bulletin 62, covering the years 1950-1959 (Helen L. Nace: 1979), 203 pp.

2469 Hansen, M. W. "Bibliography of theses and dissertations on Wyoming, by all institutions, 1968-1979." *Montana geologist* 17:2 (1980):45-56.

Unannotated.

ZOOLOGY

2470 Clark, T. W.; Saab, V. A.; and Casey, D. "A partial bibliography of Wyoming, U.S.A. mammals." *Northwest science* 54:1 (1980):55-67.

PUBLIC ADMINISTRATION

2471 Harmon, Robert B. *Government and politics in Wyoming: an information source survey.* Monticello IL: Vance Bibliographies, 1980. (Public administration series bibliography P-556) 16 pp.

Annotated.

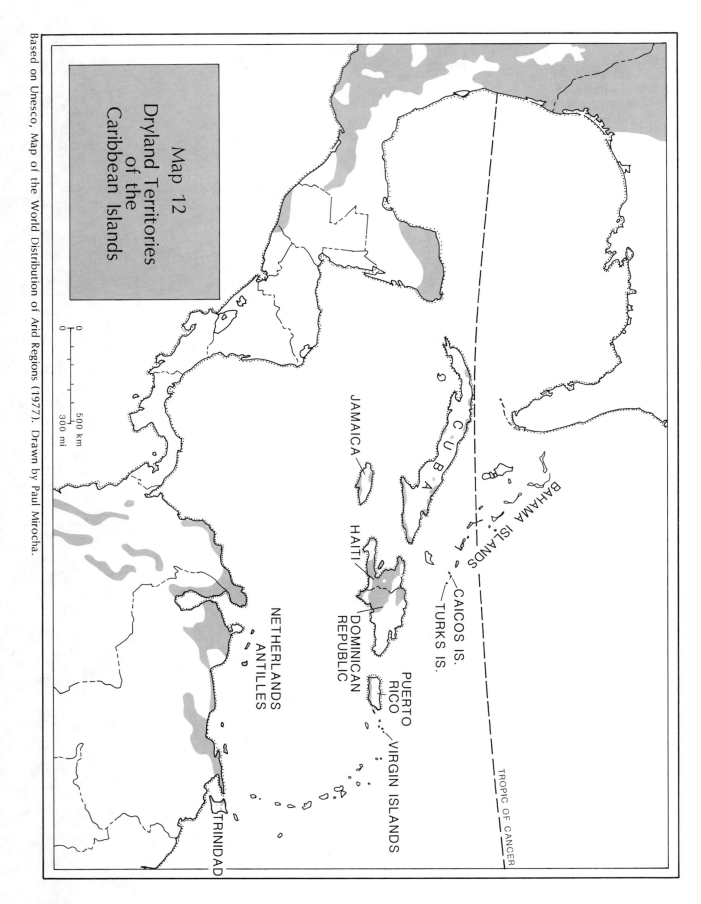

Map 12. Dryland Territories of the Caribbean Islands

Based on Unesco, Map of the World Distribution of Arid Regions (1977). Drawn by Paul Mirocha.

The Caribbean Islands

The drylands of the West Indies are the result of steady easterly trade winds that create rainshadows on the lee of mountainous islands; in other cases the islands are too low to intercept rain-bearing clouds. The driest portion of the Caribbean centers on a point just off the coast of Venezuela and includes stretches of the Venezuelan and Colombian coasts as well as the islands of the Netherlands Antilles (see section on Venezuela for explanation). Economically the Caribbean Basin has flourished in the past and has not lost the potential to do so again given sufficient capital and some enlightened solutions to the region's urgent social and political needs.

Dryland areas of THE CARIBBEAN ISLANDS

DESERTIFICATION RISK	ARIDITY									
	Hyperarid		Arid		Semiarid		Subhumid		Aridity Totals	
	km²	%	km²	%	km²	%	km²	%	km²	%
Very High	By definition, desertification does not exist in hyperarid regions.									
High										
Moderate										
Desertification Totals										
No Desertification					16,054	6.9	30,371	13.1	46,425	20.0
Total Drylands					16,054	6.9	30,371	13.1	46,425	20.0
Non-dryland									185,076	80.0
Total Area of the Territory									231,501	100.0

GENERAL BIBLIOGRAPHIES

2472 Comitas, Lambros. *The complete Caribbeana 1900-1975: a bibliographic guide to the scholarly literature.* Millwood NY: KTO Press, 1977. 4 vols.

Unannotated. Covers the non-Spanish Caribbean: Bermuda, Bahamas, Turks and Caicos, lesser Antilles, Trinidad, Netherlands Antilles, Jamaica, Caymans, Belize, the Guianas. Volume 3, covering Resources, is especially valuable to geographers.

THESES AND DISSERTATIONS

2473 Baa, Enid M. *Theses on Caribbean topics 1778-1968.* San Juan, Puerto Rico: University of Puerto Rico Press, 1970. 146 pp.

Unannotated. Contains over 1,200 entries covering theses from the United States, the United Kingdom, and France.

HANDBOOKS

2474 *Caribbean year book.* Toronto, Ontario: Caribook Ltd., 1978-. Annual.

Supersedes *The West Indies and Caribbean year book/Anuario comercial de las Antillas y paises del Caribe* (1927-77).

GEOGRAPHY

2475 Norton, A. V. *A bibliography of the Caribbean area for geographers.* Kingston, Jamaica: University of the West Indies. Dept. of Geography, 1971. (Occasional publication 7) 3 vols.

GAZETTEERS

2476 *British West Indies and Bermuda: official standard names approved by the United States Board on Geographic Names.* Washington DC: U.S. Board on Geographic Names, 1955. (Gazetteer 7) 157 pp.

List of 10,500 place-names with latitude/longitude coordinates. Includes the Bahamas and Turks and Caicos Islands, Jamaica, the Caymans, Barbados, Dominica, Grenada, St. Lucia, St. Vincent, Anguilla, Antigua, Montserrat, and other British Caribbean islands.

GEOLOGY

2477 *Abstracts of North American geology.*
See entry 1882.

2478 Rutten, Louis Martin Robert. *Bibliography of West Indian geology.* Utrecht, The Netherlands. 1938. (Geographische en geologische mededeelingen, Physiographisch-geologische reeks, no. 16).

The Caribbean Islands/2479-2496

2479 Smith, Alan L., and Weaver, John D. *Status of geological research in the Caribbean.* Mayaguez, Puerto Rico: University of Puerto Rico, 1974. (Institute of Caribbean Science 18) 126 pp.

2480 *Stratigraphic atlas of North and Central America.*

See entry 1885.

CLIMATOLOGY

2481 *Bibliography on meteorology, climatology, and physical/chemical oceanography.* Washington DC: American Meteorological Society, 1970. 614 pp.

Covers the Caribbean Sea and surrounding lands.

2482 *Climatic atlas of North and Central America. I: Maps of mean temperature and precipitation.*

See entry 1886.

2483 Portig, W. H. "The climate of Central America." In *Climates of Central and South America*, pp. 405-478. Edited by Werner Schwardtfeger. Amsterdam, The Netherlands: Elsevier Scientific, 1976. (World survey of climatology 12).

Covers the Central American isthmus and the Caribbean islands. A narrative summary with maps, charts, and an unannotated bibliography.

HYDROLOGY

2484 Beavington, C. F., and Williams, J. B. "Bibliography of water resources of Commonwealth countries in the Caribbean and Mediterranean."

See entry 123.

SOILS

2485 Cornforth, I. S. *Bibliography of soil science and fertilizer agronomy for the Commonwealth Caribbean.* St. Augustine, Trinidad: University of the West Indies. Dept. of Soil Sciences, 1969. 97 pp.

BOTANY

2486 Kuechler, August Wilhelm. *International bibliography of vegetation maps: volume 1: North America.*

See entry 1891.

2487 Rundel, Philip W. *An annotated bibliography of West Indian plant ecology.* Charlotte Amalie: Virgin Islands. Dept. of Conservation and Cultural Affairs. Bureau of Libraries and Museums, 1974. (Bibliography series 1) 70 pp.

ZOOLOGY

2488 Arnett, Ross H., ed. *Bibliography of Coleoptera of North America, north of Mexico, 1758-1948.*

See entry 1894.

2489 Hull, William B., ed. *Directory of North American entomologists and acarologists.*

See entry 1896.

AGRICULTURE

2490 *Agriculture in the economy of the Caribbean: a bibliography.* Madison WI: The University of Wisconsin. Land Tenure Center, 1974. (Training and methods 24) 84 pp.

Unannotated. Reflects the holdings of the Land Tenure Center library; includes its call numbers.

DEMOGRAPHY

2491 Johson, David M. *Population, anthropology, Caribbean: an overview and guide to the literature.* Chapel Hill NC: University of North Carolina, 1968. 154 pp.

M.A. thesis.

ANTHROPOLOGY

2492 Clermont, Norman. *Bibliographie annotee de l'anthropologie physique des Antilles [annotated bibliography of the physical anthropology of the Antilles].* Montreal, Quebec: University of Montreal. Center of Caribbean Research, 1972. 51 pp.

2493 Marshall, Trevor G. *A bibliography of the Commonwealth Caribbean peasantry, 1838-1974.* Cave Hill, Barbados: University of the West Indies. Institute of Social and Economic Research, 1975. (Occasional bibliography series 3) 47 pp.

ECONOMIC DEVELOPMENT

2494 *Quarterly economic review of the West Indies, Belize, Bahamas, Bermuda, Guyana.* London: EIU, 1974-. Quarterly.

Includes summary of political and economic news as well as charts and statistics of selected economic indicators. Annual supplement.

HISTORICAL GEOGRAPHY

2495 Ashdown, Peter. *Caribbean history in maps.* Trinidad and Jamaica: Longman Caribbean, 1979. 84 pp.

2496 Lux, William. *Historical dictionary of the British Caribbean.* Metuchen NJ: Scarecrow, 1975. (Latin American historical dictionaries 12) 266 pp.

Entries cover major topics, events, and personalities in these nations' history; includes extensive unannotated bibliography. Covers Barbados, Belize, Guyana, Jamaica, Leeward Islands, Trinidad, Tobago, Windward Islands.

BAHAMA ISLANDS
The Commonwealth of the Bahamas
(a former colony of the United Kingdom)

The Bahamas are a low-lying chain of islands stretching 800 km (500 miles) between Florida and Hispaniola. The entire chain represents the highest points of a large limestone platform similar in structure to the Florida and Yucatan Peninsulas.

Only the southeasterly islands of Long Island, Crooked Island, Acklins Island, Mayaguana Island, and the Inagua Islands are dry: all are subhumid and form part of a dryland belt that includes the Turks and Caicos group (which see). Surface water is scarce due to the porous limestone but well water can supply domestic needs. Agriculture, especially on the smaller islands, is extremely limited by the lack of both soil and water; fishing supplies much of the locally-produced food. Tourism is very important to the nation as a whole but has a more limited potential on the remote dryland islands.

Dryland areas of THE BAHAMA ISLANDS

DESERTIFICATION RISK	ARIDITY									
	Hyperarid		Arid		Semiarid		Subhumid		Aridity Totals	
	km²	%	km²	%	km²	%	km²	%	km²	%
Very High	By definition, desertification does not exist in hyperarid regions.									
High										
Moderate										
Desertification Totals										
No Desertification					1,705	12.3	1,663	12.0	3,368	24.3
Total Drylands					1,705	12.3	1,663	12.0	3,368	24.3
Non-dryland									10,496	75.7
Total Area of the Territory									13,864	100.0

GEOGRAPHY

2497 Gillis, W. T.; Byrne, R.; and Harrison, W. *Bibliography of the natural history of the Bahama Islands.* Washington DC: Smithsonian Institution, 1975. (Atoll research bulletin 191) 123 pp.

2498 Posnett, N. W. *Bahamas.* Surbiton, U.K.: United Kingdom. Land Resources Division. Overseas Development Administration, 1971. (Land resource bibliography 1) 74 pp.

Bibliography on natural resources and economic conditions.

BOTANY

2499 Knapp, R. "Vegetation of the Bahamas: review and bibliography." *Excerpta botanica: sectio B: sociologica* 20:2 (1980):137-143.

Unannotated.

ANTHROPOLOGY

2500 LaFlamme, A. G. "An annotated ethnographic bibliography of the Bahama Islands." *Behavior science research* 11:1 (1976):57-66.

CUBA
Republica de Cuba
(a former colony of Spain)

The drylands of Cuba are limited to a few coastal pockets on the southeastern third of the island. These are considered subhumid and share the two rainy seasons common to many of the Caribbean islands. What little moisture deficiency exists on Cuba is due more to rapid runoff on hard soil than to a lack of precipitation. The driest area is at Guantanamo Bay, a sheltered deepwater harbour presently occupied by the United States Navy. Save for this strategic base, Cuba's drylands are of minor importance to the nation.

Only selected reference sources are provided.

Dryland areas of CUBA

DESERTIFICATION RISK	ARIDITY									
	Hyperarid		Arid		Semiarid		Subhumid		Aridity Totals	
	km²	%	km²	%	km²	%	km²	%	km²	%
Very High	By definition, desertification does not exist in hyperarid regions.									
High										
Moderate										
Desertification Totals										
No Desertification					1,030	0.9	3,321	2.9	4,351	3.8
Total Drylands					1,030	0.9	3,321	2.9	4,351	3.8
Non-dryland									110,173	96.2
Total Area of the Territory									114,524	100.0

HANDBOOKS

2501 *Area handbook for Cuba.* 2nd ed. Washington DC: U.S. Dept. of Defense, 1976. (American University foreign area studies) 550 pp.

Narrative summary of the nation's history, politics, geography, economy, social life; includes unannotated bibliography.

STATISTICS

2502 Schroeder, Susan. *Cuba: a handbook of historical statistics.* Boston MA: G. K. Hall, 1982. 589 pp.

Covers climate, demography, labor, economic development, tourism, and other subjects.

ATLASES

2503 *Atlas de Cuba.* La Habana, Cuba: Instituto Cubano de Geodesia y Cartografia, 1978. 143 pp.

Spanish text. Maps of resources, economy, population, history, and geography.

2504 *Atlas nactional de Cuba, en el decimo aniversario de la Revolucion [national atlas of Cuba, on the tenth anniversary of the Revolution].* La Habana, Cuba: Cuba. Academia de Cienicias. Instituto de Geografia, 1970. 132 pp.

In Spanish.

GAZETTEERS

2505 *Cuba: official standard names approved by the United States Board on Geographic Names.* 2nd ed. Washington DC: U.S. Board on Geographic Names, 1963. (Gazetteer 30) 619 pp.

List of 44,000 place-names with latitude/longitude coordinates.

BOTANY

2506 Howard, R. A. "Current work on the flora of Cuba: a commentary." Harvard University. *Arnold Arboretum journal* 26:4 (1977):417-423.

AGRICULTURE

2507 **Kononkov, Petr Fedorovich, and Ustimenko, G. V.** *Selskoe khoziaistvo Respubliki Kuba: obzor literatury [agriculture of Cuba].* Moskva: Selkhoz MSKh SSSR: 79:1, 1970.

In Russian.

AGRICULTURAL ECONOMICS

2508 *Cuban agrarian economy: a bibliography.* Madison WI: The University of Wisconsin. Land Tenure Center, 1974. (Training and methods 23) 47 pp.

Unannotated. Reflects the holdings of the Land Tenure Center library; includes its call numbers.

ECONOMIC DEVELOPMENT

2509 *Quarterly economic review of Cuba, Dominican Republic, Haiti, Puerto Rico.* London: Economist Intelligence Unit, 1976-. Quarterly.

Includes summary of political and economic news as well as charts and statistics of selected economic indicators. Annual supplement.

DOMINICAN REPUBLIC
Republica Dominicana
(a former colony of Spain)

The eastern two thirds of the island of Hispaniola (Santo Domingo) are occupied by the Dominican Republic. Although not as dry overall as Haiti to the west, the nation has two extensive dryland regions.

The Cibao Valley. Inland from the port of Monte Cristi, this semiarid to subhumid region parallels the north coast and follows the Rio Yaque del Norte upstream. Natural vegetation is dryland savanna; rice is grown on irrigated land at the river's mouth and tobacco and cattle are found inland. Santiago, the nation's second city, occupies the center of the valley at the edge of the dryland area.

The Enrequillo Plain. To the south, a structural trough extends across the island from Haiti's capital of Port-au-Prince to the mouth of the Rio Yaque del Sur in the Dominican Republic. The salty lakes of Saumatre and Enrequillo lie on either side of the international border. With a semiarid climate the land supports a natural cover of cactus and thorn scrub; salt and gypsum are mined and irrigation has opened up the Yaqui delta to sugarcane.

Dryland areas of THE DOMINICAN REPUBLIC

DESERTIFICATION RISK	Hyperarid km²	Hyperarid %	Arid km²	Arid %	Semiarid km²	Semiarid %	Subhumid km²	Subhumid %	Aridity Totals km²	Aridity Totals %
Very High	By definition, desertification does not exist in hyperarid regions.									
High										
Moderate										
Desertification Totals										
No Desertification					8,380	17.3	7,169	14.8	15,549	32.1
Total Drylands					8,380	17.3	7,169	14.8	15,549	32.1
Non-dryland									32,893	67.9
Total Area of the Territory									48,442	100.0

GAZETTEERS

2510 *Dominican Republic: official standard names approved by the United States Board on Geographic Names.* 2nd ed. Washington DC: U.S. Defense Mapping Agency Topographic Center, 1972. (Gazetteer 128) 477 pp.

List of 28,400 place-names with latitude/longitude coordinates.

ECONOMIC DEVELOPMENT

2511 *Quarterly economic review of Cuba, Dominican Republic, Haiti, Puerto Rico.*

See entry 2509.

HAITI
Republique d'Haiti
(a former colony of France)

On the rugged and complex terrain of Hispaniola are to be found several small dryland areas on the Haitian third of the island. All are semiarid to subhumid with the two rainy seasons characteristic of the Caribbean drylands.

The Cul de Sac. This is a deep structural depression running eastwards from Port-au-Prince across the nation into the Dominican Republic; the sea once covered the valley floor and left two brackish lakes on either side of the international border. High evapotranspiration leaves even Port-au-Prince—at 1,400 mm (54 inches) of rain per year—too dry for unirrigated sugarcane. With generally less than 1,300 mm (50 inches) of rain, most of the Cul de Sac supports a native vegetation of thorny scrub and giant cacti.

Although initially avoided by the eighteenth century French sugar planters, the region soon became the center of cane production. The French used slaves to construct an elaborate irrigation scheme; some of which survives in use today despite the collapse of the sugar economy after Haiti's 1804 independence. Rapid depopulation of the Cul de Sac followed as the irrigation system fell into disrepair. Subsequent deforestation and soil erosion have left their scars here and elsewhere in the nation. Recently, the western end of the Cul de Sac has been reopened for sugarcane.

The Plain Centrale. On the international border is an intermontane basin about 300 meters (1,000 feet) above sea level; its drier northwestern section was originally covered by open savanna. This dry section was ignored by both the colonial French and the Haitians after independence.

The Artinobite Plain. This is a lowland area drained by the river of the same name and shares a similar climate and natural vegetation with the Cul de Sac. Rice is cultivated at the river mouth, while cotton is grown inland.

Dryland areas of HAITI

DESERTIFICATION RISK	\multicolumn{2}{Hyperarid}	Arid		Semiarid		Subhumid		Aridity Totals	
	km²	%	km²	%	km²	%	km²	%	km²	%
Very High	By definition, desertification does not exist in hyperarid regions.									
High										
Moderate										
Desertification Totals										
No Desertification					3,302	11.9	14,541	52.4	17,843	64.3
Total Drylands					3,302	11.9	14,541	52.4	17,843	64.3
Non-dryland									9,907	35.7
Total Area of the Territory									27,750	100.0

GENERAL BIBLIOGRAPHIES

2512 Bissainthe, Max. *Dictionnaire de bibliographie haitienne [dictionary of Haitian bibliography].* Washington DC: Scarecrow, 1951. 1,052 pp.

In French. *Supplement,* 1973-. Vol. 1 covers period 1950-1970, and includes appendix to main work.

2513 Laguerre, Michel. *The complete Haitiana: a bibliographic guide to the scholarly literature, 1900-1980.* Millwood NY: Kraus-Thomson, (forthcoming). 2 vols.

HANDBOOKS

2514 Weil, Thomas E. *Area handbook for Haiti.* Washington DC: U.S. Dept. of Defense, 1973. (American University foreign area studies) 189 pp.

Narrative summary of the nation's history, politics, geography, economy, social life; includes unannotated bibliography.

ATLASES

2515 *Haiti: mission d'assistance technique integree [Haiti: mission of integrated technical aid].* Washington DC: Organization of American States. Office of Regional Development, 1972. 3 vols. in 1 plus atlas.

In French with English summaries.

GAZETTEERS

2516 *Haiti: official standard names approved by the United States Board on Geographic Names.* 2nd ed. Washington DC: U.S. Board on Geographic Names, 1973. (Gazetteer 129) 211 pp.

List of 13,000 place-names with latitude/longitude coordinates.

AGRICULTURE

2517 Lespinasse, Raymonde. *Bibliographie agricole haitienne, 1950-77 [agriculture bibliography Haiti, 1950-77].* Port-au-Prince: Institut interamericain des sciences agricoles de l'OEA. Representation en Haiti, 1978. 127 pp.

2518 **Zuvekas, Clarence.** *An annotated bibliography of agricultural development in Haiti.* Washington DC: Agency for International Development. Bureau for Latin America. Rural Development Division, 1977. (Working document series. Haiti. General working document 1) 106 pp.

ECONOMIC DEVELOPMENT

2519 *Quarterly economic review of Cuba, Dominican Republic, Haiti, Puerto Rico.*

See entry 2509.

HISTORICAL GEOGRAPHY

2520 **Perusse, Roland I.** *Historical dictionary of Haiti.* Metuchen NJ: Scarecrow, 1977. (Latin American historical dictionaries 15) 124 pp.

Entries cover major topics, events, and personalities in the nation's history; includes extensive unannotated bibliography.

JAMAICA

(a former colony of the United Kingdom)

The drylands of Jamaica mirror Puerto Rico's: a dry leeward south and southwest coast giving way quickly to wet interior mountains. The capital city of Kingston receives only 750 mm (29 inches) of rain a year and surrounding lands require irrigation for agriculture. The area is considered subhumid and displays the typical Caribbean two rainy seasons with a winter drought.

Kingston and the lowland to the west was the heart of Jamaica's sugar industry when the island was one of Great Britain's most valuable colonies. The abolition of slavery in 1838 so disrupted the labor supply that the industry went into sharp decline, never to recover. Widespread abandonment of the coastal sugar lands left the region relatively underpopulated, though population density all over the island continued, and remains, very high.

Steps taken in the 1950's by the government to improve agricultural practices parallel those in Puerto Rico and have given Jamaica one of the stronger economies in the Caribbean though violence, political instability, and scarce capital are major hurdles to overcome.

Dryland areas of JAMAICA

DESERTIFICATION RISK	ARIDITY									
	Hyperarid		Arid		Semiarid		Subhumid		Aridity Totals	
	km²	%	km²	%	km²	%	km²	%	km²	%
Very High	By definition, desertification does not exist in hyperarid regions.									
High										
Moderate										
Desertification Totals										
No Desertification					395	3.6	1,263	11.5	1,658	15.1
Total Drylands					395	3.6	1,263	11.5	1,658	15.1
Non-dryland									9,333	84.9
Total Area of the Territory									10,991	100.0

HANDBOOKS

2521 Kaplan, Irving. *Area handbook for Jamaica.* Washington DC: U.S. Dept. of Defense, 1976. (American University foreign area studies) 332 pp.

Narrative summary of the nation's history, politics, geography, economy, social life; includes unannotated bibliography.

ATLASES

2522 Clarke, Colin G. *Jamaica in maps: graphic perspectives of a developing country.* New York: Africana Publishing, 1974. 104 pp.

GEOLOGY

2523 Kinghorn, Marion, ed. *Bibliography on Jamaican geology.* Norwich, U.K.: GeoAbstracts Ltd., 1977. 150 pp.

AGRICULTURE

2524 Erickson, Frank A., and Erickson, Elizabeth. *An annotated bibliography of agricultural development in Jamaica.* Washington DC: U.S. Dept. of Agriculture. Office of International Cooperation and Development. Development Planning Group, 1979. (Working document series, Jamaica) 197 pp.

2525 Steer, Edgar S. *A select bibliography of reference material providing an introduction to the study of Jamaican agriculture.* Hope, Kingston, Jamaica: Jamaica. Ministry of Agriculture and Fisheries. Agricultural Planning Unit, 1970. 40 pp.

THE NETHERLANDS ANTILLES
De Nederlandse Antillen
(an internally self-governing unit of The Netherlands)

These six islands comprise an autonomous state in the Kingdom of the Netherlands. The six cluster into two groups: Curacao, Aruba, and Bonaire lie off the coast of Venezuela; St. Martin, Saba, and St. Eustatius are parts of the leeward chain.

Curacao, Aruba, and Bonaire. These islands are located within the unusually dryland climatic belt centered on the southern Caribbean and adjacent coast of Venezuela (which see). The islands are arid with a single summer rainy season and winter drought.

Original vegetation consists of scrub woodland and tree cacti on thin-soiled limestone. Little agriculture is practised except on Curacao, where dwarf orange trees produce a fruit used in manufacturing the well-known liquor. Sisal and divi-divi are also cultivated.

The modern wealth of the islands rests on oil refineries on Curacao and Aruba and their attendant commerce. Willemstad, the capital city of the group, is a major trading center for the Caribbean.

St. Martin, Saba, and St. Eustatius. These three lie between the Virgin Islands and St. Kitts and are marginally dry on their leeward slopes. Except for tourism and local fishing their economies are moribund; aid from The Netherlands and Curacao as well as remittances from emigrants help support the islands. At one time Statia (St. Eustatius) had a population some 15 times greater than today's when its sugar industry flourished.

Dryland areas of THE NETHERLANDS ANTILLES

DESERTIFICATION RISK	ARIDITY									
	Hyperarid		Arid		Semiarid		Subhumid		Aridity Totals	
	km²	%	km²	%	km²	%	km²	%	km²	%
Very High	By definition, desertification does not exist in hyperarid regions.									
High										
Moderate										
Desertification Totals										
No Desertification					929	93.2	68	6.8	997	100.0
Total Drylands					929	93.2	68	6.8	997	100.0
Non-dryland									0	0.0
Total Area of the Territory									997	100.0

GENERAL BIBLIOGRAPHIES

2526 Nagelkerke, Gerard A. *Netherlands Antilles: a bibliography 17th century-1980.* Leiden, The Netherlands: Royal Institute of Linguistics and Anthropology. Dept. of Caribbean Studies, 1982. 422 pp.

Unannotated list of 2,852 items.

ENCYCLOPEDIAS

2527 Hoetink, H., ed. *Encyclopedie van de Nederlandse Antillen.* Amsterdam, The Netherlands: Elsevier, 1969. 708 pp.

GAZETTEERS

2528 *Netherlands Antilles: official standard names approved by the United States Board on Geographic Names.* Washington DC: U.S. Board on Geographic Names, 1952. (Gazetteer 157) 17 pp.

List of 600 place-names with latitude/longitude coordinates.

ECONOMIC DEVELOPMENT

2529 **Kok, Michiel.** "Selected bibliography of the Netherlands Antilles economy after 1940." (2 parts). *Economische notities* (Willemstad, Curacao) 2:4 (Oct 1973) and 3:1 (Feb 1974), 2-15, 2-5.

2530 *Quarterly economic review of Venezuela, Netherlands Antilles, Suriname.* London: Economist Intelligence Unit, 1976-. Quarterly.

Includes summary of political and economic news as well as charts and statistics of selected economic indicators. Annual supplement.

HISTORICAL GEOGRAPHY

2531 **Gastmann, Albert L.** *Historical dictionary of the French and Netherlands Antilles.* Metuchen NJ: Scarecrow, 1978. (Latin American historical dictionaries 18) 162 pp.

Entries cover major topics, events, and personalities in the nation's history; includes extensive unannotated bibliography.

PUERTO RICO
The Commonwealth of Puerto Rico
(a former colony of Spain; now a self-governing Commonwealth of the United States)

The drylands of Puerto Rico mirror those of Jamaica: the leeward south coast contains a narrow belt of subhumid territory with the characteristic Caribbean two-rainy seasons and a winter drought. Ponce, the island's second city, receives only 900 mm (36 inches) of rain a year (compare to north coast San Juan's 1,500 mm or 60 inches), an amount adequate only for a sparse scrub vegetation.

Most of this dryland area has supported sugarcane, cattle pasture, coastal coconut palms, and fallow brush. Too dry and hot for coffee and tobacco, the region has historically contributed relatively little to the island's economic wealth. A large population in Ponce and the surrounding countryside, however, has encouraged the establishment of manufacturing industries not tied to the land.

Note: Many additional citations can be found under United States above.

Dryland areas of PUERTO RICO

DESERTIFICATION RISK	ARIDITY									
	Hyperarid		Arid		Semiarid		Subhumid		Aridity Totals	
	km²	%	km²	%	km²	%	km²	%	km²	%
Very High	By definition, desertification does not exist in hyperarid regions.									
High										
Moderate										
Desertification Totals										
No Desertification					168	1.9	1,280	14.4	1,448	16.3
Total Drylands					168	1.9	1,280	14.4	1,448	16.3
Non-dryland									7,443	83.7
Total Area of the Territory									8,891	100.0

GENERAL BIBLIOGRAPHIES

2532 Vivo, Paquita. *The Puerto Ricans: an annotated bibliography.* New York: R. R. Bowker, 1973. 299 pp.

Annotated. Covers most major subject areas.

GEOGRAPHY

2533 Lugo Lugo, Herminio. "Puerto Rico." In *Handbook of contemporary developments in world ecology*, pp. 53-65. Edited by Edward J. Kormondy and J. Frank McCormick. Westport CT: Greenwood, 1981.

A good summary of recent research, most valuable for its mention of research organizations. Includes a short annotated bibliography.

GAZETTEERS

2534 *U.S. possessions in the Caribbean: official standard names approved by the United States Board on Geographic Names.* Washington DC: U.S. Board on Geographic Names, 1958. (Gazetteer 163) 116 pp.

List of 8,500 place-names with latitude/longitude coordinates. Includes Puerto Rico and the Virgin Islands.

GEOLOGY

2535 Hooker, Marjorie. *Bibliography and index of the geology of Puerto Rico and vicinity, 1866-1968.* San Juan, Puerto Rico: Geological Society of Puerto Rico, 1969. 53 pp.

ECONOMIC DEVELOPMENT

2536   Goldsmith, William W.; Clavel, Pierre; and Roth, Deborah. "A bibliography on public planning in Puerto Rico." *Latin American research review* 9:2 (1974):143- 169.

Unannotated.

2537   *Quarterly economic review of Cuba, Dominican Republic, Haiti, Puerto Rico.*

See entry 2509.

HISTORICAL GEOGRAPHY

2538   Faur, Kenneth R. *Historical dictionary of Puerto Rico and the U.S. Virgin Islands.* Metuchen NJ: Scarecrow, 1973. (Latin American historical dictionaries 9) 148 pp.

Entries cover major topics, events, and personalities in the nation's history; includes extensive unannotated bibliography.

TRINIDAD
Trinidad and Tobago

(a former colony of the United Kingdom)

The island of Trinidad contains scattered pockets of subhumid dryland; the largest area, extending into the semiarid zone, exists at the northwestern tip as a climatic extension of Venezuela's Paria Peninsula. The only importance of the region to the nation is its leasing to the United States as a naval base.

Note: Only selected reference sources are provided.

Dryland areas of TRINIDAD

DESERTIFICATION RISK	ARIDITY									
	Hyperarid		Arid		Semiarid		Subhumid		Aridity Totals	
	km²	%	km²	%	km²	%	km²	%	km²	%
Very High	By definition, desertification does not exist in hyperarid regions.									
High										
Moderate										
Desertification Totals										
No Desertification					133	2.6	564	11.0	697	13.6
Total Drylands					133	2.6	564	11.0	697	13.6
Non-dryland									4,431	86.4
Total Area of the Territory									5,128	100.0

HANDBOOKS

2539   Black, Jan Knippers. *Area handbook for Trinidad and Tobago.* Washington DC: U.S. Dept. of Defense, 1976. (American University foreign area studies) 301 pp.

Narrative summary of the nation's history, politics, geography, economy, social life; includes unannotated bibliography.

STATISTICS

2540   *Statistical activities of the American nations: Trinidad and Tobago.* Washington DC: Organization of American States. General Secretariat, 1979. 75 pp.

Prepared by the Central Statistical Office of Trinidad and Tobago. Includes lists of reports published through 1977.

THE TURKS AND CAICOS ISLANDS
(a dependent territory of the United Kingdom)

This island group is a physiographic extension of the Bahama Islands (which see) and thus consists of a limestone platform with low, scrub-covered islands barely rising above the level of the sea. The climate is subhumid and—coupled with scarce water supplies—limits agriculture to home gardening and some sisal cultivation. Fishing is a mainstay and salt evaporation provides some exports. Despite their relatively remote location, the islands offer good tourism potential with their wide beaches, reefs, fishing, and sunny climate.

See sources covering the Caribbean in general for information on this territory.

Dryland areas of THE TURKS AND CAICOS ISLANDS

DESERTIFICATION RISK	ARIDITY									
	Hyperarid		Arid		Semiarid		Subhumid		Aridity Totals	
	km²	%	km²	%	km²	%	km²	%	km²	%
Very High	By definition, desertification does not exist in hyperarid regions.									
High										
Moderate										
Desertification Totals										
No Desertification					12	2.8	418	97.2	430	100.0
Total Drylands					12	2.8	418	97.2	430	100.0
Non-dryland									0	0.0
Total Area of the Territory									430	100.0

THE VIRGIN ISLANDS

(administered as territories of the United States and the United Kingdom)

This small chain of islands between Puerto Rico and the broad Anegada Passage into the Caribbean has small pockets of subhumid dryland on each of the major islands. In general, the drylands occur on the lee side (southwest coast) of each.

The United States Virgin Islands. The largest islands of St. Thomas, St. John, and St. Croix belong to the United States and dominate the chain economically. Sugarcane, cattle, and other crops supplement an economy based primarily on tourism. The drylands are of no particular significance except as locations of the sunniest beaches and sailing grounds. Lack of water is a continuing problem throughout the islands; especially so at the height of the tourist season.

The British Virgin Islands. Smaller and much less populated than the U.S. group, the British islands of Tortola, Virgin Gorda, and nearby islets once had a flourishing cane and cotton plantation economy; they are now economic dependencies of the U.S. group though tourism potential is large and beginning to be developed.

Note: For the U.S. Virgin Islands, many additional sources can be found under United States above.

Dryland areas of THE VIRGIN ISLANDS

DESERTIFICATION RISK	ARIDITY									Aridity Totals	
	Hyperarid		Arid		Semiarid		Subhumid			
	km²	%	km²	%	km²	%	km²	%	km²	%
Very High	By definition, desertification does not exist in hyperarid regions.									
High										
Moderate										
Desertification Totals										
No Desertification								84	17.4	84	17.4
Total Drylands								84	17.4	84	17.4
Non-dryland										400	82.6
Total Area of the Territory										484	100.0

GENERAL BIBLIOGRAPHIES

2541 Reid, Charles Frederick. *Bibliography of the Virgin Islands of the United States.* New York: H.W. Wilson, 1941. 225 pp.

Though old, a comprehensive annotated listing covering most subject areas.

GAZETTEERS

2542 *U.S. possessions in the Caribbean: official standard names approved by the United States Board on Geographic Names.*

See entry 2534.

HISTORICAL GEOGRAPHY

2543 Faur, Kenneth R. *Historical dictionary of Puerto Rico and the U.S. Virgin Islands.*

See entry 2538.

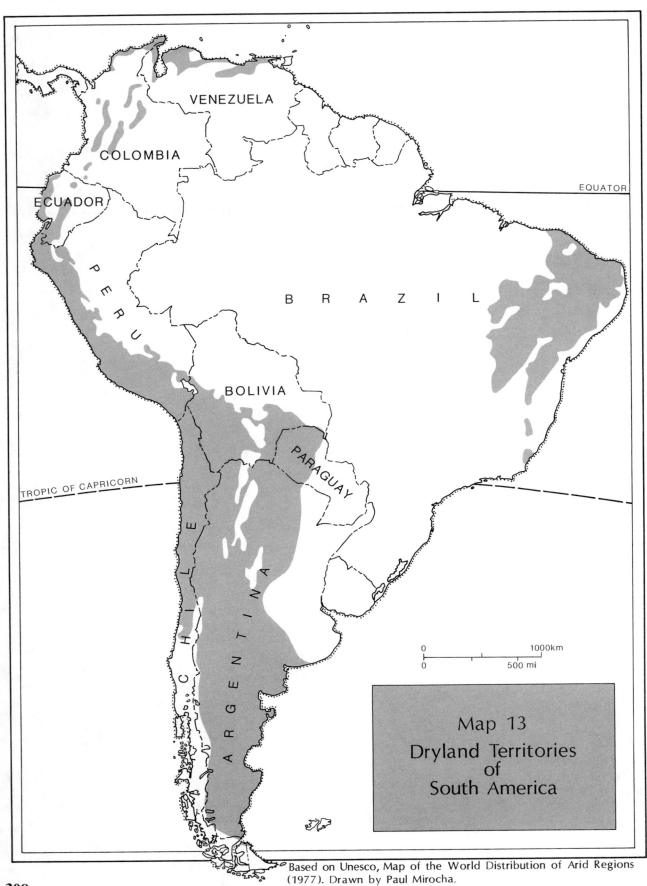
Based on Unesco, Map of the World Distribution of Arid Regions (1977). Drawn by Paul Mirocha.

South America

South America's drylands can be grouped into three regions based on their locations around the great rainforest center of the continent. On the north, along the Caribbean coast of Colombia and Venezuela, exists a small anomalous dryland of little economic importance. To the west the mighty Andes Mountains serve as a formidable control over a number of dryland environments ranging from coastal fog deserts to interior basins to broad steppes beyond the mountain base. To the east at the shoulder of Brazil the third dryland is an extensive semiarid to subhumid region where rainfall is extremely unreliable rather than non-existent.

South America has a number of landscape elements in common with North America yet their arrangement has always made development more difficult and less successful. Though both continents are drained by impressive navigable river systems, those in the South often fail to reach into primary agricultural areas. The Pacific mountain front is geologically similar but, in the South, rises much higher and effectively cuts off the coast from the interior. South America lacks the North's great Arctic wastes yet also loses the humid temperate lands as the continent attenuates to the tip of Cape Horn. Dryland environments are more important to the national economies of the South yet are also more arid and less accessible than those of the North.

Dryland areas of SOUTH AMERICA

DESERTIFICATION RISK	Hyperarid km²	Hyperarid %	Arid km²	Arid %	Semiarid km²	Semiarid %	Subhumid km²	Subhumid %	Aridity Totals km²	Aridity Totals %
Very High	By definition, desertification does not exist in hyperarid regions.		41,792	0.2	347,765	1.9			389,557	2.1
High			874,891	4.9	369,609	2.0	63,341	0.4	1,307,841	7.3
Moderate			72,606	0.4	891,701	5.0	700,646	3.9	1,664,943	9.3
Desertification Totals			989,289	5.5	1,609,075	8.9	763,977	4.3	3,462,331	18.7
No Desertification	198,366	1.1			35,692	0.2	816,119	4.5	1,050,177	5.8
Total Drylands	198,366	1.1	989,289	5.5	1,644,767	9.1	1,580,096	8.8	4,412,518	24.5
Non-dryland									13,569,201	75.5
Total Area of the Territory									17,981,719	100.0

GEOLOGY

2544 Cramer, Howard Ross. "Selected bibliography of South American geology." Tulsa Geological Society. *Digest* 31 (1963):213-239.

Unannotated. Contains only titles referring to geological maps, bibliographies, and complete areal geological studies of various South American nations.

CLIMATOLOGY

2545 Hoffmann, Jose A. J., dir. *Atlas climatico de America del Sur/Climatic atlas of South America.* Ginebra, Switzerland: WMO, 1975-.

In Spanish, English, French, and Portuguese. Contains temperature and precipitation maps. Data refer to periods 1931-60 and 1951-60. Scale 1:10,000,000 and 1:5,000,000.

BIOLOGY

2546 Hurlbert, Stuart H., ed. *Aquatic biota of southern South America being a compilation of taxonomic bibliographies for the fauna and flora of inland waters of southern South America.* San Diego CA: San Diego State University, 1977. 342 pp.

Covers Chile, Argentina, and Uruguay and claims to be useful for neighboring regions of Peru, Bolivia, Paraguay, and Brazil. In Spanish and English.

BOTANY

**2547	Kuechler, August Wilhelm.** *International bibliography of vegetation maps. Section 1. South America.* 2nd ed. Lawrence KS: University of Kansas Libraries, 1980. (Library series 45) 324 pp.

ZOOLOGY

**2548	Vanzolini, Paulo Emilio.** *An annotated bibliography of the land and freshwater reptiles of South America (1758-1975).* Sao Paulo: Universidade de Sao Paulo. Museu Zoologia, 1977-1978. 2 vols.

Unusually well-informed annotations. In English.

ANTHROPOLOGY

**2549	O'Leary, Timothy J.** *Ethnographic bibliography of South America.* New Haven CT: HRAF, 1963, (Behavior science bibliographies) 387 pp.

Unannotated. No coverage of Panama; coverage of Caribbean islands restricted to those belonging to Colombia or Venezuela.

ARGENTINA
Republica Argentina
(a former colony of Spain)

Argentina was Latin America's first industrialized nation and still retains a strong economic base. The economic center gravitates toward Buenos Aires, the capital, chief city, and major port located in the humid zone. Virtually the entire country outside this region is, however, dry to some extent.

Gran Chaco. Argentina occupies significant portions of this semiarid to subhumid tropical scrubland that extends across northern Argentina into Paraguay and Bolivia. Floodwaters spreading across these lowlands permit natural irrigation of maize, wheat, flax, and cotton; the croplands so watered are called *banados*. Lands unsuitable for *banado* cultivation are grazed with cattle. A peculiar resource—the hardwood quebracho scrub tree—grows in salty soils and is exploited for its tannin.

The Andean Piedmont. This complex region of rugged mountains and deep basins is an extension of Bolivia's physiography. A string of oasis cities—Jujuy, Salta, Tucuman, Catamarca, La Rioja, San Juan, Mendoza—are based on irrigated cultivation of sugarcane, wine grapes, alfalfa, maize, and other crops. The region was an important extension of the Incan Empire and, later, was the site of Argentina's first European settlements. It still retains an economic and agricultural importance second only to Buenos Aires and the humid pampa. The nation's relatively small mining industry is also concentrated here; asbestos and beryllium are both imported to the United States in considerable quantities.

The Dry Pampa. This is the drier portion of the great grassland that extends east of the Andes and south of the Chaco to the Atlantic coast. Cattle raising is predominant, especially in the drier western lands where maize and wheat yields fall sharply. Estates are large and population centers few.

Patagonia. This region forms one-quarter of Argentina's territory yet holds less than three percent of its population. It is generally a cold, dry, windswept plateau cut deeply by canyons that provide shelter for men and livestock. Here water is relatively plentiful but outside the canyons the land is often too dry even for sheep. Irrigated croplands in the canyons and giant sheep ranches are the principal land uses. Mining is becoming increasingly important.

Dryland areas of ARGENTINA

DESERTIFICATION RISK	Hyperarid km²	Hyperarid %	Arid km²	Arid %	Semiarid km²	Semiarid %	Subhumid km²	Subhumid %	Aridity Totals km²	Aridity Totals %
Very High					132,955	4.8			132,955	4.8
High	By definition, desertification does not exist in hyperarid regions.		652,163	23.5	192,432	6.9			844,595	30.4
Moderate			22,857	0.8	553,596	19.9	295,182	10.7	872,805	31.4
Desertification Totals			675,020	24.3	878,983	31.6	295,182	10.7	1,849,955	66.6
No Desertification							92,708	3.3	92,708	3.3
Total Drylands			675,020	24.3	878,983	31.6	389,060	14.0	1,943,063	69.9
Non-dryland									834,752	30.1
Total Area of the Territory									2,777,815	100.0

HANDBOOKS

2550 Weil, Thomas E. *Area handbook for Argentina*. 2nd ed. Washington DC: U.S. Dept. of Defense, 1974. (American University foreign area studies) 404 pp.

Narrative summary of the nation's history, politics, geography, economy, social life; includes unannotated bibliography.

GEOGRAPHY

2551 Aparicio, F. de, and Difrieri, H. A., eds. *La Argentina, suma de geografia [Argentina: a geographic summary]*. Buenos Aires: Ediciones Pevser, 1958-1963. 9 vols.

An extensive geographical encyclopedia, with bibliographies.

Argentina/2552-2566

2552 Creasi, Vincent J. *A selected annotated bibliography of environmental studies of Argentina, Chile, and Uruguay.* Washington DC: U.S. Air Force. Environmental Technical Applications Center, 1969. (Technical note ETAC-TN-70- 10; NTIS: AD 717-196) 46 pp.

2553 *Evaluacion de los recursos naturales de la Argentina [evaluation of the natural resources of Argentina].* Buenos Aires: Argentina. Consejo Federal de Inversiones, 1962-63. 9 vols.

2554 Rey Balmaceda, Raul C. *Bibliografia geografica referida a la Republica Argentina [geographical bibliography of the Republic of Argentina].* Buenos Aires: GAEA (Sociedad Argentina de Estudios Geograficos), 1975. (Serie especial 2) 648 pp.

2555 Tuya, O. H., and d'Andrea, F. N. "Bibliografia sobre zonas aridas y semiaridas de la Republica Argentina" [bibliography on arid and semi-arid zones of the Argentine Republic]. Argentina. Bibliografias. Ministerio de Agricultura y Ganaderia. Instituto Nacional de Tecnologia Agropecuaria. Estacion Experimental Regional Agropecuaria, Anguil. *Serie bibliografias* 2 (1972) 16 pp.

In Spanish with English summary.

ATLASES

2556 *Atlas de la Republica Argentina [atlas of the Republic of Argentina].* 4th ed. Buenos Aires: Argentina. Instituto Geographico Militar, 1970. 28 pp.

In Spanish. Scale 1:2,500,000.

2557 Randle, Patricio H. *Atlas del desarrollo territorial de la Argentina [atlas of the territorial development of Argentina].* Buenos Aires: OIKOS, 1981. 3 vols.

In Spanish. Large atlas covers a variety of subjects with separate volumes of explanatory text and historical statistics.

GAZETTEERS

2558 *Argentina: official standard names approved by the United States Board on Geographic Names.* Washington DC: U.S. Board on Geographic Names, 1968. (Gazetteer 103) 699 pp.

List of 48,300 place-names with latitude/longitude coordinates.

GEOLOGY

2559 Barrello, Angel V., ed. *Indice bibliografico de estratigrafia Argentina [bibliographic index of the stratigraphy of Argentina].* La Plata: Argentina. Provincia de Buenos Aires. Comision de Investigacion Cientifica, 1965. 638 pp.

In Spanish. Contains about 4,500 entries.

2560 Leanza, Armando F., ed. *Geologia regional argentina; resultados del Primer Simposio de Geologia Regional realizado en Cordoba (11-15 de septiembre de 1969) [regional geology of Argentina; results of the 1st Symposium of Regional Geology (Cordoba: 11-15 September 1969)],* .86900.

Bibliography: pp. 815-861.

CLIMATOLOGY

2561 Prohaska, F. "The climate of Argentina, Paraguay and Uruguay." In *Climates of Central and South America,* pp. 13-112. Edited by Werner Schwerdtfeger. Amsterdam, The Netherlands: Elsevier Scientific, 1976. (World survey of climatology 12).

A narrative summary with maps, charts, and an unannotated bibliography.

2562 Sparn, Enrique. "Cuarta contribucion al conocimiento de la bibliografia meteorologica y climatologica de la Republica Argentina. Anos 1949-1955" [fourth contribution to the science of meteorological and climatological bibliography of the Republic of Argentina]. *Miscelanea* 44 (1964):1-39.

SOILS

2563 *Soils of Argentina (annotated bibliography 1936-1975).* Harpenden U.K.: Commonwealth Agricultural Bureaux, 1977. (Annotated bibliography) 17 pp.

BOTANY

2564 Eskuche, V. "Bibliographia phytosociologica: Argentina" [bibliography of plant ecology: Argentina]. *Excerpta botanica: sectio B: sociologica* 8:4 (1967):290-315; 19:3 (1979):193-200.

Unannotated.

2565 Morello, J. "Estudios botanicos en las regiones aridas de la Argentina" [botanical studies in the arid regions of Argentina]. *Revista agronomica del Noroeste Argentina* 1 (1955):301-370, 385-524; 2 (1956):79-152.

Contains 160 citations. In Spanish with English summaries. Also appears as: Universidad Nacional de Tucuman. Publicacion 632.

2566 Perez Moreau, Roman A. *Bibliografia geobotanica patagonica; contribucion a la bibliografia botanica argentina contenental patagonico [geobotanical bibliography of Patagonia; contribution to the bibliography of Argentine botany].* Buenos Aires: Argentina. Instituto Nacional del Hielo, 1965. (Publicacion 8) 110 pp.

AGRICULTURE

2567　Hemsy, Victor, ed. *Aporte a la bibliografia agricola argentina, 1909-1979 [contribution to the Argentine agricultural bibliography, 1909-1979].* Tucuman, Argentina: Estacion Experimental Agro-Industrial "Obispo Colombres," 1980. (Publicacion miscelanea 68) 127 pp.

SOLAR ENERGY

2568　Hawkins, Donna. *Solar energy in Argentina: a profile of renewable energy activity in its national context.* Golden CO: U.S. Solar Energy Research Institute, 1981. 44 pp.

A summary and directory of solar energy research activities and organizations in Argentina.

ECONOMIC DEVELOPMENT

2569　*Directory: American business in Argentina.* Buenos Aires: Chamber of Commerce of the United States of America in the Argentine Republic, 1973-. Annual.

A directory of members of the Chamber of Commerce, resident and nonresident.

2570　Longrigg, S. J., ed. *Major companies of Argentina, Brazil, Mexico and Venezuela.* London: Graham & Trotman, 1981-. Annual.

Directory of 3,500 companies in the four nations, giving addresses, officers, principal activities, and, sometimes, financial information.

2571　*Quarterly economic review of Argentina.* London: Economist Intelligence Unit, 1955-. Quarterly.

Includes summary of political and economic news as well as charts and statistics of selected economic indicators. Annual supplement.

HISTORICAL GEOGRAPHY

2572　Wright, Ione Stuessy, and Nekhom, Lisa M. *Historical dictionary of Argentina.* Metuchen NJ: Scarecrow, 1978. (Latin American historical dictionaries 17) 1,113 pp.

Entries cover major topics, events, and personalities in the nation's history; includes extensive unannotated bibliography.

BOLIVIA
Republica de Bolivia
(a former colony of Spain)

Bolivia occupies a poorly defined national territory embracing the central Andes and a vast expanse of the Amazon basin. Despite a considerable natural wealth of minerals, oil, and potentially arable land, the nation suffers from lack of capital, inadequate transportation, and no seaport (which it once had but lost after an 1879-1883 war with Chile). Most of the population is concentrated in the dry rugged highlands of the Andes.

The Andes. This great mountain chain reaches its greatest width in Bolivia. The Eastern Cordillera bordering the Amazon basin is the wetter range; its gentler slopes and valleys contain the best soils and are homes to the highest concentration of rural people. Deeply entrenched rivers provide water for irrigation; maize is the principal crop.

The Titicaca Basin and Altiplano form a generalized intermountain plateau between the Eastern and Western Cordilleras. Though still heavily dissected, the region offers fertile agricultural lands around the Lake and along alluvial fans at the mountains' margins.

The Western Cordillera borders the Atacama Desert and is the driest and least settled portion of the nation. Both altitude and aridity combine to limit agriculture to a few ribbons of irrigated valley soil supporting potatoes and hay. Save for occasional nomadic herds the remainder of the land is empty.

Gran Chaco. Bolivia's portion of this dry scrub forest/savanna once extended considerably farther to the southeast but after a war in 1932-1935 the nation lost much land to Paraguay that it had never effectively controlled. The region, though, offers a crucial resource to Bolivia—oil and natural gas. Extensive fields are located along the Andean piedmont on the eastern slopes of the Eastern Cordillera. Pipelines and refineries have been constructed to give Bolivia an important, if non-renewable, source of income.

Dryland areas of BOLIVIA

DESERTIFICATION RISK	Hyperarid km²	Hyperarid %	Arid km²	Arid %	Semiarid km²	Semiarid %	Subhumid km²	Subhumid %	Aridity Totals km²	Aridity Totals %
Very High	By definition, desertification does not exist in hyperarid regions.									
High			77,999	7.1	57,126	5.2	61,520	5.6	196,645	17.9
Moderate			2,197	0.2	64,816	5.9	21,971	2.0	88,984	8.1
Desertification Totals			80,196	7.3	121,942	11.1	83,491	7.6	285,629	26.0
No Desertification							42,844	3.9	42,844	3.9
Total Drylands			80,196	7.3	121,942	11.1	126,335	11.5	328,473	29.9
Non-dryland									770,108	70.1
Total Area of the Territory									1,098,581	100.0

HANDBOOKS

2573 Weil, Thomas E. *Area handbook for Brazil.* 2nd ed. Washington DC: U.S. Dept. of Defense, 1974. (American University foreign area studies) 417 pp.

Narrative summary of the nation's history, politics, geography, economy, social life; includes unannotated bibliography.

GEOGRAPHY

2574 Munoz Reyes, Jorge. *Bibliografia geografica de Bolivia [geographical bibliography of Bolivia].* La Paz: Bolivia. Academia Nacional de Ciencias, 1967. (Publicacion 16) 170 pp.

In Spanish.

GAZETTEERS

2575 *Bolivia: official standard names approved by the United States Board on Geographic Names*. Washington DC: U.S. Board on Geographic Names, 1955. (Gazetteer 4) 269 pp.

List of 18,800 place-names with latitude/longitude coordinates.

GEOLOGY

2576 **Barth, W.** "Die Gewissenschaftliche literatur Boliviens in der Jahren 1960-1971: ein Uberblick" [the geologic literature of Bolivia from 1960-1971: an overview]. *Zentralblatt fur Geologie und Palaeontologie* 1:1-2 (1972):100-130.

In German with English summaries.

2577 **Munoz Reyes, Jorge; Branisa, Leonardo; and Freile, Alfonso J.** *Bibliografia geologica, mineralogica y paleontologica de Bolivia [geological, mineralogical, and paleontological bibliography of Bolivia]*. La Paz: Bolivia. Ministerio de Minas y Petroleo. Departamento de Geologia, 1962. (Boletin 4) 186 pp.

In Spanish.

CLIMATOLOGY

2578 **Johnson, A. M.** "The climate of Peru, Bolivia and Ecuador."

See entry 2686.

AGRICULTURE

2579 **Altaga de Vizcarra, Irma.** *Bibliografia agricola boliviana [Bolivian agricultural bibliography]*. La Paz: Bolivia. Ministerio de Agricultura. Biblioteca, 1967. 137 pp.

PLANT CULTURE

2580 **Altaga de Vizcarra, Irma.** *Bibliografia boliviana de pastos y forrages [Bolivian bibliography of pastures and fodders]*. La Paz: Ministerio de Asuntos Campesinos y Agricultura. Division de Investigaciones, 1971. 10 leaves.

AGRICULTURAL ECONOMICS

2581 **Anderson, Theresa J., comp.** *Bolivia: agricultura, economia, y politica: a bibliography [Bolivia: agriculture, economics, and politics: a bibliography]*. Madison WI: University of Wisconsin. Land Tenure Center, 1968. (Training and methods 7) 21 pp.

Unannotated. Reflects the holdings of the Land Tenure Center library; includes its call numbers. Continued by *Supplement* one (1970) and *Supplement* two (1972).

2582 **Zuvekas, Clarence.** *An annotated bibliography of agricultural development in Bolivia*. Washington DC: Agency for International Development. Bureau for Latin America. Rural Development Division; U.S. Dept. of Agriculture. Economic Research Division. Foreign Development Division. Sector Analysis Internalization Group, 1977. (Working document series. Bolivia General Working Document U.S. Agency for International Development, Bureau for Latin America, Rural Development Division 1) 162 pp.

ECONOMIC DEVELOPMENT

2583 *Quarterly economic review of Peru, Bolivia*. London: Economist Intelligence Unit, 1976-. Quarterly.

Includes summary of political and economic news as well as charts and statistics of selected economic indicators. Annual supplement.

HISTORICAL GEOGRAPHY

2584 **Heath, Dwight B.** *Historical dictionary of Bolivia*. Metuchen NJ: Scarecrow, 1972. (Latin American historical dictionaries 4) 324 pp.

Entries cover major topics, events, and personalities in the nation's history; includes extensive unannotated bibliography.

BRAZIL
Republica Federativa do Brasil
(a former colony of Portugal)

Brazil is Latin America's largest and most populous nation and possesses, perhaps, the greatest potential for development. Most of the nation is adequately to exceptionally watered but a sizeable dryland area exists in the northeast covering all or part of ten states. Although long-term precipitation records indicate no area is drier than semiarid, short-term drought and flood cycles make the region extremely vulnerable to aridity and desertification. Natural vegetation cover is known as *caatinga*, a tropical scrub woodland ranging from dense thorn forest to open savanna. Several species of trees have economic importance for their fiber, oil, wax, or fruit, though much forest has been cleared for firewood and to open lands to grazing and cultivation.

Land tenure is held by large landowners who lease their acreage to tenants who, in turn, clear the land and plant a variety of crops—maize, beans, cotton—for short term gain. When droughts periodically occur, the tenants leave the region for work elsewhere in Brazil. The landowners then bring in cattle and goats to graze on grass and *palma* (a thornless South African succulent high in water content). Scrub trees invade the pastures and, in time, the cycle is repeated.

Large scale development along the Rio Sao Francisco has been slow to occur, primarily because the river is blocked 160 km (100 miles) from its mouth by falls and its narrow upstream floodplain provides little flatland for irrigation. Nonetheless the river has great potential for hydropower and water diversion given sufficient capital and careful design.

Dryland areas of BRAZIL

DESERTIFICATION RISK	ARIDITY									
	Hyperarid		Arid		Semiarid		Subhumid		Aridity Totals	
	km²	%	km²	%	km²	%	km²	%	km²	%
Very High	By definition, desertification does not exist in hyperarid regions.				214,810	2.5			214,810	2.5
High					35,905	0.4			35,905	0.4
Moderate					148,710	1.7			148,710	1.7
Desertification Totals					399,425	4.6			399,425	4.6
No Desertification							408,195	4.8	408,195	4.8
Total Drylands					399,425	4.6	408,195	4.8	807,620	9.4
							Non-dryland		7,704,345	90.6
							Total Area of the Territory		8,511,965	100.0

HANDBOOKS

2585 Weil, Thomas E. *Area handbook for Brazil.* 3rd ed. Washington DC: U.S. Dept. of Defense, 1975. (American University foreign area studies) 482 pp.

Narrative summary of the nation's history, politics, geography, economy, social life; includes unannotated bibliography.

STATISTICS

2586 *Anuario estatistico do Brasil [statistical yearbook of Brazil].* Rio de Janeiro: Brazil. Instituto Brasileiro de geografia e estatistica, 1936-. Annual.

In Portuguese.

2587 *Indice do Brasil [Brazilian index yearbook].* Rio de Janeiro: Indice-o Banco de Dados, 1945-. Annual.

In English and Portuguese.

GEOGRAPHY

2588 *Dicionario de geografia do Brasil com terminologia geografica [dictionary of the geography of Brazil with geographical terminology].* 2nd ed. Sao Paulo, Brazil: Edicoes Melhoramentos, 1976. 544 pp.

In Portuguese.

2589 *Dicionario geografico Brasileiro [geographical dictionary of Brazil]*. Rio de Janeiro: Editora Globo, 1966. (Enciclopedias e dicionarios Globo) 559 pp.

In Portuguese.

2590 *Geografia do Brasil: regiao nordeste [geography of Brazil: Northwest region]*. Rio de Janeiro: Fundacao Instituto Brasileiro de geografia e estatistica. Directoria tecnica, 1977. 454 pp.

In Portuguese. Excellent narrative survey of the region's physical and economic geography by a number of contributors. Includes charts, maps, unannotated bibliographies. Volume 2 of a five volume set.

2591 *Nordeste Brasileiro: catalogo da exposicao [northeast Brazil: catalog of interpretation]*. Rio de Janeiro: Biblioteca Nacional. Divisao de Publicacoes e Divulgacao, 1970. 86 pp.

In Portuguese.

2592 Tundisi, Jose G. "Brazil." In *Handbook of contemporary developments in world ecology*, pp. 3-22. Edited by Edward John Kormondy and J. Frank McCormick. Westport CT: Greenwood, 1981.

A recent survey of research on Brazilian ecology, most valuable for its lists of research organizations and a good annotated bibliography.

ATLASES

2593 *Atlas nacional do Brasil [national atlas of Brazil]*. Rio de Janeiro: Instituto Brasileiro de Geografia e Estatistica, 1966. Unpaged.

In Portuguese. Part one contains political-demographic-economic maps (50 leaves); part two, not yet published, will contain regional maps. Prepared by the Conselho Nacional de Geografia (Brasil).

GAZETTEERS

2594 *Brazil: official standard names approved by the United States Board on Geographic Names*. Washington DC: U.S. Board on Geographic Names, 1963. (Gazetteer 71) 915 pp.

List of 62,500 place-names with latitude/longitude coordinates.

SCIENCE AND TECHNOLOGY

2595 Erber, Fabio Stefano. "Science and technology in Brazil: a review of the literature." *Latin American research review* 16:1 (1981):3-56.

Narrative discussion with extensive notes and unannotated bibliography.

GEOLOGY

2596 Cruz, Paulo Roberto et al. *Bibliografia comentada e indice da geologia da Bahia [annotated bibliography and index of the geology of Bahia]*. Rio de Janeiro: Brazil. Divisao de Geologia e Mineralogia, 1968. (Boletim 242) 175 pp.

In Portuguese. Bahia state includes the Rio Sao Francisco drainage.

2597 Iglesias, Dolores, and Meneghezzi, Maria de Lourdes. *Bibliografia e indice da geologia do Brasil, 1642- 1940 [bibliography and index of the geology of Brazil, 1641- 1940]*. Rio de Janeiro: Brazil. Ministerio das Minas e Energia. Departamento Nacional da Producao Mineral. Divisao da Geologia e Mineralogia, 1943. (Boletim 111) 323 pp.

In Portuguese. Updated by later bibliographies in the same series. Unannotated. Volume for 1641-1940 is a new edition, completely revised and brought up-to-date, of *Bibliografia da geologia, mineralogia e paleontologia do Brasil*, by A. D. Gonzales (1928), Boletim 27 of the Divisao. Boletim 117 (1955) covers period 1941-1942, 35 pp.; Boletim 131 (1949) covers period 1943-1944, 45 pp.; Boletim 164 (1957) covers period 1945-1950, 128 pp. Volumes for 1951-1960 have been issued as a cumulative volume, Boletim 238 (1967), 203 pp. Boletim 220 (1964) covers period 1960-1961, 71 pp.; Boletim 244 (1969) covers period 1962- 1963, 134 pp.; and Boletim 254 (1970) covers period 1964-1965, 129 pp.

CLIMATOLOGY

2598 *Bibliografia sobre meteorologia agricola do Brasil, 1963-1973 [bibliography on agricultural meteorology in Brazil, 1963-1973]*. Brasilia: Brazil. Ministerio da Agricultura. Departamento Nacional de Meteorologia. Biblioteca, 1974. 22 leaves.

In Portuguese.

2599 *Bibliografia sobre meteorologia do nordeste [bibliography on the meteorology of the northeast]*. Rio de Janeiro: Brazil. Ministerio da Agricultura. Departamento Nacional de Meteorologia. Biblioteca, 1973. 33 leaves.

In Portuguese.

2600 Nimer, Edmon. *Climatologia do Brasil [climatology of Brazil]*. Rio de Janeiro: Instituto Brasileiro de geografia e estatistica, 1979. (Serie recursos naturais e meio ambiente 4) 421 pp.

In Portuguese.

2601 Ratisbona, L. R. "The climate of Brazil." In *Climates of Central and South America*, pp. 219-293. Edited by Werner Schwerdtfeger. Amsterdam, The Netherlands: Elsevier Scientific, 1976. (World survey of climatology 12).

A narrative summary with maps, charts, and an unannotated bibliography.

BOTANY

2602 Pereira, J. F.; Valente, M. da C.; and Silva, N. M. F. da. "Bibliografia botanica Brasileira (taxinomia e anatomia das angiospermae) e levantamento dos 'Tipos' do herbario do Jardim Botanico do Rio de Janeiro" [Brazilian botany bibliography (taxonomy and anatomy of angiospermae) and survey of types in the herbarium of the Botanical Gardens of Rio de Janeiro, Brazil]. *Rodriguesia* 31:50 (1979):269-273.

ZOOLOGY

2603 *Bibliografia brasileira de zoologia [Brazilian bibliography of zoology]*. Rio de Janeiro: Instituto Brasileiro de Informacao em Ciencia e Tecnologia. Instituto Brasileiro de Bibliografia e Documentacao.

AGRICULTURE

2604 *Bibliografia agricola do Brasil [agricultural bibliography of Brazil]*. Rio de Janeiro: Sociedade Nacional de Agricultura, 1968-.

2605 *Bibliografia brasileira de agricultura [Brazilian bibliography of agriculture]*. Brasilia: EMBRATER, 1978-. Annual.

EMBRATER = Empresa Brasileira de Assistencia Tecnica e Extenao Rural. In Portuguese. Supersedes: *Bibliografia brasileira de ciencias agricolas [Brazilian bibliography of agricultural sciences]* (Rio de Janeiro: Instituto Brasileiro de Bibliografia e Documentacao, 1968-75), 8 volumes.

ANIMAL CULTURE

2606 *Bibliografia brasileira de medicina veterinaria e zootecnia [Brazilian bibliography of veterinary medicine and zootechnology]*. Sao Paulo, Brazil: EMBRAPA, Departamento de Informacao e Documentacao for Universidade. Facultade de Medicina Veterinaria e Zootecnia, 1977.

AGRICULTURAL ECONOMICS

2607 *Agrarian reform in Brazil: a bibliography*. Madison WI: The University of Wisconsin. Land Tenure Center, 1972. (Training and methods 18/19).

In two parts, with one supplement for each part. Unannotated. Reflects holdings of the Land Tenure Center library; includes its call numbers.

2608 Santos, M. C. D. "Economia rural e planejamento economico: factores que dificultam o planejamento economico da producao rural brasileira—Bibliografia" [rural economy and economic planning: factors which hamper economic planning of Brazilian rural production: a bibliography]. *Brasil Acucareiro* 80:4 (1972):88-93.

2609 Schuh, George Edward. *Research on agricultural development in Brazil*. New York: Agricultural Development Council, 1970. (Research monographs 7) 302 pp.

HUMAN GEOGRAPHY

2610 Levine, Robert M. *Brazil since 1930: an annotated bibliography for social historians*. New York: Garland, 1980. (Reference library of social science 59) 336 pp.

DEMOGRAPHY

2611 *Sinopse preliminar do censo demografico: IX recensenmento geral do Brasil 1980 [preliminary synopsis of the demographic census: IX general census of Brazil 1980]*. Rio de Janeiro: Instituto Brasileiro de Geografia e Estatistica, 1981. 25 vols.

MEDICINE

2612 Campos, E. C., and Uehara, B. "Meningites no Brasil. Levantamento bibliografico de 1880 a 1975"/"Meningitis in Brazil. Bibliographical survey from 1880 to 1975." Sao Paulo. Instituto Adolfo Lutz. *Revista* 37 (1977):95-130.

In Portuguese with English summary.

2613 De Lacerda, E. S. B., and De Lacerda, J. P. "Poliomielite no Brasil. Levantamento bibliografico de 1911 a 1977."/"Poliomielitis in Brazil. Bibliographical survey from 1911 to 1977." Sao Paulo. Instituto Adolfo Lutz. *Revista* 39:2 (1979):99-115.

In Portuguese with English summary.

2614 Mearim, A. B., and Correa, M. A. A. "Leptospiroses no Brasil. Levantamento bibliografico de 1971 a 1977."/"Leptospiroses in Brazil. Bibliographical survey from 1971 to 1977." Sao Paulo. Instituto Adolfo Lutz. *Revista* 37 (1977):131-140.

In Portuguese with English summary.

ECONOMIC DEVELOPMENT

2615 Ferreira, Carmosina N.; Ferreira, Lieny do Amaral; and Moletta, Elizabeth Tolomei. *Bibliografia selectiva sobre desenvolvimento economico no Brasil [selective bibliography on Brazilian economic development]*. Rio de Janeiro: Ministerio do Planejamento e Coordenacao Geral. IPEA. Sector de Documentacao, 1972. (Serie bibliografica 1) 96 pp.

2616 *Kompass: register of industry and commerce in Brazil/Kompass: sinopse comercial do Brasil*. Sao Paulo, Brazil: Publicacoes Kompass, 1975-. Annual.

In Portuguese, Spanish, and English.

2617 Longrigg, S. J., ed. *Major companies of Argentina, Brazil, Mexico and Venezuela 1982*.

See entry 2570.

2618 *Quarterly economic review of Brazil*. London: Economist Intelligence Unit, 1976-. Quarterly.

Includes summary of political and economic news as well as charts and statistics of selected economic indicators. Annual supplement.

HISTORICAL GEOGRAPHY

2619 Levine, Robert M. *Historical dictionary of Brazil*. Metuchen NJ: Scarecrow, 1979. (Latin American historical dictionaries 19) 297 pp.

Entries cover major topics, events, and personalities in the nation's history; includes extensive unannotated bibliography.

CHILE
Republica de Chile

(a former colony of Spain)

The Atacama Desert. Chile's dryland core is the Atacama Desert, a coastal fog desert so dry that at some stations many years pass without any recorded precipitation. The coast itself is extremely dangerous and inaccessible; frequent fog, a great escarpment, and no natural harbors limit navigation while earthquakes periodically destroy man's precarious footholds. Inland the cliffs give way to low hills and then to a narrow structural trench dotted with dry lake beds. Further east the Western Cordillera of the Andes rises up; where streams cut through the mountains steep alluvial fans have formed at the desert margin. These fans hold the only reliable water supplies and thus are the location of scattered agricultural areas developed originally by native Indians and occupied by them today. One river, the Loa, crosses the desert to reach the sea while still holding water; all others dry up in mid-course depending on season and floodstage.

The Atacama's modern development began with the mining of silver, copper, and sodium nitrate. Each mineral has produced a cycle of boom and bust in its exploitation; presently, copper is paramount in importance. Attempts to stabilize the workforce and the economic base have led to construction of other industrial plants including refineries, fish meal plants, and hydroelectric installations.

Central Chile. South of the Atacama lies a narrow semiarid to subhumid belt of Mediterranean-type climate, roughly comparable to coastal California in its relationship to the coastal desert. Also comparable is the general topography: low coast ranges give way to a long central valley and then to a great mountain range. The combination of adequate water supplies for irrigation and a mild climate make this region an attractive core for the nation; most of the population lives here and is concentrated in the cities of Valparaiso and Santiago. Land tenure follows the typical Latin American pattern with large estates given over to livestock and smallholding tenants producing grains (in Chile, principally wheat), fruits, and other food crops. Vineyards exist on the steeper slopes. Despite similarities to California, agricultural production has not kept pace with population, due more to the antiquated land tenure system and lack of capital improvements than to any inherent deficiencies in resources.

Dryland areas of CHILE

DESERTIFICATION RISK	ARIDITY									
	Hyperarid		Arid		Semiarid		Subhumid		Aridity Totals	
	km²	%	km²	%	km²	%	km²	%	km²	%
Very High	By definition, desertification does not exist in hyperarid regions.		41,792	5.6					41,792	5.6
High			92,702	12.3					92,702	12.3
Moderate			24,315	3.2	19,721	2.6	26,215	3.5	70,251	9.3
Desertification Totals			158,809	21.1	19,721	2.6	26,215	3.5	205,745	27.2
No Desertification	117,398	15.6					30,394	4.0	147,792	19.6
Total Drylands	117,398	15.6	158,809	21.1	19,721	2.6	56,609	7.5	352,537	46.8
Non-dryland									399,089	53.2
Total Area of the Territory									751,626	100.0

HANDBOOKS

2620 Weil, Thomas E. *Area handbook for Chile.* Washington DC: U.S. Dept. of Defense, 1969. (American University foreign area studies) 509 pp.

Narrative summary of the nation's history, politics, geography, economy, social life; includes uannotated bibliography.

GEOGRAPHY

2621 Creasi, Vincent J. *A selected annotated bibliography of environmental studies of Argentina, Chile, and Uruguay.*

See entry 2552.

Chile/2622-2636

2622 Gligo Viel, N. et al. *Bibliografia de recursos naturales (1966-1971)* [bibliography of natural resources]. Santiago: Chile. Instituto de Investigacion de Recursos Naturales, 1973. 266 pp.

Continues entry 2623.

2623 *Informacion bibliografica de recursos naturales, 1945-1965* [bibliographic information on natural resources, 1945-1965]. Santiago: Chile. Instituto de Investigacion de Recursos Naturales, 1967. 294 pp.

Continued by entry 2622.

2624 Risopatron Sanchez, Luis. *Diccionario jeografico de Chile* [geographical dictionary of Chile]. Santiago: Imprenta Universitaria, 1924. 958 pp.

In Spanish.

ATLASES

2625 *Atlas de la Republica de Chile* [atlas of the Republic of Chile]. Santiago: Chile. Instituto Geografico Militar, 1966. 240 pp.

In Spanish, English, and French.

GAZETTEERS

2626 *Chile: official standard names approved by the United States Board on Geographic Names.* 2nd ed. Washington DC: U.S. Board on Geographic Names, 1967. (Gazetteer 6) 591 pp.

List of 39,700 place-names with latitude/longitude coordinates.

GEOLOGY

2627 Munoz Cristi, Jorge, and Karot, Juan K. *Bibliografia geologica de Chile (1927-1953)* [geological bibliography of Chile (1927-1953)]. Santiago: Universidad de Chile. Instituto de Geologica, 1955. (Publicacion 5) 121 pp.

Contains some 600 annotated citations.

2628 Srytrova, D. *Bibliografia geologica del norte de Chile: apartados, informes ineditos y memorias de titulo del Departamento de Geologia de la Universidad de Chile en Santiago* [geological bibliography of northern Chile: separates, unpublished reports and articles from the Dept. of Geology, University of Chile at Santiago]. Antofagasta: Chile. Universidad del Norte. Centro de Documentacion y Informacion, 1975-76. 2 parts. 52 pp. and 234 pp.

In Spanish.

CLIMATOLOGY

2629 *Bibliography on the climate of Chile.* Washington DC: U.S. Weather Bureau, 1956. (NTIS: PB 176-732; AD 670-036) 22 pp.

Supplement 1 (1967), 25 pp. (NTIS: PB 176-739; AD 670-043).

2630 Miller, A. "The climate of Chile." In *Climates of Central and South America*, pp. 113-145. Edited by Werner Schwerdtfeger. Amsterdam, The Netherlands: Elsevier Scientific, 1976. (World survey of climatology 12).

A narrative summary with maps, charts, and an unannotated bibliography.

BOTANY

2631 Oberdorfer, Erich. "Bibliographia phytosociologica: Chile" [bibliography of plant ecology: Chile]. *Excerpta botanica* B 3(1):79-80.

Unannotated. Includes journal articles from 1905 through 1959.

2632 Ramirez, C. "Bibliographia phytosociologica et scientiae vegetationis: Chile" [bibliography of plant ecology and vegetation: Chile]. *Excerpta botanica: sectio B: sociologica* 19:1 (1979):63-92; 20:1 (1980):61-64; 20:4 (1980):305-319.

Unannotated, with author and geographical indexes.

AGRICULTURAL ECONOMICS

2633 Ammon, Alf. *Probleme der Agrarreform in Chile: Ubersichtsstudie und Bibliographie* [problems of the agrarian reform in Chile: review study and bibliography]. Bonn: Verlag Neue Gesellschaft, 1971. 125 pp.

Bibliography: pp. 90-125.

2634 *Chile's agricultural economy: a bibliography.* Madison WI: The University of Wisconsin. Land Tenure Center, 1970. (Training and methods 12) 65 pp.

Unannotated. Reflects holdings of the Land Tenure Center library; includes its call numbers. Continued by Supplement 1 (1971) and Supplement 2 (1974).

ECONOMIC DEVELOPMENT

2635 *Quarterly economic review of Chile.* London: Economist Intelligence Unit, 1976-. Quarterly.

Includes summary of political and economic news as well as charts and statistics of selected economic indicators. Annual supplement.

HISTORICAL GEOGRAPHY

2636 Bizzarro, Salvatore. *Historical dictionary of Chile.* Metuchen NJ: Scarecrow, 1972. (Latin American historical dictionaries 7) 309 pp.

Entries cover major topics, events, and personalities in the nation's history; includes extensive unannotated bibliography.

COLOMBIA
Republica de Colombia
(as Nueva Granada, a former colony of Spain)

Colombia's drylands consist of a semiarid Caribbean coastal strip and a number of deep subhumid valley bottoms inland along the Cauca and Magdalena Rivers.

The Guajira Peninsula. This is Colombia's share of the Caribbean coastal dryland anomaly that extends across Venezuela (which see). Composed of a platform of crystalline rock, the peninsula has little economic importance beyond a poor goat range.

The Interior Valleys. The great mountain-building forces that shaped Colombia's landscape left several deep structural troughs that, at their bottoms, hold subhumid lands along the major rivers. Beef cattle and sugarcane are the chief products; in addition, a dryland variety of the coca plant (*Erythroxylum novogranatense*) is produced for illegal export.

Dryland areas of COLOMBIA

DESERTIFICATION RISK	Hyperarid km²	Hyperarid %	Arid km²	Arid %	Semiarid km²	Semiarid %	Subhumid km²	Subhumid %	Aridity Totals km²	Aridity Totals %
Very High	By definition, desertification does not exist in hyperarid regions.									
High			2,277	0.2	11,389	1.0			13,666	1.2
Moderate										
Desertification Totals			2,277	0.2	11,389	1.0			13,666	1.2
No Desertification					2,277	0.2	84,279	7.4	86,556	7.6
Total Drylands			2,277	0.2	13,666	1.2	84,279	7.4	100,222	8.8
Non-dryland									1,038,692	91.2
Total Area of the Territory									1,138,914	100.0

HANDBOOKS

2637 Blutstein, Howard I. *Area handbook for Colombia.* 3rd ed. Washington DC: U.S. Dept. of Defense, 1977. (American University foreign area studies) 508 pp.

Narrative summary of the nation's history, politics, geography, economy, social life; includes unannotated bibliography.

STATISTICS

2638 *General features of Colombia.* Bogota: Colombia. Fondo de Promocion de Exportaciones, 1979. 108 pp.

Consists of a short narrative text on the nation's economy and international trade plus numerous statistical tables.

GEOGRAPHY

2639 Banderas, Pedro Antonio. *Diccionario geografico, industrial y agricola del Valle del Cauca [geographical, industrial and agricultural dictionary of the Valle del Cauca].* Buenos Aires: Instituto del Libro, 1944. 421 pp.

In Spanish. Dated but still a good source of information on the region.

2640 Martinson, Tom L. *Research guide to Colombia.* Pan American Institute of Geography and History, 1975. (Publicacion 341) 58 pp.

Concentrates on map coverage of the nation.

ATLASES

2641 *Atlas basico de Colombia [basic atlas of Colombia].* Bogota: Instituto Geografico Agustin Codazzi, 1970. 106 pp.

In Spanish.

2642 *Atlas de Colombia [atlas of Colombia].* Bogota: Instituto Geografico Agustin Codazzi, 1967. 203 pp.

In Spanish.

GAZETTEERS

2643 *Colombia: official standard names approved by the United States Board on Geographic Names*. Washington DC: U.S. Board on Geographic Names, 1964. (Gazetteer 86) 396 pp.

List of 27,000 place-names with latitude/longitude coordinates.

GEOLOGY

2644 *Bibliografia de la biblioteca del Instituto Geofisico de Los Andes Colombianos sobre geologica y geofisica de Colombia [bibliography of the library of the Geophysical Institute of the Colombian Andes, on geology and geophysics of Colombia]*. Bogota: Instituto Geofisico de Los Andes Colombianos, 1951. (Boletin, serie C. Geologica 2) 267 pp.

In Spanish. Supplemented by: Jesus Emilio Ramirez, *Primer suplemento a la bibliografia de la biblioteca...* (1973), Publicacion, serie C 18, 436 pp.

CLIMATOLOGY

2645 Snow, J. W. "The climate of northern South America." In *Climates of Central and South America*, pp. 295-403. Edited by Werner Schwerdtfeger. Amsterdam, The Netherlands: Elsevier Scientific, 1976. (World survey of climatology 12).

Covers Colombia, Venezuela, Surinam, Guyana, French Guiana. A narrative summary with maps, charts, and an unannotated bibliography.

SOILS

2646 *Soils of Colombia (annotated bibliography 1941-1975)*. Harpenden: U.K. Commonwealth Agricultural Bureaux, 1977. (Annotated bibliography) 16 pp.

AGRICULTURE

2647 Perez Cordero, Luis de J. *Bibliografia del sector agropecuario colombiano [bibliography of Colombian agriculture]*. Bogota: Instituto Colombiana de la Reforma Agraria. Division de Estudios Tecnicos, 1967. 399 pp.

In Spanish.

AGRICULTURAL ECONOMICS

2648 *Colombia: background and trends: a bibliography*. Madison WI: University of Wisconsin. Land Tenure Center, 1969. (Training and methods 9) 56 pp.

Unannotated. Reflects holdings of the Land Tenure Center library; includes its call numbers. Continued by *Supplement* one (1971) and *Supplement* two (1973).

2649 Uribe C., Maruja, comp. *Bibliografia selectiva sobre desarrollo rural en Colombia [selective bibliography on rural development in Colombia]*. Bogota: Biblioteca IICA-CIRA, 1978. (Documentacion e informacion agricola 57) 207 pp.

In Spanish, English, French, and Portuguese.

ANTHROPOLOGY

2650 Bernal Villa, Segundo. *Guia bibliografica de Colombia de interes para el antropologo [bibliographic guide to Colombia of interest to the anthropologist]*. Bogota: Ediciones Universidad de los Andes, 1969. 782 pp.

In Spanish. Extensive unannotated listing covering anthropology, archeology, geography, history, linguistics.

MEDICINE

2651 Cantillo, Jaime et al. "Leichmaniasis visceral (Kala-azar) en Colombia" [visceral leishmaniasis (Kala azar) in Colombia]. *Revista Latinomericana de patologia* 9:3-4 (1970):163-171.

In Spanish with English summary. It includes a review of the literature, including the complete National Medical Bibliography.

ECONOMIC DEVELOPMENT

2652 *Quarterly economic review of Colombia, Ecuador*. London: Economist Intelligence Unit, 1976-. Quarterly.

Includes summary of political and economic news as well as charts and statistics of selected economic indicators. Annual supplement.

HISTORICAL GEOGRAPHY

2653 Davis, Robert Henry. *Historical dictionary of Colombia*. Metuchen NJ: Scarecrow, 1977. (Latin American historical dictionaries 14) 280 pp.

Entries cover major topics, events, and personalities in the nation's history; includes extensive unannotated bibliography.

ECUADOR
Republica del Ecuador
(a former colony of Spain)

The Atacama Desert. Ecuador occupies the northernmost extension of the Atacama and its scrubland continuation along the coast. The transition from wet tropical forest to arid desert is unusually abrupt. Indian settlements predominate (as opposed to black African villagers in the wetter north); cotton is a major commercial crop and fishing is becoming increasingly important.

The Interior Basins. Several of the deep interior basins between the cordilleras of the Andes have lands that range from subhumid to semiarid. Settlement is dense, mainly mestizo and Indian, and supported by subsistence farming and cattle-raising. In some areas can be found irrigated sugarcane and cotton. A rail line and the Pan American Highway connect each basin so transportation is not as difficult as it is in other Andes valleys.

The Galapagos Islands. These islands lie some 1,050 km (650 miles) west of Ecuador and are of great ecological importance. Although a territory of Ecuador, they are considered separately in this Guide.

Dryland areas of ECUADOR (EXCLUDES THE GALAPAGOS ISLANDS)

DESERTIFICATION RISK	ARIDITY								Aridity Totals	
	Hyperarid		Arid		Semiarid		Subhumid			
	km²	%	km²	%	km²	%	km²	%	km²	%
Very High	By definition, desertification does not exist in hyperarid regions.									
High					1,366	0.3	1,821	0.4	3,187	0.7
Moderate			15,029	3.4	4,554	1.0	61,486	13.7	81,069	18.1
Desertification Totals			15,029	3.4	5,920	1.3	63,307	14.1	84,256	18.8
No Desertification							5,009	1.1	5,009	1.1
Total Drylands			15,029	3.4	5,920	1.3	68,316	15.2	89,265	19.9
Non-dryland									358,345	80.1
Total Area of the Territory									447,610	100.0

HANDBOOKS

2654 Weil, Thomas E. *Area handbook for Ecuador.* Rev. ed. Washington DC: U.S. Dept. of Defense, 1973. (American University foreign area studies) 403 pp.

Narrative summary of the nation's history, politics, geography, economy, social life; includes unannotated bibliography.

GEOGRAPHY

2655 Navarro Andrade, Ulpiano. *Geografia economica del Ecuador [economic geography of Ecuador].* Quito: Editorial Santo Domingo, 1965-66. 2 vols.

In Spanish. A general geography text with extensive bibliographies; not limited strictly to economic geography.

ATLASES

2656 *Atlas geografico de la Republica del Ecuador [geographical atlas of the Republic of Ecuador].* Quito: Ecuador. Instituto Geografico Militar, 1978. 82 pp.

In Spanish. Scale of base map 1:2,000,000. The Galapagos Islands are covered at 1:4,000,000.

2657 Sampedro V., Francisco. *Atlas geografico del Ecuador [geographical atlas of Ecuador].* Rev. ed. Quito: Offset Editorial Colon, 1975-76. (Atlas geografico escolar del Ecuador "SAM" 1975) 71 pp.

In Spanish. The Galapagos Islands are minimally covered at a much reduced scale.

GAZETTEERS

2658 *Ecuador: official standard names approved by the United States Board on Geographic Names.* Washington DC: U.S. Board on Geographic Names, 1957. (Gazetteer 36) 189 pp.

List of 14,850 place-names with latitutde/longitude coordinates. Includes the Galapagos Islands.

Ecuador/2659-2665

GEOLOGY

2659 **Bristow, Clement Roger.** *An annotated bibliography of Ecuadorian geology*. London: United Kingdom. Her Majesty's Stationery Office, 1981. (Overseas geology and mineral resources 58) 38 pp.

CLIMATOLOGY

2660 **Johnson, A. M.** "The climate of Peru, Bolivia and Ecuador."

See entry 2686.

AGRICULTURAL ECONOMICS

2661 *Economic aspects of agricultural development in Ecuador: a bibliography*. Madison WI: The University of Wisconsin. Land Tenure Center, 1972. (Training and methods 21) 28 pp.

Unannotated. Reflects the holdings of the Land Tenure Center library; includes its call numbers.

HUMAN GEOGRAPHY

2662 **Larrea, Carlos Manuel.** *Bibliografia cientifica del Ecuador: antropologia, etnografia, arqueologia, prehistoria, linguistica [scientific bibliography of Ecuador: anthropology, ethnography, archeology, prehistory, linguistics]*. 3rd ed. Quito: Corporacion de Estudios y Publicaciones, 1968. 289 pp.

In Spanish. Unannotated. Contains 2,234 entries covering monographs and journal articles.

ECONOMIC DEVELOPMENT

2663 *Quarterly economic review of Colombia, Ecuador*.

See entry 2652.

HISTORICAL GEOGRAPHY

2664 **Bork, Albert William, and Maier, Georg.** *Historical dictionary of Ecuador*. Metuchen NJ: Scarecrow, 1973. (Latin American historical dictionaries 10) 192 pp.

Entries cover major topics, events, and personalities in the nation's history; includes extensive unannotated bibliography.

2665 **Norris, Robert E.** *Guia bibliografica para el estudio de la historia ecuatoriana [bibliographic guide to the study of Ecuadorian history]*. Austin TX: The University of Texas at Austin. Institute of Latin American Studies, 1978. (Guides and bibliographies series 11) 295 pp.

In Spanish. Partially annotated.

PARAGUAY
Republica del Paraguay
(a former colony of Spain)

Paraguay is a land of broad rivers, forests, a mild climate, and a single large dryland region known as the Gran Chaco. This is a vast alluvial plain formed by sediments washed down from the Andes far to the west. The natural vegetation is a dense dry thorn forest with occasional openings onto grassy savannas. Surface streams are rare but groundwater—often alkaline—lies within a few feet of the surface. Economic exploitation of the region has been minimal. Quebracho trees are harvested for their tannin and cattle are run on the grassy areas; Mennonite agricultural settlements in the interior are the most ambitious attempts at colonization.

Dryland areas of PARAGUAY

DESERTIFICATION RISK	ARIDITY									
	Hyperarid		Arid		Semiarid		Subhumid		Aridity Totals	
	km²	%	km²	%	km²	%	km²	%	km²	%
Very High	By definition, desertification does not exist in hyperarid regions.									
High										
Moderate					52,877	13.0	111,450	27.4	164,327	40.4
Desertification Totals					52,877	13.0	111,450	27.4	164,327	40.4
No Desertification										
Total Drylands					52,877	13.0	111,450	27.4	164,327	40.4
Non-dryland									242,425	59.6
Total Area of the Territory									406,752	100.0

GENERAL BIBLIOGRAPHIES

2666 Jones, David Lewis. *Paraguay, a bibliography*. New York: Garland, 1979. (Garland reference library of social science 51) 499 pp.

2667 Moscov, Stephen C. *Paraguay: an annotated bibliography*. Buffalo NY: State University of New York at Buffalo. Council on International Studies, 1972. (Special studies 10) 103 leaves.

Includes 884 annotated citations to items published through 1968. Many Latin American publications are included.

HANDBOOKS

2668 Weil, Thomas E. *Area handbook for Paraguay*. Washington DC: U.S. Dept. of Defense, 1972. (American University foreign area studies) 316 pp.

Narrative summary of the nation's history, politics, geography, economy, social life; includes unannotated bibliography.

ATLASES

2669 Emategui, Federico. *Atlas Hermes: compendio geografico del Paraguay [Hermes atlas: geographical compendium of Paraguay]*. Asuncion: Hermes Editorial Pedagogica, 1977. 62 pp.

In Spanish.

GAZETTEERS

2670 *Paraguay: official standard names approved by the United States Board on Geographic Names*. Washington DC: U.S. Board on Geographic Names, 1957. (Gazetteer 35) 32 pp.

List of 2,300 place-names with latitude/longitude coordinates.

CLIMATOLOGY

2671 Prohaska, F. "The climate of Argentina, Paraguay and Uruguay."

See entry 2561.

Paraguay/2672-2674

AGRICULTURE

2672 **Sanchez, Maria Elsa Bareiro de.** *Tesis de tecnicos paraguayos sobre temas agropecuarios y forestales (bibliografia) [theses of Paraguayan technicians on agricultural and forestry subjects (bibliography)]*. Asuncion: Paraguay. Ministerio de Agricultura y Ganaderia. Instituto Interamericano de Ciencias Agricolas, 1979. 21 pp.

In Spanish.

ECONOMIC DEVELOPMENT

2673 *Quarterly economic review of Uruguay, Paraguay*. London: Economist Intelligence Unit, 1976-. Quarterly.

Includes summary of political and economic news as well as charts and statistics of selected economic indicators. Annual supplement.

HISTORICAL GEOGRAPHY

2674 **Kolinski, Charles J.** *Historical dictionary of Paraguay*. Metuchen NJ: Scarecrow, 1973. (Latin American historical dictionaries 8) 282 pp.

Entries cover major topics, events, and personalities in the nation's history; includes extensive unannotated bibliography.

PERU
Republica del Peru
(former colony of Spain)

The drylands of Peru occupy two of the nation's three great physiographic regions and hold one of the world's largest arid cities.

The Atacama Desert. Peru's portion of this narrow coastal desert is typically cool, foggy, barren of vegetation, and very dry. Although the land offers little in exploitable resources the cool ocean water teems with microscopic life that draws great schools of fish and flocks of birds. Together with irrigated oasis agriculture along streams crossing the desert, the Atacama and its offshore waters offer a surprisingly rich resource base. Disaster, however, occurs irregularly when warmer tropical waters invade the coast, drive off the fish, starve the birds, and bring torrential rains that destroy homes and irrigation works; this phenomenon (called El Nino for its appearance around Christmas) occurred most recently during 1982-83.

Lima is the nation's capital and principal city and, next to Cairo, is the world's largest urban area in a hyperarid dryland. It exists due to its location on one of the larger agricultural oases and near one of the few protected coastal anchorages; it flourishes now as the center of manufacturing, trade, and administration.

The Andes. The complex topography of the Andes ranges makes generalization about the dryland areas difficult. Northern and eastern slopes tend to be wet while those facing the Pacific are drier. The deeper canyons, especially that of the Rio Maranon, are also dry. Vegetation on these lands is typically mountain grassland given over to sheep. Crops are grown depending on altitude as much as water supply; bananas, oranges, sugarcane, maize, wheat, barley, and potatoes are favored in ascending order.

Dryland areas of PERU

DESERTIFICATION RISK	ARIDITY									
	Hyperarid		Arid		Semiarid		Subhumid		Aridity Totals	
	km²	%	km²	%	km²	%	km²	%	km²	%
Very High	By definition, desertification does not exist in hyperarid regions.									
High			48,838	3.8	66,831	5.2			115,669	9.0
Moderate					28,274	2.2	143,944	11.2	172,218	13.4
Desertification Totals			48,838	3.8	95,105	7.4	143,944	11.2	147,799	22.4
No Desertification	80,968	6.3			33,415	2.6	114,384	8.9	228,767	17.8
Total Drylands	80,968	6.3	48,838	3.8	128,520	10.0	258,328	20.1	516,654	40.2
Non-dryland									768,561	59.8
Total Area of the Territory									1,285,215	100.0

HANDBOOKS

2675 Nyrop, Richard F. *Peru: a country study.* 2nd ed. Washington DC: U.S. Dept. of Defense, 1972. (American University foreign area studies) 330 pp.

Narrative summary of the nation's history, politics, geography, economy, social life; includes unannotated bibliography.

GEOGRAPHY

2676 *Guide to cartographic and natural resources information of Peru.* Mexico DF: Instituto Panamericano de Geografia e Historia, 1979. 99 pp.

Prepared by the staff of ONERN (National Office for the Evaluation of Natural Resources, Republic of Peru). Consists of descriptions and index maps of various carto- and photo-graphic series covering geology, forests, conservation, highways, soils, hydrology, and other topics.

2677 Moreyra y Paz Soldan, Carlos. *Bibliografia regional Peruana [Peruvian regional bibliography].* Lima: Libreria Internacional del Peru, 1967. 518 pp.

In Spanish.

2678 Psuty, Norbert P.; Beckwith, Wendell; and Craig, Alan L. *1000 selected references to the geography, oceanography, geology, ecology, and archaeology* *of coastal Peru and adjacent areas.* 3rd ed. Washington DC: U.S. Office of Naval Research. Geography Branch, 1968. (Paracas papers 1:1; NTIS: AD 671-870) 52 pp.

2679 Richardson, James B., III. "A bibliography of archaeology, Pleistocene geology, and ecology of the Departments of Piura and Tumbes, Peru." *Latin American research review* 12:1 (1977):122-137.

Unannotated. Covers the extreme northwest corner of the nation; the departments occupy portions of the Atacama Desert along the Pacific coast.

ATLASES

2680 *Atlas historico y geografico de paisajes Peruanos [historical and geographical atlas of the Peruvian countryside].* Lima: Peru. Instituto Nacional de Planificacion. Asesoria Geografica, 1970. 737 pp.

Includes facsimile maps, transparent overlays and bibliographical references. Spanish text.

GAZETTEERS

2681 *Peru: official standard names approved by the United States Board on Geographic Names.* Washington DC: U.S. Board on Geographic Names, 1955. (Gazetteer 149) 609 pp.

List of 24,100 place-names with latitude/longitude coordinates.

SCIENCE AND TECHNOLOGY

2682 Zevallos y Muniz, Marco Aurelio. *Ciencia y tecnologia para el desarrolla: una bibliografia [science and technology for development: a bibliography].* Lima: Universidad del Pacifico. Centro de Investigacion, 1980. (Cuadernos ensayo 17) 270 pp.

In Spanish.

GEOLOGY

2683 Castro Bastos, Leonidas. *Bibliografia geologica del Peru [geological bibliography of Peru].* Lima: The Author, 1960. 317 pp.

In Spanish.

CLIMATOLOGY

2684 *Bibliography on the climate of Peru.* Washington DC: U.S. Weather Bureau, 1955. (NTIS: PB 176-726; AD 670-031) 13 pp.

2685 Indacochea G., and Angel, J. *Bibliografia climatologica del Peru [climatological bibliography of Peru].* Lima: Peru. Instituto Geologico del Peru, 1946. (Boletin 4) 81 pp.

Includes some 750 unannotated citations.

2686 Johnson, A. M. "The climate of Peru, Bolivia and Ecuador." In *Climates of Central and South America*, pp. 147-218. Edited by Werner Schwerdtfeger. Amsterdam, The Netherlands: Elsevier Scientific, 1976. (World survey of climatology 12).

A narrative summary with maps, charts, and an unannotated bibliography.

BIOLOGY

2687 "Bibliografia para el Peru" [bibliography for Peru]. *Biota* 5:43 (1964):307-314.

Covers Peruvian botany and zoology, 1962-65. In Spanish.

AGRICULTURE

2688 *Boletin bibliografico/Bibliographical bulletin.* Lima: C.E.D.S.A., 1974-. 2 per year.

CEDSA = Centro de Documentacion del Sector Agrario.

PLANT CULTURE

2689 Crosby, Martha. *La papa en la bibliografia peruana de 1965 a 1980 [the potato in Peruvian bibliographies from 1965 to 1980].* Lima: Centro Internacional de la Papa, 1981. 172 pp.

In Spanish.

2690 Werge, Robert W. *Aspectos socio-economicos de la produccion y la utilizacion de la papa en el Peru/ Socio-economic aspects of the production and utilization of potatoes in Peru.* Lima: International Potato Center. Socio-Economic Unit, (1976?). 71 pp.

In Spanish and English.

AGRICULTURAL ECONOMICS

2691 Phillips, Beverly, and Saupe, Barbara, comps. *Peru, land and people: a bibliography (with supplement).* Madison WI: University of Wisconsin. Land Tenure Center, 1971. (Training and methods, 15, 15 supple.) 73 pp.

Unannotated. Reflects the holdings of the Land Tenure Center library; includes its call numbers.

ANTHROPOLOGY

2692 Webster, Steven S. "The contemporary Quechua indigenous culture of highland Peru: an annotated bibliography." *Behavior science notes* 5:2 (1970):71-96; 5:3 (1970):213-247.

Extensively annotated.

ECONOMIC DEVELOPMENT

2693 Graber, Eric S. *An anotated (sic) bibliography of rural development, urbanization, and levels of living in Peru.* Washington DC: Agency for International Development. Bureau for Latin America and the Caribbean, 1979. (Peru, general working document 1) 109 pp.

2694 *Quarterly economic review of Peru, Bolivia.*
See entry 2583.

HISTORICAL GEOGRAPHY

2695 Alisky, Marvin. *Historical dictionary of Peru.* Metuchen NJ: Scarecrow, 1979. (Latin American historical dictionaries 20) 157 pp.

Entries cover major topics, events, and personalities in the nation's history; includes extensive unannotated-bibliography.

VENEZUELA
Republica de Venezuela
(as part of Nueva Granada, a former colony of Spain)

The drylands of Venezuela exist because of an unusual set of meteorological and topographic conditions that prevail along the nation's Caribbean coast. The center of this region lies in fact over the sea off the Guajira and Paraguana Peninsulas; the region embraces a long stretch of Colombia's coastline, nearly all of Venezuela's eastward to Trinidad, and the offshore islands belonging to both Venezuela and the Netherlands Antilles. Normally, in an area with onshore trade winds located only 10 to 12 degrees from the Equator one would expect abundant rainfall and vegetation; here, cool sea water meets extremely rugged topography to create windflow patterns that force air to diverge and subside along the coast, and thus inhibit precipitation. The region ranges from subhumid to arid and usually displays two rainy seasons and a winter drought; average annual temperatures are the highest in the American tropics.

Land use along the nation's arid fringe is generally limited to goat-grazing in the scrub woodlands and limited coconut cultivation at the mouth of Lake Maracaibo. The close proximity of well-watered lands in the interior discouraged development of this marginal region. Venezuela's strong economy—based as it is on petroleum extraction and refining—makes little demand on the drylands except as locations for seaports.

Dryland areas of VENEZUELA

DESERTIFICATION RISK	ARIDITY									
	Hyperarid		Arid		Semiarid		Subhumid		Aridity Totals	
	km²	%	km²	%	km²	%	km²	%	km²	%
Very High	By definition, desertification does not exist in hyperarid regions.									
High					5,472	0.6			5,472	0.6
Moderate			4,560	0.5	20,065	2.2	71,139	7.8	95,764	10.5
Desertification Totals			4,560	0.5	25,537	2.8	71,139	7.8	101,236	11.1
No Desertification			2,736	0.3					2,736	0.3
Total Drylands			7,296	0.8	25,537	2.8	71,139	7.8	103,972	11.4
Non-dryland									808,078	88.6
Total Area of the Territory									912,050	100.0

HANDBOOKS

2696 Blutstein, Howard I. *Area handbook for Venezuela*. 3rd ed. Washington DC: U.S. Dept. of Defense, 1977. (American University foreign area studies) 448 pp.

Narrative summary of the nation's history, politics, geography, economy, social life; includes unannotated bibliography.

GEOGRAPHY

2697 Chaves, Luis Fernando, and Vivas, Leonel. *Geografia regional de Venezuela [regional geography of Venezuela]*. Rio de Janeiro: Instituto Panamericana de Geografia y Historia, 1968. (Publicacao 275) 267 pp.

In Spanish.

ATLASES

2698 *Atlas de Venezuela [atlas of Venezuela]*. 2nd ed. Caracas: Venezuela. Ministerio del Ambiente y de los Recursos Naturales Renovables, 1979. 331 pp.

In Spanish. Includes gazetteer index. Prepared by the Direccion de Cartografia Nacional (Venezuela).

2699 *Venezuela: inventario nacional de recursos [national inventory of resources]*. Washington DC: U.S. Engineer Agency for Resources Inventories, 1968. (AID/EARI atlas 8) various pagings.

In Spanish and English.

GAZETTEERS

2700 *Venezuela: official standard names approved by the United States Board on Geographic Names.* Washington DC: U.S. Board on Geographic Names, 1961. (Gazetteer 56) 245 pp.

List of 17,200 place-names with latitude/longitude coordinates.

GEOLOGY

2701 "Bibliografia e indice de publicaciones oficiales referentes a geologia venezolana" [bibliography and index of official publications relating to Venezuelan geology]. Venezuela. Direccion de Geologia. *Boletin de geologia* 11:21 (1970):439-504.

In Spanish.

CLIMATOLOGY

2702 **Snow, J. W.** "The climate of northern South America."

See entry 2645.

SOILS

2703 **Butters, B.** *Soils of Venezuela (1974-1965).* Slough, U.K.: Commonwealth Bureau of Soils, 1976. (Annotated bibliography 1727) 7 pp.

AGRICULTURE

2704 *Atlas agricola de Venezuela [agricultural atlas of Venezuela].* Caracas: Venezuela. Direccion de Planificacion Agropecuaria, 1960. 51 pp.

Contains 107 colored maps.

2705 **Badillo, Victor M., comp.** *Indice bibliografico agricola de Venezuela [bibliographic index of agriculture in Venezuela].* Caracas: Fundacion Eugenio Mendoza, 1957. 305 pp.

Annotated. In Spanish. Covers material from 1935-1955. Supplement 1 (1962); supplement 2 (1967).

PLANT CULTURE

2706 **Ramirez, Ricardo, and Marquez, Orfilia.** *Bibliografia selectiva de maiz en Venezuela [selective bibliography of corn in Venezuela].* Maracay, Venezuela: Fondo Nacional de Investigaciones Agropecuarias. Direccion de Investigaciones. Oficina de Comunicaciones Agricolas. Ministerio de Agricola y Cria, 1974. 49 pp.

In Spanish.

AGRICULTURAL ECONOMICS

2707 *Rural development in Venezuela and the Guianas: a bibliography.* Madison WI: The University of Wisconsin. Land Tenure Center, 1972. (Training and methods 20) 67 pp.

Unannotated. Reflects holdings of the Land Tenure Center library; includes its call numbers.

ECONOMIC DEVELOPMENT

2708 **Longrigg, S. J., ed.** *Major companies of Argentina, Brazil, Mexico and Venezuela 1982.*

See entry 2570.

2709 *Quarterly economic review of Venezuela, Netherlands Antilles, Suriname.*

See entry 2530.

HISTORICAL GEOGRAPHY

2710 **Rudolph, Donna Keyse.** *Historical dictionary of Venezuela.* Metuchen NJ: Scarecrow, 1971. (Latin American historical dictionaries 3) 142 pp.

Entries cover major topics, events, and personalities in the nation's history; includes extensive unannotated bibliography.

PART TWO:

Subjects of Drylands Research

BIBLIOGRAPHY AND INFORMATION SCIENCE

National bibliographies are defined as those attempting to provide more or less complete records of publication activity for their nations. Most nations have some form of national bibliography but coverage and reliability varies widely. Space forbids listing each national bibliography separately in Part One; the following are useful guides to these valuable sources.

NATIONAL BIBLIOGRAPHIES

2711 *Commonwealth national bibliographies: an annotated bibliography.* London: Commonwealth Secretariat, 1977. 97 pp.

2712 **Pomassl, Gerhard, comp.** *Synoptic tables concerning the current national bibliographies.* Berlin; Leipzig, German Democratic Republic: Bibliotheksverband der Deutschen Demokratischen Republik; Deutsche Bucherei, 1975. 3 pp. plus 25 folding leaves.

ONLINE DATABASES

2713 *Directory of online databases.* Santa Monica CA: Cuadra Associates, 1979-. Quarterly.

Provides basic information on available online databases worldwide including subject, producer, content, coverage, and updating. Pricing policies are explained in the introduction. Includes a variety of indexes: by subject, producer, online service, and name. The most useful and timely source for information on this rapidly changing field.

THESES AND DISSERTATIONS

2714 **Borchardt, D. H., and Thawley, J. D.** *Guide to the availability of theses.* Munchen; New York: K. G. Saur, 1981. (IFLA publication 17) 443 pp.

Provides information on consulting and obtaining theses and dissertations from 698 institutions worldwide.

2715 *Comprehensive dissertation index.* Ann Arbor MI: University Microfilms International, 1973-. Annual.

A massive current and retrospective index of doctoral dissertations from institutions in the United States, Canada, and selected other locations awarded since 1861. Includes no abstracts and covers more titles than *Dissertation abstracts international.* Also available online (vendors: BRS, DIALOG, SDC).

2716 *Dissertation abstracts international.* Ann Arbor MI: University Microfilms International, 1938-. Monthly.

Provides abstracts of selected doctoral dissertations from many institutions in the United States, Canada, and Europe; most major subject areas are covered. For broader coverage without abstracts, use *Comprehensive dissertation index.*

2717 *Masters abstracts: a catalog of selected masters theses on microfilm.* Ann Arbor MI: University Microfilms International, 1962-. Quarterly.

Provides abstracts of selected theses from a selected number of institutions worldwide; most major subject areas are covered.

2718 **Reynolds, Michael M.** *A guide to theses and dissertations: an annotated, international bibliography of bibliographies.* Detroit MI: Gale Research, 1975. 599 pp.

Covers most major subject areas.

CONFERENCE PUBLICATIONS

2719 *Conference papers index (CPI).* Bethesda MD: Cambridge Scientific Abstracts, 1973-. Updated monthly.

An online database corresponding to printed *Conference papers index.* Covers major conferences in scientific subjects worldwide. Vendors: DIALOG, ESA-IRS.

2720 *Index of conference proceedings received.* Boston Spa, U.K.: United Kingdom. National Lending Library for Science and Technology, June 1973-. Monthly.

Covers all subjects but emphasizes the sciences. Also available online (as *Conference proceedings index*; vendor BLAISE). Formerly: *Index of conference proceedings received by the NLL* (1964-May 1973), nos. 1-68.

GOVERNMENT DOCUMENTS

2721 *Catalog of government publications in the research libraries.* Boston MA: G. K. Hall, 1972. 40 vols.

Prepared by the New York Public Library. Supplemented annually by *Bibliographic guide to government publications: foreign* (1975-).

SUBJECT COLLECTIONS

2722 **Ash, Lee.** *Subject collections: a guide to special book collections and subject emphases as reported by university, college, public, and special libraries and museums in the United States and Canada.* 5th ed. New York: Bowker, 1978. 1,184 pp.

A standard reference to special book collections and subject emphases in the United States and Canada. Arranged by subject; has no indexes. Weak on science and technical topics.

2723 **Lenroot-Ernt, Lois, ed.** *Subject directory of special libraries and information centers.* 7th ed. Detroit MI: Gale, 1982. 5 vols.

Covers literature in the United States and Canada containing materials on a variety of subjects. Arranged by general subject area; includes index to alternate names only. This material is a rearrangement by subject classification of *Directory of special libraries and information centers* (entry 2725) which contains full subject, geographic, and personal indexes. Volume 1 covers business and law libraries; volume 2 covers education and information sciences libraries; volume 3 covers healthy sciences libraries; volume 4 covers social sciences and humanities libraries, and volume 5 covers science and technology libraries.

2724 **Lewanski, Richard Casimir.** *Subject collections in European libraries.* 2nd ed. London; New York: Bowker, 1978. 495 pp.

Includes index.

2725 **Young, Margaret Labash, and Young, Harold C., eds.** *Directory of special libraries and information centers.* 5th ed. Detroit MI: Gale, 1979. 2 vols.

Covers libraries and information centers in the United States and Canada containing materials on a variety of subjects. Volume 1 arranged by name of institution; volume 2 contains geographic and personnel indexes. Each entry gives address, phone, staff, holdings, and other information. See *Subject directory of special libraries and information centers* (1981) (entry 2733) for rearrangement by general subject area.

STATISTICS

2726 *American statistics index.* Washington DC: U.S. Congressional Information Service, 1973-. Monthly.

A comprehensive guide and index to the statistical publications of the United States governemnt. Also available online (vendors: DIALOG, SDC).

2727 **Pieper, Frank C., comp.** *SISCIS: subject index to sources of comparative international statistics.* Beckenham, Kent, U.K.: CBD Research, 1978. 745 pp.

Indexes a variety of international statistical sources. Best used as supplement to *American statistics index* (1973-) (entry 2726).

2728 *Statistical yearbook/Annuaire statistique.* New York: U.N. Statistical Office, 1948-. Annual.

In English and French.

GEOGRAPHY

RESEARCH GUIDES

2729 **Brewer, James Gordon.** *The literature of geography: a guide to its organisation and use.* 2nd ed. Hamden CT: Linnet Books, 1978. 264 pp.

Narrative guide arranged by topic. Emphasis on English language materials recently published.

2730 **Lock, Clara Beatrice Muriel.** *Geography and cartography: a reference handbook.* 3rd ed. London; Hamden CT: Clive Bingley; Linnet Books, 1976. 762 pp.

A combined and revised edition of *Geography: a reference handbook* (1968) and of *Modern maps and atlases* (1969).

2731 **Wood, David N.** *Use of earth sciences literature.* London: Butterworths, 1973. 459 pp.

GENERAL BIBLIOGRAPHIES

2732 American Geographical Society of New York. *Research catalogue.* Boston MA: G. K. Hall, 1962. 16 vols.

Supplements 1 (1974) and 2 (1978). Updated by listings in *Current geographical publications* since 1938.

2733 *Bibliographie geographique internationale [international geographical bibliography].* Paris: France. Centre National de la Recherche Statistique, 1891-. Quarterly.

Annotated. Explanations in French and English; abstracts in French.

2734 *Current geographical publications.* Milwaukee WI: University of Wisconsin-Milwaukee. American Geographical Society Collection, 1938-. Monthly.

Unannotated. Additions to the research catalog of the American Geographical Society collection of the University of Wisconsin- Milwaukee.

2735 **Heise, Jon.** *The travel book: guide to the travel guides.* New York: Bowker, 1981. 319 pp.

A bibliography and critical review of travel guides, the best of which are excellent sources of geographical information.

2736 **Hewes, Leslie, et al.** "Recent geography dissertations and theses completed and in preparation." American Society for Professional Geographers. *Professional geographer* 16:6 (1964-) with irregular cumulations.

Unannotated. Available in microform. Generally printed on an annual basis, beginning in 1964.

ONLINE DATABASES

2737 *EDE (Environmental data and ecological parameters).* Vaerloese, Denmark: International Society of Ecological Modelling (ISEM), 1970-. Updated quarterly.

Corresponds to printed *Handbook of environmental data and ecological parameters.* Provides both bibliographic and numeric data on the environment and ecology. Vendor: Datacentralen.

2738 *ENVIROLINE.* New York: Environment Information Center, 1971-. Updated monthly.

Corresponds to printed *Environment abstracts.* Covers the literature of environmental studies and natural resource management. Vendors: BRS, DIALOG, SDC, ESA-IRS.

2739 *Environmental bibliography.* Santa Barbara CA: Environmental Studies Institute, 1973-. Updated 6 times per year.

Corresponds to printed *Environmental periodicals bibliography.* Covers environmental sciences worldwide. Vendor: DIALOG.

2740 *NASA.* Washington DC: U.S. National Aeronautics and Space Administration, 1962-. Updated monthly.

Corresponds to printed *Scientific and technical aerospace reports* (STAR) and *International aerospace abstracts* (IAA). Covers space and aerospace literature; includes remote sensing of earth resources. Vendor: NASA/RECON.

INDEXES AND ABSTRACTS

2741 *Environment index; a guide to the key environmental literature of the year.* New York: Environment Information Center, 1972-. Monthly.

Indexes English language publications.

2742 *Environmental periodicals: bibliography.* Santa Barbara CA: International Academy at Santa Barbara, 1972-. 6 per year.

Covers human ecology, energy, land resources, agriculture, marine resources, water management, nutrition, and health.

ENCYCLOPEDIAS

2743 *McGraw-Hill encyclopedia of environmental science.* 2nd ed. New York: McGraw-Hill, 1980. 858 pp.

Covers physical geography generally, including meteorology, land and soil resources, pollution, energy supplies, biology, and ecology.

DIRECTORIES

2744 *Country experts in the Federal government.* 1981 ed. Washington DC: Washington Researchers, 1981. 35 pp.

Lists name, phone, and address of specialists in various federal departments and agencies, with their nation and topic of specialization.

2745 *World environmental directory.* 4th ed. Silver Spring MD: Business Publishers, 1980. 964 pp.

Gives names, addresses, and specializations of manufacturers, organizations, and government bodies involved in environmental matters worldwide.

MAP COLLECTIONS

2746 *Bibliographia cartographica.* Munchen, Federal Republic of Germany: Verlag Dokumentation, 1974-. Annual.

In German, English, and French. Unannotated. Continues: *Bibliotheca cartographica* (1957-72).

2747 *Geo-Katalog: internationale Ausgabe.* Stuttgart, Federal Republic of Germany: Internationales Landkartenhaus. Geo Center, 1973-. Annual.

Explanatory notes in German, English, and French. This complex multivolume catalog is the best source for currently available mapping worldwide.

2748 *International yearbook of cartography/ Annuaire international de cartographie/Internationales Jahrbuch fur Kartographie.* Zurich, Switzerland, Drell Fuessli, 1961-. Annual.

In English, French, or German with summaries in languages other than the language of the article. Includes translated Asian articles.

2749 **Ristow, Walter W., comp.** *World directory of map collections.* Munchen, Federal Republic of Germany: Verlag Dokumentation, 1976. (International Federation of Library Associations publication 8) 326 pp.

ATLASES

2750 *CartActual: topical map service.* Budapest: Kultura, 1965-. 2 per month.

In English, French, German, and Hungarian. A period updating service presenting maps to illustrate changes in world geography.

2751 *The Times atlas of the world: comprehensive edition.* 6th ed. New York: Times Books, 1980. 227 pp.

Perhaps the best single volume world atlas.

REMOTE SENSING

2752 **Bryan, M. Leonard.** *Remote sensing of earth resources: a guide to information sources.* Detroit MI: Gale Research, 1979. (Geography and travel information guide series 1) 188 pp.

An excellent bibliographic guide to remote sensing literature.

2753 *Earth resources: a continuing bibliography with indexes.* Washington DC: U.S. National Aeronautics and Space Administration. Scientific and Technical Information Office, 1974-. Quarterly.

An abstracting journal covering remote sensing references drawn from *Scientific and technical aerospace reports* (STAR) and *International aerospace abstracts* (IAA.

2754 *Geo abstracts. G: Remote sensing and cartography.* Norwich, U.K.: University of East Anglia, 1974-. Annual.

2755 **Krumpe, Paul F., comp.** *The world remote sensing bibliographic index: comprehensive geographic index bibliography to remote sensing site investigations of natural and agricultural resources throughout the world.* Fairfax VA: Tensor Industries, 1976. 619 pp.

2756 *Landsat data users handbook.* 3rd ed. Arlington VA: U.S. Geological Survey, 1979. 195 pp.

Handbook includes detailed descriptions of the Landsat system, a glossary of remote sensing terms, and appendixes.

2757 *Quarterly literature review of the remote sensing of natural resources.* Albuquerque NM: University of New Mexico. Technology Application Center, 1974-. Quarterly.

2758 *Remote sensing of earth resources: a literature survey with indexes.* Washington DC: U.S. National Aeronautics and Space Administration. Scientific and Technical Information Facility, 1970. (NTIS: N70-41047). Various pagings.

A selection of annotated references to unclassified reports and journal articles introduced into the NASA information system between January 1962 and February 1970, in *Technical publication announcements* (TPA), *Scientific and technical aerospace reports* (STAR), and *International aerospace abstracts* (IAA). Supplemented in 1975, covering period 1970-73; also supplemented by *Earth resources* (entry 2753).

2759 *Skylab earth resources data catalog.* Houston TX: U.S. National Aeronautics and Space Administration, 1974. (NTIS: N75-20798) 359 pp.

Index of EREP photographs is provided as well as information on how EREP may be obtained. Suggestions are presented for utilization of data in various fields.

2760 Thompson, W. I., III. *Earth survey bibliography: a KWIC index of remote sensing information.* Washington DC: U.S. National Aeronautics and Space Administration, 1971. (DOT-TSC-NASA-70-1; NTIS: N70-42766) 265 pp.

Contains 1,650 citations.

SCIENCE AND TECHNOLOGY

GENERAL BIBLIOGRAPHIES

2761 *Science citation index.* Philadelphia PA: Institute for Scientific Information, 1961-. 2 per month.

An international interdisciplinary index to the literature of science. Includes enhanced citation indexing.

ONLINE DATABASES

2762 *COMPENDEX (Computerized engineering index).* New York NY: Engineering Information, 1970-. Updated monthly.

Corresponds to printed *Engineering index*. Covers engineering and technology. Vendors: BRS, DIALOG, SDC, CISTIC, ESA-IRS.

2763 *ISI/ISTP&B (Index to scientific and technical proceeding and books).* Philadelphia PA: Institute for Scientific Information, 1978-. Updated monthly.

Corresponds to printed *Index to scientific and technical proceedings and books*. Provides worldwide coverage of conference proceedings and annual review series for the sciences. Vendor: ISI.

2764 *NTIS.* Springfield VA: U.S. National Technical Information Service, 1964-. Updated 2 per month.

Corresponds to printed *Government reports announcements and index*. Provides coverage of the NTIS document collection, which includes unrestricted research reports on many topics in the pure and applied sciences. Vendors: BRS, DIALOG, SDC, DATA-STAR, ESA-IRS.

2765 *SCISEARCH.* Philadelphia PA: Institute for Scientific Information, 1974-. Updated monthly.

Corresponds to printed *Science citation index*. Covers the literature of science and technology worldwide; noted for coverage of basic research and timely indexing. Vendors: DIALOG, ISI.

INDEXES AND ABSTRACTS

2766 *Applied science and technology index.* New York: H. W. Wilson Co., 1958-. Monthly.

Indexes English-language periodicals covering space sciences, chemistry, computers, energy, foods, geology and mining, oceanography, petroleum, etc.

2767 *CTI: Current technology index.* London: Library Association Publishing, 1981-. Monthly.

Formerly: *British technology index*. Covers technology, engineering, architecture, agricultural machinery, mining, chemicals, manufacturing, etc.

DOCUMENT COLLECTIONS

2768 *Government reports: announcements and index.* Springfield VA: U.S. National Technical Information Service, 1946-. 2 per month.

Provides current access to documents deposited into the NTIS system. Also available online.

2769 *A reference guide to the NTIS bibliographic data file: an NTIS database reference aid.* Reston VA: U.S. National Technical Information Service, 1978. (NTIS: PR 253) 22 pp.

Provides background information on the database helpful to those working with NTIS files or other computer services. This guide is scheduled to become a modular part of the *NTIS users manual*.

2770 *Technical abstract bulletin.* Washington DC: U.S. Defense Documentation Center, 1979-. 2 per month.

Provides current access to Defense Technical Information Center documents, which partially overlap the NTIS system.

2771 *TRANSDEX: bibliography and index to the United States Joint Publications Research Service (JPRS) translations.* New York: Macmillan, 1962-. Monthly.

Bibliography of the approximately 30,000 articles and books translated each year by JPRS. Of particular interest are articles from Communist, Near Eastern, and developing nations. Topics covered include meteorology, geomorphology, water resources, agriculture, remote sensing of environment, etc.

DIRECTORIES

2772 Fifield, Richard J., ed. *Guide to world science*. 2nd ed. Guernsey, U.K.: Frances Hodgson, 1974-76. 24 vols.

A multi-volume set of directories of scientific research organizations and government agencies arranged geographically. (Individual volumes are listed separately in this Guide.)

2773 *World guide to scientific associations*. 3rd ed. Munchen; New York: K. G. Saur, 1982.

Includes names, addresses, and directors of more than 18,000 national and international organizations in science, technology, social sciences, and the arts. Arranged by country, with subject index.

APPROPRIATE TECHNOLOGY

2774 *Appropriate technology in the Commonwealth: a directory of institutions*. London: United Kingdom. Commonwealth Secretariat. Food Production and Rural Development Division, 1977. 64 pp.

"Appropriate technology" is defined as labor-intensive processes which makes use of renewable resources. This directory was developed for use by organizations working on some aspect of appropriate technology as well as for economic agencies in the development business (i.e., aid agencies, government officials, academics).

2775 Brighton, J. Noyce. *ATINDEX*. Brighton U.K.: John L. Noyce, 1980-. Quarterly.

Quarterly indexing service to the literature of appropriate technology and related fields.

2776 Bulfin, Robert L. and Weaver, Harry L. *Appropriate technology for natural resources development: an overview, annotated bibliography, and a guide to sources of information*. Tucson AZ: University of Arizona, 1977. (NTIS: PB 279-193) 166 pp.

2777 Carr, Marilyn. *Economically appropriate technologies for developing countries: an annotated bibliography*. London: Intermediate Technology Publications, 1976. 101 pp.

2778 *Directory of institutions and individuals active in environmentally-sound and appropriate technologies*. Oxford; New York: Pergamon Press, 1979. (U.N. Environment Programme. Reference series 1) 152 pp.

Lists about 2,000 institutions worldwide which are able to supply information on some aspect of environmentally-sound technology.

2779 Jequier, Nicolas. *Appropriate technology directory*. Paris: Organisation for Economic Co-operation and Development. Development Centre, 1979. 361 pp.

An international "who's doing what" in appropriate technology, arranged by country. Includes organizations involved in the development or promotion of appropriate technology, international banks, intergovernmental organizations, and availability of information.

2780 Mathur, Brij. *International directory of appropriate technology resources*. Mt. Rainier MD: VITA, 1979. Approximately 600 pp.

Result of 1977 questionnaire. Lists organizations and publications by those organizations.

2781 *Selected appropriate technologies for developing countries; abstracts from the NTIS data file*. 2nd ed. Reston VA: U.S. National Technical Information Service, 1979. (NTIS: PB 291-573) 201 pp.

Contains over 1,000 entries with abstracts pertinent to "small scale, employment generating, and energy conserving technologies relevant to needs and resources of developing countries."

EARTH SCIENCES

GENERAL BIBLIOGRAPHIES

2782 *Land and water development*. Wageningen, The Netherlands: International Institute for Land Reclamation and Improvement, 1975. (Bibliography 11) 80 pp.

Contains approximately 750 entries with brief annotations.

ENCYCLOPEDIAS

2783 Fairbridge, Rhodes W. *The encyclopedia of geochemistry and environmental sciences*. New York: Van Nostrand Reinhold, 1972. (Encyclopedia of earth sciences IV-A) 1,321 pp.

Covers the chemistry of the earth, atmosphere, and oceans as well as pollution and other environmental topics.

2784 Smith, David G., ed. *The Cambridge encyclopedia of earth sciences*. Cambridge U.K.: Cambridge University Press, 1982. 496 pp.

A well-illustrated general account of geology, plate tectonics, the atmosphere, climate, and remote-sensing, valuable chiefly for its non-specialist summary of current research. Short bibliography.

DIRECTORIES

2785 *A directory of information resources in the United States: geosciences and oceanography*. Washington DC: U.S. Library of Congress. National Referral Center, 1981. 375 pp.

Provides names, addresses, and areas of interest of research centers in the United States covering earth sciences, geology, mineral and energy resources, meteorology, etc.

GEOLOGY

RESEARCH GUIDES

2786 Mackay, John W. *Sources of information for the literature of geology: an introductory guide.* 2nd ed. Edinburgh, U.K.: Scottish Academic Press for the Geological Society of London, 1974. 59 pp.

GENERAL BIBLIOGRAPHIES

2787 Bergquist, Wenonah E. *Bibliography of reports resulting from U.S. Geological Survey technical cooperation with other countries, 1967-1974.* Washington DC: U.S. Geological Survey, 1976. (Bulletin 1426) 68 pp.

2788 *Catalog of the United States Geological Survey library.* Boston MA: G. K. Hall, 1964. 25 vols.

Includes primarily maps and monographic materials. *Supplement* one, 11 volumes, is a record of all publications added between late 1964 through 1971 (1972); *Supplement* two, four volumes, lists titles added in 1972-73 (1974); and *Supplement* three, six volumes, lists titles added between 1974 and March 1976 (1976).

2789 *Comprehensive index of the publications of the American Association of Petroleum Geologists.* Tulsa OK: American Association of Petroleum Geologists, 1947-. Irregular.

Covers the world but strongest on North American geology. Includes all formal publications of the AAPG except maps. Volume 1 covers period 1917-45 (1947), 603 pp.; volume 2 covers period 1946-55 (1957), 301 pp.; volume 3 covers period 1956-65 (1967), 792 pp.; volume 4 covers period 1966-70 (1975), 608 pp.; volume 5 covers period 1971-75 (1979), 906 pp.

2790 Hall, Vivian S. *Environmental geology: a selected bibliography.* Boulder CO: Geological Society of America, 1975. 331 pp.

2791 *Publications of the Geological Survey.* Washington DC: U.S. Geological Survey, 1879-. Monthly.

Covers the world, with strong emphasis on the United States.

2792 Ward, Dederick C. *Geologic reference sources: a subject and regional bibliography of publications and maps in the geological sciences.* 2nd ed. Metuchen NJ: Scarecrow, 1981. 560 pp.

Contains 4,324 citations—some lightly annotated—to worldwide literature on geology.

ONLINE DATABASES

2793 *Geoarchive.* London: Geosystems, 1969-. Updated monthly.

Corresponds in part to the following printed indexes: *Geoscience documentation, Bibliography of vertebrate paleontology, Bibliography of economic geology, Bibliography of engineering geology, Bibliography of Afro-Asian geology, Bibliography of American geology.* Vendor: DIALOG. A printed thesaurus (*Geosaurus: geosystems' thesaurus of geoscience*) is available.

2794 *Georef (geological reference file).* Falls Church VA: American Geological Institute, 19—. Updated monthly.

Corresponds to the following printed indexes: *Bibliography and index of geology* (1969-), *Bibliography and index of North American geology* (1785-1970), *Bibliography and index of geology exclusive of North America* (1933-1968). Vendors: DIALOG, SDC. A printed thesaurus is available.

INDEXES AND ABSTRACTS

2795 *Bibliography and index of geology.* Falls Church VA: American Geological Institute, 1969-. Monthly.

Supersedes *Bibliography and index of geology exclusive of North America* (Washington DC: Geological Society of America, 1933-68), 32 vols. See entry 2794 regarding online access.

2796 *Geographical abstracts: A - Landforms and the quaternary.* Norwich, U.K.: University of East Anglia, 1960-. 6 per year.

Formerly titled *Geographical abstracts: A - Geomorphology* (1960-73).

2797 *Geographical Abstracts: E - Sedimentology.* Norwich, U.K.: University of East Anglia, 1972-. 2 per month.

2798 *Geophysical abstracts.* Norwich, U.K.: Geo Abstracts, 1977-. 2 per month.

2799 *Mineralogical abstracts.* London: Mineralogical Society of Great Britain and Mineralogical Society of America, 1920-. Quarterly.

2800 *Zentralblatt fur Geologie und Palaeontologie.* Stuttgart, Federal Republic of Germany: E. Schweizerbart, 1950-.

Teil [part] 1 appears monthly; Teil [part] 2 appears 6 per year. In German. Primarily German sources. Supersedes *Neues Jahrbuch fur Mineralogie, Geologie und Palaeontologie.* 1830-1949. Teil 1 covers geology and marine sciences, Teil 2 paleontology.

ENCYCLOPEDIAS

2801 Fairbridge, Rhodes W. *Encyclopedia of geomorphology.* New York: Reinhold Books, 1968. (Encyclopedia of earth science series 3) 1,295 pp.

Contains concise discussions of major topics in geomorphology; pertinent entries include "arid cycle," "deserts and desert land forms," "sand dunes," and "wind action." Each entry concludes with a brief bibliography and cross references to related topics.

2802 Fairbridge, Rhodes W. *The encyclopedia of world regional geology.* New York: Halsted Press, 1975. (Encyclopedia of earth science series) 704 pp.

To be completed in two volumes. Vol. 1 covers the Western Hemisphere; Vol. 2 for the Eastern Hemisphere, has not yet appeared.

Earth Sciences

2803 Fairbridge, Rhodes W., and Bourgeois, Joanne, eds. *The encyclopedia of sedimentology.* New York: Academic, 1978. 901 pp.

Contains about 500 entries.

2804 Lapedes, Daniel N., ed. *McGraw-Hill encyclopedia of the geological sciences.* New York: McGraw-Hill, 1978. 915 pp.

2805 Simkin, Tom, et al. *Volcanoes of the world: a regional directory, gazetteer, and chronology of volcanism during the last 10,000 years.* Stroudsburg PA: Hutchison Ross, 1981. 232 pp.

DIRECTORIES

2806 *Financial Times mining international yearbook.* London: Longman, 1979-. Annual.

A directory of over 700 mining companies worldwide, with addresses, officers, and detailed statements on their properties and finances.

2807 *The geologists' year book.* Poole, U.K.: Delphin, 1977-. Annual.

A directory of universities, museums, companies, and other geology-related organizations worldwide.

2808 Ricaldi, Victor, ed. *Directory of geoscience departments in universities in developing countries 1979-80.* 2nd ed. St. John's, Newfoundland: Association of Geoscientists for International Development, 1980. 28 pp.

Provides addresses, names of personnel, and some information on research facilities.

2809 *World mines register.* San Francisco CA: Miller Freeman, 1975/76-. Freq. unknown.

Provides names, addresses, and details of mining firms and active operations world-wide. Supersedes *Mines register* (1937-71), *The mines handbook: an enlargement of The copper handbook; a manual of the mining industry of North America* (1916-31), and *The copper handbook* (1900-14).

STATISTICS

2810 Lofty, G. J. et al. *World mineral statistics 1976-80: production: exports: imports.* London: U.K. Institute of Geological Sciences, 1982. 267 pp.

Provides production and trade statistics for many commodities and many nations of the world. Updated periodically.

2811 *Minerals yearbook.* Washington DC: U.S. Bureau of Mines, 1861-. Annual.

ATLASES

2812 Derry, Duncan R. *World atlas of geology and mineral deposits.* London: Mining Journal Books, 1980. 110 pp.

Lists sources of additional information by region.

2813 Snead, Rodman E. *World atlas of geomorphic features.* Rev. ed. Huntington NY: R. E. Krieger, 1980. 301 pp.

Includes unannotated bibliography: pp. 271-290.

CLIMATOLOGY

GENERAL BIBLIOGRAPHIES

2814 Brierly, William B. *Bibliography on atmospheric (cyclic) sea-salts.* Natick MA: U.S. Army. Natick Laboratories, 1970. (Technical report 70-63-ES; NTIS: AD 718-613) 70 pp.

Contains over 600 citations.

2815 Butson, Keith D., and Hatch, Warren L. *Selective guide to climatic data sources.* Asheville NC: U.S. National Climatic Center, 1979. (Key to meteorological records documentation 4.11) 142 pp.

The best introduction to United States government climatological publications, emphasizing data series currently available but including access information for several non-published series as well. Facsimile examples for many titles are included to illustrate the type and format of information. No geographic index is provided but an extensive index does key particular data elements (e.g., barometric pressure, precipitation, dew point, etc.) to individual titles. Geographical coverage naturally concentrates on North America and its surrounding oceans but often ranges worldwide.

2816 *Catalog of the Atmospheric Science collection.* Boston MA: G. K. Hall, 1978. 24 vols.

Subtitled: "In the Library and Information Services Division, Environmental Science Information Center, Environmental Data and Information Service, National Oceanic and Atmospheric Administration." First 20 volumes list acquisitions from 1890-1971. Volumes 21-24 list acquisitions from 1971-1977.

2817 Hacia, Henry. *An annotated bibliography of climatic atlases and charts of the world.* Silver Spring MD: U.S. Environmental Data Service, 1970. (NTIS: PB 193-287) 2 vols.

Volume 1 covers land areas; volume 2 covers oceans, seas, and islands.

2818 Ownbey, J. W. *Guide to standard weather summaries and climatic services.* Washington DC: Naval Air Systems Command, 1980. (NTIS: AD A078-660/8) 211 pp.

Description of published and unpublished climatological summaries available from the National Climatic Center in Asheville NC.

2819 Rigby, M., and Rice, M. L. "Annotated bibliography on textbooks and monographs containing subject bibliographies." American Meteorological Society. *Meteorological abstracts and bibliography* 6:2 (1955):217-264.

Covers the period 1856-1954.

ONLINE DATABASES

2820 *Guide to NOAA's computerized information retrieval services.* Washington DC: U.S. Dept. of Commerce. National Oceanic and Atmospheric Administration. Environmental Data and Information Service, 1979. 34 pp.

Serves as a reference to available data files, and the procedure for conducting computerized searches. Covers environmental data and atmospheric and marine sciences. Supersedes *User's guide to ENDEX/OASIS* (1976).

2821 *Meteorological and geoastrophysical abstracts (MGA).* Boston MA: American Meteorological Society, 1972-. Updated irregularly.

Corresponds to printed index of same title. Provides worldwide coverage of meteorological, climatological, and geoastrophysical research; includes related work in oceanography, hydrology, etc. Vendor: DIALOG.

2822 *Real-time weather.* Bedford MA: Weather Services International, 19—. Updated continuously.

A nonbibliographic database providing weather data of all types (including forecasts) from some 4,000 stations worldwide; data may be displayed in either tabular or cartographic form. Some historical data is available back to 1890. Vendor: Weather Services International.

2823 Ropelewski, C. F.; Predoehl, M. C.; and Platto, M. *The interim climate data inventory: a quick reference to selected climate data.* Washington DC: U.S. Center for Environmental Assessment Services, 1980. 176 pp.

A descriptive inventory of climate data covering much of the world; gives type, location/accessibility, period of observation, etc. Also available online for specialized searches. Developed from information presented at the 1979 Climate Data Management Workshop held at Harpers Ferry, West Virginia.

INDEXES AND ABSTRACTS

2824 *Bibliography of meteorological literature.* London: Royal Meteorological Society, 1920-. 2 per year.

2825 *Geographic Abstracts: B-Climatology and hydrology.* Norwich U.K.: University of East Anglia, 1974-. 2 per month.

2826 *Meteorological and geoastrophysical abstracts.* Boston MA: American Meteorological Society, 1950-. Monthly.

ENCYCLOPEDIAS

2827 Fairbridge, Rhodes W. *Encyclopedia of atmospheric sciences and astrogeology.* New York: Reinhold Publishing, 1967. (Encyclopedia of earth sciences series 2) 1,200 pp.

2828 Parker, Sybil P., ed. *McGraw-Hill encyclopedia of ocean and atmospheric sciences.*

See entry 67.

DIRECTORIES

2829 *Composition of the WMO.* Geneva, Switzerland: WMO, 1969-. Quarterly.

Provides names and addresses of WMO members and thus serves as an international directory of climatologists and organizations.

2830 *Meteorological services of the world.* Geneva, Switzerland: WMO, 1971. (WMO/OMM 2) various pagings.

List of meteorological organizations and institutes by country. Publications and directors are listed when reported.

STATISTICS

2831 *Catalogue of meteorological data for research.* Geneva, Switzerland: World Meteorological Organization, 1965-. (WMO 174 TP 86). Irreg.

Contains nation by nation summaries of climatic data.

2832 *International meteorological tables.* Geneva, Switzerland: WMO, 1966-. Irreg.

Compendium of formulae and tables intended to standardize reporting and documentation of meteorological data. Based in part on the Smithsonian Meteorological Tables.

2833 Landsberg, H. E. *World survey of climatology.* New York: Elsevier Scientific, 1969-.

An outstanding reference source reviewing the state of knowledge for every region of the world including the oceans and the polar regions. Bibliographies accompany text. Individual volumes and chapters are listed separately in this Guide. To be completed in 15 volumes.

2834 List, Robert J. *Smithsonian meteorological tables.* 6th ed. Washington DC: Smithsonian Institution, 1958. (Smithsonian miscellaneous collections 14) 527 pp.

2835 Nelson, Herman L. *Climatic data for representative stations of the world.* Lincoln NE: University of Nebraska Press, 1968. 81 pp.

Consists of tables of temperature and precipitation data from stations throughout the world selected to provide representative data for each country or region.

2836 Rudloff, Willy. *World-climates with tables of climatic data and practical suggestions.* Stuttgart, Federal Republic of Germany: Wissenschaftliche Verlagsgesellschaft, 1981. 632 pp.

2837 *Tables of temperature, relative humidity and precipitation for the world.* London: United Kingdom. Meteorological Office, 1958; reprint ed., London: United Kingdom. Meteorological Office, 1966-68. 6 vols.

Tables provide a variety of averaged values by month for reporting stations throughout the world. Volume 1 *North America, Greenland and the North Pacific Ocean (1968)*, 84 pp.; volume 2 *Central and South America, the West Indies and Bermuda* (1967), 53 pp.; volume 3 *Europe and the Atlantic Ocean north of 35° N* (1967), 153 pp.; volume 4 *Africa, the Atlantic Ocean south of 35° N and the Indian Ocean* (1967), 208 pp.;

volume 5 *Asia* (1966), 126 pp.; and volume 6 *Australasia and the Southern Pacific Ocean, including the corresponding sectors of Antarctica* (1967), 54 pp.

2838 **Wernstedt, Frederick L.** *World climatic data.* Lemont PA: Climatic Data Press, 1972. 522 pp.

A compilation of average long-term monthly and annual temperature and/or precipitation data for a network of nearly 19,000 world climatic stations. Organized first by continent and then by country. Bibliographies at the end of each section document data sources.

2839 *World weather records.* Asheville NC: U.S. National Climatic Center, 1927-. (Publication). Every 10 years.

Records covering 1750-1920 (1927), 1,199 pp.; records covering 1921-1930 (1934), 616 pp.; records covering 1931-1940 (1947), 646 pp.; records covering 1941-1950 (1959), 535 pp.; records covering 1951-1960 (1966), 6 volumes; records covering 1961-1970 (1979), 6 volumes (volumes 1, 2, and 6 available only).

CLIMATE CHANGE

2840 **Mitchell, J. M., Jr., and Kiss, E.** "Annotated bibliography on climatic changes: general works - theories; and annotated bibliography on climatic changes in historical times: synthetic studies." *Meteorological and geoastrophysical abstracts*) 15:11 (1964): 2236-2284; 15:12 (1964):2434-2478.

DROUGHT

2841 **Palmer, Wayne C., and Denny, Lyle M.** *Drought bibliography.* Silver Spring MD: U.S. Environmental Data Service, 1971. (NTIS: COM 71-00937) 236 pp.

Covers the period through 1968.

EVAPORATION

2842 **Panara, R., and Thuronyi, G.** "Annotated bibliography on evaporation measurement;... on evapotranspiration (-1955);...on evapotranspiration, 1956-1959; and...on evaporation." *Meteorological abstracts and bibliography* 10:8 (1959): 1234-1262; 10:9 (1959): 1394- 1426; 10:10 (1959): 1552-1595; 10:11 (1959): 1725-1765.

WEATHER MODIFICATION

2843 *An annotated bibliography on weather modification, 1960-1969.* Springfield VA: U.S. National Technical Information Service, 1972. (NTIS: NOAA-TM-EDS-ESIC-1; COM-72-11287) 416 pp.

2844 *Weather modification effects and management: 1964-January 1981 (citations from the NTIS data bank).* Springfield VA: U.S. National Technical Information Service, 1981. (NTIS: PB 81-805-236) 290 pp.

Includes 282 citations.

2845 *Weather modification: precipitation inducement: a bibliography.* Washington DC: U.S. Water Resources Scientific Information Center, 1973. (WRSIC 73-212) 246 pp.

HYDROLOGY

GENERAL BIBLIOGRAPHIES

2846 **Giefer, Gerald J.** *Sources of information in water resources: an annotated guide to printed materials.* Port Washington NY: Water Information Center, 1976. 290 pp.

2847 **Llaverias, Rita K.** *Bibliography of remote sensing of earth resources for hydrological applications, 1960- 1967; a working bibliography of selected references.* Springfield VA: U.S. National Technical Information Service, 1968. (Interagency report NASA-134) 127 pp.

2848 **Margat, Jean.** *Guide bibliographique d'hydrogeologic: ouvrage et articles en langue francaise [bibliographic guide to hydrogeology: works and articles in French].* Paris: Bureau de Recherche Geologiques et Minieres, 1964. (Suite hydrogeologie 113) 113 pp.

In French. Covers the period 1950-1964.

2849 **Van der Leeden, Frits.** *Ground water: a selected bibliography.* 2nd ed. Port Washington NY: Water Information Center, 1974. 146 pp.

ONLINE DATABASES

2850 *Aqualine.* Marlow, U.K.: Water Research Centre, 1974-. Updated monthly.

Corresponds to printed *WRC Information*. Covers all aspects of hydrology, water resources, and other water-related subjects. Vendors: DIALOG, ESA-IRS.

2851 *Water resources abstracts.* Washington DC: U.S. Dept. of the Interior, 1968-. Updated monthly.

Corresponds to printed *Selected Water Resources Abstracts*. Covers hydrology and water resources worldwide. Vendors: DIALOG, SDC. A printed thesaurus (*Water Resources Thesaurus*)is available.

2852 *WATERLIT.* Pretoria, South Africa: South African Water Information Center, 1976-. Updated monthly.

Available online only. Covers water resources and hydrology worldwide, though African coverage is especially strong; complements *Water resources abstracts*. Vendor: SDC.

2853 *WATERNET.* Denver CO: American Water Works Association, 1971-.

Available online only. Covers water quality, re-use, testing, economics, other applied aspects of water resources. Vendor: DIALOG.

INDEXES AND ABSTRACTS

2854 *Geographical abstracts: B - Climatology and hydrology.*

See entry 2825.

2855 *Selected water resources abstracts.* Washington DC: U.S. Dept. of the Interior, 1968-. Monthly.

THESES AND DISSERTATIONS

2856 *Doctoral dissertations on hydrology and desalination, 1962-1976.* London: University Microfilm International, 1977. 24 pp.

Contains dissertations submitted to North American universities during the years 1962-1976.

ENCYCLOPEDIAS

2857 Todd, David Keith. *The water encyclopedia: a compendium of useful information on water resources.* Port Washington NY: Water Information Center, 1970. 559 pp.

Presents a range of information including climatic data, surface and ground water statistics, water use and water quality. World wide coverage but emphasis on the United States.

ATLASES

2858 *Atlas of world water balance.* Paris: Unesco Press, 1977. 36 pages and 65 maps.

Identical to the original Russian version published in 1974. Contains explanatory text.

CONSERVATION

2859 Baumann, Duane E., et al. *An annotated bibliography on water conservation.* Fort Belvoir VA: U.S. Army Corps of Engineers. Institute for Water Resources, 1979. 181 pp.

DEVELOPMENT AND MANAGEMENT

2860 *Bibliography of water management.* Logan UT: Utah State University, 1974. 352 pp.

Unannotated.

2861 Cohen, Sanford F. *Cost-benefit analysis for water project evaluation: a selected bibliography.* Monticello IL: Vance Bibliographies, 1980. (Public administration series bibliography P-578) 5 pp.

Unannotated.

2862 *EPA environmental modeling catalogue: abstracts of environmental models.* Washington DC: U.S. Environmental Protection Agency, 1979. 229 pp.

A compendium of descriptions of environmental models used by the EPA. Most (about 25) cover water quality and runoff. Technical contacts and bibliographic references are provided for each.

2863 Gysi, Marshall. *A selected annotated bibliography on the analysis of water resource systems.* Ithaca NY: Cornell University. Dept. of Water Resources Engineering, 1969-1976. 8 vols.

2864 *Public participation in water resource development: a bibliography.* Washington DC: U.S. Water Resources Scientific Information Center, 1976. (OWRT/WRSIC 76-205) 166 pp.

2865 *Water resources, planning and management: a select bibliography/Les ressources en eau, leau planification, et leur gestion: bibliographie selective.* New York: United Nations. Dag Hammarskjold Library, 1977. (Bibliographical series 23) 117 pp.

Unannotated. In English and French.

2866 Wellisch, Hanan. *Water resources development: an international bibliography, 1950-1965.* Jerusalem: Israel Program for Scientific Translations, 1967. 144 pp.

Includes some 2,000 citations.

EVAPORATION SUPPRESSION

2867 *Evaporation suppression: a bibliography.* Washington DC: U.S. Office of Water Resources Research, 1973. (WRSIC 73-216; NTIS: PB 231-200) 478 pp.

Contains 320 citations from the SWRA database.

2868 Magin, George B., and Randall, Lois E. "Review of literature on evaporation suppression (with bibliography)." U.S. Geological Survey. *Professional paper* 272-C (1960):53-69.

GROUNDWATER

2869 Knapp, George L. *Artificial recharge of groundwater: a bibliography.* Washington DC: U.S. National Technical Information Service, 1973. (WRSIC 73-202) 309 pp.

2870 Lehmann, Edward J. *Studies of saline soils and ground water: a bibliography with abstracts.*

See entry 2903.

INTER-BASIN DIVERSION

2871 Whetstone, G. A. *Interbasin diversion of water: an annotated bibliography.* Lubbock TX: Texas Tech University. Water Resources Center, 1970-71. 2 vols.

Contains some 2,000 citations from 1855.

LEGAL AND SOCIAL ASPECTS

2872 Hamilton, H. R. et al. *Bibliography on socio-economic aspects of water resources.* Washington DC: U.S. Office of Water Resources Research, 1966. 453 pp.

PHREATOPHYTE CONTROL

2873 Horton, Jerome S. *Evapotranspiration and watershed research as related to riparian and phreatophyte management, an abstract bibliography.* Washington DC: U.S. Forest Service, 1973. 192 pp.

2874 Paylore, Patricia, ed. *Phreatophytes: a bibliography, revised.* Washington DC: U.S. Office of Water Resources Research, 1974. (NTIS: PB 232-259) 277 pp.

Contains 183 citations from the SWRA database.

QUALITY, POLLUTION, RE-USE

2875 *Annotated bibliography for water quality management.* 6th ed. Washington DC: U.S. Environmental Protection Agency, 1979. 87 pp.

2876 Carberry, Michael E. *Remote sensing applications to water quality assessment: an annotated bibliography of selected literature, 1970 to 1975.* Monticello IL: Council of Planning Librarians, 1976. (CPL exchange bibliography 1121) 36 pp.

Extensively annotated.

2877 Kepinski, Alfred. *Water supply and wastewater disposal: international almanac.* Essen, Federal Republic of Germany: Vulkar-Verlag, 1976-78. 5 vols.

In English, French, and German. Contains a directory of organizations and unannotated bibliography.

2878 Knapp, George L. *Aeration of natural waters: a bibliography.* Washington DC: U.S. Water Resources Scientific Information Center, 1973. (WRSIC 73-206) 358 pp.

2879 Knight, Allen W., and Simmons, Mary Ann. *Water pollution: a guide to information sources.* Detroit MI: Gale, 1980. (Man and the environment guide series 9) 278 pp.

2880 *Land application of wastewater: a bibliography.* Washington DC: U.S. Water Resources Scientific Information Center, 1977. (WRSIC 77-204) 408 pp.

2881 Ogg, Clayton W.; Christensen, Lee A.; and Heimlich, Ralph E. *Economics of water quality in agriculture: a literature review.* Washington DC: U.S. Dept. of Agriculture. Economics, Statistics, and Cooperatives Service. 39 pp.

2882 Rogoff, Marc Jay. *Use of the universal soil loss equation in water quality assessment: an annotated bibliography.* Monticello IL: Council of Planning librarians, 1978. (CPL exchange bibliography 1498) 16 pp.

Extensive annotation.

2883 Tchobanoglous, George; Smith, Robert; and Crites, Ronald. *Wastewater management: a guide to information sources.* Detroit MI: Gale, 1976. (Man and the environment guide series 2) 202 pp.

2884 *Use of naturally impaired water: a bibliography.* Washington DC: U.S. Water Resources Scientific Information Center, 1973. (WRSIC 73-217; NTIS: PB 220-350) 364 pp.

Contains 246 citations derived from the SWRA database.

2885 *Water quality instructional resources information system (IRIS): a compilation of abstracts to water quality and water resources materials.* Cincinnati OH: U.S. Environmental Protection Agency. Office of Water Program Operations, 1979. (EPA 431/1-79-002) 357 pp.

2886 *Water reuse: a bibliography.* Washington DC: U.S. Water Resources Scientific Information Center, 1973-76. (WRSIC 73-215) 5 vols.

SALINE WATER AND DESALINATION

2887 *Bibliography of desalting literature, 1969.* Washington DC: U.S. Office of Saline Water, 1970. (Research and development progress report 552) 465 pp.

Prepared by Columbia Software, Inc.

2888 DePay, G. W. *Disposal of brine effluents from inland desalting plants: review and bibliography.* Washington DC: U.S. Dept. of the Interior, 1969. 209 pp.

2889 *Desalination abstracts.* Tel Aviv: Israel. National Council for Research and Development. Center for Scientific and Technological Information, 1966-. Quarterly.

2890 *Desalting plants inventory report.* (Washington DC): U.S. Office of Saline Water, 1968-. Annual.

2891 *Literature on saline water conversion available in the library of the Sea-Water Conversion Commission as of January 1970.* 5th ed. Tel Aviv: Israel. National Council for Research and Development. Center of Scientific and Technical Information, 1970. Various pagings.

Covers period 1959-1969.

2892 Lockerby, Robert W. *Desalination technology.* Monticello IL: Vance Bibliographies, 1981. (Public administration series bibliography P-668) 12 pp.

Unannotated.

2893 Schamus, J. J. *Bibliography of saline water conversion literature.* Washington DC: U.S. Dept. of the Interior, 1963. 262 pp.

SOILS

GENERAL BIBLIOGRAPHIES

2894 Greenwood, L. Larry, and Rohrer, Richard L. *KWIC index to the Commonwealth Bureau of Soils annotated bibliographies on soils and fertilizers, 1956-1972.* Manhattan KS: Kansas State University, 1973. (Bibliography 13) 198 pp.

A listing and keyword index to 1,553 annotated bibliographies.

2895 Kamphorst, A. *Annotated bibliography on tropical and subtropical alluvial and organic soils.* Wageningen, The Netherlands: International Institute for Land Reclamation and Improvement, 1961. (Bibliography 3) 2 vols.

2896 Orvedal, Arnold Clifford. *Bibliography of the soils of the Tropics.* Washington DC: U.S. Agency for International Development. Technical Assistance Bureau. Office of Agriculture, 1975. (Technical series bulletin 17) 225 pp.

INDEXES AND ABSTRACTS

2897 *Soils and fertilizers.* Farnham Royal, U.K.: Commonwealth Agricultural Bureau, 1938-. Monthly.

Includes subsections on aridic soils, irrigation, and land reclamation. Supersedes Commonwealth Bureau of Soils *Monthly letter.*

MINED LAND RECLAMATION

2898 Gifford, Gerald F.; Dwyer, D. D.; and Norton, B. E. *A bibliography of literature pertinent to mining reclamation in arid and semi-arid environments.* Logan UT: Utah State University. Dept. of Range Science, 1972. (SWRA-W73- 00912) 23 pp.

Contains 312 citations.

2899 *SEAMALERT.* Tucson AZ: University of Arizona. Office of Arid Lands Studies, 1977-. Quarterly.

Quarterly index with abstracts. SEAMALERT = Surface environment and mining: bibliographic reference to published materials.

SALINE AND ALKALINE SOILS

2900 Abell, L. F., and Gelderman, W. J. *Annotated bibliography on reclamation and improvement of saline and alkali soils (1957-1964).* Wageningen, The Netherlands: International Institute for Land Reclamation and Improvement, 1964. (Bibliography 4) 59 pp.

Updated by Bibliography 6 (1967) (see entry 2901).

2901 van Alphen, J. G., and Abell, L. F., comps. *Annotated bibliography on reclamation and improvement of saline and sodic soils (1966-1960).* Wageningen, The Netherlands: International Institute for Land Reclamation and Improvement, 1967. (Bibliography 6) 43 pp.

Updated version of Bibliography 4 (1964) (entry 2900), which covers the period 1957-1964.

2902 Goudie, A. S. "A regional bibliography of calcrete." In *Food, fiber, and the arid lands,* pp. 421-427. Edited by William G. McGinnies, Bram J. Goldman, and Patricia Paylore. Tucson AZ: University of Arizona Press, 1971.

2903 Lehmann, Edward J. *Studies of saline soils and ground water: a bibliography with abstracts.* Springfield VA: U.S. National Technical Information Service, 1973. (NTIS: COM 73-11799/6) 60 pp.

Contains 55 citations.

2904 Pahwa, K. N., and Gupta, I. C. *World literature on reclamation and management of salt affected soils (1950-1981).* New Delhi: Associated Publishing, 1982. 352 pp.

Extensively annotated; 811 citations cover the world with greatest emphasis on India, the Soviet Union, and the United States.

2905 Varallyay, G. "Bibliography on salt-affected soils." In *Review of research on salt-affected soils,* pp. 99-127. Edited by Istvan Szaboks. Paris: Unesco, 1979.

SAND DUNES AND CONTROL

2906 Hagedorn, Horst et al. *Dune stabilization: a survey of literature on dune formation and dune stabilization.* Eschborn, Federal Republic of Germany: Deutsche Gesellschaft fur Technische Zusammenarbeit, 1977. 223 pp.

2907 Mainguet, M., comp. *Dunes et autres edifices sableux eoliens [dunes and other aeolian sand formations].* Paris: Delegation generale a la Recherche Scientifique et Technique, 1978. 344 pp.

2908 "Special retrospective bibliography: sand control." *Arid lands abstracts* 7 (1975): items 199- 357.

2909 Warren, Andrew. "A bibliography of desert dunes and associated phenomena." In *Arid lands in perspective,* pp. 75-99. Edited by William G. McGinnies and Bram J. Goldman. Washington DC: American Association for the Advancement of Science; Tucson AZ: University of Arizona Press, 1969.

Contains some 700 citations covering the period to 1967.

SOIL FORMATION AND PROPERTIES

2910   Brook. Ronald H. *Soil survey interpretation: an annotated bibliography (selected references 1972-1960).* Wageningen, The Netherlands: International Institute for Land Reclamation and Improvement, 1975. (Bibliography 10) 64 pp.

2911   Smalley, Ian J. *Loess: a partial bibliography.* Norwich, U.K.: Geo Books, 1980. (Geo Abstracts bibliography 7) 103 pp.

2912   *Soil nitrogen cycle; a bibliography.* Washington DC: U.S. Office of Water Resources Research. Water Resources Scientific Information Center, 1972. (WRSIC 72-208; NTIS: PB 209-931) 288 pp.

Contains 201 citations.

2913   Thuronyi, G. "Annotated bibliography of recent literature on soil moisture." American Meteorological Society. *Meteorological abstracts and bibliography* 10:3 (1959): 423-461.

2914   Zikeev, Nikolai Tikhonovich. "A selective annotated bibliography on soil temperature." American Meteorological Society. *Meteorological abstracts and bibliography* 2:3 (1951): 207-232.

BIOLOGY

RESEARCH GUIDES

2915   Blanchard, J. Richard, and Farrell, Lois, eds. *Guide to sources for agricultural and biological research.*

See entry 2949.

2916   Davis, Elisabeth B. *Using the biological literature: a practical guide.* New York: M. Dekker, 1981. (Books in library and information science 35) 286 pp.

Partially annotated. Primary and secondary resources, aimed at undergraduate and graduate students. Emphasis on current English-language materials. Applied areas (medicine, clinical psychology) not included.

2917   Smith, Roger Cletus. *Smith's guide to the literature of the life sciences.* 9th ed. Minneapolis MI: Burgess, 1980. 223 pp.

Introduction to the most useful materials for biological research and suggestions for reporting research to the scientific community. Designed for undergraduate and graduate students. Includes bibliographies, journals, library searching and thesis preparation.

GENERAL BIBLIOGRAPHIES

2918   Wood, Don A. *A bibliography on the world's rare, endangered and recently extinct wildlife and plants.* Stillwater OK: Oklahoma State University, 1977. (Environmental series 3) 85 pp.

ONLINE DATABASES

2919   *BIOSIS previews.* Philadelphia PA: BioSciences Information Service, 1969-. Updated monthly.

Corresponds to printed *Biological abstracts* and *Biological abstracts/ RRM*. Covers the literature of the life sciences. Vendors: BRS, DIALOG, SDC, DATA-STAR, DIMDI, CISTI. A printed thesaurus (*BIOSIS search guide*) is available.

2920   *Life sciences collection.* Bethesda MD: Cambridge Scientific Abstracts, 1978-. Updated monthly.

Corresponds to 17 printed CSA abstracting journals, including *Animal behavior abstracts*, *Ecology abstracts*, *Entomology abstracts*, etc. Covers the literature of the life sciences worldwide. Vendor: DIALOG. Formerly *IRL life sciences collection*.

INDEXES AND ABSTRACTS

2921   *Berichte biochemie und biologie.* Berlin, Federal Republic of Germany: Springer Verlag, 1969-. 19 per year.

In German. Cumulative indexes for every two volumes. Continues *Berichte uber die Wissenschaftliche Biologie* (1926- 69).

2922   *Biological abstracts.* Philadelphia PA: BioSciences Information Service of Biological Abstracts, 1926-. Annual.

2923   *Biological abstracts/Reports, reviews, meetings.* Philadelphia PA: BioSciences Information Service, 1980-. Monthly.

Continues *BioResearch index* (1965-1979).

2924   *Biology digest.* Louisville KY: Data Courier, 1974-. 9 per year.

Organizes, summarizes and indexes worldwide scientific literature in the life sciences. Designed for use by secondary school and undergraduate students. Each issue also contains a feature article.

2925   *Ecology abstracts.* London: Information Retrieval, 1980-. Monthly.

Prepared in cooperation with Unesco program on man and the biosphere. Continues *Applied ecology abstracts* (1975- 79).

2926   Goetz, Rita D., ed. *Biological and agricultural index: a cumulative subject index to periodicals in the fields of biology, agriculture, and related sciences.* New York: H. W. Wilson Co., 1916-. 11 per year.

ENCYCLOPEDIAS

2927 Gray, Peter. *The encyclopedia of the biological sciences.* 2nd ed. New York: Van Nostrand Reinhold, 1970. 1,027 pp.

2928 Kormondy, Edward J., and McCormick, J. Frank, eds. *Handbook of contemporary developments in world ecology.* Westport CT: Greenwood, 1981. 776 pp.

34 major countries are each covered by narratives written by regional experts. There are many major omissions (e.g., only five African countries are included, and India and most of China are omitted), but for those countries included, the information is quite useful for background and coverage of basic texts. Special section on Man and the Biosphere Programme (MAB). Chapters on individual nations are listed separately in this Guide.

2929 Lapedes, Daniel N., ed. *McGraw-Hill dictionary of the life sciences.* New York: McGraw-Hill, 1976. 907 pp.

DIRECTORIES

2930 Gunn, C. R. et al. *Systematic collections of the Agricultural Research Service.* Washington DC: U.S. Dept. of Agriculture, 1977. (Miscellaneous publication 1343) 84 pp.

A directory of collections of plant and animal species giving addresses, personnel, history, extent, and bibliographic references. Holdings in these collections cover the world's flora and fauna.

BOTANY

GENERAL BIBLIOGRAPHIES

2931 *Excerpta botanica: sectio B: sociologica.* New York: G. Fischer, 1959-. Quarterly.

A running series of botanical bibliographies, mostly unannotated. Selected titles are included individually in this Guide.

2932 Swift, Lloyd H. *Botanical bibliographies: a guide to bibliographic materials applicable to botany.* Minneapolis MI: Burgess, 1970. 804 pp.

INDEXES AND ABSTRACTS

2933 *Excerpta botanica: sectio A: taxonomica et chorologica.* New York: G. Fischer Verlag, 1959-. 7 per year.

A briefly annotated index to recent botanical literature. Includes taxonomy, geographic areas, paleobotany, prehistory, and ethnobotany. Primarily English and German.

2934 *Medicinal and aromatic plants abstracts.* New Delhi: India. Council of Scientific and Industrial Research. Publications and Information Directorate, 1979-. 6 per year.

An abstracting index of current literature; covers the world but emphasis is on South Asia.

DIRECTORIES

2935 Holmgren, Patricia K.; Keuken, Wil; and Schofield, Eileen K., comps. "The herbaria of the world." In *Index herbariorum: a guide to the location and contents of the world's public herbaria*, pt. 1. 7th ed. Edited by Frans Antonie Stafleu. Utrecht, The Netherlands: Bohn, Schelteme and Holkeme, 1981. (Regnum vegetabile 106).

Part of a multi-volume set.

2936 Kiger, Robt. William. *International register of specialists and current research in plant systematics.* Pittsburgh PA: Hunt Institute for Botanical Documentation, 1981. 346 pp.

REMOTE SENSING

2937 Pettinger, Lawrence R. *A selected bibliography: remote sensing applications for tropical and subtropical vegetation analysis.* Sioux Falls SD: Technicolor Graphic Services, 1978. (NTIS: PB 284-683) 50 pp.

Contains 425 citations covering the period 1924-1978.

VEGETATION MAPPING

2938 Kuechler, August Wilhelm. *International bibliography of vegetation maps.* Lawrence KS: University of Kansas Libraries, 1965-. (Library series 21, 26, 29, 36, 45).

An extensive annotated bibliography arranged and published as follows. First edition: volume 1 covers North America (1965), 453 pp.; volume 2 covers Europe (1966), 584 pp.; volume 3 covers the U.S.S.R., Asia, and Australia (1968), 389 pp.; and volume 4 covers Africa, South America and the world (1970), 561 pp. Second edition: section 1 covers South America (1980), 324 pp. Volumes are listed separately in this Guide.

2939 Phillips. W. Louis. *Index to plant distribution maps in North American periodicals through 1972.* Boston MA: G. K. Hall, 1976. 686 pp.

Covers the world.

ZOOLOGY

GENERAL BIBLIOGRAPHIES

2940 Harmon, Ralph Winter. *Bibliography of animal venoms.* Gainesville FL: University of Florida Press, 1948. 340 pp.

Contains 4,157 citations covering 1875-1946.

INDEXES AND ABSTRACTS

2941 *Keyword index of wildlife research.* Zurich, Switzerland: Swiss Wildlife Information Service, 1974-. Annual.

2942 *The zoological record.* London: The Zoological Society, 1864-. Annual.

ENCYCLOPEDIAS

2943 Grzimek, Bernhard. *Grzimek's animal life encyclopedia.* New York: Van Nostrand Reinhold, 1972. 13 vol.

ENTOMOLOGY

2944 *Acrida.* Paris: Association d'Acridologie, 1972-. Quarterly.

In French and English. Continues *Acridological abstracts.* Covers the literature of the grasshopper and locust (Orthoptera).

2945 Bonnet, Pierre. *Bibliographia araneorum: analyse methodique de toute la litterature araneologique jusqu'en [bibliography of araneology: methodical analysis of all the araneological literature until 1939].* Toulouse, France: Impremerie Douladoure, 1945-61. 3 vols. in 7.

In French. On spiders. Includes brief biographies of principal araneologists and general bibliographies.

2946 *Entomology abstracts.* London: Information Retrieval Ltd., 1969-. Monthly.

2947 Hammack, Gloria Mae. *The serial literature of entomology: a descriptive study.* College Park MD: Entomological Society of America, 1970. 85 pp.

Includes annual reports, proceedings, government reports, and abstracting publications which appear in serial form, as well as scientific journals.

HERPETOLOGY

2948 Russell, Findlay E., and Scharffenberg, R. *Bibliography of snake venoms and venomous snakes.* West Covina CA: Bibliographic Associates, 1964. 220 pp.

Includes 5,829 citations.

AGRICULTURE

RESEARCH GUIDES

2949 Blanchard, J. Richard, and Farrell, Lois, eds. *Guide to sources for agricultural and biological research.* Berkeley CA: University of California Press, 1981. 735 pp.

A massive annotated bibliography covering the full range of subjects connected with research in biology, agriculture, food science, environmental science, and agricultural economics. Contains long narrative introductions to each major topic highlighting the most useful sources. A separate chapter considers online databases. Covers literature appearing from about 1958 through 1979. Librarians will appreciate the elaborate arrangement of entries; researchers should turn to the author, subject, and title indexes. An essential tool.

2950 Lilley, G. P. *Information sources on agriculture and food science.* London: Butterworths, 1981. 603 pp.

Very good narrative treatment of agricultural information sources, with brief sections on specific topics.

2951 Ryan, Mary, and O'Shea, John. *Information retrieval in agricultural engineering.* Toowoomba, Queensland: Darling Downs Institute Press, 1979. 50 pp.

Covers agriculture generally with some emphasis on engineering aspects.

GENERAL BIBLIOGRAPHIES

2952 Bush, Ernest Alfred Radford. *Agriculture: a bibliographical guide.* London: Macdonald, 1974. 2 vols. (Macdonald bibliographical guides).

Covers the period 1958 to 1971. Primarily English-language items.

2953 Buttel, Frederick H. *Energy and agriculture: a bibliography of social science literature.* Monticello IL: Council of Planning Librarians, 1977. (CPL exchange bibliography 1430) 39 pp.

Unannotated.

2954 *Dictionary catalog of the National Agricultural Library: 1862-1965.* New York: Rowman and Littlefield, 1967. 73 vols.

Updated by monthly issues since 1966; cumulated annually. Five year cumulation (1972) of 12 volumes covers period 1966-1970. Lists all books and serials held by NAL including a large collection of materials on most aspects of arid lands agriculture and related sciences.

2955 Dworaczek, Marian. *Sources of information in agriculture.* Monticello IL: Vance Bibliographies, 1982. (Public administration series P-1032) 25 pp.

Contains 259 unannotated citations.

2956 *FAO documentation: current bibliography.* Rome: FAO, 1967-. Monthly.

Text in English, French and Spanish; English summaries. Annual cumulative author and subject indexes.

2957 **Greenwood, L. Larry, and Rohrer, Richard L.** *KWIC index to the Commonwealth Bureau of Soils annotated bibliographies on soils and fertilizers 1956-1972.*

See entry 2894.

2958 *Index of FAO technical assistance reports 1951-1965.* Rome: FAO, 1966. 273 pp.

Index to over 2,000 projects in some 100 different nations.

2959 *Land and water; index. 1945-1966.* Rome: FAO, 1968. (Special indexes DC/Sp. 8) various pagings.

Contains approximately 1,300 entries. In English, French, and Spanish. Covers most documents published from 1945-66 by the Land and Water Division. Covers watershed management, soil resources, fertilizer, irrigation, and agricultural engineering.

ONLINE DATABASES

2960 *AGLINE.* St. Louis MO: Doane-Western, 1977-. Updated monthly.

Available online only. Covers agriculture worldwide. Formerly *DOANE*. Vendor: SDC.

2961 *AGRICOLA.* Beltsville MD: U.S. Dept. of Agriculture, 1970-. Updated monthly.

Corresponds to printed *Bibliography of agriculture* in part. Covers all aspects of agriculture. Vendors: BRS, DIALOG, DIMDI. A manual (*AGRICOLA user's guide*) is available.

2962 *AGRIS (International information system for the agricultural sciences and technology).* Rome: U.N. FAO, 1975-. Updated monthly.

Corresponds to printed *Agrindex*. Covers all aspects of agriculture. Vendors: ESA-IRS, IAEA, DIMDI.

2963 *Aquaculture.* Kansas City MO: U.S. National Oceanic and Atmospheric Administration, 1970-. Updated quarterly.

Available only online. Covers the cultivation of marine, brackish, and freshwater organisms; excludes hydroponics. Vendor: DIALOG.

2964 *CAB abstracts.* Slough U.K.: Commonwealth Agricultural Bureaux, 19—. Updated monthly.

Corresponds to the 28 printed abstracting journals of the CAB. Covers agriculture and related fields worldwide. Vendors: DIALOG, DIMDI, ESA-IRS.

2965 *CRIS/USDA (Current research information system).* Washington DC: U.S. Dept. of Agriculture. Cooperative State Research Service, 1974-. Updated quarterly.

Available online only. Contains descriptions of agriculture-related research performed in the United States. Vendor: DIALOG.

2966 *TROPAG.* Amsterdam, The Netherlands: Royal Tropical Institute, 1975-. Updated monthly.

Corresponds to printed *Abstracts on tropical agriculture*. Vendor: SDC.

INDEXES AND ABSTRACTS

2967 *Agrindex.* Rome: AGRIS Coordinating Centre, 1975-. Freq. unknown.

In English, French and Spanish.

2968 *Bibliography of agriculture.* Phoenix AZ: The Oryx Press, 1942-. Monthly.

Unannotated. Monthly index to literature of agriculture and related sciences. Includes journal articles, government documents, reports, conference proceedings. Prepared by the U.S. Dept. of Agriculture, National Agriculture Library. A thesaurus (*Agricultural terms as used in the Bibliography of Agriculture*) is available.

2969 **Goetz, Rita D., ed.** *Biological and agricultural index: a cumulative subject index to periodicals in the fields of biology, agriculture, and related sciences.*

See entry 2926.

2970 *Soils and fertilizers.*

See entry 2897.

ENCYCLOPEDIAS

2971 **Crabbe, David, and Lawson, Simon.** *The world food book: an A-Z, atlas and statistical source book.* London: Kogan Page; New York: Nichols, 1981. 240 pp.

An encyclopedia of food production and agriculture, with maps and statistical appendices.

DIRECTORIES

2972 *Who's who in world agriculture.* Longman House, U.K.: Francis Hodgson, 1979-. 2 vols.

STATISTICS

2973 *FAO production yearbook.* Rome: FAO, 1976-. Annual.

In English, French, and Spanish. Data based on calendar-year (period in which the bulk of the harvest took place). Includes data on crops, livestock, pesticides, agricultural machinery and land use and irrigation.

ATLASES

2974 *World atlas of agriculture.* Novara, Italy: Istituto Geografico De Agostini, 1969-76. 4 vols.

IRRIGATION

GENERAL BIBLIOGRAPHIES

2975 *Agricultural runoff: a bibliography.* Washington DC: U.S. Water Resources Scientific Information Center, 1972. (WRSIC 72-204; NTIS: PB 207-514) 252 pp.

Contains 158 citations from the SWRA database.

2976 Bebee, Charles N. *Drip irrigation, 1970-1977.* Beltsville MD: U.S. Science and Education Administration. Technical Information Systems, 1978. (Quick bibliography series; NAL-bib 78-08; NTIS: PB 294-000) 31 pp.

Contains 265 citations.

2977 Floss, Ludmilla. *Sprinkler irrigation: a bibliography selected from foreign literature, 1964-1969.* Washington DC: U.S. Dept. of the Interior. Office of Library Services, 1970. (Bibliography series 15; SWRA W70-07879) 54 pp.

Contains 251 citations.

2978 *Irrigation book list.* Bet Dagan, Israel: International Irrigation Information Center, 1977. (Publication 3) 60 pp.

Unannotated. Covers period 1960-76.

2979 *Irrigation efficiency: a bibliography.* Washington DC: U.S. Water Resources Scientific Information Center, 1973-78. (WRSIC 73-214; NTIS: PB 220-349, PB 263-155, PB 287-626/655) 3 vols.

2980 *Irrigation return flow: a bibliography.* Washington DC: U.S. Water Resources Scientific Information Center, 1975. (WRSIC 75-209; NTIS: PB 249-585/1) 207 pp.

Contains 132 citations from the SWRA database.

2981 MacLean, Jayne. *Irrigation scheduling, 1969-November 1978.* Beltsville MD: U.S. Dept. of Agriculture. Science and Education Administration. Technical Information Systems. National Agricultural Library, 1979. (NAL-BIBL-79-04) 78 pp.

Unannotated. Contains 183 English-language citations from the AGRICOLA data base.

2982 Raadsma, S., and Schrale, G. *Annotated bibliography on surface irrigation methods.* Wageningen, The Netherlands: International Institute for Land Reclamation and Improvement, 1971. (Bibliography 9) 72 pp.

2983 Smith, Stephen W. *Annotated bibliography on trickle irrigation.* Ft. Collins CO: Colorado State University. Environmental Resources Center, 1975. (Information series 6) 61 pp.

Contains 182 references.

2984 *Water for agriculture: FAO publications and documents (1945-Sept. 73), annotated bibliography, author and subject index.* Rome: FAO, 1973. 400 pp.

INDEXES AND ABSTRACTS

2985 *Irricab: current annotated bibliography of irrigation.* Bet Dagan, Israel: International Irrigation Information Centre, 1976-. Quarterly.

2986 *Irrigation and drainage abstracts.* Farnham Royal, U.K.: Commonwealth Agricultural Bureaux, 1975-. Quarterly.

Worldwide coverage of periodicals, monographs, and reports on all aspects of irrigation and drainage problems.

DIRECTORIES

2987 *Irrigation: international guide to organizations and institutions.* Bet Dagan, Israel: International Irrigation Information Center, 1980. 165 pp.

PLANT CULTURE

GENERAL BIBLIOGRAPHIES

2988 Beale, Helen Purdy, comp. *Bibliography of plant viruses and index to research.* New York: Columbia University Press, 1976. 1,495 pp.

Unannotated. Covers the period 1892-1970.

2989 Bebee, Charles N. *Aquaculture and hydroponics, 1968-1978.* Beltsville MD: U.S. Dept. of Agriculture, 1979. (Bibliographies and literature of agriculture BLA-2) 71 pp.

Citations from AGRICOLA database of articles and monographs.

2990 Francois, L. E., and Maas, E. V. *Plant responses to salinity: an indexed bibliography.* Berkeley CA: U.S. Dept. of Agriculture, 1978. (Science and Education Administration. Agricultural Reviews and Manuals. Western series 6) 369 pp.

Contains citations from 1900-1977.

2991 *Plants, index. 1945-1966.* Rome: FAO, 1967. (Special indexes DC/Sp. 5) Various pagings.

Contains approximately 2,400 entries. Indexes FAO publications and documents. In English, French, and Spanish.

INDEXES AND ABSTRACTS

2992 *Crop physiology abstracts*. Farnham Royal, U.K.: Commonwealth Agricultural Bureaux, 1975-. Monthly.

Worldwide bibliography on germination, enzymes, metabolism and other aspects of crop physiology.

2993 *Pesticides abstracts.*. Washington DC: U.S. Environmental Protection Agency, 1974-. Monthly.

Formerly *Health aspects of pesticides bulletin*. Includes worldwide literature pertaining to the effects of pesticides.

2994 *Plant growth regulator abstracts*. Farnham Royal, U.K.: Commonwealth Agricultural Bureaux, 1975-. Monthly.

DIRECTORIES

2995 *Horticultural research international: directory of horticultural research institutes and their activities in 61 countries*. Wageningen, The Netherlands: PUDOC/ Centre for Agricultural Publishing and Documentation, 1981. 698 pp.

2996 *World list of seed sources/Liste mondiale de sources de semences/Lista mundial de fuentes de semillas*. Rome: FAO, 1980. 102 pp.

In English, French, Spanish. A directory of organizations and government agencies that supply seeds for food, fiber, and ornamental crops. Listed by nation.

CEREAL CROPS

2997 *Bibliography of wheat*. Metuchen NJ: Scarecrow, 1971. 3 vols.

Covers period 1959-1968. Includes works in 47 languages. Prepared by George Washington University (Washington DC) Biological Sciences Communication Project.

2998 *The millets: a bibliography of the world literature covering the years 1930-1963*. Metuchen NJ: Scarecrow, 1967. 154 pp.

Covers 7 species of millets. Prepared by George Washington University (Washington DC) Biological Sciences Communication Project.

2999 **Rachie, Kenneth O.** *The millets and minor cereals: a bibliography of the world literature on millets pre- 1930 and 1964-69; and of all literature on other minor cereals*. Metuchen NJ: Scarecrow, 1974. 202 pp.

Covers seven species of millet, included in entry 2998, in addition to a further three millets and teff (*Eragrostis*), fonio (*Digitaria* sp.), and adlay(*Coix*). Supplements entry 2998.

3000 *Sorghum: a bibliography of the world literature covering the years 1930-1963*. Metuchen NJ: Scarecrow, 1967. 301 pp.

Unannotated. Supplemented by entry 3001. Prepared by George Washington University (Washington DC) Biological Sciences Communication Project.

3001 *Sorghum: a bibliography of the world literature, 1964-1969*. Metuchen NJ: Scarecrow, 1973. 393 pp.

Unannotated. Supplement to entry 3000.

3002 *Sorghum and millets abstracts*. Farnham Royal, U.K.: Commonwealth Agricultural Bureaux, 1976-. Monthly.

Abstracts cover various aspects of sorghum and millet.

3003 *Sorghums and millets bibliography, April 1976-August 1978*. Beltsville MD: U.S. Dept. of Agriculture, 1979. (Bibliographies and literature of agriculture 4) 186 pp.

Unannotated. Compiled from AGRICOLA data base.

FIBER CROPS

3004 **Abell, L. F., and Brouwer, C. J.** *Bibliography on cotton irrigation*. Wageningen, The Netherlands: International Institute for Land Reclamation and Improvement, 1970. (Bibliography 8; SWRA W71-05712) 41 pp.

Unannotated. Worldwide coverage. Covers the period 1959 through 1968. Includes information on work in field operation, utilization, and economic and other aspects. Prepared by the Biological Sciences Communication Project at George Washington University (Washington DC).

3005 *Cotton and tropical fibres abstracts*. Farnham Royal, U.K.: Commonwealth Agricultural Bureaux, 1976-. Monthly.

Covers cotton, sisal, jute, kenaf, roselle, and other fibers.

3006 **Dunn, Henry Arthur.** *Cotton boll weevil: abstracts of research publications 1843-1960*. Washington DC: U.S. Dept. of Agriculture, 1964. (Miscellaneous publication 985) 194 pp.

Includes items from U.S. Dept. of Agriculture publications, professional journals, and biological abstracts. Continued by entry 3007.

3007 **Mitlin, Luceille Liston.** *Boll weevil (Anthonomus grandis Boh), abstracts of research publications, 1961-65*. Washington DC: U.S. Dept. of Agriculture, 1968. (Miscellaneous publication 1092) 32 pp.

Continuation of entry 3006.

FORESTRY

3008 **Evans, Peter A., and Skupa, Gary L.** *International directory of documentation services concerning forestry and forest products*. Berkeley CA: U.S. Forest Service. Pacific Southwest Forest and Range Experiment Station, 1981. (General technical report PSW-47) 71 pp.

3009 *Forestry abstracts.* Farnham Royal, U.K.: Commonwealth Agricultural Bureaux, 1939/40-. Monthly.

3010 Myers, Brian J., and Craig, Ian E. *Bibliography of remote sensing in forestry 1950-1978.* Canberra A.C.T.: CSIRO. Division of Forest Research, 1980. (Divisional report 5) 150 pp.

An unannotated list of some 840 citations of English-language sources. Coverage is worldwide.

3011 Pert, Mary. *Pinus radiata: a bibliography.* Canberra A.C.T.: Australia. Forestry and Timber Bureau, 1963. 145 pp.

"A bibliography to 1963." Unannotated. *Pinus radiata* = the Monterey pine. *Pinus radiata* is native to California and is of major importance in New Zealand, Australia, South Africa, and Chile. *Supplement* (1965) covers period 1963-64, 70 pp.; *Supplement* (1966) covers period 1965-66, 53 pp.; *Supplement* (1969) covers period 1967-68, 58 pp.; and *Supplement* (1972) covers period 1969-70, 56 pp.

3012 Torunsky, Richard, ed. *Weltforstatlas./World forestry atlas./Atlas des forets du monde./Atlas forestal del mundo.* Berlin, Federal Republic of Germany: Fritz Haller Verlag, 1951-. Loose-leaf (not yet complete).

In German, English, French, and Spanish.

3013 *Yearbook of forest products/Annuaire des produits forestiers/Anuario de productos forestales.* Rome: FAO, 1967-. Annual.

In English, French, and Spanish.

FRUIT, NUT AND LEGUME CROPS

3014 Cheek, Emory, comp. *List of theses and dissertations on peanuts and peanut related research.* Athens GA: University of Georgia. College of Agriculture. Experiment Stations, 1969. (Research report 54) 38 pp.

Unannotated. Covers masters theses and doctoral dissertations issued worldwide. *Research report* 54 contains 234 titles; *Research report* 107 (1971) contains 97 titles, 20 pp.; *Research report* 164 (1973) contains 83 titles, 20 pp.; *Research report* 265 (1977) contains 87 titles, 21 pp.; *Research report* 346 (1980) contains 76 titles, 18 pp.; and *Research report* 392 (1982) contains 77 titles, 18 pp.

3015 Dowson, Valentine, and Wilfred, Hugh, comps. *Bibliography of the date palm.* Miami FL: Field Research Projects, 1976. (Study 102) 139 pp.

Unannotated.

3016 *Faba bean abstracts.* Slough, U.K.: Commonwealth Agricultural Bureaux, 1981-. Monthly.

3017 *Lentil abstracts.* Farnham Royal, U.K.: Commonwealth Agricultural Bureaux, 1981-. Annual.

Annotated citations on lentils drawn from the CAB database.

3018 Nene, Y. L. et al. *An annotated bibliography of chickpea diseases, 1915-1976.* Hyderabad, India: International Crops Research Institute for the Semi-arid Tropics (ICRISAT), 1978. (Information bulletin 1; NTIS: PB 81-220-766 or PB 81-214-462) 52 pp.

3019 Poulter, N. H., and Dench, J. E. *The winged bean (Psophocarpus tetragonolobus (L.) DC): an annotated bibliography.* Croydon, U.K.: Tropical Products Institute, 1981. 233 pp.

Comprehensive for period since 1978 but includes major earlier works.

3020 Singh, K. B., and Van der Maeson, L. J. G. *Chickpea bibliography, 1930 to 1974.* Hyderabad, India: International Crops Research Institute for the Semi-arid Tropics (ICRISAT), 1977. (NTIS: PB 81-216-541) 229 pp.

Also includes literature prior to 1930.

OIL AND RUBBER CROPS

3021 Elias-Cesnik, Anna. *Jojoba: guide to the literature.*

See entry 2247.

3022 McGinnies, William G., and Haase, Edward F. *Guayule: a rubber-producing shrub for arid and semi-arid regions.*

See entry 2027.

3023 Sherbrooke, Wade C. *Jojoba: an annotated bibliography.*

See entry 2250.

3024 Sherbrooke, Wade C., and Haase, E. F. *Jojoba: a wax-producing shrub of the Sonoran Desert: literature review and annotated bibliography.*

See entry 2251.

3025 *Tropical oil seeds abstracts.* Farnham Royal, U.K.: Commonwealth Agricultural Bureaux, 1976-. Monthly.

Covers the groundnut, the safflower, the coconut, the oil palm, the castor, the sesame, and others.

ANIMAL CULTURE

GENERAL BIBLIOGRAPHIES

3026 *Animals index 1945-1966.* Rome: FAO, 1967. Approximately 400 pp.

Covers most documents produced from 1945-1966 by the Animal Production and Health Division.

3027 Rowe, M. L., and Merryman, Linda. *Livestock and the environment: a bibliography with abstracts.* Ada OK: U.S. Environmental Protection Agency, 1971-. (Environmental protection technology series). Annual.

Annotated references deal with the impact of animal wastes on the environment.

INDEXES AND ABSTRACTS

3028 *Animal disease occurrence.* Farnham Royal, U.K.: Commonwealth Agricultural Bureaux, 1980-. 2 per year.

3029 *Poultry abstracts.* Farnham Royal, U.K.: Commonwealth Agricultural Bureaux, 1975-. Monthly.

Contains abstracts on types of poultry as well as on disorders affecting poultry.

3030 *Small animal abstracts.* Farnham Royal, U.K.: Commonwealth Agricultural Bureaux, 1975-. Quarterly.

Contains abstracts concerning disease, genetics, nutrition, surgery, etc. of small animals.

RANGE MANAGEMENT

3031 Vallentine, John F. *Range science: a guide to information sources.* Detroit MI: Gale, 1980. (Natural world information guide series 2) 231 pp.

AGRICULTURAL ECONOMICS

GENERAL BIBLIOGRAPHIES

3032 *Food and agricultural industries: annotated bibliography, author and subject index.* Rome: FAO, 1970. Approximately 450 pp.

Covers FAO publications produced from 1945-1970, by the Agricultural Services Division. In English, French, and Spanish.

3033 Martin, Lee R., ed. *A survey of agricultural economics literature.* Minneapolis MN: University of Minnesota Press, 1977. 3 vols., to be completed in 4 vols.

Thoroughly-documented articles on aspects of agricultural economics. Vol. I covers farm management, vol. II covers statistical inference in economics, vol. III covers rural population.

3034 Vondruska, John. *Aquacultural economics bibliography.* Washington DC (?): U.S. Dept. of Commerce, 1976. (NOAA technical report NMFS SSRF-703) 123 pp.

Includes recent published and some unpublished U.S. and foreign literature.

INDEXES AND ABSTRACTS

3035 *World agricultural economics and rural sociology abstracts.* Oxford, U.K.: Commonwealth Agricultural Bureaux. Commonwealth Bureau of Agricultural Economics, 1959-. Monthly.

Covers publications on agricultural economics, policy, supply, demand and prices, marketing and distribution, and international trade.

STATISTICS

3036 *FAO commodity review and outlook.* Rome: FAO, 1968/69-. (FAO committee on commodity problems. Document). Every other year.

A narrative and statistical review of agricultural commodities and marketing for the world.

3037 *FAO trade yearbook.* Rome: FAO, 1958-. Annual.

Provides statistics on agricultural production and trade for the world. Explanatory text in English, French, and Spanish.

3038 McFall, Jane Buzby, ed. *Agricultural commodities index: ready-reference index to USDA statistical series.* Phoenix AZ: The Oryx Press, 1978. 1,956 pp.

Covers all situation and outlook reports, national and selected local market news and livestock reports and other sources, from 1965 to 1977.

EXTENSION AND TRAINING

3039 *Rural extension, education and training abstracts.* Oxford, U.K.: Commonwealth Bureau of Agricultural Economics, 1978-. Quarterly.

Contains abstracts on extension and training, arranged by geographical region or by subject.

FOOD AID

3040 Cadet, Melissa Lawson. *Food aid and policy for economic development: an annotated bibliography and directory.* Sacramento CA: Trans Tech Management Press, 1981. 178 pp.

Annotated. Covers period 1964-1980. Emphasis on role of food aid in development. Includes international directory of food aid in development agencies.

LAND TENURE

3041 *Land reform: annotated bibliography, author and subject index.* Rome: FAO, 1971. Approximately 250 pp.

Covers FAO publications and documents produced since 1945 on land reform and related subjects. Titles given in English. Primarily documents produced by the Rural Institutions Division.

3042 *Land reform, land settlement and cooperatives.* Rome: Food and Agriculture Organization of the United Nations, 1972-. 2 per year.

Consists of country review papers and bibliographies on land reform and related issues worldwide. Issued in English, French, and Spanish.

MARKETS AND COOPERATIVES

3043 *Agricultural cooperation: annotated bibliography, author and subject index.* Rome: FAO, 1971. Approximately 250 pp.

Covers FAO publications produced from 1945-1970, on agricultural cooperation by the Rural Institutions Division.

3044 Bromley, Ray. *Periodic markets, daily markets, and fairs: a bibliography supplement to 1979.* Swansea, U.K.: University College of Swansea. Centre for Development Studies, 1979. 69 pp.

An unannotated list of 1,195 citations covering the world.

3045 Kubal, Gene J. *Cooperation in agriculture 1954-1964, a list of selected references.* Washington DC: U.S. Dept. of Agriculture, 1966. (Library list 41, supplement 2) 115 pp.

Contains English-language citations only.

ENERGY

GENERAL BIBLIOGRAPHIES

3046 Buttel, Frederick H. *Energy and agriculture: a bibliography of social science literature.*

See entry 2953.

3047 *Energy bibliography and index.* Houston TX: Gulf Publishing for Texas A & M University. College Station Library, 1978. 4 vols.

Bibliography and index to all energy-related materials in the Texas A & M University Library. Monographs only; no journal articles.

3048 Morrison, Denton E. et al. *Energy: a bibliography of social science and related literature.* New York: Garland, 1975. (Garland reference library of social science 9) 173 pp.

Updated by *Energy II: a bibliography of 1975-76 social science and related literature*, 1977. (Garland reference library of social science 42) 256 pp.

3049 Yanarella, Ernest J., and Yanarella, Ann-Marie. *Energy and the social sciences: a bibliographic guide to the literature.* Boulder CO: Westview Press, 1982. 347 pp.

Unannotated. Covers a broad spectrum of documents on energy policy, politics, conventional and alternative sources, national surveys, and more. Includes an annotated list of essential books for an energy library.

ONLINE DATABASES

3050 *DOE energy: energy database.* Oak Ridge TN: U.S. Dept. of Energy. Technical Information Center, 1974-. Updated monthly.

Corresponds to printed *Energy research abstracts*, *Energy abstracts for policy analysis*, *Atomindex*. Covers energy-related literature processed by the Department's Technical Information Center; includes alternative energy and the social impact of energy policy. Vendors: BRS, DIALOG, SDC, INKA. A printed thesaurus (*Energy information database subject thesaurus*) is available.

3051 *EBIB (Energy bibliography and index).* College Station TX: Texas A & M University. Library.; Gulf Publishing, 19__-.

Corresponds to printed *Energy bibliography and index*. Covers non-journal literature on energy held by Texas A & M University, from 1919 to present. Vendor: SDC.

3052 *P/E NEWS (Petroleum/energy business news index).* New York NY: American Petroleum Institute, 1975-. Updated weekly.

Available online only. Covers energy and environmental issues, including government actions, economics, supplies. Vendor: SDC.

3053 *POWER.* Washington DC: U.S. Dept. of Energy. Energy Library, 19__- Updated 2 per month.

Available online only. Accesses the book collection of the Department's main library; provides good coverage of energy, environmental, and water resources literature. Vendor: SDC.

INDEXES AND ABSTRACTS

3054 *ASSET: abstracts of selected solar energy technology.* Tokyo: United Nations University, 1979-. 10 to 12 per year.

Despite the title, includes abstracts of current literature on solar, biomass, wind, and other new and renewable energy sources.

3055 *Energy: a continuing bibliography with indexes.* Washington DC: U.S. National Aeronautics and Space Administration. Scientific and Technical Information Office, 1974-. Quarterly.

3056 *Energy research abstracts.* Oak Ridge TN: U.S. Dept. of Energy. Technical Information Center, 1976-. 2 per month.

Continues: *ERDA energy research abstracts.*

3057 *Fossil energy update: a current awareness journal.* Oak Ridge TN: U.S. Energy Research and Development Administration. Technical Information Center, 1976-. Monthly.

Covers coal, petroleum, natural gas, oilshale, biomass energy, and other subjects.

DIRECTORIES

3058 **Banly, J. A.,** ed. *World energy directory: a guide to organizations and research activities in non-atomic energy.* Detroit MI: Gale, 1981. 567 pp.

STATISTICS

3059 **Balachandran, Sarojini.** *Energy statistics: a guide to information sources.* Detroit MI: Gale, 1980. (Natural world information guide 1) 272 pp.

3060 *Energy interrelationships, a handbook of tables and conversion factors for combining and comparing international energy data.* Springfield VA: U.S. National Technical Information Service, 1977. (FEA B-77/166) 60 pp.

3061 *1979 yearbook of world energy statistics/ Annuaire des statistiques mondiales de l'energie.* New York: United Nations, 1981. 1,209 pp.

In English and French. Statistics as of November 1980. This initial *Yearbook* expands upon and updates the statistical series shown in previous issues of *World energy supplies*.

Solar Energy

3062 **Lockerby, Robert W.** *Solar ponds.* Monticello IL: Vance Bibliographies, 1982. (Architecture series A-765) 15 pp.

Unannotated.

3063 *Passive solar design: an extensive bibliography.*

See entry 3163.

3064 **Seeley, Dwight** et al. *Solar energy legal bibliography, final report.* Golden CO: U.S. Dept. of Energy, 1979. 160 pp.

Updated in *Solar law reporter*.

3065 *Solar energy index.* Tempe AZ: Arizona State University, 1978-. Irreg.

Indexes the extensive Solar Energy Collection at the Hayden Library, Arizona State University.

3066 *Solar energy update: a current awareness journal.* Oak Ridge TN: U.S. Energy Research and Development Administration. Office of Public Affairs. Technical Information Center, 1976-. Monthly.

3067 "Special retrospective bibliography: solar energy." *Arid lands abstracts* 6 (1974): items 201- 314.

Geothermal Energy

3068 *Geothermal energy update: a current awareness journal.* Oak Ridge TN: U.S. Energy Research and Development Administration. Technical Information Center, 1977-. Monthly.

Supplement to the bibliography *Geothermal resources*. Information included is in the energy data base of the U.S. Dept. of Energy's Technical Information Center.

3069 *Geothermal resources explanation and exploitation: a bibliography.* Oak Ridge TN: U.S. Energy Research and Development Administration. Technical Information Center, 1976. 631 pp.

Contains 5,476 annotated citations. Updated in *Geothermal energy update* and available online through ERDA's RECON database.

3070 *Legal and institutional impediments to geothermal energy resource development: a bibliography.* Washington DC: U.S. Dept. of Energy. Technical Information Center, 1978. 47 pp.

3071 **Meadows, Katherine F.** *Geothermal world directory.* Glendora CA: Katherine F. Meadows, 1972-. Annual.

Contains articles on current developments, names and addresses of firms and individuals worldwide, and much advertising.

3072 **Summers, W. Kelly,** comp. *Annotated and indexed bibliography of geothermal phenomena.* Socorro NM: New Mexico. State Bureau of Mines and Mineral Resources, 1971. Various pagings.

Provides 14,177 citations—many annotated—appearing through 1969. The major retrospective bibliography.

Wind Energy

3073 **Burke, Barbara L.,** and **Meroney, Robert N.** *Energy from the wind: annotated bibliography.* Fort Collins CO: Colorado State University. Solar Energy Applications Laboratory, 1975. 179 pp.

First supplement (1977), Second supplement (1979), and Third supplement (1982) each add additional citations. The entire file is computer-searchable.

3074 **Otawa, Toru.** *Wind energy planning: a bibliography.* Chicago IL: CPL Bibliographies, 1979. (CPL bibliography 16) 25 pp.

Unannotated.

3075 Randall, Paul. *Wind power: a bibliography.* London: Institution of Electrical Engineers. Library, 1980. 56 pp.

3076 *Wind energy information directory.* Golden CO: U.S. Dept. of Energy, 1979. 28 pp.

3077 *Wind energy utilization: a bibliography.* 7th ed. East Lansing MI: Michigan State University. Dept. of Agricultural Engineering, 1979. (Energy in agriculture collection).

Prepared at the University of New Mexico.

3078 *Wind energy utilization: a bibliography with abstracts (cumulative volume, 1944-74).* Albuquerque NM: U.S. National Aeronautics and Space Administration, 1975. 496 pp.

Prepared by the Technology Application Center, University of New Mexico.

Biomass Energy

3079 Bente, Paul F., Jr., ed. *International bioenergy directory.* Washington DC: Bio-Energy Council, 1981. 770 pp.

3080 Goldstein, Edward L., and Gross, Meier. *Land use and planning implications of wood energy development: a bibliography.* Chicago, IL: CPL Bibliographies, 1981. (CPL bibliography 42) 27 pp.

Unannotated.

3081 *Information sources on bioconversion of agricultural wastes.* New York: U.N. Industrial Development Organization, 1979. (UNIDO guides to information sources 33) 84 pp.

Unannotated. Lists research organizations by country as well as handbooks, series, proceedings and other potential sources of information.

3082 Lockerby, Robert W. *Gasohol: energy from agriculture.* Monticello IL: Vance Bibliographies, 1980. (Public administration series bibliography P-533) 17 pp.

Unannotated.

3083 McCarl, Henry N., and Handlin, Jayne. *Bibliography on energy from biomass.* Monticello IL: Vance Bibliographies, 1981. (Public administration series bibliography P-806) 16 pp.

Unannotated.

3084 *A selected bibliography on alcohol fuels (1901-November 1981).* Golden CO: U.S. Solar Energy Research Institute. Technical Information Office, 1982. 475 pp.

Includes over 2,000 annotated references to methyl and ethyl alcohol fuel produced from biomass. Excludes citations to butyl, propyl, methane, or methanol fuels. Prepared by the Solar Energy Information Data Bank of the Solar Energy Research Institute. Supplements *Alcohol fuel bibliography (1901-March 1980)* (1981), 458 pp.

Petroleum

3085 *Basic petroleum data book: petroleum industry statistics.* Washington DC: American Petroleum Institute, 1977-. Quarterly.

3086 *Oil shales and tar sands: a bibliography.* Washington DC: U.S. Department of Energy, 1977. 312 pp.

Includes 5,049 annotated references covering the world.

3087 *The whole world oil directory.* Deerfield IL: Whole World, 1977-. Annual.

A 2-volume set issued annually. Volume 1 is a personnel index, volume 2 a company index.

HUMAN GEOGRAPHY

ONLINE DATABASES

3088 *PAIS International.* New York NY: Public Affairs Information Service, 1972-. Updated quarterly.

Corresponds to printed *PAIS bulletin* and *PAIS foreign language index.* Covers the social sciences worldwide. Vendors: BRS, DIALOG.

3089 *Social SciSearch.* Philadelphia PA: Institute for Scientific Information, 1972-. Updated monthly.

Corresponds to printed *Social sciences citation index.* Covers social sciences generally, indexing some 1,400 journals. Vendors: BRS, DIALOG, DIMDI, ISI.

3090 *Sociological abstracts.* San Diego CA: Sociological Abstracts, 1963-. Updated quarterly.

Corresponds to printed *Sociological abstracts.* Covers social sciences generally. Vendor: DIALOG.

INDEXES AND ABSTRACTS

3091 *Arts and humanities citation index.* Philadelphia PA: Institute for Scientific Information, 1977-. Annual.

Coverage includes architecture, history.

3092 *British humanities index.* London: The Library Association, 1962-. Quarterly.

Indexes English-language journals from the United Kingdom and Commonwealth nations, covering the social sciences generally.

3093 *Geo Abstracts: D-Social and historical geography.* Norwich, U.K.: University of East Anglia, 1974-. 6 per year.

3094 *Humanities index.* New York: H. W. Wilson, 1974-. Quarterly.

Indexes English-language journals covering archeology, area studies, history, and the arts.

3095 *Index to social sciences and humanities proceedings.* Philadelphia PA: Institute for Scientific Information, 1979-. Quarterly.

3096 *Public affairs information service bulletin.* New York: Public Affairs Information Service, 1914-. 2 per month.

Indexes English-language books, pamphlets, journal articles, and government documents, covering economics and public affairs in the broadest sense. An excellent first-step reference source.

3097 *Social sciences citation index.* Philadelphia PA: Institute for Scientific Information, 1973-. 3 per year.

Indexes 1,500 journals and 200 monographs. Covers the full range of social science literature, including but not limited to anthropology, archeology, area studies, business and finance, demography, environmental sciences, ethnic studies, geography, information and library science, political science, sociology, urban studies. Provides enhanced citation indexing.

3098 *Social sciences index.* New York: H. W. Wilson, 1974-. Quarterly.

Indexes English-language journals covering anthropology, area studies, economics, environmental sciences, geography, political sciences, and other social sciences.

DICTIONARIES

3099 *Dictionary of human geography.* New York: Free Press, 1981. 411 pp.

DIRECTORIES

3100 *Government sponsored research on foreign affairs; a quarterly report of project information.* Washington DC: U.S. Dept. of State. Office of External Research, 1977-. Quarterly.

Lists research contracts and grants that involve social sciences or humanities as these bear substantively on foreign affairs.

DEMOGRAPHY

GENERAL BIBLIOGRAPHIES

3101 Driver, Edwin D. *World population policy: an annotated bibliography.* Lexington MA: Lexington Books, 1972. 1,280 pp.

Annotated.

3102 Goyer, Doreen S., comp. *The international population census bibliography, revision and update, 1945-1977.* New York: Academic Press, 1980. (Studies in population; Texas bibliography 2) 570 pp.

Includes sovereign nations and some territories which have either changed administrative jurisdiction and/or taken their own census. Covers period 1945-1977. Titles in all languages which do not use the Roman alphabet have been transliterated.

3103 Goyer, Doreen S., comp. *National population censuses, 1945-1976: some holding libraries.* Clarion PA: Association for Population/Family Planning Libraries and Information Centers, International, 1979. (APLIC special publication 1) 44 pp.

Contains responses to a 1978 questionnaire about foreign census holdings in 53 libraries in the United States.

3104 Lyle, Katherine Ch'iu, and Segal, Sheldon J. *International family-planning programs, 1966-1975.* New York: Population Council, 1977. 207 pp.

3105 Waltisperger, Dominique, and Canedo, D. A. *Mortality project: annotated bibliography on the sources of demographic data.* Paris: OECD. Development Centre, 1977. 3 vols.

Covers: Vol. 1-Africa, Near East, Vol. 2-Latin America, Caribbean area, Vol. 3-Asia.

ONLINE DATABASES

3106 *International demographic data directory; a computerized information retrieval system for demographic and family planning data, December 1972.* Washington DC: U.S. Bureau of the Census, 1972. 54 pp.

Contains a retrieval system to access 1,000 tables in the International Data Directory; 2,500 added annually. Covers the major areas of the world, component regions, countries, major subdivisions of countries (e.g., states or provinces), urban/rural/metropolitan settings. Emphasis on Africa, Asia, and Latin America. Includes United Nations data and national censuses.

3107 *Population bibliography.* Chapel Hill NC: University of North Carolina. Carolina Population Center, 1966-. Updated 6 times per year.

Available online only. Covers demography, migration, and other areas of population research worldwide. Vendor: DIALOG. a printed thesaurus (*Population/family planning thesaurus*) is available.

Human Geography

INDEXES AND ABSTRACTS

3108 *Population index.* Princeton NJ: Princeton University. School of Public Affairs, 1935-. Quarterly.

Includes articles and an annotated bibliography on demography (urban-rural affairs, past trends, projections, mortality, etc.).

ENCYCLOPEDIAS

3109 *International encyclopedia of population.* New York: The Free Press, 1982. 2 vols.

A collection of long, signed essays on various aspects of demography, fertility, national and regional census data, etc. Bibliographies appear after each article.

DIRECTORIES

3110 **Trzyna, Thaddeus C.** *Population: an international directory of organizations and information resources.* Claremont CA: Public Affairs Clearinghouse, 1976. (Who's doing what series 3) 132 pp.

Contains a listing of organizations concerned with population and family planning. Emphasis on organizations of an international character, and those in the United States and Canada.

STATISTICS

3111 *Demographic yearbook/Annuaire demographique.* New York: U.N. Statistical Office, 1948-. Annual.

In English and French.

3112 *Demographic yearbook: historical supplement/Annuaire demographique, supplement retrospectiv.* New York: United Nations. Statistical Office, 1979. 1,171 pp.

In English and French.

3113 *World population, 1979: recent demographic estimates for the countries and regions of the world.* Washington DC: U.S. Bureau of the Census, 1980. 2 vols.

Contains basic demographic information. Lists sources for each country.

ANTHROPOLOGY

GENERAL BIBLIOGRAPHIES

3114 **Douglass, Barbara.** *A planner's guide to anthropology.* Monticello IL: Council of Planning Librarians, 1978. (CPL exchange bibliography 1519) 36 pp.

Annotated.

3115 **Murdock, George Peter.** *Atlas of world cultures.* Pittsburgh PA: University of Pittsburgh Press, 1981. 151 pp.

Not an atlas of maps but rather a scheme of organizing human societies based on linguistic and cultural affinities. This volume extracts the most useful bibliographic references for 563 representative societies worldwide.

INDEXES AND ABSTRACTS

3116 *Abstracts in anthropology.* Westport CT: Greenwood Periodicals for Harvard University. Peabody Museum of Archeology and Ethnology, 1970-. Quarterly.

Covers archeology, ethnology, linguistics, physical anthropology, and cultural anthropology.

3117 *Anthropological index to current periodicals in the Museum of Mankind Library.* London: United Kingdom. Royal Anthropological Institute, 1963-. Quarterly.

3118 *Anthropological literature: an index to periodical articles and essays.* Pleasantville NY: Redgrave, 1979-. Quarterly.

Unannotated.

THESES AND DISSERTATIONS

3119 **McDonald, David R.** *Master's theses in anthropology: a bibliography of theses from United States colleges and universities.* New Haven CT: HRAF, 1977. 453 pp.

Unannotated. Covers theses up to 1975.

ENCYCLOPEDIAS

3120 **Hunter, David E., and Whitten, Phillip, eds.** *Encyclopedia of anthropology.* New York: Harper and Row, 1976. 411 pp.

Contains about 1,400 entries.

DIRECTORIES

3121 *Fifth international directory of anthropologists.* Chicago IL: The University of Chicago Press, 1975. (Current anthropology resource series) 496 pp.

Lists addresses, academic background, and special interests and projects.

WOMEN'S STUDIES

3122 **Danforth, Sandra C.** *Women and national development.* Monticello IL: Vance Bibliographies, 1982. (Public administration series bibliography P-916) 35 pp.

Unannotated. Contains approximately 500 references, most since 1960.

3123 **Rihani, May, and Joy, Jody.** *Development as if women mattered: an annotated bibliography with a Third World focus.* Washington DC: Overseas Development Council and New Transcentury Foundation, 1978. (Occasional paper 10) 137 pp.

Contains 63 items. Emphasis on period 1976-77. Includes unpublished reports. Updates Maria Buvinic' *Women and world development: an annotated bibliography* (1976), 162 pp.

3124 *Women and family in rural development/La femme et la famille dans le developpement rural/La mujer y la familia en el desarrollo rural: annotated bibliography, author and subject index.* Rome: United Nations. FAO Documentation Centre and Population Documentation Centre, 1977. 66 pp.

In English, French, and Spanish. Contains 276 items. Covers FAO publications and documents from 1966 to 1976, primarily Africa- related. Arranged by FAO accession number.

MEDICINE

RESEARCH GUIDES

3125 *Human and animal ecology: reviews of research.*

See entry 46.

3126 Morton, Leslie Thomas. *Use of medical literature.* 2nd ed. London; Boston MA: Butterworths, 1977. (Information sources for research and development) 462 pp.

Guide to general and specialist literature. Chapter X covers tropical medicine. Refers primarily to English-language works.

INDEXES AND ABSTRACTS

3127 *Index medicus.* Bethesda MD: U.S. National Library of Medicine, 1960-. Monthly.

The standard indexing journal.

STATISTICS

3128 *Statistiques epidemiologiques et demographiques annuelles/World health statistics annual.* Geneva, Switzerland: World Health Organization, 1965-. Annual.

Supersedes *Rapport epidemiologique annuel/Annual epidemiological and vital statistics* (1951-1964). Contains data on vital statistics, causes of death, causes of infectious diseases, and health personnel and establishments. In French and English.

FOOD, HUNGER, NUTRITION

3129 *Food and nutrition bibliography.* 9th ed. Phoenix AZ: The Oryx Press, 1980. 345 pp.

Includes print and non-print materials. Annotated. Compiled by the U.S. Dept. of Agriculture, National Agriculture Library.

3130 Freedman, Robert L., comp. *Human food uses: a cross-cultural comprehensive annotated bibliography.* Westport CT: Greenwood, 1981. 552 pp.

3131 *Nutrition abstracts and reviews: Ser. A: human and experimental.* Farnham Royal, U.K.: Commonwealth Agricultural Bureaux, 1977-. Monthly.

Worldwide multi-lingual coverage. Covers foods, physiology, nutrition and disease.

3132 Trzyna, Thaddeus C.; Smith, Joan Dickson; and Ruggles, Judith, eds. *World food crisis: an international directory of organizations and information resources.* Claremont CA: Public Affairs Clearinghouse, 1977. (Who's doing what series 40) 140 pp.

HEALTH CARE SERVICES

3133 Elling, Ray H. *Cross-national study of health systems, concepts, methods, and data sources: a guide to information sources.* Detroit MI: Gale Research, 1980. (Health affairs information guide series 2) 293 pp.

3134 Elling, Ray H. *Cross-national study of health systems, countries, world regions, and special problems: a guide to information sources.* Detroit MI: Gale Research, 1980. (Health affairs information guide series 3) 687 pp.

3135 Kohn, R., and Rodius, S. "International comparison of health services systems: an annotated bibliography." *International journal of health services* 3 (1973):295-309.

3136 Miller, L. D.; Walthall, C. L.; and Mathews, M. L. *Remote sensing and geoinformation systems as related to the regional planning of health services: a bibliography.* Chicago IL: CPL Bibliographies, 1981. (CPL bibliography 51) 76 pp.

Unannotated. References here were selected from the RESENA database.

3137 Rylko-Bauer, Barbara, and Bletzer, Keith V. *National health care systems: a bibliographic tool for comparative and cross-national research into health care.* Monticello IL: Vance Bibliographies, 1982. (Public administration series P-1009) 36 pp.

Contains narrative and 277 unannotated citations.

HEAT STRESS

3138 Chapman, Carleton B., and Reinmiller, Elinor C. *The physiology of physical stress: a selective bibliography, 1500-1964.* Cambridge MA: Harvard University Press, 1975. 369 pp.

Multilingual references.

3139 *Human reactions to high temperatures annotated bibliography (1927-1962).* Fort Clayton, Panama Canal Zone: U.S. Army Tropic Test Center, 1967. (NTIS: AD 651- 940) 115 pp.

Contains 352 citations from three sources: *Psychological abstracts*, the Defense Documentation Center, and Tufts University Human Engineering Service.

3140 "Physiological response to heat stress: selected references." *Arid lands abstracts* 1 (1972): items 100-133.

Contains 34 annotated references arranged by author.

NATIVE MEDICINES

3141 **Spinelle, William B.** *The primitive therapeutic use of natural products: a bibliography*. Pittsburgh PA: Duquesne University Library, 1971. 106 pp.

Includes literature world-wide dealing with pre-scientific use of natural products to treat disease.

TROPICAL MEDICINE

3142 **Ajello, Libero, ed.** *Coccidioidomycosis papers*. Tucson AZ: University of Arizona Press, 1967. 434 pp.

From the 2nd Symposium on Coccidioidomycosis held in Phoenix AZ, December 8-10, 1965. Includes bibliography of 543 items, which updates Fiese (entry 3143).

3143 **Fiese, Marshall J.** *Coccidioidomycosis*. Springfield IL: Thomas, 1958. 253 pp.

Includes a 968-item bibliography covering 1892-1957.

3144 **Hoogstraal, Harry.** *Bibliography of ticks and tickborne diseases from Homer (about 800 B.C.) to 31 December 1969*. Cairo: U.S. Navy Medical Research Unit 3, 1970-1972. (NAMRU - 3) 4 vols.

Unannotated. *Supplement* (1974-) covers period to 31 December 1973.

3145 **Neumann, Hans H.** *Foreign travel and immunization guide*. 10th ed. Oradell NJ: Medical Economics Co. Book Division, 1981. 72 pp.

Practical guide to required and recommended health-care. Based on data from the World Health Organization, U.S. Center for Disease Control, and sources within various countries.

3146 *Review of medical and veterinary mycology*. Kew, U.K.: Commonwealth Mycological Institute, 1943-. Quarterly.

Includes recent world literature on diseases of humans and animals (including fish and some large invertebrates, but excluding insects) which are caused by fungi and actinomysetes. Examines more than 360 periodicals. Includes references in non- Western languages.

3147 *Tropical diseases bulletin*. London: United Kingdom. Bureau of Hygiene and Tropical Diseases, 1912-. Monthly.

3148 *Trypanosomiasis: bibliography*. Hyattsville MD: U.S. Dept. of Agriculture. Animal and Plant Inspection Service, 1978, 172 pp.

Part of the Emergency Programs Foreign Animal Disease Data Bank. All articles are in English, including some translated pieces.

3149 **Warren, Kenneth S., and Newill, Vann A.** *Schistosomiasis: a bibliography of the world's literature from 1852 to 1962*. Cleveland OH: Press of Western Reserve University, 1967. 2 vols.

VENOM DISEASES

3150 **Harmon, Ralph Winter.** *Bibliography of animal venoms*.

See entry 2940.

3151 **Minton, Sherman A., Jr.** *Venom diseases*. Springfield IL: Thomas, 1974. (American lecture series. Publication 937) 235 pp.

Bibliography: pp. 189-222.

URBAN GEOGRAPHY

GENERAL BIBLIOGRAPHIES

3152 **Lambert, Claire M., ed.** *Village studies: data analysis and bibliography*. Epping, U.K.: Bowker for the Institute of Development Studies at the University of Sussex, 1976-78. 2 vols.

Volume one: *India, 1950-1975*. Volume two: *Africa, the Middle East and North Africa, Asia (excluding India), Pacific Islands, Latin America, West Indies and the Caribbean, 1950-1975*. 320 pp.

INDEXES AND ABSTRACTS

3153 *Geo abstracts: F-Regional and community planning*. Norwich, U.K.: University of East Anglia, 1972-. 2 per month.

Includes sections on urban and rural planning, transportation planning and environmental planning.

3154 *Urbanism, past and present*. Milwaukee WI: University of Wisconsin-Milwaukee. Dept. of History, 1975/76-. 2 per year.

ARCHITECTURE

GENERAL BIBLIOGRAPHIES

3155 **Rivera de Figueroa, Carmen A.** *Architecture for the tropics: a bibliographical synthesis (from the beginnings to 1972)*. Rio Piedras, Puerto Rico: Universidad de Puerto Rico, 1980. 203 pp.

Unannotated; text in English and Spanish.

INDEXES AND ABSTRACTS

3156 *Architectural periodicals index*. London: RIBA, 1972-. Quarterly.

Indexes some 450 worldwide architectural journals.

3157 *Art index: a cumulative author and subject index to a selected list of fine arts periodicals and museum bulletins.* New York: H. W. Wilson, 1929-. Quarterly.

Coverage includes architecture, city planning, landscape architecture.

ADOBE AND EARTHEN MATERIALS

3158 **Hopson, Rex C.** *Adobe: a comprehensive bibliography.* Santa Fe NM: The Lightning Tree, 1979. 127 pp.

An unannotated listing of 1,321 references covering earthen construction worldwide. Updated in *Adobe today.*

3159 **Vance, Mary.** *Adobe building: a short bibliography.* Monticello IL: Vance Bibliographies, 1979. (Architecture series bibliography A76) 3 pp.

Unannotated.

LANDSCAPING

3160 **White, Anthony G.** *Consumptive water use by landscape plants: a brief source list for landscape architects.* Monticello IL: Vance Bibliographies, 1980. (Architecture series bibliography A319) 4 pp.

Unannotated.

SOLAR BUILDING DESIGN

3161 **Eggers-Lura, A.** *Solar energy for domestic heating and cooling: a bibliography with abstracts and a survey of literature and information sources.* Oxford: Pergamon Press, 1979. (Pergamon European heliostudies 2) 229 pp.

Contains a directory of organizations and an annotated bibliography covering solar collectors, thermal storage, space heating and cooling, and other domestic applications.

3162 **Lee, Kaiman.** *Bibliography of energy conservation in architecture: keyword searched.* Boston MA: Environmental Design and Research Center, 1977. 428 pp.

Includes primarily works from 1973 to 1977.

3163 *Passive solar design: an extensive bibliography.* Washington DC: U.S. Dept. of Energy, 1978. (HCP/M-4113-3) 218 pp.

Unannotated. Relates to heating and cooling techniques. Includes material in major computerized data banks as of January 1978, as well as *Proceedings* of the Second National Passive Conference (1968), 3 volumes. Prepared by AIA Research Corporation. Available from NTIS.

3164 *Solar thermal heating and cooling.* Albuquerque NM: University of New Mexico, 1977-. Quarterly.

ECONOMIC DEVELOPMENT

GENERAL BIBLIOGRAPHIES

3165 **Brooke, Michael Z.; Black, Mary; and Neville, Paul.** *International business bibliography.* New York: Garland, 1977. 480 pp.

Includes books published within past 20 years, articles published within the past 5 years.

3166 **Brownstone, David M.** *Where to find business information.* 2nd ed. New York: Wiley, 1982. 632 pp.

3167 **Daniells, Lorna M., comp.** *Business reference sources: an annotated guide for Harvard Business School students.* Rev. ed. Cambridge MA: Harvard University. Graduate School of Business Administration. Baker Library, 1979. (Reference list 30) 133 pp.

Annotated.

3168 **Dicks, G. R.** *Sources of world financial and banking information.* Westmead, U.K.: Gower, 1981. 720 pp.

Annotated.

3169 *Economic analysis: index.* Rome: FAO, 1969. (Special indexes DC/Sp,12) approximately 250 pp.

In English, French, and Spanish. Includes FAO publications produced from 1945 to 1966 by the Economic Analysis Division. Over 900 entries.

3170 **Field, Barry C., and Willis, Cleve E.** *Environmental economics: a guide to information sources.* Detroit MI: Gale Research, 1979. (Man and environment information guide series 8) 243 pp.

A guide to significant information sources on selected topics of reference material, legal cases, and ideas. Selected collection of periodical literature. Annotated.

3171 *International business publications from ITA: a geographical index.* Washington DC: U.S. Dept. of Commerce. Industry and Trade Administration, 1978. 13 pp.

An index of four series of international business publications by the ITA (foreign economic trends, overseas business reports, global market surveys, and country market sectoral surveys). Indexed by country.

3172 **Wasserman, Paul, ed.** *Encyclopedia of geographic information sources: a detailed listing of publications and agencies of interest to managerial personnel....* 3rd ed. Detroit MI: Gale Research, 1978. 167 pp.

Unannotated listings of reference sources valuable chiefly as a guide to business and commerce worldwide. Nation-by-nation coverage.

Human Geography / 3173-3191

ONLINE DATABASES

3173 *Economic abstracts international.* Oxford, U.K.: Learned Information, 1974-. Updated monthly.

Corresponds to printed *Economic titles/abstracts* and *Key to economic science and managerial science*. Covers information on markets, international economics, country-specific data. Vendors: DIALOG, BELINOIS.

INDEXES AND ABSTRACTS

3174 *Business periodicals index.* New York: H.W. Wilson, 1958-. 11 per year.

Indexes current English-language journals in the fields of business, economics, commerce, and commercial industries and technologies; the best general index for these fields.

3175 *Geo abstracts: C-Economic geography.* Norwich, U.K.: University of East Anglia, 1966-. 2 per month.

Annotated.

3176 *Geo abstracts: F-Regional and community planning.*

See entry 3153.

3177 *Human resources abstracts.* Beverly Hills CA: Sage, 1975-. Quarterly.

Formerly *Poverty and human resources abstracts*. Contains abstracts on human, social and manpower problems and solutions ranging from slum rehabilitation to job training.

3178 *Journal of economic literature.* Cambridge MA: American Economic Association, 1963-. Quarterly.

Contains articles and abstracts on world economics.

3179 *Rural development abstracts.* Oxford, U.K.: Commonwealth Bureau of Agricultural Economics, 1978-. Quarterly.

Covers physical resources, human resources, economic utilization of resources, public services and projects.

DIRECTORIES

3180 Angel, J. L., comp. *Directory of American firms operating in foreign countries.* 9th ed. New York: Uniworld, 1979. 3 vols.

Volume 1 covers United States firms operating overseas with domestic addresses. Volumes 2 and 3 list every U.S. firm under the country in which it has subsidiaries. Includes only those firms which have substantial direct capital investment. Non-commercial enterprises omitted.

3181 *Bottin international business register.* Paris: Annuaire du Commerce. Didot Botton, 1947-. Annual.

In English, German, Spanish, and French. Consists largely of a directory of addresses of manufacturers, exporters, and others engaged in foreign trade. Volume 1 covers the world outside Europe; volume 2 covers Europe only. Arranged by country; includes an index to products, services, and advertisements as well as extensive geographical index to countries and towns.

3182 Cyriax, George, ed. *World index of economic forecasts.* Farnborough, U.K.: Gower, 1978. 379 pp.

An explanatory directory of the extent and coverage of various economic forecasting organizations for the world. Includes information on national development plans and planning agencies.

STATISTICS

3183 *Monthly bulletin of statistics/Bulletin mensuel de statistique.* New York: U.N. Statistical Office, 1947-. Monthly.

In English and French. Statistics obtained by United Nations from official sources in the various countries.

3184 *World casts.* Cleveland OH: Predicasts, 1960-. 8 per year.

Looseleaf index to economic statistics accessible by nation and commodity.

3185 *World economic survey.* New York: U.N. Statistical Office, 1977-. Annual.

Intended to provide the basis for a synthesized appraisal of current trends in world economy, particularly as they affect developing countries.

3186 *Yearbook of industrial statistics.* New York: United Nations. Statistical Office, 1974-. Annual.

Supersedes *The growth of world industry/Croissance de l'industrie mondiale* (1961-73). Volume 1 covers basic data for each country in the form of separate chapters. Primarily based on replies to the United Nations General Industrial Statistics Questionnaire; and global and regional trends in industrial activity. Volume 2 provides detailed information on world production of industrial commodities.

3187 *Yearbook of international trade statistics.* New York: U.N. Statistical Office, 1950-. Annual.

Provides basic information for individual countries' external trade performance and analyzes the flow of trade between countries.

3188 *Yearbook of labour statistics.* Geneva, Switzerland: International Labour Office, 1935/6-. Annual.

In English, French, and Spanish. Summary of labor statistics for approximately 180 countries or territories. Data drawn primarily from information sent by national statistical services.

3189 *Yearbook of national accounts statistics.* New York: U.N. Statistical Office, 1957-. Annual.

ATLASES

3190 *Oxford economic atlas of the world.* 4th ed. London: Oxford University Press, 1972. 239 pp.

Covers environment, agriculture, energy, natural resources, industry, demography and political maps, with statistics by political region.

TOURISM AND RECREATION

3191 *Leisure, recreation and tourism abstracts.* Farnham Royal, U.K.: Commonwealth Agricultural Bureaux, 1976-. Annual.

HISTORICAL GEOGRAPHY

RESEARCH GUIDES

3192 **Poulton, Helen J.** *The historian's handbook: a descriptive guide to reference works.* Norman OK: University of Oklahoma Press, 1972. 304 pp.

GENERAL BIBLIOGRAPHIES

3193 *The combined retrospective index set to journals in history 1838-1974.* Washington DC: Carrollton Press, 1977-78. 11 vols.

The standard retrospective bibliography to the journal literature of history.

3194 **Cox, Edward Godfrey.** *A reference guide to the literature of travel, including voyages, geographical descriptions, adventures, shipwrecks and expeditions.* Seattle WA: University of Washington, 1935-49. (Publications in language and literature 9-10, 12) 3 vols.

Annotated. Covers the Old World in volume 1, the New World in volume 2, and Great Britain in volume 3. Covers the earliest times to 1800. The major retrospective bibliography for monographs on exploration.

INDEXES AND ABSTRACTS

3195 *Geo abstracts: D-Social and historical geography.*

See entry 3093.

3196 *Historical abstracts: bibliography of the world's historical literature.* Santa Barbara CA: Clio Press, 1955-. Annual.

ENCYCLOPEDIAS

3197 **Henze, Dietmar.** *Enzyklopaedie der Entdecker und Erforscher der Erde [encyclopedia of the discoverers and explorers of the earth].* Graz, Austria: Akademische Druck und Verlagsanstalt, 1975-.

In German. Issued in fascicles.

ATLASES

3198 **Barraclough, Geoffrey, ed.** *The Times atlas of world history.* London: Times Books, 1978. 360 pp.

Unlike most historical atlases this colorful and readable volume aspires to cover the world, giving equal weight to non-European peoples.

3199 *Rand McNally historical atlas of the world.* Chicago IL: Rand McNally, 1981. 192 pp.

APPENDIX A:

THE AREAL EXTENT OF THE DRYLANDS OF THE WORLD

The charts appearing throughout Part One of this Guide represent a second attempt to quantify the areal extent of the world's drylands. The first compilation appeared in 1981.[1]

The present effort is based on two complementary maps designed for the 1977 Nairobi Conference on World Desertification which summarize the extent of the world's arid regions and their degree of desertification. The first map—*World Distribution of Arid Regions*[2]—illustrates several sets of data on aridity, temperature, and periods of drought. The second—*Desertification Map of the World*[3]—presents estimates of the degree of desertification hazard as an overlay upon the aridity pattern of the first map. An explanation of these categories of data is given below.

Although the entire land surface of the earth (excluding Antarctica) is fitted onto a single sheet for each map at the relatively small scale of 1:25,000,000, the resolution was considered sufficiently precise to allow a nation-by-nation analysis of the areas of aridity and desertification. For the 1981 attempt a K & E Compensating Polar Planimeter was used; this was later found to be less accurate than the more tedious method of counting squares on a grid overlaying each map. Because the map projections are not of the equal area type, for the present effort each nation (or national subunit in the case of larger nations) was counted separately and then compared to published figures for total area.[4] For example, if a territory had 10 grid units total, two of which were arid, the percentage of arid dryland would be 20 percent; if the territory's total area was 100,000 square kilometers then 20,000 square kilometers was determined to be arid.

Using the two Unesco maps made most of the data compilation a largely mechanical exercise; however, there were a sufficient number of inconsistencies and omissions to require calculating additional aridity values from raw climatic data. The charts for Hawaii, Kiribati, Thailand, and several other small island and mainland areas were based on these newly calculated values.

Degrees of Aridity. The first Unesco map was compiled using the ratio P/ETP to define aridity (where P equals the mean value of annual precipitation and ETP equals the mean annual evapotranspiration). Data from 1,600 stations around the world were fed through the Penman formula to determine ETP while the pattern for intervening areas was developed from the FAO/Unesco *Soil Map of the World* and from sources depicting vegetation cover; all this was originally drafted onto a set of 1:5,000,000 base maps and then photo-reduced to the present scale of 1:25,000,000. The ratio P/ETP was generalized into four degrees of aridity, each defined by a range of values and a set of environmental characteristics, as follows:

Hyperarid. With an aridity value of less than 0.03, this indicates an extreme desert where rainfall can be nonexistent for a year or more and natural vegetation cover is almost totally lacking. Water is available only below the surface or where exotic sources (such as the Nile River through the Sahara) pass into the area. Sensitivity to desertification is virtually nil because the hyperarid land has no biomass to be reduced.

Arid. Aridity ranges from 0.03 to 0.20, supporting xerophytic vegetation with some seasonal grasses. Without special efforts to trap floodwaters rainfed agriculture is usually impossible. Lands can be grazed but unit yields are low. Susceptibility to desertification can be high because the biomass is fragile and carrying capacities are easily exceeded.

Semiarid. Aridity ranges from 0.20 to 0.50, supporting scrub forest and perennial grassland steppe. Grazing is often the most productive land use, though rain-fed agriculture is possible. Susceptibility to desertification is also high.

Subhumid. Aridity ranges from 0.50 to 0.75, supporting dry forest, grassland steppe, mediterranean chaparral, and tropical savanna. Grazing and dry farming are both possible, though desertification risk remains. The point where subhumid land becomes humid is difficult to define because seasonal and annual rainfall variability tends to blur the mathematical boundary; this is especially true of the Asian monsoon climates where a very dry season alternates with a very wet, and annual rainfall totals tend to obliterate dry season aridity.

A more extended discussion of these categories can be found in the Explanatory Note accompanying the map.

Degrees of Desertification Hazard. For the Unesco desertification map a looser set of criteria was used to depict the risk of land to the processes of desertification. Such factors as climate, aridity, surface cover, and the impact of man's land use patterns were all considered in making the assessment. Three degrees of desertification hazard were defined, as follows (quoting from the map):

Appendix A

Very high. "The region will be subject to very rapid desertification if existing conditions do not change."

High. The desertification hazard "lies between these two conditions."

Moderate. "The region will change only slowly from its present stage to a more degraded stage if existing conditions do not change."

Again, a more detailed discussion of these categories can be found in the Explanatory Note accompanying the map.

It must be mentioned that any attempt at bioclimatic mapping of continuously variable phenomena is open to broad differences in definition and interpretation; the precise delineation of areas is often impossible. At least one author[5] has criticized the Unesco maps for alleged inaccuracies in their depiction of Australia; as the creators of the Unesco maps admit, their effort is at best a provisional one at a small scale. In calculating the areas of aridity and desertification we relied heavily on the Unesco data, except in the specific instances mentioned above. The authors hope to publish a fuller tabulation including subunits of the larger nations in a later publication.

References:

[1] John A. Rogers, "Fools Rush In, Part 3: Selected Dryland Areas of the World" *Arid Lands Newsletter* 14 (July 1981), pp. 24-5.

[2] Laboratoire de Cartographie Thematique, *World Distribution of Arid Regions* (Paris: Unesco, 1977), map and accompanying text.

[3] Food and Agriculture Organization of the United Nations and Unesco, *Desertification Map of the World/Carte mondiale de la Desertification* (Nairobi: UNEP, 1977), map and accompanying text.

[4] Derived from John Paxton, ed. *The Statesman's Yearbook* (New York: St. Martin's Press, 1981).

[5] D. J. Carder, "Desertification in Australia: a muddled concept" *Search* 12:7 (1981), pp. 217-21.

APPENDIX B:

SOURCES USED IN THE PREPARATION OF THIS GUIDE

Best, Alan C. G., and de Blij, Harm J. *African survey.* New York: John Wiley & Sons, 1977. 626 pp.

Body, Alexander, comp. *Annotated bibliography of bibliographies on selected government publications and supplementary guides to the Superintendent of Documents classification system.* Kalamazoo MI: Western Michigan University, 1967. 181 pp.

Carter, John, ed. *Pacific Islands year book.* 14th ed. Sydney; New York: Pacific Publications, 1981. 560 pp.

Cressey, George B. *Asia's lands and peoples: a geography of one-third of the earth and two-thirds of its people.* 3rd ed. New York: McGraw-Hill, 1963. 663 pp.

Cressey, George B. *Crossroads: land and life in Southwest Asia.* Chicago IL: J. B. Lippincott, 1960. 593 pp.

Cumberland, Kenneth B. *Southwest Pacific: a geography of Australia, New Zealand, and their Pacific Island neighbors.* New York: Praeger, 1968. 423 pp.

Dewdney, John C. *A geography of the Soviet Union.* 3rd ed. Oxford; New York: Pergamon, 1979. 175 pp.

Directory of online databases. Santa Monica CA: Cuadra Associates, 1979-. Quarterly.

Evans, F. C. *The West Indies.* London: Cambridge University Press, 1973. 128 pp.

Government reference books. 6th biennial volume. Littleton CO: Libraries Unlimited, 1980. 517 pp.

Hoffman, George W., ed. *A geography of Europe.* New York: Ronald, 1969. 671 pp.

James, Preston E. *Latin America.* 4th ed. New York: Odyssey, 1969. 947 pp.

Jeans, D. N., ed. *Australia: a geography.* London: Routledge and Kegan Paul, 1978. 571 pp.

Kanely, Edna A. *Cumulative subject guide to U.S. government bibliographies, 1924-1973.* Arlington VA: Carrollton, 1976- 1977. 7 vols.

MacPherson, John. *Caribbean lands: a geography of the West Indies.* 2nd ed. London: Longmans, Green, 1967. 181 pp.

Musiker, Reuben. *South Africa.* Oxford U.K.; Santa Barbara CA: Clio Press, 1979. 194 pp.

Pacific Islands. London: United Kingdom. Naval Intelligence Division, 1943-45. 4 vols.

Paxton, John, ed. *The statesman's year-book.* 118th ed. New York: St. Martin's, 1981. 1,696 pp.

Robinson, Harry. *Monsoon Asia: a geographical survey.* New York: Praeger, 1967. 561 pp.

Sale, Colin. *Australia: the land and its development.* Canberra, A.C.T.: Australian Government Publishing Service, 1975. 147 pp.

Scheven, Yvette. *Bibliographies for African Studies, 1970- 1975.* Waltham MA: Crossroads, 1977. (Archival and Bibliographic Series 103) 159 pp.

Scheven, Yvette. *Bibliographies for African studies, 1976- 1979.* Waltham MA: Crossroads, 1980. 142 pp.

Scull, Roberta A. *A bibliography of United States government bibliographies, 1968-1973.* Ann Arbor MI: Pierian, 1975. 353 pp.

Appendix B

Ward, Dederick C.; Wheeler, Marjorie W.; and Bier, Robert A., Jr. *Geologic reference sources: a subject and regional bibliography of publications and maps in the geological sciences.* Metuchen NJ: Scarecrow, 1981. 560 pp.

Warketin, John, ed. *Canada: a geographical interpretation.* Toronto: Methuen, 1968. 608 pp.

White, Charles Langdon et al. *Regional geography of Anglo-America.* 5th ed. Englewood Cliffs NJ: Prentice-Hall, 1979. 585 pp.

AUTHOR INDEX

Abdul-Rasoul, M. S., 1265
Abdullaev, I. K., 1622
Abell, L. F., 2900-2901, 3004
Abul-Hab, J., 1265
Acciaivoli, L. de Menezes Correa, 189
Achema, John A., 652
Adamec, Ludwig W., 1211, 1234
Adams, John, 1955
Adika, G. H., 1058
Agboola, S., 663
Agnew, Swanzie, 880
Aguilar y Santillan, Rafael, 2012-2015
Aguolu, Christian Chukwunedu, 537, 665
Ahmed, Abdel Ghaffar M., 706
Ahmed, Zaki, 1498
Aitro, Vincent P., 2293
Aiyepeku, Wilson O., 646
Ajaegbu, Hyacinth I., 357
Ajello, Libero, 3142
Akhtar, Shahid, 88, 1578
al-Barbar, Aghil M., *see* Barbar, Aghil M.
Al-Qazzaz, Ayad, 414, 1180
Alam, S. Manzoor, 1430
Alexandrov, Eugene A., 1641
Aliev, S. D., 1623
Alisky, Marvin, 2053, 2695
Allen, Peter Sutton, 134, 166
Allouse, Bashir E., 1168-1169, 1262
Allworth, Edward, 1662
Alonso, Patrica Ann Greechie, 1730, 1732
Alpatiev, S. M., 1658
Alsinawi, S. A., 1255
Altaga de Vizcarra, Irma, 2579-2580
Alvear, Alfredo, 1849
Ambrose, David P., 861-862, 864
Amedekey, E. Y., 538
Amiet, Carolyn F., 873
Aminullah, 1428
Amiran, David H. K., 68, 1279
Ammon, Alf, 2633
Anderson, A. Grant, 1828
Anderson, Anders H., 2254
Anderson, Ian Gibson, 111
Anderson, Margaret, 1141
Anderson, Theresa J., 1406, 1539, 1853, 2581
Andriot, Donna, 2088
Andriot, John L., 2188
Angel, J. L., 2685, 3180
Ansari, Mary B., 2354, 2358
Anthony, John Duke, 1142, 1338, 1385
Anthony, John W., 2229
Anwar, Muhammad, 1490
Aparicio, F. de, 2551
Arbingast, Stanley Alan, 1976, 1985, 2006, 2422
Archer, R. W., 1782
Ardizonne, G. D., 323
Armstrong, Frederick H., 1966
Armstrong, R. Warwick, 1813
Arnett, Ross H., 1894, 2488
Arnould, Eric, 922
Arone, G., 175

Artibise, Alan F. J., 1966
Asamani, J. O., 278
Ash, Lee, 2722
Ashdown, Peter, 2495
Asiedu, Edward Seth, 522, 539, 666
Askow, Catherine, 2080
Ataeva, E. Sh., 1657
Atiyeh, George Nicholas, 1143
Atkins, A. G., 2058
Aubriot, Bernard, 592
Auerbach, Devoira, 98
Avnimelech, Moshe A., 127, 491, 1157
Axelton, Elvera A., 2141-2143
Ayensu, Edward S., 2144
Baa, Enid M., 2473
Baark, Erik, 1562
Babcock, H. M., 2238
Badillo, Victor M., 2705
Bagur, Jacques D., 2420
Baines, John, 511
Bainey, S., 1951
Baiou, M. A., 563-564
Baker, Melvin, 1966
Baker, N. M., 1952, 2119
Baker, R., 51
Baker, Simon, 2079
Bako, Elemer, 221
Balachandran, Sarojini, 3059
Baldwin, John L., 2102
Balima, Mildred Grimes, 784, 860, 997
Balkan, Behire, 1374
Balogh, Janos, 224
Bambrick, S., 1734
Banderas, Pedro Antonio, 2639
Banerjee, S., 572
Banks, R. C., 2365
Banly, J. A., 3058
Banta, B. H., 2366
Bantje, Han, 589, 602, 742
Baranov, A. N., 1628
Barbar, Aghil M., 413, 417, 419-420, 424, 428-429, 510, 578-581, 584-585, 735, 738, 1181, 1184-1186, 1190-1191, 1198-1199, 1364, 1325
Barbour, K. Michael, 653
Bardsley, Elaine, 1681
Baretje, Rene, 371, 1092
Barlow, Alex, 1798
Barnard, Joseph D., 2052
Barnett, A. Doak, 1581
Barr, Charles W., 765
Barraclough, Geoffrey, 3198
Barrello, Angel V., 2559
Barrett, Ellen C., 1999
Barth, Hans Karl, 486
Barth, W., 2576
Bartov, Yosef, 1283
Bartsch, William H., 1246
Basinski, J. J., 1771
Bastani-Parizi, Mohammad Ebrahim, 1250
Batanouny, K. H., 1346
Bates, Margaret L., 1007
Battistini, Rene, 1098
Bauer, A. J. H., 99
Baumann, Duane E., 2859

Baumer, Michel, 405
Bayoumi, A. A., 705
Beale, Helen Purdy, 2988
Beavington, C. F., 123, 2484
Bebee, Charles N., 2976, 2989
Beck, Martinus Adrianus, 1266
Beck, Warren A., 2307
Beckwith, Wendell, 2678
Bederman, Sanford Harold, 305
Beeley, Brian W., 1376
Beetle, Alan Ackerman, 1889
Beg, Abdur Rahman, 1494
Bela, James, 2408, 2453
Bellot-Courdec, Beatrice, 394
Bender, Gordon Lawrence, 1879-1880
Bender, Thomas A., Jr., 460, 656
Benneh, George, 530
Bennett, J. D., 792
Bennett, Norman Robert, 716
Bennett, William Alfred Glen, 2454
Benson, Lyman David, 2145, 2245, 2286
Bente, Paul F., Jr., 3079
Berberian, Manuel, 1236
Berger, Thomas J., 2155
Bergquist, Wenonah E., 2787
Beriel, Marie-Magdeleine, 455, 472, 635
Bernal, Ignacio, 1993, 2041
Bernal Villa, Segundo, 2650
Bernus, Edmond, 405, 638
Berquist, Richard J., 2295
Berry, George W., 2111
Berry, Leonard, 1016
Besairie, Henri, 1100
Beshir, M. E., 693
Bespalov, N. D., 1590
Bettis, M. G., 2322
Beudot, Francoise, 395
Bharier, Julian, 1246
Bialor, Perry A., 134, 166
Bichet, E., 398
Billington, Ray Allen, 2208
Bindagji, Hussein Hamza, 1347
Birkenmayer, Sigmund S., 1638
Bishop, Olga B., 1907
Bissainthe, Max, 2512
Bisset, Ronald, 2081
Bizzarro, Salvatore, 2636
Black, Jan Knippers, 2539
Black, Mary, 3165
Blair, Patricia W., 1581
Blake, Gerald Henry, 509
Blanchard, J. Richard, 2915, 2949
Blanchard, Wendell, 1586
Blankhart, D. M., 853
Blasquez Lopez, Luis, 2020-2021
Blaudin de The, Bernard Marie Samuel, 378
Blauvelt, Evan, 368, 1193
Bletzer, Keith V., 3137
Blutstein, Howard I., 1975, 2637, 2696
Boe, Kenneth N., 2153
Boeseken, A. J., 992
Bogomolov, G. V., 1608
Bogusch, E. R., 1842
Bonine, Michael E., 1217, 1245
Bonnet, Pierre, 2945

Author Index

Booke, Bradley L., 2130
Borchardt, Dietrich Hans, 1689, 1704, 2714
Borchert, John R., 2334
Bord, Ernest W., 2343
Borisov, Anatolii Aleksandrovich, 1644
Bork, Albert William, 2664
Borton, R. L., 2380
Borza, Alexander, 238-239
Bose, Ashish, 1474
Bose, P. R., 1465
Botha, Laurette Isabella, 905
Botz, Maxwell K., 2343
Bouchard, D., 361
Boughton, Valerie H., 1774
Bourgeois, Joanne, 2803
Bovee, Gladys G., 2468
Bovy, Philippe H., 90
Bowden, Charles, 27, 56
Bowers, J. E., 2246
Boyer, Dennis L., 1438
Bradley, Raymond S., 2103
Branisa, Leonardo, 2577
Branstedt, Wayne G., 2297
Brasseur, Paule, 587-598, 634, 674, 740
Brewer, James Gordon, 2729
Bricault, Giselle C., 426, 1194
Bridge, Peter J., 1736
Bridgman, Jon, 450, 711, 901, 1008
Brien, J. P., 1770
Brierly, William B., 2814
Briggs, Donald C., 2053
Brighton, J. Noyce, 2775
Bristow, Clement Roger, 2659
Broadbent, Kieran P., 1574
Broekhuizen, Simon, 114, 118, 397, 1649, 1654
Bromley, Ray, 3044
Brook, Ronald H., 2910
Brooke, Michael Z., 3165
Brooks, M., 542
Brooks, Wahner F., 2216
Brooks, Walter R., 62
Brouwer, C. J., 3004
Brown, Barbara E., 1968
Brown, Clifton F., 803
Brown, Edward E., 875
Brownstone, David M., 3166
Bruchet, Susan, 1904
Brunn, Stanley D., 91
Bryan, M. Leonard, 2752
Bryant, Mark, 2323, 2409, 2455
Brygoo, Edouard R., 1104-1105
Buick, Barbara, 92
Bulfin, Robert L., 2776
Bullock, A. A., 760
Bullwinkle, Davis A., 331
Bunge, Frederica M., 1528, 1546
Burford, James B., 2112
Burge, Frederica M., 126
Burgess, Robert L., 1230, 2063, 2146
Burke, Barbara L., 3073
Burke, Hubert D., 2135
Busch, Lawrence, 684
Bush, Ernest Alfred Radford, 2952
Buss, W. R., 2441
Butler, A. C., 514
Butrum, Ritva Rauanheimo, 355
Butson, Keith D., 2815
Buttel, Frederick H., 2953, 3046

Butters, B., 1321, 1359, 2703
Byrne, R., 2497
Bystriakov, O. V., 1658
Caballero Deloya, Miguel, 2035
Cabot, Jean, 474
Cadet, Melissa Lawson, 3040
Calvo, Susanna, 2261
Cameron, R. J., 1710
Campbell, R. Wayne, 1956, 2462
Campos, E. C., 2612
Canedo, D. A., 3105
Cantillo, Jaime, 2651
Cantu, V., 174
Carberry, Michael E., 2876
Carlson, Helen S., 2355
Carr, Marilyn, 2777
Carraway, D. M., 776, 1304
Carrigan, Philip Hadley, 2113
Carrington, David K., 1918, 2073
Carrozza, A., 178
Carter, John, 1674
Case, Glenna L., 1784
Casebeer, R. L., 761
Casey, D., 2470
Casey, Hugh E., 53
Castagno, Margaret, 936
Castellanos Jarquin, S., 2035
Castro Bastos, Leonidas, 2683
Cathcart, Richard Brook, 60
Cavan, Ann, 763
Chalmers, John R., 2125
Chambers, Frances, 136
Chang, Chi'-yun, 1555, 1587
Chang, Raymond C., 1575-1576
Chapman, Carleton B., 3138
Chaves, Luis Fernando, 2697
Cheek, Emory, 3014
Chekki, Danesh A., 1467
Cheney, Roberta Carkeek, 2340
Chimutengwende, Chen, 299
Chimutengwende, Mary, 299
Chimwano, A. M. P., 1066
Chonchol, Maria-Edy, 888
Choudhari, J. S., 1458
Christensen, Earl M., 2445
Christensen, Lee A., 2881
Christian, C. J., 1754
Chronic, Felicie, 2315
Cigar, Norman, 422, 442, 445, 502, 586, 627-629, 633, 703, 730, 736, 739, 1201-1202, 1222, 1251, 1263, 1300, 1308, 1323, 1361, 1380
Clark, Brian Drummond, 2081
Clark, Charles Manning Hope, 1810
Clark, John M., 2112
Clark, T. W., 2470
Clarke, Colin G., 2522
Clarke, David G., 450, 901, 1008, 1083
Clarke, Walter Sheldon, 830
Clarke, David G., 901, 1008
Clary, W. P., 2148
Clavel, Pierre, 2536
Clements, Frank A., 1326, 1343
Clermont, Norman, 2492
Clogg, Mary Jo, 163
Clogg, Richard, 163
Cochran, Anita, 2136
Cochrane, T. W., 534
Codarcea, Alexandru, 235
Coger, Dalvan M., 373
Cohen, John M., 818

Cohen, Phyllis, 2300
Cohen, Sanford F., 2174-2177, 2260, 2296, 2387, 2861
Colhoun, Eric A., 1737
Colling, Gene, 2098
Collins, Kay, 2257, 2291
Collins, Michael Owen, 1076
Collison, Robert Lewis, 833, 1036
Colvin, Lucie Gallisted, 686
Combe, Michel, 620-621
Comitas, Lambros, 2472
Condarco Morales, Ramiro, 1877
Conover, Helen F., 280
Cook, Alison, 1022
Coons, William, 2212
Cooper, C. F., 1756
Cooper, E. Nathan, 2241, 2282
Coppell, W. G., 1790
Corbin, John Boyd, 2090
Corcoran, R. E., 2410
Cornforth, I. S., 2485
Cornwallis, L., 1230
Correa, M. A. A., 2614
Cortese, Charles F., 2178
Cortese, Jane Archer, 2178
Corwin, Arthur F., 2040
Coulanges, M., 1106
Coulanges, Pierre, 1106
Coult, Lyman H., 506
Coumans, Cheryl Louise, 2226
Court, Arnold, 2105
Cowling, S. S., 1761
Cox, Edward Godfrey, 3194
Coxon, Howard, 1705
Crabbe, David, 2971
Craig, Alan L., 2678
Craig, Beryl F., 1791-1795
Craig, Ian E., 3010
Cramer, Howard Ross, 2544
Creasi, Vincent J., 497, 1275, 1745-1746, 2552, 2621
Crippen, J. R., 2283
Crites, Ronald, 2883
Crocker, R. L., 1754
Crockfield, W. E., 1933
Crosby, Cynthia A., 887
Crosby, Martha, 2689
Crout, Robert R., 444
Crowder, Michael, 297
Crump, Ian A., 1706-1707
Crush, Jonathan S., 797, 800, 870-871, 1003-1004
Cruz, Paulo Roberto, 2596
Cuff, David J., 2179
Culey, Alma G., 1777-1778
Curran, Brian Dean, 601
Curtis, Neville M., Jr., 2399
Cyriax, George, 3182
d'Andrea, F. N., 2555
Danforth, Sandra C., 3122
Daniells, Lorna M., 3167
Darch, Colin, 804, 822, 924
Das Gupta, S. P., 1448, 1462
Daubenmire, Rexford, 2461
Daveau, S., 392
Davey, Lois, 1749
David, Tannatt William Edgeworth, Sir, 1738
Davies, D. Hywel, 1059
Davies, Harold Richard John, 308
Davies, J. L., 1722
Davis, Elisabeth B., 2916

Author Index

Davis, Lenwood G., 329, 333, 346, 358, 365-366, 370, 421, 1187
Davis, Peter Hadland, 160, 1372
Davis, Robert Henry, 2653
Dazy, Jean, 435
de Burlo, Charles, 110
de Gennaro, Nat, 2214
de Graff, Gerrit, 956
de Kock, C. I., 990
de Kok, David A., 2217
De Lacerda, J. P., 2613
De Lacerda, E. S. B., 2613
De Silva, A., 1524
de Vries, C. A., 84
De Vries, James., 1024
Decalo, Samuel, 449, 481, 642, 715
Dedet, J-P., 440, 732
Deike, R. G., 1952, 2119
Delancey, Mark W., 451
Delancey, Virginia H., 451
Delorme, Robert L., 1861
Dench, J. E., 3019
Denevan, William M., 1878
Denham, Woodrow W., 1796
Denman, J. M., 493
Denny, Lyle M., 2841
DePay, G. W., 2888
Derry, Duncan R., 2812
Dey, Nundo Lal, 1480, 1514
Dick, Trevor J. O., 1970
Dicks, G. R., 3168
Dickson, Bertram Thomas, 1
Difrieri, H. A., 2551
Dill, Henry W., Jr., 2079
Dillon, John L., 1781
Dimaras, C. Th., 164
Dinstel, Marion, 291
Dione, Josue, 682
Dippold, Max F., 452
Dombrowski, John, 1983
Donley, Michael W., 2274
Donque, G., 1096, 1097
Dorward, D. C., 514
dos Passos, Cyril F., 1957, 2156
Dosaj, N. P., 846
Dossick, Jesse John, 1594-1595
Dost, H., 1166
Doucet, Cyril D., 1910
Douglass, Barbara, 3114
Douglass, John F., 1895
Doumani, George A., 1650
Doumato, Lamia, 1351
Dowson, Valentine, 3015
Dregne, Harold E., 37
Driver, Edwin D., 3101
Droulia, L., 164
Dubreuil, Lorraine, 1919
Dudai, Yadin, 1273
Duffield, Christopher, 57-58
Duffy, David, 79, 303
Duffy, James, 78, 300
Duignan, Peter, 281, 375
Dujarier, M., 446
Dunford, Christopher, 15
Dunn, Henry Arthur, 3006
Dunn, James E., Jr., 2065
Dupuy, Trevor N., 1586
Durrenberger, Robert W., 2017, 2275
Duster, Joachin, 1327
Dutt, Ashok K., 1431
Dworaczek, Marian, 2955
Dwyer, D. D., 2898

Dzotsenidze, G. S., 1629
Eaton, T. J., 2381
Ebolo, Josue E., 406, 411
Eby, P. A., 2099
Eckert, Jerry Bruce, 869
Edie, Milton J., 2339
Edmondson, J. R., 160, 1372
Edmonston, Barry, 1864
Efrat, E., 1276
Eggers-Lura, A., 87, 3161
Egunjobi, James E., 647
Ehlers, Eckart, 1227
Eicher, Carl K., 521, 536, 664
Eichler, George R., 2310
Eisenman, Eric, 2138
Eister, Margaret F., 2092
el Awamy, Alad Musa, 576
el Nasri, Abdel Rahman, 689-690
El-Kassas, Mohamed, 488
Elias, Joel N., 1900
Elias-Cesnik, Anna, 2023, 2247, 2287, 3021
Elling, Ray H., 3133-3134
Ellis, Harold H., 2126, 2131
Elsasser, Albert B., 2299
Emanuel, Muriel, 1268
Emategui, Federico, 2669
Emerson, John Philip, 1577
Emery, L. A., 2352
Emery, Sarah Snell, 2036
Emezi, Herbert O., 667
Enciso de la Vega, Salvador, 2016
Ene, Ngozi, 657
Enevoldsen, Thyge, 899
Erber, Fabio Stefano, 2595
Eribes, Richard A., 2201
Erickson, Elizabeth, 2524
Erickson, Frank A., 2524
Erickson, Judith B., 1294
Erinc, Sirri, 1369
Eskuche, V., 2564
Ettalhi, J. Azzoyz, 572
Evans, A. Kent, 2099
Evans, Gwynneth, 1904
Evans, Peter A., 2165, 3008
Evenson, J. P., 1771
Ewusi, Kodwo, 540
Fage, J. D., 374
Faillace, C., 175
Fairbridge, Rhodes W., 72, 2783, 2801-2803, 2827
Farley, Albert L., 1922
Farrell, Lois, 2915, 2949
Faur, Kenneth R., 2538, 2543
Feierman, Steven, 352
Ferguson, Donald S., 463
Ferreira, Carmosina N., 2615
Ferreira, H. Amorim, 777
Ferreira, Lieny do Amaral, 2615
Ferrier, Walter Frederick, 1935
Ffolliott, Peter Frederick, 2148, 2244, 2384
Field, Barry C., 3170
Field, Henry, 1471
Field, William Dewitt, 1957, 2156
Fiese, Marshall J., 3143
Fifield, Richard J., 2772
Findlay, Allan M., 717
Findlay, Anne M., 717
Finlayson, Jennifer A. S., 1711
Fischer-Galati, Stephen A., 230
Fitzpatrick, Lilian Linder, 2346

Fleischer, Beverly, 1026
Fleming, Quentin W., 1196
Flemion, Philip F., 1982
Florescano, Enrique, 2034
Floss, Ludmilla, 2977
Fontana, Bernard L., 2024, 2248
Fontolliet, Micheline, 917
Foster, Stephanie, 2047, 2202
Fowler, Catherine S., 2193
Francois, L. E., 2990
Franz, Erhard, 1377
Fraser, John Keith, 1915
Freedman, Robert L., 3130
Freile, Alfonso J., 2577
Freitag, H., 1216
Freitag, Ruth S., 334
French, Norman R., 1611
Frey, Mitsue, 300, 303
Fuerer-Haimendorf, Elizabeth von, 1409
Fuller, S. C., 1399, 1817, 1820
Furlan, D., 159, 218, 237, 248
Gabriel, Baldur, 473
Gabrovska, Svobodozarya, 119, 1661
Gagner, Lorraine, 741
Gaignebet, Wanda, 292
Gailey, Harry A., 525
Gakenheimer, Ralph A., 90
Galazii, G. I., 1630
Gamble, David P., 515
Gann, Lewis H., 375
Ganssen, Robert, 915
Garcia de Miranda, Enriqueta, 2007, 2018
Gastmann, Albert L., 2531
Gaston, M. P., 2324
Gat, Z., 1291
Geddes, Charles L., 483, 687, 1144-1145
Geelan, P. J. M., 1556
Geiger, H. Kent, 103
Gelderman, W. J., 2900
Gentilli, J., 1747, 1831
Geraghty, James J., 2115
Gerakis, Pantazis A., 154
Gerhard, Peter, 2054
Gerteiny, Alfred G., 609
Geyser, O., 995
Gibb, H. A. R., 388, 1128
Gibson, G. D., 347
Gibson, Lay James, 2217
Gibson, Mary Jane, 260, 551, 770, 889
Gidwani, N. N., 1420
Giefer, Gerald J., 2116-2117, 2846
Gifford, Gerald F., 2898
Gill, Dhara S., 1965
Gillis, W. T., 2497
Girard, Roselle M., 2425
Girno, Aristides de Amorim, 187
Gischler, Christiaan, 399, 1167
Gleason, V. E., 1938, 2100
Gleeson, Thomas Alexander, 122, 313, 1161, 1404
Glenn, C. R., 493
Gligo Viel, N., 2622
Glore, Denise, 1890
Goehlert, Robert, 1673
Goeltz, N. S., 2441
Goetz, Rita D., 2926, 2969
Goins, Charles R., 2403

Author Index

Golany, Gideon, 1298
Gold, H. K., 1241, 1284
Goldman, Bram J., 2
Goldsmith, William W., 2536
Goldstein, Edward L., 3080
Goodhew, Barbara, 1805
Goodman, Lowell Robert, 2394
Goodwin, Scheila McMillan, 979
Goonetileke, H. A. I., 1515
Gordon, Mary S. J., 2060
Gorter, G. J. M. A., 983
Goudie, Andrew S., 13, 2902
Gouvernet, C., 140
Goyer, Doreen S., 3102-3103
Graber, Eric S., 1991, 2693
Graber, M., 400
Graham, Alan, 1843-1844
Grainger, A., 42
Grandidier, Guillaume, 1093
Grant, A. Paige, 590
Gray, Beverly Ann, 1037
Gray, Peter, 2927
Grayson, Donald K., 1887
Green, Christine R., 26
Green, Muriel M., 146
Greenway, John, 1797
Greenway, M. E., 1821
Greenwood, L. Larry, 2894, 2957
Greig, Doreen E., 989
Grey, D. R. C., 1503
Greyeris, Harry A., 63
Griffith, J. F., 314, 658, 812-813, 825, 847, 896, 929, 1020, 1046, 1101
Grimes, Annie E., 202, 496, 573, 619, 726, 814, 826, 930, 1162, 1319-1320, 1415, 1416, 1439-1441, 1500-1501, 1521
Grimwood-Jones, Diana, 1146
Gritzner, Charles F., 2339
Gross, Meier, 3080
Grossman, Jorge, 1862
Grotpeter, John J., 1006, 1069
Grover, Ray, 1824
Gruenberg-Fertig, I., 1288
Grzimek, Bernhard, 2943
Guclu, Meral, 1365
Gudde, Erwin Gustav, 2276
Guiell, G., 140
Guillarmod, A. J., 321
Guliani, T. D., 1451
Gunn, C. R., 2930
Gunn, Clare A., 2421
Gupta, I. C., 1444, 2904
Gupta, Raj Kumar, 1216, 1445-1446, 1505, 1570
Gustafson, Neil C., 2334
Gustafson, W. Eric, 1508
Guttadauro, Guy J., 2294
Gysi, Marshall, 2863
Haacke, W. D., 916
Haase, Edward F., 2027, 2029, 2251, 2289, 2428, 3022, 3024
Haber, Stephen, 1838
Hacia, Henry, 497, 2817
Hackbarth, D. A., 1951
Hadac, E., 1261
Hadley, D. G., 1086
Hagedorn, Horst, 2906
Hahn, Lorna, 583
Hailu, Alem Seged, 595, 849
Haislip, T. W., Jr., 2415

Haliburton, Gordon MacKay, 874
Hall, Arthur Lewis, 969
Hall, E., 1933
Hall, S. A., 1049
Hall, Vivian S., 2790
Hallaron, Shirley Anderson, 93
Hamblin, L. S., 2444
Hamidon, Sidikon A., 638
Hamilton, H. R., 2872
Hamilton, Michael S., 2182
Hammack, Gloria Mae, 2947
Hammond, Nicholas G. L., 125
Handlin, Jayne, 3083
Hanelt, P., 1571
Hanifi, M. Jamil, 1220
Hansen, Alfred, 273
Hansen, Gerda, 1175
Hansen, M. W., 2325, 2395, 2442, 2469
Haralambous, Diomedes, 129-130, 157-158
Hare, Frederick Kenneth, 1945-1946
Harmon, Ralph Winter, 2940, 3150
Harmon, Robert B., 513, 1218, 1354, 1585, 2051, 2306, 2318, 2326, 2333, 2337, 2345, 2351, 2367, 2392, 2397, 2404, 2416, 2418, 2434, 2449, 2463, 2471
Harniss, Roy O., 2149
Harris, Chauncy Dennison, 1610
Harrison, Colin, 116, 401, 1170
Harrison, John P., 1832
Harrison, W., 2497
Hartnig, Charles W., 852
Harvey, Joan M., 304, 1130, 1675, 1712, 1837
Harvey, Stephen J., 2149
Hasemeier, Amanda N., 2236
Hastenrath, Stefan, 250-251, 1090, 1683-1684
Hatch, Warren L., 2815
Hatcher, Suzanne, 2183
Hattingh, Phillipus Stefanus, 982
Haug, Peter T., 2140
Haughton, S. H., 968
Havinden, M. A., 403
Hawkes, Clifford L., 39
Hawkins, Donna, 2039, 2568
Hay, John E., 1945
Hayes, R. O., 2386
Hayward, C. L., 2447
Heaser, Eileen, 2269, 2301
Heath, Dwight B., 2584
Hecht, Melvin E., 2227
Heggoy, Alf Andrew, 444
Heijnen, J. E., 1021
Heimlich, Ralph E., 2881
Heise, Jon, 2735
Heizer, Robert Fleming, 2298-2299
Heller, Thomas, 2257, 2291
Hemsy, Victor, 2567
Henderson, F. I., 788
Hendrickson, Lupe P., 2031, 2253
Henrickson, James, 2010, 2228, 2371, 2423
Henze, Dietmar, 3197
Herbst, J. F., 949, 1075
Hernandez, John W., 2381
Herrick, Allison Butler, 1039
Herrmann, Albert, 1584
Hershfield, David M., 2107
Hess, Robert L., 373

Hevelin, John, 78
Hewes, Leslie, 2736
Hicks, Margaret Rowland, 1753
Hidaru, Alula, 805
Highsmith, Richard M., 2319, 2405, 2450
Hill, Dorothy, 1739
Hill, Marji, 1798
Hill, R. W., 565
Hill, Richard H., 2237
Hill, Richard Leslie, 688
Hilty, Steven L., 863
Hinckley, Dexter A., 2140
Hitchcock, R. K., 795
Hodge, Edwin T., 2411
Hodges, Tony, 754
Hodson, Warren G., 2330
Hoeller, Erich, 315
Hoetink, H., 2527
Hoffmann, Jose A. J., 2545
Hogan, Edward Patrick, 2417
Holdsworth, Mary, 282
Holmes, Beatrice H., 2126
Holmgren, Patricia K., 2935
Honkala, Barbara H., 2151
Hoogstraal, Harry, 3144
Hooker, Marjorie, 2535
Hopson, Rex C., 3158
Hopwood, Derek, 1146
Horecky, Paul Louis, 148, 208
Horowitz, Michael M., 410
Horrell, Muriel, 996
Horton, Jerome S., 2873
Horton, John Joseph, 242
Hottin, G., 1100
Houghton, John G., 2360
Houston, Carol A., 1799
Houts, Didier van, 372
Howard, R. A., 2506
Howell, John Bruce, 834-835, 1009, 1038
Hsieh, Chiao-min, 1557
Huber, Otto, 1845
Huerta, Fernando, 203
Hull, William B., 1896, 2489
Hulme, M. I., 1775
Hundsdorfer, Volkhard, 1029
Hunter, David E., 3120
Huq, A. M. Abdul, 80
Hurlbert, Stuart H., 2546
Hynes, Mary C., 1915
Ibrahim, A., 689
Iglesias, Dolores, 2597
Illes, Doris, 48
Iltis, J. C., 1427, 1492, 1517
Imperato, Eleanor M., 588
Imperato, Pascal James, 588, 597
Inch, Peter, 1666
Indacochea G., 2685
Indesha, N., 733, 1197, 1247
Inglis, Michael, 1890
Isaac, G., 839
Isaza Velez, Guillermo, 1857
Ita, Nduntuei O., 670
Ives, Alan, 1714
Izco, J., 205
Jackson, Stanley Percival, 316
Jacobs, Barbara, 79
Jacobstein, J. Myron, 2127
Jadhav, P. S., 1461
Jaeger, Edmund Carroll, 1881, 1897
Jain, M. K., 1421, 1451

Author Index

Jain, Tara Chand, 1450, 1453
Jamail, Milton Henry, 1997, 2022, 2059, 2128, 2242, 2285
James, Dilmus D., 1841
Jansen, Anne, 318
Jequier, Nicolas, 2779
Jermy, Tibor, 224
Jett, John H., 2231
Johnsen, Vibe, 899
Johnson, A. F., 526
Johnson, A. M., 2578, 2660, 2686
Johnson, W. Carter, 1611
Johnston, A. G., 1940
Johson, David M., 2491
Jones, David Lewis, 2666
Jones, E. G., 209
Jones, George Neville, 2025
Jonsen, Roar, 1562
Jordan, Jeffrey L., 353
Joseph, Richard M., Jr., 105
Joseph, Timothy W., 2309, 2338, 2368, 2393, 2435, 2464
Joy, Jody, 3123
Joyce, Stephen J., 395
Kabdebo, Thomas, 222
Kabeel, Soraya M., 1223, 1313, 1328, 1339
Kaestner, Hermann, 1214
Kafe, Joseph Kofi, 528
Kagan, Alfred, 294
Kahrl, William L., 2284
Kai-Samba, Ibrahim B., 1025
Kalck, Pierre, 470
Kaldani, Elias H., 494
Kalia, D. R., 1421, 1451
Kalley, Jaqueline Audrey, 938, 984
Kamm, Antony, 1208, 1229, 1400, 1548
Kamphorst, A., 2895
Kane, F., 683
Kaplan, Irving D., 529, 772, 807, 837, 891, 926, 946, 1010, 1055, 2521
Karot, Juan K., 2627
Kartawinata, Kuswata, 1483
Kasapligil, Baki, 1307
Kassahun, Checole, 806
Katz, Zev, 1663
Kayser, Bernard, 155
Kazmer, Daniel R., 1667
Kazmer, Vera, 1667
Kazmi, S. M. A., 1506
Keast, Allen, 1754
Keefe, Eugene K., 150, 168, 185, 196, 213, 223, 231, 1596
Keith, Susan Jo, 28
Kepinski, Alfred, 2877
Kerst, Erna W., 97
Kerven, Carol, 798
Ketso, L. Victor, 767
Keuken, Wil, 2935
Khan, M. N. G. A., 1208, 1229, 1400, 1548
Kharbas, Datta Shankarrao, 1468
Khukin, N. V., 1631
Kiger, Robt. William, 2936
Killick, Tony, 855, 1030, 1050
Kimhi, Israel, 1279
Kimmerling, Jon, 2319, 2405, 2450
Kinch, M., 51
King, John Wucher, 512
King, Russell, 1329

Kinghorn, Marion, 2523
Kirkpatrick, Meredith, 210-211, 1612, 1660
Kish, George, 1668
Kiss, E., 2840
Kitanov, Boris Pavlov, 219
Klock, Glen O., 2413-2414, 2459-2460
Kloot, T., 1762
Klosterman, Richard E., 2190
Kluckhohn, Clyde, 2263
Knabe, Danota T., 659
Knapp, George L., 2869, 2878
Knapp, R., 462, 476-477, 501, 535, 575, 594, 607, 640, 660, 680, 702, 748, 816, 829, 848, 934, 1023, 1047, 1065, 1124, 2499
Knaster, Meri, 1867
Knight, Allen W., 2879
Kocher, James E., 1026
Koehler, Jochen, 293
Kogan, Sh. I., 1657
Kohn, R., 3135
Kok, Michiel, 2529
Kolinski, Charles J., 2674
Kong, Les, 2301
Kononkov, Petr Fedorovich, 2507
Kormondy, Edward J., 2928
Korson, J. Henry, 1509
Kostinko, Gail A., 380, 682
Kotze, D. A., 939
Koumarianou, D., 164
Kowal, Jan M., 659
Kraehenbuehl, D. N., 1757
Kramer, H. P., 256, 272, 396, 498, 697, 815, 827, 931, 1134, 1163
Kramers, J. H., 388, 1128
Kreeb, Karlheinz, 1261
Krieg, P., 1980, 1992
Kroenke, Loren W., 1681
Krokovic, D., 572
Krummes, Daniel C., 430, 769
Krumpe, Paul F., 2755
Kubal, Gene J., 3045
Kuechler, August Wilhelm, 115, 322, 1136, 1651, 1758, 1891, 2331, 2486, 2547, 2938
Kumerloeve, Hans, 1322, 1360, 1373
Kuper, Wolfgang, 1029
Kupsch, Walter Oscar, 1941
Kurian, George Thomas, 77, 1481
Kurtz, Laura S., 1034
Laclavere, Georges, 458, 604
Lacombe, Bernard, 1103
LaFlamme, A. G., 2500
Laguerre, Michel, 2513
Laird, Edith M., 1471
Laird, W. David, 2264
Laking, Phyllis W., 214, 232, 1368, 1613
Lal, Chhotey, 1452
Lamar, Howard R., 1901
Lamb, Peter J., 250-251, 1090, 1683-1684
Lambert, Claire M., 1473, 3152
Lamprecht, Sandra J., 1614, 2270
Lancaster, Henry Oliver, 1787-1788
Land, Brian, 1971
Landsberg, H. E., 2833
Landy, Marc, 2084
Lang, A. R. G., 1753
Langer, Sylvie, 523, 544, 671

Langlands, B. W., 1040, 1049
Langman, I. K., 2026
Lapedes, Daniel N., 2804, 2929
Lappalainen, T. N., 1231, 1609
LaPray, Barbara A., 2444
Larrea, Carlos Manuel, 2662
Laughton, C. A., 794
LaVoie, Joseph R., 2218
Lavrentiades, G. J., 161
Lawani, S. M., 328
Lawless, Richard I., 431, 717
Lawson, Merlin P., 2347, 2349
Lawson, Simon, 2971
Le Bourdiee, Francoise, 1098
Le Bourdiee, Paul, 1098
Le Vire, Victor T., 465
Leanza, Armando F., 2560
LeBaron, Allen, 1856
Lebrun, J. P., 681
Lee, David, 423
Lee, Kaiman, 3162
Lee, Lawrence B., 2209
Lehmann, Edward J., 2870, 2903
Leistner, O. A., 790, 866, 907, 963, 1000
Lenroot-Ernt, Lois, 2723
Lespinasse, Raymonde, 2517
Leteure, P. C., 354
Leung, Woot-Tsuen (Wu), 355
Leupen, A. H. A., 412
Levine, Robert M., 2610, 2619
Levi-Provencal, E., 388, 1128
Lewanski, Richard Casimir, 2724
Lewin, Meredith, 944
Lewis, John van Dusen, 410
Li, Hsiao-Fang, 1563
Liakhova, A. G., 1632
Lilley, G. P., 2950
Lime, David W., 2206
Linden, George L., 2000
Lines Escardo, A., 190, 204
List, Robert J., 2834
Little, Elbert L., Jr., 2150-2151
Little, Peter D., 762
Littlefield, David W., 381, 1147
Llaverias, Rita K., 2847
Lobban, Richard, 269, 557
Lock, Clara Beatrice Muriel, 2730
Lockerby, Robert W., 2892, 3062, 3082
Lockwood, Sharon Burdge, 527, 643
Loehnberg, Alfredo, 2021
Lofty, G. J., 2810
Logan, Richard F., 906
Logan, William J. C., 1573
London, Elizabeth B. H., 2108
Longley, Richard Wilberforce, 1947
Longrigg, S. J., 2048, 2570, 2617, 2708
Lorimer, John Gordon, 1200
Louis, Andre, 731
Low, Donald A., 1427, 1492, 1517
Low, Jessop, 2435
Lowe, Charles H., 47
Loy, William G., 2406
Loya, Yossef, 1277
Lucas, David, 668-669
Lugo Lugo, Herminio, 2533
Lundgren, Bjorn, 840, 1012
Lustig, Lawrence K., 29
Lutsey, Ira A., 2359
Lux, William, 2496

Author Index

Luz, Maria Cira Padilha da, 54
Lydolph, Paul E., 1646-1647
Lyle, Katherine Ch'iu, 3104
Lynott, Bill, 2393
Lytle, Elizabeth Edith, 499, 700, 1178-1179, 1209, 1253, 1293, 1375, 1476, 1664
Ma, Laurence J. C., 1579
Maas, E. V., 2990
Mabbutt, J. A., 18
MacDonald, Gordon A., 1814
Mackay, John W., 2786
Mackenzie, Donald R., 97
MacLean, Jayne, 2981
Macro, Eric, 1386
Magin, George B., 2868
Mahayni, Riad C., 109, 1188
Mahdi, Ali-Akbar, 1243, 1248
Maier, Georg, 2664
Mainguet, M., 2907
Malan, Stephanus Immelman, 943
Malek, Jaromir, 511
Maloney, Frank Edward, 2129
Mamoun, Izz Eldin, 704
Manson, C., 2456
Manu, Comfort Henrietta, 545
Marcus, Harold G., 820, 832, 937
Margat, Jean, 622, 2848
Mariam, Mesfin Wolde, 809
Mark, Alan F., 1826
Marquez, Orfilia, 2706
Marris, Bernice, 1776
Marsden, B. S., 1677, 1715
Marshall, Mary, 2004, 2225, 2273
Marshall, Trevor G., 2493
Martin, Henno A., 911
Martin, Lee R., 3033
Martin, Phyllis M., 398, 783
Martinez Rios, Jorge, 2038
Martinson, Tom L., 2640
Marton, Shraga T., 1291
Mascarenhas, Ophelia C., 858, 1033, 1053
Massing, Andreas, 408
Massola, Aldo, 1800
Massoni, Colette, 591, 637, 648
Masters, John H., 1957, 2156
Mathews, M. L., 3136
Mathot, G., 324
Mathur, Brij, 2780
Mathur, U. B., 1472
Matthes, Hubert, 325
Matulic, Rusko, 243
Mbwana, Salum S., 1025
McAllister, Paul E., 2370
McArthur, Lewis Ankeny, 2407
McCain, John R., 2128
McCarl, Henry N., 3083
McCarn, Davis B., 75
McCarthy, Joseph M., 261, 552
McCarthy, Justin, Jr., 152, 1154
McClain, Thomas, 2327
McClymont, D. S., 1082
McCormick, J. Frank, 2928
McCone, Gary K., 1541
McCrossen, Robert G., 1942
McDonald, David R., 3119
McDonald, Gordon C., 246
McDonald, J. L., 2255
McFall, Jane Buzby, 3038
McFarland, Daniel Miles, 750
McGarvin, Thomas G., 2218

McGill, John T., 71
McGinnies, William Grovenor, 2, 24, 41, 2027, 2219, 2428, 3022
McHenry, Dean E., Jr., 1027
Mchomba, Vallery G., 1025
McKiernan, Gerard, 2164
McLoughlin, Peter F. M., 850, 1028, 1048
McReynolds, Edwin C., 2403
McWilliam, John, 668-669
Meadows, Katherine F., 3071
Mearim, A. B., 2614
Medellon-Leal, Fernando, 1879
Medler, John T., 661
Meghdessian, Samira Rafidi, 415, 1182
Meher-Homji, V. M., 1216, 1447, 1505, 1523
Meigs, Peveril, 69
Mekkawi, Mod M., 364
Melone, Theodore G., 2335
Memmi, L., 723
Meneghezzi, Maria de Lourdes, 2597
Merabet, Omar, 393
Meroney, Robert N., 3073
Merrett, Christopher Edward, 960
Merrill, Elmer Drew, 1537, 1686
Merryman, Linda, 3027
Mersky, Roy M., 2127
Merz, Robert W., 2336
Miller, A., 2630
Miller, E. Willard, 283, 382, 755
Miller, James D., 2200
Miller, John F., 2109
Miller, L. D., 3136
Miller, Lester L., Jr., 64
Miller, Ruby M., 283, 382, 755
Miller, Tom, 2008
Milligan, G. A., 949, 1075
Mingkov, N. I., 1648
Mings, Robert C., 1876
Mink, John F., 1815
Minter, Ella S. G., 1958
Mintier, J. Laurence, 2303
Minton, Sherman A., Jr., 3151
Miska, John P., 1954, 1959
Mitchell, J. M., Jr., 2840
Mitlin, Luceille Liston, 3007
Moe, Christine E., 2382, 2390
Mohome, Paulus, 786
Mokhtarzadeh, A., 1231
Moletta, Elizabeth Tolomei, 2615
Molnos, Angela, 856, 1031, 1051
Mondesir, Simone L., 1387, 1393
Montague, Joel, 734
Monteiro, Palmyra V. M., 1840
Moodie, Peter M., 1801
Moody, Elize, 982
Moore, Elizabeth T., 2426-2427
Moore, Howard, 169
Moore, Richard Thomas, 2232
Morehead, Joe, 2056
Morello, J., 2565
Moreyra y Paz Soldan, Carlos, 2677
Morgan, John, 2157
Morin, Philippe, 617, 724
Moritz, Walter, 781, 919
Morris, John W., 790, 866, 907, 963, 1000, 2403
Morrison, Denton E., 3048
Mortensen, B. Kim, 15
Morton, Leslie Thomas, 3126

Moscov, Stephen C., 2667
Mosino Aleman, Pedro A., 2018
Moskowitz, Harry, 1542
Mostyn, Trevor, 1344, 1381
Moulds, Maxwell Sydney, 1763-1764
Mowery, M. Kay, 2303
Moyal, Ann Mozley, 1691
Moyer, D. David, 2067
Muller, C. F. J., 993
Munoz Cristi, Jorge, 2627
Munoz Reyes, Jorge, 2574, 2577
Murdock, George Peter, 1963, 2042, 2194, 3115
Murray, Robert B., 2149
Musgrave, A., 1765
Musiker, Reuben, 941
Myers, Brian J., 3010
Nagelkerke, Gerard A., 2526
Nankivell, P. S., 1780
Naqash, A. B., 1255
Narzikulov, I. K., 1633
Naseva, V., 219
Nash, H. G., 701
Nathan, Andrew J., 1543
Naude, M. H., 951
Navalani, K., 1420
Navarro Andrade, Ulpiano, 2655
Nazarevskiy, A., 1615
Nede, G. J., 1709, 1825
Nekhom, Lisa M., 2572
Nelson, Harold D., 432, 454, 471, 547, 569, 611, 644, 676, 692, 719, 878, 1074, 1095
Nelson, Herman L., 2835
Nelson, Michael C., 2130
Nene, Y. L., 3018
Neser, L., 988
Neumann, Hans H., 3145
Neville, Paul, 3165
Neville-Rolfe, Edmund, 335
Newill, Vann A., 3149
Newman, Mark D., 356
Nicholson, Norman L., 1920
Nielsen, G. L., 1951
Nielsen, L. T., 2386
Nieuwolt, S., 1417, 1530, 1535
Nilsen, Odd, 1019
Nimer, Edmon, 2600
Norgaard, Ole, 851
Norris, Robert E., 2665
Norton, A. V., 2475
Norton, B. E., 2898
Norwell, M. H., 1955
Nuhn, H., 1980, 1992
Nuttonson, Michael Y., 433, 489, 570, 613, 721, 958, 1716
Nye, Roger P., 465
Nyrop, Richard F., 484, 1224, 1228, 1252, 1271, 1301, 1314, 1332, 1340, 1345, 1355, 1366, 1382, 1426, 1491, 1516, 2675
O'Connor, Anthony Michael, 360
O'Farrell, T. P., 2352
O'Leary, Timothy J., 124, 1963, 2042, 2194, 2549
O'Shea, John, 2951
O'Toole, Thomas E., 550
Oakeshott, Gordon B., 2277
Obeid, M., 694
Oberdorfer, Erich, 2631
Obudho, Constance E., 854
Obudho, Robert A., 854

Author Index

Odimuko, C. L., 361
Offield, Tarry W., 1499
Ofori, Patrick, 340
Ogden, Gerald R., 2166
Ogg, Clayton W., 2881
Ogot, Bethwell A., 859
Ohman, Howard L., 1418, 1531, 1536
Okolowica, W., 227
Oliver, Roland, 297
Olsen, Wallace C., 2159
Olton, G. S., 2255
Opheim, Lee A., 2417
Opschoor, J. B., 788
Oren, O. H., 1287
Orni, E., 1276
Orsi, Richard J., 2292
Ortiz, Alfonso, 2043, 2265, 2388
Orvedal, Arnold Clifford, 2896
Osterreicher, Suzanne, 78
Otawa, Toru, 3074
Ownbey, J. W., 2818
Pahwa, K. N., 1444, 2904
Painter, Tom, 410
Palia, Aspy P., 107
Pallier, Ginette, 744
Palmer, Wayne C., 2841
Pampe, William R., 2095
Panara, R., 2842
Panofsky, Hans E., 284
Pantazis, Th. M., 131
Parker, Susan, 614
Parker, Sybil P., 67, 2828
Patterson, D. S., 519-520
Patterson, Karl David, 351
Patterson, Maureen L. P., 1469, 1510, 1525
Patton, David R., 2243-2244, 2383-2384
Paylore, Patricia, 2, 5-9, 14, 17-18, 20, 1551, 1879, 2220, 2874
Pearce, Thomas Matthews, 2372
Pearl, Richard Howard, 2317
Pearson, J. D., 1204, 1398, 1544
Pederson, B. O., 42
Pedersen, E. B., 1801
Peebles, Patrick, 1518
Peltier, J. P., 625
Pereira, Benjamim Enes, 193
Pereira, J. F., 2602
Perez Moreau, Roman A., 2566
Perez Cordero, Luis de J., 2647
Perkins, David L., 1545
Perkins, Lee, 1665
Pernet, Ann, 1601
Peron, Yves, 745
Perry, J. W. B., 864
Pert, Mary, 3011
Perusse, Roland I., 2520
Pesoa, Lillian, 1874
Peterson, A. D., 1240
Petersen, J. C. Briand, 761
Peterson, Nicolas, 1802
Petrov, Mikhail Platonovich, 1616-1619, 1643
Petrovich, Michael Boro, 244
Pettinger, Lawrence R., 2937
Pfund, Rose T., 1816
Phillips, Beverly, 2691
Phillips, David W., 1948-1949
Phillips, James Wendell, 2452
Phillips. W. Louis, 1892, 2939

Pickard, J., 1759
Pieper, Frank C., 2727
Pierce, Richard A., 1592
Pignatti, Erika, 176
Pignatti, Sandro, 176
Pinto, Justine, 1545
Pinto da Silva, A. R., 191, 265, 274, 779, 894, 1485
Platto, M., 2823
Plomley, Norman James Brian, 1803
Plummer, Thomas F., 1634
Pollak, Karen, 758, 1070-1071
Pollak, Oliver B., 758, 1070-1071
Polushkin, V. A., 1231, 1609
Pomassl, Gerhard, 2712
Pool, William C., 2433
Porges, Laurence, 675
Portig, W. H., 1979, 1989, 2483
Porto, Everaldo Rocha, 54
Porzecanski, Leopoldo, 1870
Posnett, N. W., 649, 1056, 2498
Post, George, 1898
Poulter, N. H., 3019
Poulton, Helen J., 3192
Powell, Donald M., 2213
Powell, John M., 1947
Powell, Margaret S., 1932, 2077
Powell, S. C., 1779
Powelson, John P., 106
Powys, J. J., 1780
Praite, R., 1727
Prasannalakshmi, S., 1453, 1461
Predoehl, M. C., 2823
Price, Lynn, 1890
Prohaska, F., 2561, 2671
Prokhorov, Aleksandr Mikhailovich, 1598
Prouty, Chris, 821
Prussin, Labelle, 423
Psuty, Norbert P., 2678
Pundeff, Marin V., 212
Punyasiri Perera, N., 1523
Puri, R. K., 928
Pyhala, Mikk, 553
Quello, Steve, 1876
Raadsma, S., 2982
Rabin, Albert I., 1295
Raccagni, Michelle, 416, 1183
Rachie, Kenneth O., 2999
Rada, Edward L., 2295
Rahman, Mushtaqur, 1495
Rahmato, Dessalegn, 805
Rajemisa-Raolison, Regis, 1109
Rakowski, Cathy A., 350, 1868
Ralston, Sally, 2101
Ramirez, C., 2632
Ramirez, Ricardo, 2706
Ramirez de Amaya, Ruth, 1990
Ranaivoson, R., 1101
Randall, Lois E., 2868
Randall, Paul, 3075
Randle, Patricio H., 2557
Randolph, J. R., 1952, 2119
Rao, K. N., 1091
Rao, Y. P., 1442, 1502, 1522
Raper, P. E., 964
Rapp, John S., 2278
Rasmussen, R. Kent, 1085
Rasmussen, Randall, 2052
Ratisbona, L. R., 2601
Rauchle, Nancy M., 1732
Rautio, S. A., 2344

Rau, William E., 1067
Ravera, Oscar, 171
Reed, Alexander Wyclif, 1728-1729
Reid, Charles Frederick, 2541
Reilly, P. M., 649, 808, 1056
Reiner, Ernst, 1717
Reining, Priscilla, 19
Reinmiller, Elinor C., 3138
Reitan, Clayton H., 26
Rey Balmaceda, Raul C., 2554
Reynolds, Hudson Gillis, 2259
Reynolds, Michael M., 2718
Rhoades, Marjorie, 32, 38, 50
Ribeiro, Marilia da Cunha Ferro, 774
Ricaldi, Victor, 2808
Rice, M. L., 2819
Richard, D., 326, 1137
Richards, Alan, 505
Richards, Horace G., 72
Richards, J. Howard, 1924
Richards, Robert A. C., 186, 197, 1835
Richardson, James B., III., 2679
Richardson, R. A., 1780
Rickert, Jon E., 1899
Ridge, John Drew, 132, 312, 618, 655, 725, 912, 973, 1061, 1078, 1133, 1237, 1371, 1414, 1436, 1520, 1682, 1741
Ridge, Martin, 2208
Riehl, S. K., 51, 52
Rigby, M., 2819
Rihani, May, 3123
Riley, Carroll L., 1353
Riley, Peter, 86
Rishworth, Susan Kroke, 610, 751
Risopatron Sanchez, Luis, 2624
Ristow, Walter W., 2749
Rivera de Figueroa, Carmen A., 3155
Riviere, Lindsay, 1114
Roberts, Albert E., 2279
Roberts, Jack, 1542
Roberts, Robert S., 2330
Roberts, T. D., 558, 1411
Robertson, J. H., 2363
Robinove, Charles Joseph, 22
Robinson, Anthony, 944
Robinson, G. D., 2198
Robson, John R. K., 1900
Rodier, J., 319
Rodius, S., 3135
Rodrigues, Jose Joaquim, 270
Roe, Arthur L., 2153
Rogers, Dilwyn J., 306, 532
Rogers, Earl M., 2160-2161
Rogoff, Marc Jay, 2069, 2882
Rohrer, Richard L., 2894, 2957
Roman, David, 235
Roman, S. J., 1258
Romney, E. M., 2353, 2364
Ronayne, J., 1709, 1825
Rondinelli, Dennis A., 107
Ropelewski, C. F., 2823
Rosen, Norman C., 1238
Rosenfeld, Eugene, 821
Rosenthal, Eric, 756
Rossi, Georges, 1041
Roth, Deborah, 2536
Roukens de Lande, E. J., 902
Rousset, C., 140
Roussine, N., 438, 626, 729
Rowe, M. L., 3027

Author Index

Ruder, Ruth L., 2221
Rudloff, Willy, 2836
Rudolph, Donna Keyse, 2710
Ruggles, Judith, 3132
Rundel, Philip W., 2487
Russell, Findlay E., 2948
Rutten, Louis Martin Robert, 2478
Rweyemamu, A. H., 1035
Ryan, Mary, 2951
Ryckman, L. Frederick, 2393
Ryder, Dorothy E., 1902
Rydjord, John, 2328
Rylander, Kristina, 553
Rylko-Bauer, Barbara, 3137
Saab, V. A., 2470
Sable, Martin Howard, 1869
Saghayroun, Atif A. Rahman, 707
Sahab, Abbas, 1210
Said, Rushdi, 495
Sakala, Carol, 1477, 1512, 1526
Salad, Mohamed Khalief, 925
Salas de Leon, Sonia, 1847
Saltzmann, E. J., 2019
Samaan, A. G., 1189
Sampedro V., Francisco, 2657
Samuelson, Ann-Marie, 840, 1012
Sanchez, Maria Elsa Bareiro de, 2672
Santos, M. C. D., 2608
Sarig, S., 1286
Sarukhan, Jose, 2002
Satyaprakash, 1454, 1488
Saulniers, Suzanne Smith, 350, 1868
Saupe, Barbara, 2691
Sauvage, C., 438, 626, 729
Sawchuk, John P., 1916
Schamus, J. J., 2893
Schapera, Isaac, 986
Scharffenberg, R., 2948
Schlick, W., 1980, 1992
Schlueter, Hans, 567, 568
Schmeckebier, Laurence Frederick, 2210
Schmidt, D. L., 1086
Schmutz, Ervin M., 2249
Schoeman, Elna, 918
Schofield, Eileen K., 1812, 2935
Scholz, Fred, 1327
Schrale, G., 2982
Schreiber, Joseph F., Jr., 70
Schrock, Joan, 601
Schroeder, Peter B., 1573
Schroeder, Susan, 2502
Schuh, George Edward, 2609
Schulze, B. R., 796, 868, 913, 974-975, 1002
Schumacher, August, 108
Schuster, Joseph L., 1893
Schwartzberg, Joseph E., 1221, 1410
Scott, J. M., 2415
Scott, Mary Woods, 2396
Scoville, Sheila A., 1156
Sealock, Margaret M., 1932, 2077
Sealock, Richard B., 1932, 2077
Seeley, Dwight, 3064
Segal, Sheldon J., 3104
Self, Huber, 2332
Selim, George Dimitri, 386, 1149
Sellers, William D., 2237
Semenova, M. I., 1635
Sen, N. K., 1430
Senn, Harold Archie, 1955
Shabad, Theodore, 1603

Shackar, Arie, 1279
Shaffer, Ray, 2311
Shah, Khawaja T., 151, 389, 1153
Shalhevet, J., 1290
Shannon, Michael Owen, 1330
Sharma, Prakash C., 1455, 1466, 1470
Sharpe, Jane, 1751
Shaw, Robert Baldwin, 376
Shaw, Robert Blaine, 2429
Shea, Carol A., 179
Shea, Donald R., 1875
Sherbrooke, Wade C., 9, 20, 2028-2029, 2250-2251, 2288-2289, 3023-3024
Shihabi, Mustafa, 402, 1171
Shinn, Rinn-Sup, 1546
Shirk, George H., 2398
Shreve, Forrest, 2030, 2252, 2290
Shrimali, D. S., 1457
Shur, Shimon, 1296-1297
Siddiqi, Akhtar Husain, 83, 1507
Silva, Aderaldo de Souza, 54
Silva, A. T. S. Ferreira da, 774-775
Silva, N. M. F. da, 2602
Simkin, Tom, 2805
Simmons, Mary Ann, 2879
Simon, Joan C., 543
Simon, Reeva S., 1176
Simons, George G., 2126
Simpson, Eugene S., 33
Sims, Michael, 76, 294, 300
Singer, Philip, 89
Singh, Ganda, 1429
Singh, Jasbir, 1456
Singh, K. B., 3020
Singhvi, M. L., 1457
Sklar, Richard L., 376
Skoglund, C., 2032
Skupa, Gary L., 3008
Skurnik, W. A. E., 285
Slamecka, Vladimir, 75
Slankey, George, 2206
Slobodkin, Lawrence B., 1277
Sloggett, Gordon, 2167
Smalley, Ian J., 2911
Smeins, Fred E., 2429
Smith, A. L., Jr., 1438
Smith, Alan L., 2479
Smith, Craig C., 1079
Smith, David G., 2784
Smith, Dwight L., 1974, 2211
Smith, Ernest Linwood, 2222
Smith, Harold E., 1533
Smith, Harvey H., 1207, 1317
Smith, Joan Dickson, 3132
Smith, Roger Cletus, 2917
Smith, Robert, 2883
Smith, Robert H. T., 341, 1138, 1854
Smith, Stephen W., 2983
Smith-Sanclare, Shelby, 2369
Snead, Rodman E., 1232, 2813
Snodgrass, Marjorie P., 2204
Snow, J. W., 2645, 2702
Snowball, George J., 1062
Sochava, V. B., 1636
Socolofsky, Homer Edward, 2332
Solomon, Alan C., 377
Sommer, John W., 307
Sonderegger, J. L., 2344
South, Aloha, 286
Spamer, Earle E., 2224

Sparn, Enrique, 2562
Specht, M. M., 1760
Specht, Raymond L., 1718, 1760
Speece, Mark W., 695, 1057, 1278, 1335, 1389
Spencer, Katherine, 2263
Spencer, William, 632
Sperling, Louise, 515
Spieker, Andrew M., 2198
Spinelle, William B., 3141
Srytrova, D., 2628
Stahler, Gerald, 2262
Stanley, Janet L., 673
Stanukovich, K. V., 1633
Steenberg, Warren F., 2031, 2253
Steer, Edgar S., 2525
Steinitz, H., 1287
Steller, Dorothy L., 1816
Stengel, H. W. von, 914
Stevens, J. H., 1329
Stevens, Pamela, 787
Stevens, Richard P., 802
Stevenson, Merritt R., 1679
Stewart, Omer C., 2195
Stoddard, Theodore L., 1111, 1116, 1121
Stoner, J. D., 2400
Stork, Karen E., 2348
Stow, D. A. V., 81
Strand, Rudolph G., 2280
Stranz, Dietrich, 315
Straw, Richard M., 2010, 2228, 2371, 2423
Strohmeyer, Eckhard, 781, 919
Stubbs, Michael, 880
Stuckey, Ronald L., 1892
Sturtevant, William C., 1964, 2044, 2196
Sukhwal, B. L., 1087, 1401-1402, 1552-1553
Summers, W. Kelly, 3072
Sunding, Per, 258, 266
Swearingen, W. D., 509
Sweet, H. R., 2412
Sweet, Louise Elizabeth, 124
Swift, Lloyd H., 2932
Tadmor, R., 1289
Taha, M. F., 133, 1165
Talbot, A. M., 865, 962, 999
Talbot, W. J., 865, 962
Talbot, W. S., 999
Tamayo, Jorge L., 2009
Tanis, Norman E., 1545
Tanoglu, Ali, 1369
Tatarsky, I. V., 1619
Taylor, Beth Ann., 404
Taylor, David Ruxton Fraser, 844
Taylor, Donald C., 1172
Taylor, George F., II, 1463
Taylor, George F., III, 404
Taylor, Robert L., 2339
Taylor, Suzanne N., 15
Tchalenko, J. S., 1236
Tchobanoglous, George, 2883
Teesdale-Smith, E. N., 1742
Teles, A. N., 191-192, 265, 267, 274-275, 779-780, 894-895, 1485-1486
Templer, Otis Worth, 10
Teplova, S. N., 1628
Terry, G. M., 245

Author Index

Tesdell, Lee S., 441, 1309-1311, 1362
Thauvin, Jean Pierre, 621
Thawley, J. D., 2714
Thomas, Morley K., 1946, 1949
Thomas, Robert N., 1865
Thompson, Anthony., 1593
Thompson, B. W., 317
Thompson, Leonard M., 994
Thompson, M., 2415
Thompson, W. I., III., 2760
Thomsen, Nyla R., 2348
Thran, P., 114, 118, 397, 1649, 1654
Thuronyi, G., 2842, 2913
Tindale, Norman Barnett, 1804
Titus, Elizabeth A., 89
Tobias, Phillip V., 799, 920
Todd, David Keith, 2857
Tolley, J. C., 1727
Torrance, J. D., 882, 1063, 1080
Torunsky, Richard, 3012
Toth, J., 1953
Toupet, Charles, 599, 603-604
Townsend, Ruth J., 2182
Townshend, J., 1013
Toye, Beatrice Olukemi, 662
Traore, Issa Baba, 592
Trotta, Victoria K., 2304
Troughton, Michael J., 1961
Trzyna, Thaddeus C., 3110, 3132
Tseng, T. C., 1564
Tueller, P. T., 2363
Tugby, E. E., 1677, 1715
Tulupnikov, A. I., 1655
Tumertekin, Erol, 1369
Tundisi, Jose G., 2592
Turner, P. V., 884
Turney, Jack R., 2131
Tuya, O. H., 2555
Twindale, C. R., 1719
Twitchett, D. C., 1556
Tyler, M. J., 1719
Udo, Reuben K., 650
Uehara, B., 2612
Uhleman, E. W., 2099
Ulfstrand, Stattan, 841, 1014, 1042
Ulibarri, George S., 1832
Ullery, Scott J., 2022, 2128, 2242, 2285
Urbanek, Mae Bobb, 2465
Uribe Contreras, Maruja, 1857, 2649
Ustimenko, G. V., 2507
Valente, M. da C., 2602
Vallentine, John F., 1962, 2173, 3031
van Alphen, J. G., 2901
Van der Heyde, H. E., 1079
Van der Leeden, Frits, 2849
Van der Maeson, L. J. G., 3020
van Garsse, Yvan, 708
Van Maele, Bernard, 600
van Warmelo, Nicolaas Jacobus, 764
Vance, Mary, 2266, 3159
Vanzolini, Paulo Emilio, 2548
Varady, Robert G., 879, 1043, 1412, 1496
Varallyay, G., 2905
Vashist, V. N., 1465
Velasquez Gallardo, Pablo, 2033
Vetrov, A. S., 1637
Vidergar, John J., 439, 507, 577, 691, 718, 1177, 1205, 1244, 1264, 1292, 1379, 1511

Vidich, Charles, 2186
Visser, S. A., 651
Vitale, Charles S., 828, 932, 1566
Vithal, B. P. R., 1430
Vivas, Leonel, 2697
Vivo, Paquita, 2532
Vlachos, Evan, 165
Vogel, Harvey, 3
Vogt, Martin, 903
Voll, John Obert, 709
von Soo, Rezso, 228
Vondruska, John, 3034
Vorhis, Robert Corson, 2121
Vreeland, Nena, 1482
Vuich, J. S., 2234
Vulkov, Yordan, 215
Wagner, Donald B., 1562
Wainwright, M. D., 1427, 1492, 1517
Walker, Alta Sharon, 22
Walker, Audrey A., 516, 877, 1054, 1072
Walker, Egbert Hamilton, 1537-1538
Walker, Henry P., 2267
Wallace, J. Allen, Jr., 147, 183, 698, 699, 778, 976, 1305, 1567-1568, 1589
Wallace, A., 2364
Walsh, Gretchen, 409
Walsh, J., 846
Walthall, C. L., 3136
Waltisperger, Dominique, 3105
Ward, Dederick C., 2792
Wards, Ian McLean, 1829
Ware, Helen Ruth E., 1688
Waring, Gerald Ashley, 2122
Warren, Andrew, 2909
Warren, Kenneth S., 3149
Wasserman, Paul, 3172
Wathern, Peter, 2081
Watts, I. E. M., 1569
Wauchope, Robert, 2045
Wazaife, Rashid, 1152
Weaver, Harry L., 2776
Weaver, John D., 2479
Webb, B. P., 1719
Weber, Michael T., 86
Webster, John B., 786, 836
Webster, Steven S., 2692
Weekes, Richard V., 348, 1140
Weight, Marie L., 1241, 1358, 2019
Weil, Thomas E., 1995, 2514, 2550, 2573, 2585, 2620, 2654, 2668
Weir, Thomas R., 1925-1926
Weis, Leonard W., 2335
Weiss, Joseph E., 2192
Weiss, Marianne, 318, 987
Welch, Florette Jean, 904
Wellisch, Hanan, 2866
Welsh, S. L., 2446
Weniger, Del, 2385, 2401, 2430
Werge, Robert W., 2690
Wernstedt, Frederick L., 2838
Westfall, Gloria, 137
Wheeler, Stella E. L., 151, 389, 1153
Whelan, John, 1333
Whetstone, G. A., 2871
Whitaker, Donald P., 1702
White, Anthony G., 2132, 2169, 3160

White, Sarah, 1604
Whiteside, R. M., 121
Whitfield, P., 649
Whiting, A. F., 2154
Whittell, Hubert Massey, 1766
Whitten, Phillip, 3120
Wicks, Vera M., 1713
Wieczynski, Joseph L., 1672
Wilber, Donald N., 1206
Wilcocks, Julia Ruth Nadene, 959
Wilcox, Virginia Lee, 2308
Wilfred, Hugh, 3015
Wilken, Gene C., 873
Wilkie, James Wallace, 1838
Willet, Shelagh M., 861
Williams, Geoffrey J., 652
Williams, J. B., 123, 2484
Williams, Jerry L., 2370
Willis, Cleve E., 3170
Wilmot, B. C., 327
Wilmot, L. P., 327
Wilson, Andrew H., 68
Wilson, E. D., 2232
Wilson, Monica, 994
Wilt, J. C., 2234
Wimberly, Ronald C., 2163
Winter, Alan, 1208, 1229, 1400, 1548
Wisner, B., 1013
Witherall, Julian W., 287, 385, 453, 466, 712, 1094, 1110, 1115, 1120
Witherell, Julian, 527
Woehlcke, Manfred, 1852
Wolfenden, E. B., 1403
Wolff, T. A., 2386
Wood, David N., 2731
Wood, Don A., 2918
Wood, John, 2368
Worden, Marshall A., 2217
Worsham, John P., Jr., 2085, 2312, 2321, 2341, 2357, 2373, 2391, 2402, 2432, 2437, 2466
Wright, Ann Finley, 2235, 2316, 2376-2379, 2443
Wright, Ione Stuessy, 2572
Wright, J. F., 1933, 1943
Wydoski, Richard S., 2170
Wylie, Enid, 1694
Yadav, Shree Ram, 1453, 1461
Yanarella, Ann-Marie, 3049
Yanarella, Ernest J., 3049
Yao, Augustine Y. M., 1135
Yates, Richard, 2004, 2225, 2273
Yentsch, Anne, 1088
Young, Harold C., 2725
Young, Margaret Labash, 2725
Young, William J., 2179
Zalacain, Victoire, 745
Zarn, Mark, 2257, 2291
Zerlucke, Raul, 2033
Zevallos y Muniz, Marco Aurelio, 2682
Zieske, Scott H., 2417
Zikeev, Nikolai Tikhonovich, 1650, 2914
Zohary, M., 1288
Zubatsky, David S., 262, 554, 771, 890
Zuvekas, Clarence, 2518, 2582

SUBJECT INDEX

ADOBE AND EARTHEN MATERIALS
3158-3159
AGRICULTURAL ECONOMICS
Africa, 332-337
    Northern and Western, 408-409
The Americas, 1855-1857
Australia, 1781
Bolivia, 2581-2582
Brazil, 2607-2609
Cameroon, 463
Chile, 2633-2634
China, 1574
Colombia, 2648-2649
Cuba, 2508
Ecuador, 2661
Egypt, 505
The Gambia, 521
Ghana, 536
Guatemala, 1991
India, 1464-1466
Israel, 1291
Kenya, 849-850
Mali, 595
Mexico, 2036
Middle East, 1172
Nigeria, 664
Peru, 2691
Senegal, 682
Tanzania, 1025-1028
Third World, 83-84
Tunisia, 730
Uganda, 1048
Union of Soviet Socialist Republics, 1659
Venezuela, 2707
AGRICULTURAL MARKETS AND COOPERATIVES
Third World, 85-86
AGRICULTURE
Africa, 328
    Northern and Western, 402
The Americas, 1848-1852
Argentina, 2567
Australia, 1767-1770
Bolivia, 2579
Brazil, 2604-2605
Canada, 1958-1961
Caribbean Islands, 2490
China, 1572-1573
Colombia, 2647
Cuba, 2507
Drylands of the World, 48-52
Egypt, 502-503
Europe, 117-118
France, 143
Greece, 162
Guatemala, 1990
Haiti, 2517-2518
India, 1448-1457
Iraq, 1263
Italy, 177-178
Jamaica, 2524-2525
Jordan, 1308
Lebanon, 1323
Mexico, 2033
Middle East, 1171
Mozambique, 897-898
Nigeria, 663
Pakistan, 1507
Paraguay, 2672

Peru, 2688
Romania, 240
South Africa, 980-982
Sudan, 703-704
Syria, 1361
Tanzania, 1024
Third World, 82
Union of Soviet Socialist Republics, 1652-1656
The United States, 2158-2163
    Arizona, 2258
    California, 2292
    Nebraska, 2349
Venezuela, 2704-2705
ANIMAL CULTURE
Africa, Eastern and Southern, 762
Australia, 1777-1780
Brazil, 2606
The United States, 2170
Zambia, 1066
ANTHROPOLOGY
Africa, 347-348
    Eastern and Southern, 764
    Northern and Western, 412
The Americas, 1866
Angola, 781
Asia, 1140
    South and Southeast, 1409
Australia, 1789-1804
Bahama Islands, 2500
Botswana, 799
Canada, 1963-1964
Caribbean Islands, 2492-2493
Colombia, 2650
Cyprus, 134
Egypt, 506
Europe, Mediterranean, 124
Greece, 166
Guatemala, 1993
India, 1471-1472
Mexico, 2041-2045
Middle East, 1178-1179
Namibia, 919-920
Nigeria, 670
Peru, 2692
Portugal, 193
South Africa, 988
South America, 2549
Sudan, 708
Tunisia, 731
Union of Soviet Socialist Republics, 1662-1664
The United States, 2193-2196
    Arizona 2263-2265
    California, 2297-2299
    New Mexico, 2388
APPROPRIATE TECHNOLOGY
Science and Technology, 2774-2781
ARCHITECTURE
Afghanistan, 1218
Africa, 363-364
    Northern and Western, 422-423
Algeria, 442
Jordan, 1311
Libya, 581
Middle East, 1190
Morocco, 629
Saudi Arabia, 1351
South Africa, 989
Third World, 94

Subject Index

Tunisia, 736
The United States, 2200
  Arizona, 2266
  California, 2305-2306
  New Mexico, 2389-2390
  Texas, 2431

ATLASES
Afghanistan, 1210
Africa, 308-309
  Northern and Western, 391
Agriculture, 2974
Argentina, 2556-2557
Australia, 1720-1724
  and the Pacific, 1679
Brazil, 2593
Bulgaria, 215
Cameroon, 457-458
Canada, 1921-1926
Canary Islands, 253
Chad, 474
Chile, 2625
China, 1554-1560
Colombia, 2641-2642
Cuba, 2503-2504
Economic Development, 3190
Ecuador, 2656-2657
El Salvador, 1976
Ethiopia, 809-810
France, 138
Geography, 2750-2751
Geology, 2812-2813
Greece, 155
Guatemala, 1985-1986
Haiti, 2515
Hawaiian Islands, 1813
Historical Geography, 3198-3199
Hungary, 225
Hydrology, 2858
India, 1430-1432
Indian Ocean and Islands, 1089
Iran, 1233
Israel, 1279-1280
Ivory Coast, 559-560
Jamaica, 2522
Kenya, 842-844
Lesotho, 864-865
Madagascar, 1098
Malawi, 880
Mali, 592
Mauritania, 604
Mexico, 2006-2009
Mongolia, 1587
Morocco, 615
Mozambique, 892
New Caledonia, 1822
New Zealand, 1828-1829
Niger, 638
Nigeria, 653
Paraguay, 2669
Peru, 2680
Portugal, 187
Reunion, 1117
Romania, 233
Saudi Arabia, 1347
Senegal, 678
South Africa, 961-962
Spain, 199
Swaziland, 999
Tanzania, 1015-1016
Turkey, 1369
Uganda, 1044
Union of Soviet Socialist Republics, 1622-1637

The United States, 2075
  Arizona, 2227
  California, 2274-2275
  Idaho, 2319
  Minnesota, 2334
  Montana, 2339
  New Mexico, 2370
  North Dakota, 2394
  Oregon, 2405-2406
  South Dakota, 2417
  Texas, 2422
  Utah, 2436
  Washington, 2450-2451
Upper Volta, 745
Venezuela, 2698-2699
Zambia, 1058-1059
Zimbabwe, 1076

BIBLIOGRAPHIES
Afghanistan, 1203-1206
Africa, 277-287
  Eastern and Southern, 755
  Northern and Western, 378-385
Agricultural Economics, 3032-3034
Agriculture, 2952-2959
Algeria, 431
The Americas, 1832
Angola, 770-771
Animal Culture, 3026-3027
Anthropology, 3114-3115
Architecture, 3155
Asia, South and Southeast, 1398
Australia, 1689-1691
Bahrain, 1223
Benin, 446
Biology, 2918
Botany, 2931-2932
Botswana, 784-787
Bulgaria, 212
Cameroon, 450-453
Canada, 1902
Cape Verde Islands, 259-262
Caribbean Islands, 2472
Central African Republic, 466
China, 1541-1545
Climatology, 2814-2819
Demography, 3101-3105
Drylands of the World, 4-10
Earth Sciences, 2782
Economic Development, 3165-3172
Egypt, 482-483
Energy, 3046-3049
Ethiopia, 803-806
Europe, Eastern, 208
France, 136
The Gambia, 514-516
Geography, 2732-2736
Geology, 2787-2792
Ghana, 526-527
Gibralter, 146
Greece, 148
Guinea-Bissau, 551-554
Haiti, 2512-2513
Historical Geography, 3193-3194
Hungary, 221-222
Hydrology, 2846-2849
India, 1420-1421
Iran, 1227
Irrigation, 2975-2984
Israel, 1268
Jibuti, 822
Kenya, 833-836
Kuwait, 1313

Subject Index

Lesotho, 860-861
Libya, 563-568
Madagascar, 1093-1094
Madeira Islands, 270
Malawi, 875-877
Mali, 587
Mauritania, 598-600
Mauritius, 1110
Middle East, 1141-1147
Morocco, 610
Mozambique, 888-890
Namibia, 901-904
Netherlands Antilles, 2526
New Zealand, 1824
Niger, 634
Nigeria, 643
Oman, 1326-1330
Pakistan, 1488
Paraguay, 2666-2667
Plant Culture, 2988-2991
Puerto Rico, 2532
Qatar, 1339
Reunion, 1115
Romania, 230
Sahara, Western, 751
Saudi Arabia, 1343
Science and Technology, 2761
Senegal, 674-675
Seychelles, 1120
Soils, 2894-2896
Somalia, 924-925
South Africa, 938-942
Sri Lanka, 1515
Sudan, 687-691
Swaziland, 997-998
Tanzania, 1007-1009
Togo, 711-712
Tunisia, 716-718
Turkey, 1365
Uganda, 1036-1038
Union of Soviet Socialist Republics, 1591-1593
The United States
    Arizona, 2213
    California, 2268
    Colorado, 2308
    Texas, 2419
Upper Volta, 740-741
Urban Geography, 3152
Virgin Islands, 2541
Yemen, North, 1386-1387
Yemen, South, 1393
Yugoslavia, 242-245
Zambia, 1054
Zimbabwe, 1070-1072
Zoology, 2940
BIOLOGY
Africa, 321
Australia, 1754
Drylands of the World, 39-40
India, 1445-1446
Israel, 1286-1287
Oceans and Coasts, 74
Peru, 2687
South America, 2546
The United States, 2134-2140
    Arizona, 2243-2244
    New Mexico, 2383-2384
BIOMASS ENERGY
3079-3084
BOTANY
Afghanistan, 1216
Africa, 322
    Eastern and Southern, 760
Algeria, 438
The Americas, 1842-1847
Angola, 779-780
Argentina, 2564-2566
Asia, 1136
    Northern and Eastern, 1537-1538
Australia, 1755-1760
    and the Pacific, 1686
Bahama Islands, 2499
Brazil, 2602
Bulgaria, 219
Cameroon, 462
Canada, 1955
Canary Islands, 258
Cape Verde Islands, 265-267
Caribbean Islands, 2486-2487
Chad, 476-477
Chile, 2631-2632
China, 1570-1571
Cuba, 2506
Drylands of the World, 41-44
Egypt, 501
Ethiopia, 816
Europe, 115
Galapagos Islands, 1812
Ghana, 535
Greece, 160-161
Hungary, 228
India, 1447
Indonesia, 1485-1486
Iraq, 1261
Israel, 1288-1289
Italy, 176
Jibuti, 829
Jordan, 1307
Kenya, 848
Libya, 575
Madeira Islands, 273-275
Mali, 594
Mauritania, 607
Mexico, 2023-2031
Morocco, 625-626
Mozambique, 894-895
Niger, 640
Nigeria, 660
North America, 1889-1893
Pakistan, 1505-1506
Portugal, 191-192
Romania, 238-239
Senegal, 680-681
Socotra, 1124
Somalia, 934
South America, 2547
Spain, 205
Sri Lanka, 1523
Sudan, 702
Tanzania, 1023
Tunisia, 729
Turkey, 1372
Uganda, 1047
Union of Soviet Socialist Republics, 1651
The United States, 2141-2154
    Arizona, 2245-2253
    California, 2286-2290
    Kansas, 2331
    Nevada, 2363-2364
    New Mexico, 2385
    Oklahoma, 2401
    Texas, 2428-2430
    Utah, 2445-2446
    Washington, 2461

Subject Index

    Upper Volta, 748
    Zambia, 1065
CEREAL CROPS
    2997-3003
CLIMATE CHANGE
    2840
CLIMATOLOGY
    Africa, 313-318
        Eastern and Southern, 759
        Northern and Western, 394-397
    Angola, 776-778
    Argentina, 2561-2562
    Asia, 1134-1135
        South and Southeast, 1404
    Atlantic Ocean, 250-251
    Australia, 1743-1747
        and the Pacific, 1683-1684
    Bolivia, 2578
    Botswana, 795-796
    Brazil, 2598-2601
    Bulgaria, 218
    Burma, 1415-1418
    Cameroon, 460
    Canada, 1944-1949
    Canary Islands, 256
    Caribbean Islands, 2481-2483
    Chile, 2629-2630
    China, 1565-1569
    Colombia, 2645
    Cyprus, 133
    Drylands of the World, 25-26
    Ecuador, 2660
    Egypt, 496-498
    El Salvador, 1978-1979
    Ethiopia, 812-815
    Europe, 113-114
        Mediterranean, 122
    France, 142
    Gibralter, 147
    Greece, 159
    Guatemala, 1988-1989
    Hungary, 227
    India, 1438-1442
    Indian Ocean and Islands, 1090-1091
    Iran, 1239-1241
    Iraq, 1256-1259
    Israel, 1284
    Italy, 174
    Jibuti, 825-828
    Jordan, 1303-1305
    Kenya, 847
    Lebanon, 1319-1320
    Lesotho, 868
    Libya, 573
    Madagascar, 1101
    Madeira Islands, 272
    Malawi, 882
    Malta, 183
    Mexico, 2017-2019
    Middle East, 1159-1165
    Mongolia, 1589
    Morocco, 619
    Mozambique, 896
    Namibia, 913
    New Zealand, 1831
    Nigeria, 656-659
    North America, 1886-1887
    Oceans and Coasts, 73
    Pakistan, 1500-1502
    Paraguay, 2671
    Peru, 2684-2686
    Portugal, 190
    Romania, 236-237
    Somalia, 929-932
    South Africa, 974-976
    South America, 2545
    Spain, 202-204
    Sri Lanka, 1521-1522
    Sudan, 697-699
    Swaziland, 1002
    Syria, 1357-1358
    Tanzania, 1020
    Thailand, 1530-1531
    Tunisia, 726
    Uganda, 1046
    Union of Soviet Socialist Republics, 1644-1649
    The United States, 2102-2110
        Arizona, 2236-2237
        Nebraska, 2347
        Nevada, 2360
    Venezuela, 2702
    Vietnam, 1535-1536
    Yugoslavia, 248
    Zambia, 1063
    Zimbabwe, 1080
COASTAL DESERTS
    68-70
CONFERENCE PUBLICATIONS
    Bibliography and Information Science, 2719-2720
CONSERVATION
    Hydrology, 2859
CONSTRUCTION
    Africa, 366
COOPERATIVES
    Sudan, 707
DEMOGRAPHY
    Africa, 343-345
    The Americas, 1864-1865
    Australia, 1787-1788
        and the Pacific, 1688
    Brazil, 2611
    Caribbean Islands, 2491
    Ghana, 540-541
    India, 1475
    Libya, 578
    Madagascar, 1102-1103
    Nigeria, 667-669
    Turkey, 1374
    The United States, 2188-2191
DESERTIFICATION
    Drylands of the World, 16-22
    South Africa, 959
DEVELOPMENT AND MANAGEMENT
    Hydrology, 2860-2866
DICTIONARIES
    Africa, 298-303
    Human Geography, 3099
DIRECTORIES
    Afghanistan, 1208
    Africa, Northern and Western, 389
    Agriculture, 2972
    The Americas, 1834-1835
    Anthropology, 3121
    Asia, 1129
        South and Southeast, 1399-1400
    Australia, 1706-1709
    Biology, 2930
    Botany, 2935-2936
    Canada, 1908-1911
    China, 1547-1548
    Climatology, 2829-2830
    Demography, 3110
    Drylands of the World, 13-15
    Earth Sciences, 2785

Subject Index

Economic Development, 3180-3182
Energy, 3058
Europe, 111
   Eastern, 209
Geography, 2744-2745
Geology, 2806-2809
Greece, 151
Human Geography, 3100
Iran, 1229
Irrigation, 2987
Israel, 1273
Italy, 169
Kiribati, 1817
Mexico, 1997
Middle East, 1153
New Caledonia, 1820
New Zealand, 1825
North America, 1879
Plant Culture, 2995-2996
Portugal, 186
Science and Technology, 2772-2773
South Africa, 948-950
Spain, 197
Third World, 78-79
Union of Soviet Socialist Republics, 1599-1604
The United States, 2058-2059
Zimbabwe, 1075

DOCUMENT COLLECTIONS
Science and Technology, 2768-2771

DROUGHT
2841

ECONOMIC DEVELOPMENT
Afghanistan, 1219
Africa, 367-369
   Eastern and Southern, 767-768
   Northern and Western Africa, 426-427
Algeria, 443
The Americas, 1871-1875
Angola, 782
Argentina, 2569-2571
Australia, 1806-1808
Bahrain, 1226
Benin, 448
Bolivia, 2583
Botswana, 800-801
Brazil, 2615-2618
Bulgaria, 220
Burma, 1419
Cameroon, 464
Canada, 1968-1972
Cape Verde Islands, 268
Caribbean Islands, 2494
Central African Republic, 469
Chad, 480
Chile, 2635
China, 1580-1583
Colombia, 2652
Cuba, 2509
Cyprus, 135
Dominican Republic, 2511
Drylands of the World, 61
Ecuador, 2663
Egypt, 508
El Salvador, 1981
Ethiopia, 819
Europe, 120-121
France, 144-145
The Gambia, 523-524
Ghana, 544-546
Greece, 167
Guatemala, 1994
Guinea, 549
Guinea-Bissau, 556
Haiti, 2519
Hungary, 229
India, 1478-1479
Indonesia, 1487
Iran, 1246-1249
Iraq, 1267
Israel, 1299
Italy, 180-181
Ivory Coast, 562
Jibuti, 830-831
Jordan, 1312
Kenya, 855-857
Kuwait, 1316
Lebanon, 1324
Lesotho, 871-873
Libya, 582
Madagascar, 1108
Malawi, 885-886
Mali, 596
Malta, 184
Mauritania, 608
Mexico, 2046-2050
Middle East, 1192-1197
Morocco, 630-631
Mozambique, 900
Namibia, 921
Netherlands Antilles, 2529-2530
Niger, 641
Nigeria, 671-673
Oman, 1337
Pakistan, 1513
Paraguay, 2673
Peru, 2693-2694
Portugal, 194-195
Puerto Rico, 2536-2537
Qatar, 1342
Romania, 241
Saudi Arabia, 1352
Senegal, 684-685
Somalia, 935
South Africa, 990-991
Spain, 206-207
Sri Lanka, 1527
Sudan, 710
Swaziland, 1004-1005
Syria, 1363
Tanzania, 1030-1032
Thailand, 1532
Third World, 98-108
Togo, 714
Tunisia, 737
Turkey, 1378-1379
Uganda, 1050-1052
Union of Soviet Socialist Republics, 1666-1669
United Arab Emirates, 1384
The United States, 2202-2204
   Nebraska, 2350
   Oklahoma, 2402
   Texas, 2432
Upper Volta, 749
Venezuela, 2708-2709
Yemen, North, 1392
Yemen, South, 1397
Yugoslavia, 249
Zambia, 1068
Zimbabwe, 1083-1084

ENCYCLOPEDIAS
Africa, 297
   Eastern and Southern, 756-757
   Northern and Western, 388
Agriculture, 2971

Subject Index

Anthropology, 3120
Asia, 1128
Australia, 1703
Biology, 2927-2929
Canada, 1906
Climatology, 2827-2828
Demography, 3109
Earth Sciences, 2783-2784
Geography, 2743
Geology, 2801-2805
Greece, 149
Historical Geography, 3197
Hydrology, 2857
Israel, 1272
Mexico, 1996
Netherlands Antilles, 2527
Oceans and Coasts, 67
Third World, 77
Union of Soviet Socialist Republics, 1597-1598
Zoology, 2943
ENERGY
Africa, Eastern and Southern, 763
Asia, 1139
Australia, 1783
and the Pacific, 1687
Drylands of the World, 56-57
Europe, Eastern, 211
Union of Soviet Socialist Republics, 1660
The United States, 2174-2184
Arizona, 2260
California, 2296
New Mexico, 2387
Utah, 2448
ENTOMOLOGY
2944-2947
ENVIRONMENTAL IMPACT STATEMENTS
The United States, 2080-2085
Colorado, 2312
Idaho, 2321
Montana, 2341
Nevada, 2357
New Mexico, 2373
Utah, 2437
Wyoming, 2466
EVAPORATION
2842
EVAPORATION SUPPRESSION
2867-2868
EXTENSION AND TRAINING
Agricultural Economics, 3039
FIBER CROPS
3004-3007
FOOD AID
3040
FOOD, HUNGER, NUTRITION
3129-3132
FOODS
North America, 1900
FORESTRY, 3008-3013
Africa, 330
Northern and Western, 404
Australia, 1773-1776
India, 1462-1463
Mexico, 2035
Sudan, 705
The United States, 2165-2166
California, 2293
Minnesota, 2336
FRUIT, NUT AND LEGUME CROPS
3014-3020
GAZETTEERS
Afghanistan, 1211-1212

Africa, 310
Algeria, 434
The Americas, 1839
Angola, 773
Argentina, 2558
Asia, 1132
Australia, 1725-1729
and the Pacific, 1680
Azores Islands, 252
Bahrain, 1225
Benin, 447
Bolivia, 2575
Botswana, 789-791
Brazil, 2594
Bulgaria, 216
Burma, 1413
Cameroon, 459
Canada, 1927-1932
Canary Islands, 254
Cape Verde Islands, 264
Caribbean Islands, 2476
Central African Republic, 467
Chad, 475
Chile, 2626
China, 1561
Colombia, 2643
Cuba, 2505
Dominican Republic, 2510
Ecuador, 2658
Egypt, 490
El Salvador, 1977
Ethiopia, 811
Europe, 112
France, 139
Galapagos Islands, 1811
The Gambia, 518
Ghana, 533
Greece, 156
Guatemala, 1987
Guinea, 548
Guinea-Bissau, 555
Haiti, 2516
Hungary, 226
India, 1433
Indonesia, 1484
Iran, 1234-1235
Iraq, 1254
Israel, 1281-1282
Italy, 172
Ivory Coast, 561
Jibuti, 823
Jordan, 1302
Kenya, 845
Kiribati, 1818
Kuwait, 1315
Lebanon, 1318
Lesotho, 866-867
Libya, 571
Madagascar, 1099
Madeira Islands, 271
Malawi, 881
Mali, 593
Malta, 182
Marquesas Islands, 1819
Mauritania, 605
Mauritius, 1112
Mexico, 2010-2011
Middle East, 1155-1156
Mongolia, 1588
Morocco, 616
Mozambique, 893
Namibia, 907-908

Subject Index

Netherlands Antilles, 2528
New Caledonia, 1823
New Zealand, 1830
Niger, 639
Nigeria, 654
Oman, 1336
Pakistan, 1497
Paraguay, 2670
Peru, 2681
Portugal, 188
Puerto Rico, 2534
Qatar, 1341
Reunion, 1118
Romania, 234
Rwanda, 923
Sahara, Western, 752
Saudi Arabia, 1348
Senegal, 679
Seychelles, 1122
Somalia, 927
South Africa, 963-965
Spain, 200
Sri Lanka, 1519
St. Helena, 276
Sudan, 696
Swaziland, 1000-1001
Syria, 1356
Tanzania, 1017
Thailand, 1529
Togo, 713
Tunisia, 722
Turkey, 1370
Uganda, 1045
Union of Soviet Socialist Republics, 1638-1640
United Arab Emirates, 1383
The United States, 2076-2077
  Arizona, 2228
  California, 2276
  Colorado, 2310-2311
  Idaho, 2320
  Kansas, 2328
  Montana, 2340
  Nebraska, 2346
  Nevada, 2355-2356
  New Mexico, 2371-2372
  Oklahoma, 2398
  Oregon, 2407
  Texas, 2423
  Washington, 2452
  Wyoming, 2465
Upper Volta, 746
Venezuela, 2700
Vietnam, 1534
Virgin Islands, 2542
Yemen, North, 1391
Yemen, South, 1395
Yugoslavia, 247
Zambia, 1060
Zimbabwe, 1077

GEOGRAPHY
Afghanistan, 1209
Africa, 305-307
  Northern and Western, 390
Algeria, 433
Argentina, 2551-2555
Asia, South and Southeast, 1401-1402
Australia, 1714-1719
  and the Pacific, 1677-1678
Bahama Islands, 2497-2498
Bolivia, 2574
Botswana, 788
Brazil, 2588-2592
Bulgaria, 214
Burma, 1412
Cameroon, 455-456
Canada, 1913-1917
Cape Verde Islands, 263
Caribbean Islands, 2475
Chad, 472-473
Chile, 2621-2624
China, 1551-1553
Colombia, 2639-2640
Ecuador, 2655
Egypt, 486-489
Ethiopia, 808
Europe, Eastern, 210
The Gambia, 517
Ghana, 530-532
Greece, 154
Guatemala, 1984
Hungary, 224
India, 1428-1429
Indian Ocean and Islands, 1086-1088
Indonesia, 1483
Iran, 1230-1232
Iraq, 1253
Israel, 1275-1278
Italy, 170-171
Kenya, 838-841
Lesotho, 863
Libya, 570
Madagascar, 1096-1097
Malawi, 879
Mali, 589-591
Mauritania, 602-603
Mexico, 1998-2004
Morocco, 612-614
Namibia, 905-906
New Caledonia, 1821
New Zealand, 1826-1827
Niger, 635-637
Nigeria, 646-652
North America, 1880-1881
Oman, 1335
Pakistan, 1494-1496
Peru, 2676-2679
Puerto Rico, 2533
Romania, 232
Rwanda, 922
Saudi Arabia, 1346
Senegal, 677
South Africa, 956-958
Sudan, 694-695
Tanzania, 1011-1014
Tunisia, 720-721
Turkey, 1368
Uganda, 1040-1043
Union of Soviet Socialist Republics, 1608-1621
The United States, 2062-2072
  Arizona, 2215-2225
  California, 2269-2273
  Colorado, 2309
  Montana, 2338
  Nevada, 2352-2353
  New Mexico, 2368-2369
  North Dakota, 2393
  Texas, 2420-2421
  Utah, 2435
  Wyoming, 2464
Upper Volta, 742-744
Venezuela, 2697
Yemen, North, 1389-1390
Zambia, 1056-1057

Subject Index

GEOLOGY
Afghanistan, 1213-1214
Africa, 311-312
  Northern and Western, 392-393
Angola, 774-775
Argentina, 2559-2560
Asia, 1133
  South and Southeast, 1403
Australia, 1733-1742
  and the Pacific, 1681-1682
Bolivia, 2576-2577
Botswana, 792-793, 794
Brazil, 2596-2597
Burma, 1414
Canada, 1933-1943
Canary Islands, 255
Caribbean Islands, 2477-2480
Chile, 2627-2628
China, 1563-1564
Colombia, 2644
Cyprus, 127-132
Ecuador, 2659
Egypt, 491-495
France, 140
The Gambia, 519-520
Greece, 157-158
Hawaiian Islands, 1814
India, 1434-1437
Iran, 1236-1238
Iraq, 1255
Israel, 1283
Italy, 173
Jamaica, 2523
Jibuti, 824
Kenya, 846
Libya, 572
Madagascar, 1100
Mexico, 2012-2016
Middle East, 1157-1158
Morocco, 617-618
Namibia, 909-912
Nigeria, 655
North America, 1882-1885
Oceans and Coasts, 71-72
Pakistan, 1498-1499
Peru, 2683
Portugal, 189
Puerto Rico, 2535
Romania, 235
Sahara, Western, 753
Saudi Arabia, 1349-1350
Somalia, 928
South Africa, 966-973
South America, 2544
Spain, 201
Sri Lanka, 1520
Tanzania, 1018-1019
Tunisia, 723-725
Turkey, 1371
Union of Soviet Socialist Republics, 1641-1642
The United States, 2087-2097
  Arizona, 2229-2235
  California, 2277-2280
  Colorado, 2313-2316
  Idaho, 2322-2325
  Kansas, 2329
  Minnesota, 2335
  Nevada, 2358-2359
  New Mexico, 2374-2379
  North Dakota, 2395-2396
  Oklahoma, 2399
  Oregon, 2408-2411
  Texas, 2424-2427
  Utah, 2438-2443
  Washington, 2453-2456
  Wyoming, 2467-2469
Venezuela, 2701
Zambia, 1061-1062
Zimbabwe, 1078-1079

GEOTHERMAL ENERGY
Drylands of the World, 58-59
Energy, 3068-3072
The United States-Arizona, 2261

GOVERNMENT DOCUMENTS
Australia, 1704-1705
Bibliography and Information Science, 2721
Canada, 1907
France, 137
India, 1427
Pakistan, 1492
Sri Lanka, 1517
The United States, 2055-2057

GROUNDWATER
Hydrology, 2869-2870

HANDBOOKS
Afghanistan, 1207
Africa, 296
  Northern and Western, 387
Algeria, 432
Angola, 772
Argentina, 2550
Asia, 1127
Australia, 1695-1702
  and the Pacific, 1674
Bahrain, 1224
Bolivia, 2573
Brazil, 2585
Bulgaria, 213
Burma, 1411
Cameroon, 454
Caribbean Islands, 2474
Chad, 471
Chile, 2620
China, 1546
Colombia, 2637
Cuba, 2501
Cyprus, 126
Ecuador, 2654
Egypt, 484
El Salvador, 1975
Ethiopia, 807
Ghana, 529
Greece, 150
Guatemala, 1983
Guinea, 547
Haiti, 2514
Hungary, 223
India, 1426
Indonesia, 1482
Iran, 1228
Iraq, 1252
Israel, 1270-1271
Italy, 168
Ivory Coast, 558
Jamaica, 2521
Jordan, 1301
Kenya, 837
Kuwait, 1314
Lebanon, 1317
Lesotho, 862
Libya, 569
Madagascar, 1095
Malawi, 878
Mauritania, 601

Subject Index

Mauritius, 1111
Mexico, 1995
Middle East, 1150-1152
Mongolia, 1586
Morocco, 611
Mozambique, 891
Nigeria, 644
Oman, 1332-1333
Pakistan, 1491
Paraguay, 2668
Peru, 2675
Portugal, 185
Qatar, 1340
Reunion, 1116
Romania, 231
Saudi Arabia, 1344-1345
Senegal, 676
Seychelles, 1121
Somalia, 926
South Africa, 946-947
Spain, 196
Sri Lanka, 1516
Sudan, 692
Syria, 1355
Tanzania, 1010
Thailand, 1528
Trinidad, 2539
Tunisia, 719
Turkey, 1366
Uganda, 1039
Union of Soviet Socialist Republics, 1596
United Arab Emirates, 1381-1382
Venezuela, 2696
Yemen, North, 1388
Yemen, South, 1394
Yugoslavia, 246
Zambia, 1055
Zimbabwe, 1074

HEALTH CARE SERVICES
3133-3137

HEAT STRESS
Medicine, 3138-3140

HERPETOLOGY
2948

HISTORICAL GEOGRAPHY
Afghanistan, 1220-1221
Africa, 373-375
Algeria, 444
The Americas, 1877-1878
Angola, 783
Argentina, 2572
Asia, South and Southeast, 1410
Australia, 1810
Benin, 449
Bolivia, 2584
Botswana, 802
Brazil, 2619
Cameroon, 465
Canada, 1973-1974
Cape Verde Islands, 269
Caribbean Islands, 2495-2496
Central African Republic, 470
Chad, 481
Chile, 2636
China, 1584
Colombia, 2653
Ecuador, 2664-2665
Egypt, 511-512
El Salvador, 1982
Ethiopia, 820-821
Europe, Mediterranean, 125
The Gambia, 525

Guinea, 550
Guinea-Bissau, 557
Haiti, 2520
India, 1480-1481
Iran, 1250
Iraq, 1266
Jibuti, 832
Kenya, 859
Lesotho, 874
Libya, 583
Madagascar, 1109
Malawi, 887
Mali, 597
Mauritania, 609
Mauritius, 1114
Mexico, 2052-2054
Middle East, 1200
Morocco, 632
Netherlands Antilles, 2531
Niger, 642
North America, 1901
Oman, 1338
Pakistan, 1514
Paraguay, 2674
Peru, 2695
Puerto Rico, 2538
Sahara, Western, 754
Saudi Arabia, 1353
Senegal, 686
Somalia, 936-937
South Africa, 992-994
Sudan, 709
Swaziland, 1006
Tanzania, 1034
Thailand, 1533
Togo, 715
Union of Soviet Socialist Republics, 1670-1672
United Arab Emirates, 1385
The United States, 2207-2211
    Arizona, 2267
    California, 2307
    Kansas, 2332
    Oklahoma, 2403
    Texas, 2433
Upper Volta, 750
Venezuela, 2710
Virgin Islands, 2543
Zambia, 1069
Zimbabwe, 1085

HOUSING
Africa, 365
    Eastern and Southern, 765-766
    Northern and Western, 424-425
The Americas, 1870
Ghana, 543
Middle East, 1191
Third World, 95-97
The United States, 2201

HUMAN GEOGRAPHY
Africa, 342
    Northern and Western, 410
Algeria, 439
The Americas, 1859-1863
Australia, 1785-1786
Botswana, 797
Brazil, 2610
Chad, 478-479
China, 1577
The Gambia, 522
Ecuador, 2662
El Salvador, 1980
Europe, 119

Subject Index

Ghana, 537-539
Greece, 163-164
Guatemala, 1992
India, 1467-1470
Iran, 1243-1244
Israel, 1292
Kenya, 851
Lesotho, 870
Libya, 577
Middle East, 1175-1177
Mozambique, 899
Namibia, 917-918
Nigeria, 665-666
Pakistan, 1508-1511
South Africa, 984-987
Sri Lanka, 1525
Swaziland, 1003
Tanzania, 1029
Union of Soviet Socialist Republics, 1661
Zambia, 1067
HYDROLOGY
Africa, 319
   Northern and Western, 398-399
Algeria, 435
Australia, 1748-1751
Canada, 1950-1953
Caribbean Islands, 2484
Drylands of the World, 27-35
Egypt, 499
Europe, Mediterranean, 123
France, 141
Ghana, 534
Hawaiian Islands, 1815-1816
Italy, 175
Mexico, 2020-2021
Middle East, 1166-1167
Morocco, 620-622
Namibia, 914
North America, 1888
Pakistan, 1503
South Africa, 977
Sudan, 700-701
Tanzania, 1021
Third World, 81
The United States, 2111-2124
   Arizona, 2238-2241
   California, 2281-2284
   Colorado, 2317
   Kansas, 2330
   Montana, 2342-2344
   Nebraska, 2348
   Nevada, 2361-2362
   New Mexico, 2380-2382
   Oklahoma, 2400
   Oregon, 2412
   Utah, 2444
   Washington, 2457-2458
HYDROPOWER
Drylands of the World, 60
The United States, 2187
   Arizona, 2262
INDEXES AND ABSTRACTS
Africa, 288-290
Agricultural Economics, 3035
Agriculture, 2967-2970
The Americas, 1833
Animal Culture, 3028-3030
Anthropology, 3116-3118
Architecture, 3156-3157
Asia, 1125
Australia, 1693
Biology, 2921-2926
Botany, 2933-2934
Canada, 1903
Climatology, 2824-2826
Demography, 3108
Drylands of the World, 11-12
Economic Development, 3174-3179
Energy, 3054-3057
Geography, 2741-2742
Geology, 2795-2800
Historical Geography, 3195-3196
Human Geography, 3091-3098
Hydrology, 2854-2855
India, 1422
Irrigation, 2985-2986
Israel, 1269
Medicine, 3127
Oceans and Coasts, 66
Oman, 1331
Pakistan, 1489
Plant Culture, 2992-2994
Science and Technology, 2766-2767
Soils, 2897
Urban Geography, 3153-3154
Zimbabwe, 1073
Zoology, 2941-2942
INFORMATION SCIENCE
Australia, 1713
Third World, 80
Union of Soviet Socialist Republics, 1607
INTER-BASIN DIVERSION
Hydrology, 2871
IRRIGATION
Africa, 329
Drylands of the World, 53-54
India, 1458-1460
Union of Soviet Socialist Republics, 1657-1658
The United States, 2167-2169
   California, 2295
KIBBUTZ
Israel, 1294-1297
LAND RECLAMATION
Bulgaria, 217
The United States, 2098-2101
LAND TENURE
Africa, 338-340
Agricultural Economics, 3041-3042
The Americas, 1853
Asia, Northern and Eastern, 1539-1540
   South and Southeast, 1406-1408
Australia, 1782
Lesotho, 869
Mexico, 2037-2038
Middle East, 1173-1174
LANDSCAPING
Architecture, 3160
LEGAL AND SOCIAL ASPECTS
Hydrology, 2872
MAP COLLECTIONS
The Americas, 1840
Australia, 1730-1732
Canada, 1918-1920
Geography, 2746-2749
Mexico, 2005
South Africa, 960
The United States, 2073-2074
   Arizona, 2226
   Kansas, 2327
   Nevada, 2354
MARKETS
Africa, 341
The Americas, 1854
Asia, 1138

Subject Index

MARKETS AND COOPERATIVES
  Agricultural Economics, 3043-3045
MEDICINE
  Africa, 351
  Algeria, 440
  Brazil, 2612-2614
  Colombia, 2651
  Ghana, 542
  Iraq, 1265
  Madagascar, 1104-1106
  Third World, 88-89
  Tunisia, 732
  Uganda, 1049
  Union of Soviet Socialist Republics, 1665
MIGRATION
  Africa, 346
    Northern and Western, 411
  Botswana, 798
  Greece, 165
  Iraq, 1264
  Mexico, 2040
  The United States, 2192
MILITARY AFFAIRS
  Drylands of the World, 62-64
MINED LAND RECLAMATION
  Soils, 2898-2899
NATIONAL BIBLIOGRAPHIES
  Bibliography and Information Science, 2711-2712
NATIVE MEDICINES
  Medicine, 3141
NOMADISM
  Africa, 331
    Northern and Western, 405-407
  Sudan, 706
NUTRITION
  Africa, 354-356
    Northern and Western, 418
  Kenya, 853
OIL AND RUBBER CROPS
  Plant Culture, 3021-3025
ONLINE DATABASES
  Agriculture, 2960-2966
  Australia, 1692
  Bibliography and Information Science, 2713
  Biology, 2919-2920
  Climatology, 2820-2823
  Demography, 3106-3107
  Economic Development, 3173
  Geography, 2737-2740
  Geology, 2793-2794
  Human Geography, 3088-3090
  Hydrology, 2850-2853
  Middle East, 1148
  Oceans and Coasts, 65
  Science and Technology, 2762-2765
  Energy, 3050-3053
PETROLEUM
  The Americas, 1858
  China, 1575-1576
  Energy, 3085-3087
PHREATOPHYTE CONTROL
  2873-2874
PLANT CULTURE
  Africa, Northern and Western, 403
  Australia, 1771-1772
  Bolivia, 2580
  Egypt, 504
  India, 1461
  Israel, 1290
  Mexico, 2034
  Peru, 2689-2690
  South Africa, 983

  The United States, 2164
    California, 2294
  Venezuela, 2706
  Zimbabwe, 1082
PUBLIC ADMINISTRATION
  Afghanistan, 1222
  Africa, 376-377
  Algeria, 445
  China, 1585
  Egypt, 513
  Iran, 1251
  Israel, 1300
  Libya, 584-586
  Mexico, 2051
  Middle East, 1201-1202
  Morocco, 633
  Saudi Arabia, 1354
  South Africa, 995-996
  Tanzania, 1035
  Tunisia, 739
  Turkey, 1380
  Union of Soviet Socialist Republics, 1673
  The United States, 2212
    Colorado, 2318
    Idaho, 2326
    Kansas, 2333
    Minnesota, 2337
    Montana, 2345
    Nebraska, 2351
    Nevada, 2367
    New Mexico, 2392
    North Dakota, 2397
    Oklahoma, 2404
    Oregon, 2416
    South Dakota, 2418
    Texas, 2434
    Utah, 2449
    Washington, 2463
    Wyoming, 2471
PUBLIC HEALTH
  Africa, 352-353
    Northern and Western Africa, 417
  China, 1578
  Kenya, 852
  Libya, 579
  Middle East, 1184
  Tunisia, 733-734
RANGE MANAGEMENT
  Animal Culture, 3031
  Canada, 1962
  Drylands of the World, 55
  The United States, 2171-2173
    Arizona, 2259
REMOTE SENSING
  Botany, 2937
  Drylands of the World, 23-24
  Geography, 2752-2760
  Mauritania, 606
  The United States, 2078-2079
  Upper Volta, 747
RESEARCH GUIDES
  Agriculture, 2949-2951
  Biology, 2915-2917
  Drylands of the World, 1-3
  Geography, 2729-2731
  Geology, 2786
  Historical Geography, 3192
  Medicine, 3125-3126
  Third World, 75
RURAL GEOGRAPHY
  Canada, 1965
  Ethiopia, 818

Subject Index

India, 1473
Malawi, 884
Turkey, 1376-1377
SALINE AND ALKALINE SOILS
2900-2905
SALINE WATER AND DESALINATION
2887-2893
SAND CONTROL
   Union of Soviet Socialist Republics, 1643
SAND DUNES AND CONTROL
2906-2909
SCIENCE AND TECHNOLOGY
   The Americas, 1841
   Brazil, 2595
   China, 1562
   Peru, 2682
   The United States, 2086
SOIL FORMATION AND PROPERTIES
2910-2914
SOILS
   Afghanistan, 1215
   Africa, 320
   Algeria, 436-437
   Argentina, 2563
   Asia, South and Southeast, 1405
   Australia, 1752-1753
     and the Pacific, 1685
   Cameroon, 461
   Canada, 1954
   Canary Islands, 257
   Caribbean Islands, 2485
   Central African Republic, 468
   Colombia, 2646
   Drylands of the World, 36-38
   Egypt, 500
   Ethiopia, 817
   India, 1443-1444
   Iran, 1242
   Iraq, 1260
   Israel, 1285
   Jordan, 1306
   Lebanon, 1321
   Libya, 574
   Malawi, 883
   Mauritius, 1113
   Mongolia, 1590
   Morocco, 624-623
   Namibia, 915
   Pakistan, 1504
   Reunion, 1119
   Seychelles, 1123
   Somalia, 933
   South Africa, 978
   Syria, 1359
   Tanzania, 1022
   Tunisia, 727-728
   The United States, 2133
     Oregon, 2413-2414
     Washington, 2459-2460
   Venezuela, 2703
   Yemen, South, 1396
   Zambia, 1064
   Zimbabwe, 1081
SOLAR BUILDING DESIGN
   Architecture, 3161-3164
SOLAR ENERGY
   Argentina, 2568
   Australia, 1784
   Energy, 3062-3067
   Italy, 179
   Mexico, 2039
   Third World, 87

   The United States, 2185-2186
STATISTICS
   Africa, 304
   Agricultural Economics, 3036-3038
   Agriculture, 2973
   The Americas, 1836-1838
   Asia, 1130-1131
   Australia, 1710-1712
     and the Pacific, 1675-1676
   Bibliography and Information Science, 2726-2728
   Brazil, 2586-2587
   Canada, 1912
   China, 1549-1550
   Climatology, 2831-2839
   Colombia, 2638
   Cuba, 2502
   Demography, 3111-3113
   Economic Development, 3183-3189
   Egypt, 485
   Energy, 3059-3061
   Geology, 2810-2811
   Greece, 152-153
   Israel, 1274
   Mali, 588
   Medicine, 3128
   Middle East, 1154
   Oman, 1334
   Pakistan, 1493
   South Africa, 951-955
   Spain, 198
   Sri Lanka, 1518
   Trinidad, 2540
   Turkey, 1367
   Union of Soviet Socialist Republics, 1605-1606
   The United States, 2060-2061
     Arizona, 2214
SUBJECT COLLECTIONS
   Bibliography and Information Science, 2722-2725
THESES AND DISSERTATIONS
   Africa, 291-295
     Eastern and Southern, 758
     Northern and Western, 386
   Anthropology, 3119
   Asia, 1126
   Australia, 1694
   Bibliography and Information Science, 2714-2718
   Canada, 1904-1905
   Caribbean Islands, 2473
   Ghana, 528
   Hydrology, 2856
   India, 1423-1425
   Middle East, 1149
   Nigeria, 645
   Pakistan, 1490
   South Africa, 943-945
   Sudan, 693
   Third World, 76
   Union of Soviet Socialist Republics, 1594-1595
TOURISM AND RECREATION
   Africa, 371-372
   The Americas, 1876
   Australia, 1809
   Economic Development, 3191
   Egypt, 510
   Indian Ocean and Islands, 1092
   Kenya, 858
   Lebanon, 1325
   Syria, 1364
   Tanzania, 1033
   Third World, 110
   Tunisia, 738
   Uganda, 1053

The United States, 2205-2206
TRANSPORTATION
Africa, 370
Eastern and Southern, 769
Northern and Western, 428-430
Egypt, 509
Middle East, 1198-1199
Third World, 109
TROPICAL MEDICINE
3142-3149
URBAN GEOGRAPHY
Afghanistan, 1217
Africa, 357-362
Northern and Western, 419-421
Algeria, 441
The Americas, 1869
Australia, 1805
Canada, 1966-1967
China, 1579
Egypt, 507
India, 1474
Iran, 1245
Israel, 1298
Jordan, 1309-1310
Kenya, 854
Libya, 580
Madagascar, 1107
Middle East, 1185-1189
Morocco, 627-628
Syria, 1362
Third World, 90-93
Tunisia, 735
The United States, 2197-2199
California, 2300-2304
VEGETATION MAPPING
Botany, 2938-2939
VENOM DISEASES
Medicine, 3150-3151
WATER LAW
Mexico, 2022
The United States, 2125-2132
Arizona, 2242
California, 2285
WATER QUALITY, POLLUTION, RE-USE
2875-2886
WEATHER MODIFICATION
Climatology, 2843-2845
Union of Soviet Socialist Republics, 1650

WIND ENERGY
3073-3078
WOMEN'S STUDIES
Africa, 349-350
Northern and Western, 413-416
The Americas, 1867-1868
Anthropology, 3122-3124
India, 1476-1477
Israel, 1293
Middle East, 1180-1183
Pakistan, 1512
Senegal, 683
Sri Lanka, 1526
Turkey, 1375
ZOOLOGY
Africa, 323-327
Eastern and Southern, 761
Northern and Western, 400-401
Asia, 1137
Australia, 1761-1766
Brazil, 2603
Canada, 1957
Caribbean Islands, 2488-2489
Drylands of the World, 45-47
Europe, 116
Iraq, 1262
Lebanon, 1322
Libya, 576
Mexico, 2032
Middle East, 1168-1170
Namibia, 916
Nigeria, 661-662
North America, 1894-1899
South Africa, 979
South America, 2548
Sri Lanka, 1524
Syria, 1360
Turkey, 1373
The United States, 2155-2157
Arizona, 2254-2257
California, 2291
Nevada, 2365-2366
New Mexico, 2386
Oregon, 2415
Utah, 2447
Washington, 2462
Wyoming, 2470

DISCARDED
URI LIBRARY